International Association of Geodesy Symposia

Michael G. Sideris, Series Editor

For further volumes:
http://www.springer.com/series/1345

International Association of Geodesy Symposia

Michael G. Sideris, Series Editor

Symposium 101: Global and Regional Geodynamics
Symposium 102: Global Positioning System: An Overview
Symposium 103: Gravity, Gradiometry, and Gravimetry
Symposium 104: Sea SurfaceTopography and the Geoid
Symposium 105: Earth Rotation and Coordinate Reference Frames
Symposium 106: Determination of the Geoid: Present and Future
Symposium 107: Kinematic Systems in Geodesy, Surveying, and Remote Sensing
Symposium 108: Application of Geodesy to Engineering
Symposium 109: Permanent Satellite Tracking Networks for Geodesy and Geodynamics
Symposium 110: From Mars to Greenland: Charting Gravity with Space and Airborne Instruments
Symposium 111: Recent Geodetic and Gravimetric Research in Latin America
Symposium 112: Geodesy and Physics of the Earth: Geodetic Contributions to Geodynamics
Symposium 113: Gravity and Geoid
Symposium 114: Geodetic Theory Today
Symposium 115: GPS Trends in Precise Terrestrial, Airborne, and Spaceborne Applications
Symposium 116: Global Gravity Field and Its Temporal Variations
Symposium 117: Gravity, Geoid and Marine Geodesy
Symposium 118: Advances in Positioning and Reference Frames
Symposium 119: Geodesy on the Move
Symposium 120: Towards an Integrated Global Geodetic Observation System (IGGOS)
Symposium 121: Geodesy Beyond 2000: The Challenges of the First Decade
Symposium 122: IV Hotine-Marussi Symposium on Mathematical Geodesy
Symposium 123: Gravity, Geoid and Geodynamics 2000
Symposium 124: Vertical Reference Systems
Symposium 125: Vistas for Geodesy in the New Millennium
Symposium 126: Satellite Altimetry for Geodesy, Geophysics and Oceanography
Symposium 127: V Hotine Marussi Symposium on Mathematical Geodesy
Symposium 128: A Window on the Future of Geodesy
Symposium 129: Gravity, Geoid and Space Missions
Symposium 130: Dynamic Planet - Monitoring and Understanding …
Symposium 131: Geodetic Deformation Monitoring: From Geophysical to Engineering Roles
Symposium 132: VI Hotine-Marussi Symposium on Theoretical and Computational Geodesy
Symposium 133: Observing our Changing Earth
Symposium 134: Geodetic Reference Frames
Symposium 135: Gravity, Geoid and Earth Observation
Symposium 136: Geodesy for Planet Earth
Symposium 137: VII Hotine-Marussi Symposium on Mathematical Geodesy

Reference Frames for Applications in Geosciences

Proccedings of the Symposium
in Marne-La-Vallée,
4–8 October, 2010

Edited by

Zuheir Altamimi
Xavier Collilieux

Springer

Volume Editors
Zuheir Altamimi
Xavier Collilieux
Institut National de l'Information
Géographique et Forestière
Laboratoire de recherche en géodésie
Université Paris Diderot
Paris

Series Editor
Michael G. Sideris
Departmemt of Geomatics Engineering
University of Calgary
2500 University Drive NW
Calgary, Alberta
Canada T2N 1N4

ISSN 0939-9585
ISBN 978-3-642-32997-5 ISBN 978-3-642-32998-2 (eBook)
DOI 10.1007/978-3-642-32998-2
Springer Heidelberg New York Dordrecht London

Library of Congress Control Number: 2013930022

Printed on acid-free paper

Springer is part of Springer Science+Business Media (www.springer.com)

Associated Editors

Claude Boucher

Observatoire de Paris/SYRTE
claude-boucher@club-internet.fr

David Coulot

Institut national de l'information géographique et forestière
Laboratoire de recherche en géodésie
Université Paris Diderot
Paris, France
David.Coulot@ign.fr

Mike Craymer

Natural Resources Canada
Geodetic Survey Division
Ottawa, Ontario, Canada
Michael.Craymer@NRCan-RNCan.gc.ca

Richard S. Gross

Jet Propulsion Laboratory
Pasadena, California, USA
Richard.Gross@jpl.nasa.gov

Johaness Ihde

Federal Agency of Cartography and Geodesy
Richard-Strauss-Allee 11,
D-60598 Frankfurt am Main, Germany
johannes.ihde@bkg.bund.de

Frank Lemoine

NASA Goddard Space Flight Center
Planetary Geodynamics Laboratory
Greenbelt, Maryland U.S.A.
Frank.G.Lemoine@nasa.gov

Markus Rothacher

ETH Zurich
Institute of Geodesy and Photogrammetry
HPV G52 Schafmattstr. 34
CH-8093 Zurich, Switzerland
markus.rothacher@ethz.ch

Harald Schuh

Helmholtz-Zentrum Potsdam
DeutschesGeoForschungsZentrum GFZ
Telegrafenberg, A17
14473 Potsdam
Germany
harald.schuh@tuwien.ac.at

Michael G. Sideris

University of Calgary
Department of Geomatics Engineering
2500 University Drive N.W.
Calgary, Alberta T2N 1N4 Canada
sideris@ucalgary.ca

Peter Steigenberger

Technische Universität München
Institut für Astronomische und Physikalische Geodäsie
Munich, Germany
steigenberger@bv.tum.de

João Agria Torres

National Committee on Astronomy, Geodesy and Geophysics
Lisbon, Portugal
jatorres@iol.pt

Preface

Reference systems and frames are of primary importance for many Earth science applications, satellite navigation as well as for practical applications in geo-information. A precisely defined reference frame is needed for the quantification of, e.g. Earth rotation and its gravity field, global and regional sea level variation, tectonic motion and deformation, postglacial rebound, geocenter motion, large-scale deformation due to Earthquakes, local subsidence and other ruptures and crustal dislocations. All of these important scientific applications fundamentally depend on a truly global reference system that only space geodesy can realize.

IAG Commission 1 activities are to deal with theoretical aspects of reference systems and the practical applications for their realizations as well as applied researches. The main objectives of Commission 1 are:

- Definition, establishment, maintenance and improvement of the geodetic reference frames
- Advanced terrestrial and space observation technique development for the above purposes
- International collaboration for the definition and deployment of networks of terrestrially based space geodetic observatories
- Theory and coordination of astrometric observation for reference frame purposes
- Collaboration with space geodesy/reference frame–related international services, agencies and organizations
- Promoting the definition and establishment of vertical reference systems at the global level, considering the advances in the regional sub-commissions

In order to review the progress in the above objectives, the Commission had organized the IAG Symposium "Reference Frames for Applications in Geosciences (REFAG2010)", held in Marne la Vallée, France, during October 4–8, 2010, at the premises of Ecole Nationale des Sciences Géographiques & Université de Marne Lavallée. The primary scope of REFAG2010 was to address today's achievements on theoretical concepts of reference systems and their practical implementations by individual space geodetic techniques and their combinations, underlying limiting factors, systematic errors and novel approaches for future improvements. Additionally, reference frame requirements, usage and applications in geosciences were also addressed during the Symposium. The program of the Symposium was divided into six sessions:

1. Theory and realization of global terrestrial reference systems
2. Strengths, weaknesses, modelling standards and processing strategies of space geodetic techniques
3. Definition, establishment, maintenance and integration of regional reference frames
4. Interaction between the celestial and the terrestrial reference frames
5. Definition and establishment of vertical reference systems
6. Usage and applications of reference frames in Geosciences

The Scientific organizing Committee consisted of:

- Zuheir Altamimi (IAG Commission 1 President)
- Mike Craymer (IAG Commission 1 Vice President)
- Markus Rothacher (President SC1.1)
- Claude Boucher (President SC1.2)
- João Torres (President SC1.3)
- Harald Schuh (President SC1.4)

and the local organizing committee consisted of:

- Xavier Collilieux
- David Coulot
- Laurent Métivier
- Christiane Guerin

who are members of the Geodetic Research Laboratory (LAREG) of the Institut National de l'Information Géographique et Forestière (IGN) , France.

More than 150 scientists from 31 countries attended the Symposium. There were 43 oral and 25 poster presentations during the 5 days of the Symposium. More information is available at the REFAG2010 Symposium web site: http://iag.ign.fr/index.php?id=138. Forty papers were peer-reviewed and published in these proceedings, summarizing the main outcome of the Symposium.

The Symposium and the review process would not have been possible without the contribution of the following colleagues who acted as session conveners and associated editors, in alphabetic order: Claude Boucher, David Coulot, Mike Craymer, Richard Gross, Johaness Ihde, Frank Lemoine, Markus Rothacher, Harald Schuh, Michael Sideris, Peter Steigenberger and João Torres. I am also very grateful to all the reviewers listed in the front matter of these Proceedings for their concise reviews of the REFAG papers. My deep gratitude goes to my IGN colleagues who organized the logistics of the Symposium, and in particular to my co-editor, Xavier Collilieux, who created, managed and operated not only the Symposium website, but also the website of the Commission during its 4-year term (2007–2011).

Zuheir Altamimi
President, IAG Commission 1 (2007–2011)

List of Reviewers

Luisa Bastos
Matthias Becker
Etienne Bernard
Mario Berube
Johannes Böhm
Claude Boucher
Pierre Briole
Elmar Brockmann
Alessandro Caporali
Xavier Collilieux
Oscar Colombo
David Coulot
Michael Craymer
John Dawson
Athanasios Dermanis
Galina Dick
John Dow
Miguel Dumett
Rémi Ferland
Steve Fisher
Mathias Fritsche
Daniel Gambis
Bruno Garayt
David Gordon
Mike Heflin
Manuel Hernandez-Pajares
Thomas A. Herring
Jianliang Huang
Urs Hugentobler
Robert Heinkelmann
Sharon Kedar
Ambrus Kenyeres
Kerry Kingham
Matt King
Johannes Ihde
Sébastien Lambert
Karine Le Bail
Frank Lemoine
Gunter Liebsch
Vincenza Luceri
Brian Luzum
Dan MacMillan
Zinovy Malkin
Jürgen Müller
Jay Parker
Hans-Peter Plag
Gerard Petit
Bill Petrachenko
Markku Poutanen
Elena Rangelova
Erricos Pavlis
Jim Ray
Paul Rebischung
John Ries
Manuel Ruiz-Perez
Martina Sacher
Laura Sanchez
Pierguido Sarti
Steffen Schoen
Florian Seitz
Michael G. Sideris
Michael Soffel
Drazen Svehla
Daniela Thaller
João Torres
Paul Tregoning
Marc Véronneau
Pascal Willis
Herbert Wilmes
Hartmut Wziontek

Contents

Geodetic Reference Frames: 40 Years of Technological Progress and of International Cooperation: 1970–2010

1

C. Boucher

1 Doppler Technique in Space Geodesy

In the early 1970s, a major change occurred in geodetic activities through the operational availability of the satellite Doppler technology.

Among the various space techniques developed in the 1960s (optical, laser, radio) and mostly coordinated by the US National Geodetic Satellite Program (NGSP), this technique got a specific appeal for the following main reasons:

- Availability of a space segment provided by the Navy Navigation Satellite System (NNSS) or Transit, after its declassification for civilian users
- All weather access over the whole Earth, thanks to the polar orbits of NNSS and the radio frequencies
- Availability of portable receivers through commercial companies (Canadian Marconi, JMR, Magnavox)

At that time in Europe, a series of informal meetings, initiated by Paul Melchior (also general secretary of the IUGG), known as *Journées Luxembourgeoises de Géodynamique* (JLG), provided the frame for the design and realization of first European campaigns in space geodesy. NNSS offered the adequate technology. Two campaigns were realized:

- EDOC-1 in 1975
- EDOC-2 in 1977

This illustrated the capability in Europe to design and realize such field campaigns, as well as ensure the data analysis. These campaigns also provided for the first time an estimate of the distortions at the European level of the common reference frame available over Europe at that time, the European Datum 1950 (ED50).

At the international level, Doppler technique was also a leading technology to develop cooperation between geodesists and surveyors, such as for oil exploration or hydrography. I want at this point to mention the series of international symposia organized by the US Defense Mapping Agency (now NGA) at Las Cruces in 1974, Austin in 1979 and again Las Cruces in 1982.

The dual character of the Transit system was also an opportunity to exchange between military and civilian communities, in particular for the tracking networks (OPNET, TRANET and MEDOC) and satellite orbits (Broadcast ephemerides, Precise ephemerides). The study and use of these various data made popular within the geodetic and surveying communities the concept of terrestrial reference frame, with those specifically related to the ephemeredes: NWL9D, NWL10D, NSWC9Z2, WGS72, WGS84.

Another important domain of activity was related to data processing methods (global dynamical, short arc, point positioning). Several software packages were developed by various groups around the world, in Belgium, Canada, France, UK and USA for instance. Also some software packages were disseminated throughout the international community (DOPPLR, GEODOP).

In addition, significant works were devoted to the design, implementation and maintenance of various networks of Doppler stations. In addition to the global tracking networks necessary for satellite orbit determination (OPNET, TRANET and MEDOC), numerous local or regional campaigns were realized, particularly designed for short arc processing. Remember some of these names: EROSDOC, SEATOC, TIMEDOC, WEDOC.

Dedicated to Jean-Jacques Levallois (1911–2001) and Georges Laclavère (1906–1994)

I am grateful to the organizers of the REFAG symposium for inviting me as keynote speaker. I take this opportunity to informally review 40 years of activities related to the field of geodetic reference frames, driven by various technological progresses as well as international cooperation, and ultimately human relations around the world.

C. Boucher (✉)
Observatoire de Paris/SYRTE
e-mail: claude-boucher@club-internet.fr

Z. Altamimi and X. Collilieux (eds.), *Reference Frames for Applications in Geosciences*, International Association of Geodesy Symposia 138, DOI 10.1007/978-3-642-32998-2_1,

Multiple applications of the Doppler technique were also developed, offering to the geodetic and space communities new opportunities. We can mention some of them:

- Precise orbit determination. Doppler devices equipped several satellite missions, in particular in satellite altimetry (GEOS-3, SEASAT, GEOSAT)
- Contribution to the realization of global terrestrial reference frames (see below)
- Realization of primary reference frames of national or continental extension
- Determination of transformation parameters between global and local datums

2 Reference Systems and Earth Rotation: MERIT and the Creation of IERS

In the early 1970s, the monitoring of the rotation of the Earth in space was ensured at the international level by two services, the Bureau International de l'Heure (BIH) in Paris, France and the International Polar Monitoring Service (IPMS) in Misuzawa, Japan. Both services used optical astrometry.

Meantime, new space geodetic techniques started to provide polar motion estimates: Doppler already mentioned, but also Satellite Laser Ranging (SLR) and Very Long Baseline Interferometry (VLBI). The necessity to intercompare all these techniques led the International Astronomical Union (IAU) to establish in 1978 the MERIT project. This gave the opportunity to review and redefine the basic concepts, in particular the Earth orientation parameters (EOP: polar motion, UT, precession, nutation) as part of the transformation formula between celestial reference system (CRS) and terrestrial reference system (TRS).

Issues related to terrestrial reference systems and terrestrial reference frames (TRF) were the topic of the COTES working group established in 1980 by the International Association of Geodesy (IAG), as contributor to MERIT.

Prior to MERIT-COTES, the terrestrial reference system was the so-called CIO/BIH system, a geocentric system the orientation of which was realized by the astronomical coordinates of the network of astrometric instruments (astrolabes and PZT) used by BIH or IMPS.

A major conceptual as well as operational change occurred at that time by considering a new type of realization based on geometric positions of networks of space geodetic instruments. MERIT-COTES was the framework for the development of these activities:

- COTES provided a forum to discuss conceptual issues
- All analysis groups participating to MERIT were asked to provide a TRF consistent with the EOPs provided by them
- BIH played the role of operational center, and was in charge to collect all data, and develop related standards
- Two important standards were initiated during MERIT-COTES, which were subsequently developed: the MERIT standards and the DOMES numbering system
- BIH developed with the French *Institut Géographique National* (IGN) a joint research program, led by Bernard Guinot and Martine Feissel for BIH and myself for IGN, to implement these new concepts using data collected by MERIT, and specifically to determine a TRF by combination of submitted TRFs, consistent with EOPs (Boucher and Feissel 1984)
- Four solutions were published in the BIH annual reports during MERIT: BTS84, BTS85, BTS86 and BTS87

The conclusion of the MERIT-COTES project was the establishment in 1988, jointly by IAU and IUGG, of a new international service (IERS) implementing operationally these new procedures for Earth orientation monitoring and reference frame realizations. IERS replaced both IPMS and BIH, the time activity of the later being moved in BIPM.

3 New Challenges in Regional Reference Frames: ADOS and Africa

As I already mentioned, Doppler technology enabled the geodetic community to realize anywhere in the world reference networks at metric level. The ADOS project used this technology to establish a primary reference network over the whole African continent. This project was a joint initiative of the IAG and the *Organisation Africaine de Cartographie et de Télédétection* (OACT), and was chaired by Ivan I. Mueller. ADOS was a real success based on a wide cooperation between African and non-African organizations. Important technology transfers were achieved for field works, data processing and expertise, notably through training courses in Nairobi (Kenya) and Yamoussoukro (Ivory Coast).

4 The GNSS Revolution

The next technological revolution I was involved in was provided by the satellite navigation system designed by US in late 1970s to replace NNSS, namely the NAVSTAR (Navigation by Satellite Timing and Ranging) also called GPS (Global Positioning System).

So many things could be discussed on this matter... I will just concentrate on some key aspects

Concerning the technological developments, three major aspects need to be mentioned:

- The progressive availability of several kinds of GPS receivers (L1 C/A, L1/L2 codeless). I would just mention

the first tests in France in 1984 (TI4100, Macrometer) and the first GPS receiver bought by IGN in 1985 (SERCEL TR5S)
- The evolution and diversification of processing strategies
  - Local relative positioning
  - Global positioning with fiducial stations
  - Global dynamical processing of global tracking campaigns (remember GIG91)
  - Real time processing, PPP
- The progressive appearance of other similar systems from other countries (Russia: GLONASS, China: Beidou, Europe: Galileo) and the emerging concept of generic Global Navigation Satellite System (GNSS)

Concerning international cooperation, I recall here some events through the years:

- In years 1984–1985, IGN developed research cooperations on GPS with The US National Geodetic Survey (NGS) and the University of Bern (PhD thesis of Pascal Willis)
- The use of GPS in the 1986 Channel Tunnel project, jointly between the Ordnance Survey of Great Britain (OSGB) and IGN
- The extensive use of GPS since 1989 by the EUREF group for the realization of a common frame (see below)
- The progressive developments of global GPS tracking (CIGNET, GIG91)
- The establishment in 1993 of the International GPS Service (IGS) thanks in particular to Ivan I Mueller, Gerhard Beutler and Ruth Neilhan
- The IGEX-98 campaign for GLONASS
- More recently the establishment of the International Committee for GNSS (ICG)

Finally, I would also like to mention the tremendous range of applications of GNSS, such as

- Local surveys
- Realization of global, regional and national reference frames mostly using permanent GNSS stations
- High accuracy LEO satellite orbit determination
- Extensive use in geosciences (tectonics, glaciology, ionosphere, meteorology)

5 Developments in IERS

Returning to IERS, I was involved up to recently into its developments.

The first period ranged from 1988 to 2000. Most activities were concentrated into its Central Bureau, realized by a joint venture between the *Bureau des longitudes*, IGN and the Paris Observatory, headed by Martine Feissel. In the frame of this bureau, IGN produced nine successive solutions of the new primary reference frame known as ITRF: from ITRF88 to ITRF97. The initial solutions were based on a combination of SLR, LLR and VLBI data. GPS data were first added in the ITRF91 solution and DORIS (see below) in ITRF94.

A new organization for IERS was implemented on January 1st 2001, creating in particular several product centers. The previous work on TRF done in the frame of the central bureau was transferred to the ITRS Product center, which I was initially in charge of, up to 2003. At this time, I left this 15-year involvement in ITRS, replaced by my colleague Zuheir Altamimi, who worked with me since the beginning of IERS. Since 2000, three solutions were produced: ITRF2000, ITRF2005 and very recently ITRF2008.

I am sure that numerous presentations and discussions will show off during this meeting.

6 Evolution of European Activities: RETRIG, REUN and EUREF

Concerning reference frames, European countries had in the seventies a significant cooperative work under the umbrella of IAG through two organizations: RETRIG for horizontal control network and UELN for leveling. This was a noticeable example of effective and efficient cooperation between European scientists and operational geodesists. Nevertheless, for well known geopolitical reasons, it was restricted to western European countries, and a two step combination strategy was used, avoiding exchange of raw data between countries.

The occurrence of space techniques droved the complete modification of these activities. In a first phase, space based (mostly Doppler) networks offered an external evaluation to RETRIG activities. In a second step, RETRIG decided to provide a combined solution, and even designed for this purpose a specific Doppler campaign RETDOC. This resulted to the ED87 solution, which was the last product of RETRIG, because in the meantime, a proposal to completely reorganize the activity was done, which recommended the establishment of a new organization, EUREF, officially created in 1987 during the XIXth IUGG/IAG General Assembly in Vancouver (Canada).

EUREF adopted space techniques as primary technique, fully justified at that time by the maturity of GPS and the appearance of ITRF. The initial work of EUREF was to organize a GPS campaign over Europe, offering a primary regional frame linked to ITRF, and to adopt a new "datum" of Europe, ETRS89, tri-dimensional and rigorously linked to ITRS.

All these new developments were ensured thanks to an executive structure created by EUREF, its Technical Working Group (TWG) which meets several times per year, and the leading role of Herman Seeger. GPS campaigns were initially used for national realizations of ETRS89. Initially, the use of fiducial stations (SLR and VLBI) was a key element for the accuracy of such GPS campaigns. Thanks again to Herman Seeger, a mobile VLBI station from NGS was temporary shipped to Europe. Several stations were established, including two sites in France (Brest and Grasse).

Then the establishment of permanent GPS stations was preferred. EUREF is presently providing the EUREF Permanent Network (EPN), as link between global IGS network and national densifications as it will be mentioned below for the case of France. EPN is also planning to move from GPS to GNSS, taking the opportunity of the development of new systems such as Galileo.

Concerning the vertical information, EUREF took also this responsibility, absorbing the UELN group, and making strong connexions with geoid and tide gauge data.

Finally, I also want to emphasize the remarkable expansion in the number of country members of EUREF, as well as its recognized expertise from EuroGeographics (former CERCO) and mostly the European Commission, notably related to the recent INSPIRE directive.

7 France as Example of National Activities: RGF and RGP

I take the opportunity of this speech to just mention that my deep involvement in international activities, especially IERS and EUREF, convinced me at early stage that the implementation of a new reference frame at national level as direct densification of global infrastructures was actually technically feasible. The availability of GPS technology was increasing among users of our national grid based on triangulation (NTF), and exhibiting inaccuracies in the reference.

We therefore proposed a new national reference frame, tri-dimensional and with an accuracy compatible with GPS. The new system RGF was selected as a national realization of ETRS89.

Initially the plan was to readjust the first and second order of the French NTF triangulation. Later on, a complete use of GPS ensured the densification. In addition, the present day primary national frame is based upon permanent GNSS stations (RGP network), a densification of the EPN previously mentioned.

8 A New Generation of Doppler Technique: DORIS

Last but not least, I would like to say a few words about DORIS. In this was, I close a loop, back to NNSS initially mentioned.

As you may know, the DORIS system is a French concept based on early experience that CNES, together with IGN and GRGS, acquired, in particular by the deployment and exploitation of the MEDOC network, the GEOLE studies or the ARGOS system which implemented the concept of inverted Doppler system, receiver being onboard satellites.

The development of DORIS involved various aspects related to reference frames. IGN was and still is in charge of the deployment and maintenance of the global ground network.

DORIS progressively demonstrated to be an important contributor to the estimation of ITRF, as I guess some presentations will mention it during this meeting. Conversely, DORIS has its own realization of ITRS, based on ITRF, and used for operational activities such as Precise Orbit Determination and real time Orbit Determination using the DIODE software package, developed by CNES and available onboard these satellites.

Finally, starting from this initiative I mentioned, DORIS has now a truly international environment, thanks to the International DORIS Service (IDS).

9 Concluding Remarks

I shall finish this keynote by three concluding remarks.

Concerning IAG, I want to acknowledge the establishment of GGOS and the firm participation to the Global Earth Observation (GEO) initiative. The driving force should be the legitimate recognition of Geodesy outside our community. The operational aspects, so necessary to scientific and societal challenges, should be nevertheless ultimately handled by a proper organization, beyond the international scientific non governmental status of IAG, IAU or IUGG. Our neighboring communities, such as meteorology or oceanography, have succeeded in this way. Why not geodesy or in a broader scope Solid Earth sciences?

Concerning GNSS, we are all conscious of the tremendous impact of this technology in Geodesy as well as numerous other domains. Like Internet, there are a large number of positive aspects, but still also a few negative ones, such as addiction or vulnerability we all have to manage.

Ultimately, concerning ITRS, we can now welcome the definitive success of this product within the geodetic community and its progress within various domains of geosciences. Efforts to get recognition and the preferred adoption beyond scientists should be continued, such as by getting a relevant ISO standard.

I thank you for your attention.

Part I

Theory and Realization of Global Terrestrial Reference Systems

ITRF Combination: Theoretical and Practical Considerations and Lessons from ITRF2008

2

Z. Altamimi, X. Collilieux, and L. Métivier

Abstract

The current ITRF construction is based on a two-step approach, combining input data provided by space geodesy techniques (VLBI, SLR, GPS, DORIS) in the form of time series of station positions and Earth Orientation Parameters. In the first step, the individual technique time series are rigorously stacked (accumulated) yielding long-term secular solutions, while the second step forms the ITRF final combination of the four technique long-term solutions together with local ties at co-location sites. The combination model involves a 7- or 14-parameter similarity transformation formula, for time series stacking and multi-technique combination, respectively. Not all these parameters are necessarily estimated in the combination process, some or all of them could be eliminated from the constructed normal equation, depending on the combination purpose. The paper discusses the relevance of the combination model and its appropriateness for the ITRF combination activities, both from the theoretical and practical point of views, and in particular for the reference frame specifications (origin, scale, orientation and their time evolutions). Selected analysis tests of ITRF2008 input data and results are used to illustrate the discussion as well as to address lessons learned from ITRF2008 experience.

Keywords

Reference systems • Reference frames • Time evolution • ITRF

1 Introduction

With the advent of space geodesy in the early 1980s, the importance of reference frames has became more and more important, as a function of technological and data analysis advances. Appropriate definition of a Terrestrial Reference System (TRS) and its precise materialization through a Terrestrial Reference Frame (TRF) are fundamental to many applications in geosciences. The main TRS and TRF specifications are the origin, scale, orientation and their time evolutions. Any defect on these parameters would have an impact on the results and interpretation of geodetic and geophysical applications that require the usage of a reference frame, such as:

- Precise Orbit Determination, not only for Global Navigation Satellite Systems (GNSS), but also for other satellite missions dedicated to Altimetry, Oceanography, Gravity;
- Earth sciences applications, such as tectonic motion and crustal deformation, sea level variations and Earth rotation (Collilieux and Altamimi 2012).

Given the currently available reference frame products provided by space geodetic techniques, representations of terrestrial reference frames are divided in two categories:

- "Quasi-instantaneous" reference frame which gives access to mean station positions at "short" interval, using space geodesy observations over, e.g. 1 or several hours, 1 day, and up to 1 week. Note that over 1 month, stations may be subject to displacements of the order of

Z. Altamimi (✉) • X. Collilieux • L. Métivier
IGN/LAREG and GRGS, Université paris Diderot, 35 rue, Héléne Brion 75013 Paris, France
e-mail: zuheir.altamimi@ign.fr

Z. Altamimi and X. Collilieux (eds.), *Reference Frames for Applications in Geosciences*, International Association of Geodesy Symposia 138, DOI 10.1007/978-3-642-32998-2_2,

1–10 mm due to plate tectonic motion. Time series of such frames encompass not only station linear motion due to plate tectonic, but also non-linear motion and discontinuities due to geophysical events such as Earthquakes.

- Long-term secular frame which gives access to mean station positions (X) at a reference epoch, t_0, and station linear velocities ($\dot{X}$). The propagation of station positions (as well as their variances) at any epoch t is operated using $X(t) = X(t_0) + \dot{X} \times (t - t_0)$.

The International Terrestrial Reference Frame (ITRF) is by definition a secular frame, and therefore precise definition of its defining parameters are of interest to many Earth science applications. For more details regarding the International Terrestrial Reference System and Frame (ITRS, ITRF) description and definition, the reader may refer to Chap. 4 of the IERS Conventions (Petit and Luzum 2010). As the input data of the current ITRF construction are in the form of time series of station positions (and Earth Orientation Parameters: EOPs), it is fundamental to evaluate the temporal behavior of not only the individual station positions, but also, and equally important, the frame physical parameters, namely the scale and the origin components. As it will be quantified in this paper, any temporal discontinuity or unexpected drift of these parameters will directly impact the estimated constant station velocities. We recall here that the current scientific requirement in terms of accuracy and stability over time of the origin and scale of the ITRF are believed to be at the level of 1 mm and 0.1 mm/year (Plag and Pearlman 2009; Blewitt et al. 2010; NRC 2010). This requirement is at least 10 times higher than what is achievable today, due mainly to the degradation of the network of space geodesy techniques and their intrinsic systematic errors.

In the following, we recall in Sect. 2 the CATREF combination model used for the ITRF computation and discuss the usefulness of including the transformation parameters in that model. Section 3 is devoted to numerical applications intended to verify that the estimation of weekly transformation parameters in the time series stacking has no impact on the ITRF secular frame. In Sect. 4 we summarize the main ITRF2008 lessons regarding the usage and consistency of local ties in the ITRF combination.

2 CATREF Combination Model

Although the CATREF combination model is extensively described in several publications (see for instance Altamimi et al. 2002, 2007, 2011), we review here its main equations for the purpose of this paper, discussing the benefit of including the transformation parameters in this model. The main two equations are written as:

$$\begin{cases} X_s^i = X_c^i + (t_s^i - t_0)\dot{X}_c^i \\ \quad + T_k + D_k X_c^i + R_k X_c^i \\ \quad + (t_s^i - t_k)\left[\dot{T}_k + \dot{D}_k X_c^i + \dot{R}_k X_c^i\right] \\ \dot{X}_s^i = \dot{X}_c^i + \dot{T}_k + \dot{D}_k X_c^i + \dot{R}_k X_c^i \end{cases} \tag{2.1}$$

$$\begin{cases} x_s^p = x_c^p + R2_k \\ y_s^p = y_c^p + R1_k \\ UT_s = UT_c - \frac{1}{f} R3_k \\ \dot{x}_s^p = \dot{x}_c^p \\ \dot{y}_s^p = \dot{y}_c^p \\ LOD_s = LOD_c \end{cases} \tag{2.2}$$

where for each point i, X_s^i (at epoch t_s^i) and $\dot{X}_s^i$ are positions and velocities of technique solution s and X_c^i (at epoch t_0) and $\dot{X}_c^i$ are those of the combined solution c. For each individual frame k, as implicitly defined by solution s, D_k is the scale factor, T_k the translation vector and R_k the rotation matrix. The dotted parameters designate their derivatives with respect to time. The translation vector T_k is composed of three origin components, namely T_x, T_y, T_z, and the rotation matrix of three small rotation parameters: R_x, R_y, R_z, following the three axes, respectively X, Y, Z. t_k is a conventionally selected epoch of the seven transformation parameters. In addition to Eq. 2.1 involving station positions (and velocities), the EOPs are added by Eq. 2.2 making use of pole coordinates x_s^p, y_s^p and universal time UT_s as well as their daily rates $\dot{x}_s^p$, $\dot{y}_s^p$ and LOD_s. The link between the combined frame and the EOPs is ensured via the three rotation parameters appearing in the first three lines of Eq. 2.2. Detailed derivation of the above equations could also be found in Altamimi and Dermanis (2011), and more discussion regarding the polar motion rate equations is available in Altamimi et al. (2011).

CATREF combination model was designed to be as general as possible in order to be able dealing with reference frame solutions of different natures. In time series stacking, the first line of Eq. 2.1, nullifying its last terms involving the rates of the transformation parameters, and the entire Eq. 2.2 are used. The entire two equations are used in the combination of multi-technique long-term solutions. Note that in both combination cases, the EOPs could be included or discarded. Note also that polar motion rate equations (fourth and fifth lines of Eq. 2.2) are included in the combination model in order to be able to take into account solutions where the polar motion representation is in the form of offset and drift. For solutions where the polar motion representation is in the form of a continuous piece-wise linear function, the polar motion rate equations are of course discarded.

It should be noted here that including the transformation parameters in the combination model does not imply systematically estimating them. Depending on the application, some or all of them could/should be eliminated from the constructed normal equation system. However, from the ITRF combination perspective, we list below the main advantages of estimating the transformation parameters in the two-step procedure: time series stacking and multi-technique combination of long-term solutions. In the step of time series stacking, the main advantages are to allow:

- Evaluating the temporal behavior of the reference frame parameters and in particular the physical ones, namely the scale and origin components. In the same way as a station presenting discontinuities in its position time series (due for example to earthquakes) should not be part of the list of reference frame stations, a particular parameter having such discontinuities should be excluded from the reference frame definition. Otherwise adopting such a parameter in the ITRF definition would introduce biases in its time evolution and consequently in the estimated station velocities. In a simulated study, we found that a discontinuity of 1.5 ppb in the middle of a scale time series covering 4 years induces a bias of 3 mm/year in the vertical velocities of all the stations, and 1 mm/year over a time-span of 16 years. In the same study, a simulated scale drift of 0.15 ppb/year, induces a vertical velocity bias of 1 mm/year. Therefore estimating the transformation parameters in the time series analysis is a tool to evaluate the level of stability over time of the frame defining parameters;
- Assessing robustly the repeatability (internal precision) of the analyzed solutions by computing the Weighted Root Mean Scatter (WRMS) of each epoch (daily, weekly) solution with respect to the combined long-term solution;
- Applying the inner minimal constraints as described in Altamimi et al. (2007), having the advantage of preserving mean origin and scale of satellite technique and the mean scale of VLBI solutions;

In the second step of the ITRF combination involving multi-technique long-term solutions and local ties, estimating the transformation parameters has the following advantages:

- The ITRF defining parameters are eliminated from the normal equation system, offering in this way various possible options to define the combined reference frame among the incorporated solutions, e.g. adopting SLR origin; SLR, VLBI or their average to define the scale.
- Possible biases between the technique solutions are rigourously quantified. This is the case of the scale bias of 1.05 ppb at epoch 2005.0 and 0.049 ppb/year between VLBI and SLR solutions determined from the ITRF2008 combination (Altamimi et al. 2011);
- Assessing the uncertainties of the transfer of the SLR origin and VLBI and SLR mean scale to GPS and DORIS frames as it will be shown in this paper.

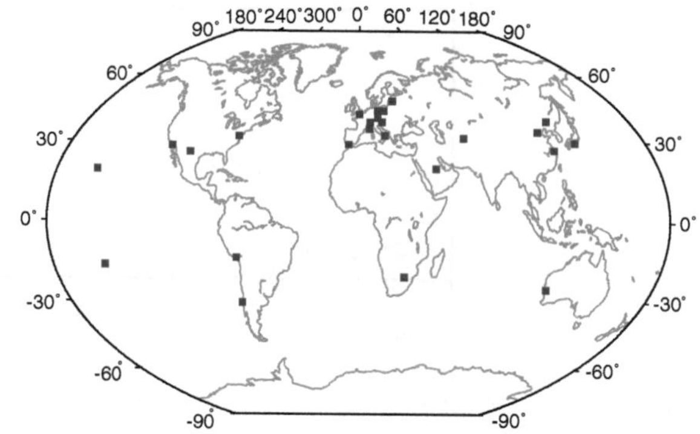

Fig. 2.1 Distribution of the 26 SLR stations used in the test combinations

In the following section devoted to numerical applications using some ITRF2008 input data, we discuss the difference in the results between estimating and not estimating the scale and origin parameters in the time series stacking and in a test combination of multi-technique long term solutions. We show in particular that including or not these parameters in the observation equations model produces the same mean origin and scale of the obtained two long-term solutions.

However, it should be noted that an obvious alternative to the two-step procedure is the one-step approach where all the technique time series are stacked together with local ties. Although this one-step procedure is computationally prohibitive, it should mathematically be equivalent to the two-step approach.

3 Numerical Applications

In order to evaluate the difference in the results between estimating and not the origin components and the scale when stacking time series provided by satellite techniques, we use hereafter, as an example, the ILRS SLR time series that contributed to the ITRF2008 (Pavlis et al. 2010). We extracted from these weekly solutions 26 stations (illustrated by Fig. 2.1) with long observation histories (from 7 to 16 years) and conducted two types of stacking: with and without including the weekly origin and scale parameters in the observation equations model. We first analyzed the results on three aspects: the weekly WRMS with respect to the combined/stacked frame, the precision of the estimated station positions and velocities and the 14

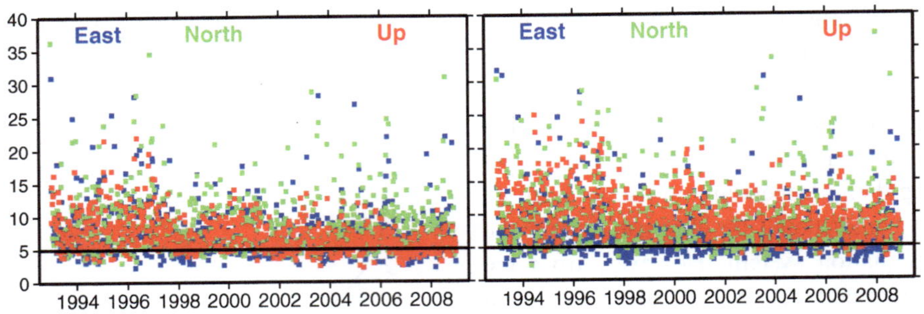

Fig. 2.2 Weekly WRMS of SLR time series as results of their stacking with (*left*) and without (*right*) estimating the weekly translations and scales

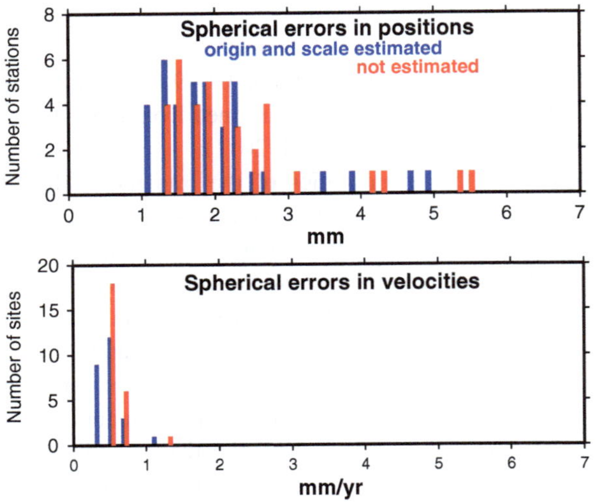

Fig. 2.3 Spherical errors in station positions and velocities of the long-term SLR solutions obtained by stacking the time series with and without including the weekly origin and scale parameters in the observation equations model

transformation parameters between the two long-term solutions.

When accumulating (rigorously stacking) time series of station positions (including or not EOPs), the constructed normal equation that includes the transformation parameters has a rank deficiency of 14, corresponding to the combined frame parameters that have to be defined. In this case we used the inner minimal constraints as described in Altamimi et al. (2007) that make use of the time series of the 7 estimated parameters. In the second stacking where the origin and scale parameters were eliminated, the normal equation has a rank deficiency of six corresponding to the three rotations and their rates. In this case we also used the inner minimal constraints over the time series of the rotation parameters.

Figure 2.2 illustrates the weekly WRMS with respect to the combined frame, with and without estimating the weekly origins and scales. This figure shows expected increase of the residuals when weekly origins and scales are not estimated, given the fact that the transformation parameters absorb part of the time series biases. These biases include technique systematic errors, as well as geocenter motion and aliased loading effects (Collilieux et al. 2009, 2010). However, as shown by Collilieux et al. (2010), applying a loading model to the station position time series before the stacking decreases the annual signal present in the residuals, but would not absorb the systematic errors in these tested SLR time series. The mean of these WRMS shown at Fig. 2.2 are 7.3, 8.7, 7.2 mm, versus 7.2, 9.5, 9.8 mm in east, north, and vertical components, with and without estimating the transformation parameters, respectively.

Another interesting feature to examine is the level of precision of the estimated station positions and velocities of the two accumulated frames, with and without including the weekly transformation parameters in the observation equations model. Figure 2.3 displays the spherical formal errors of both estimations, computed by the square sum of the formal errors resulting from the least squares adjustment, following Altamimi et al. (2002). This figure indicates that the precision of station positions and velocities is higher when the transformation parameters are estimated. This is also an expected result because the formal errors shown at Fig. 2.3 are function of the variance factor of unit weight computed with the residuals illustrated by Fig. 2.2.

We also estimated 14 transformation parameters between the two estimated long-term frames and found that all are insignificantly different from zero. The WRMS values of the 14-parameter fit are of the order of 1 mm in positions and 0.2 mm/year in velocities. These results indicate that the two estimated frames are equivalent at the level of their intrinsic uncertainties. They also confirm that the non-linear variations related to loading effects which are partly absorbed by the estimated weekly translations and scales does not affect the ITRF frame parameters as demonstrated by Collilieux et al. (2010). These results are also expected since the loading effects induce seasonal (annual or semi-annual) variations, with no impact on the estimated linear velocities for stations with long time-span, whether these variations leak to the station residuals or to the epoch transformation parameters. Blewitt and Lavallée (2002) showed for instance that in case of GPS, 2.5 years is the minimum time-span necessary to minimize the effect of annual signals on the estimated linear station velocity. This also means that despite the poor SLR network geometry (see Fig. 2.1 where only 5 stations out of 26 are located in the southern hemisphere), estimating the weekly translations and scales does not affect the estimated linear velocities. This fact was demonstrated theoretically by Collilieux (2008), using a simpler case of translation parameters only, and uniform weighting.

In order to evaluate the behavior of the two estimated long-term solutions in a multi-technique combination, two combinations were tested, involving the long-term GPS solution used in the ITRF2008 elaboration. Local ties at 21 GPS-SLR co-location sites were used, adopting the same weighting as the one used in the ITRF2008 combination. The results of these two combinations show that the estimated 14 transformation parameters between GPS and SLR solutions are consistent in both cases, their differences being within the uncertainties of the estimated parameters. The resulting two combined frames are also statistically equivalent: the differences are in average at the level of 1 mm in station positions and 0.2 mm/year in velocities. The level of their agreement with local ties is also within the noise and consistent with the results obtained in the ITRF2008 combination.

4 Lessons from ITRF2008

Detailed ITRF2008 results and discussion are published in Altamimi et al. (2011), and in particular regarding the impact of local ties on the ITRF2008 combination. Consistency between local ties and space geodesy estimates are critically analyzed. We showed in particular that 50 % of the available SLR and VLBI tie vectors to GPS exhibit residuals larger than 6 mm, and about 30 % have residuals larger than 10 mm. These discrepancies are to be understood as differences between local ties and space geodesy estimates. The reasons for these discrepancies are difficult to pinpoint, since they could be due to errors in local ties, in space geodesy estimates or in both. There are indications discussed in Altamimi et al. (2011) that these discrepancies are most likely to be due to systematic errors in space geodesy estimates, rather than in local surveys. In addition to the critical analysis discussed in Altamimi et al. (2011), we report here the impact of the usage of local ties on the level of uncertainty of the transfer of SLR origin and SLR and VLBI mean scale to GPS frame. Table 2.1 lists the levels of that uncertainty for three different cases of used local tie sets: (1) ties where the discrepancies with space geodesy estimates do not exceed 6 mm, (2) ties where the discrepancies are less than 10 mm, and (3) all available ties used in the ITRF2008 combination provided in SINEX files where 63 % of them are with full variance-covariance information. The results listed in this table indicate that the more ties used the better is the precision of the translations and scale of the ITRF2008 frame and their transfer to GPS frame. This kind of assessment is possible thanks to the inclusion of the transformation in the CATREF combination model.

Table 2.1 Uncertainties (in mm) of the transfer of SLR origin and SLR and VLBI mean scale to GPS frame, as a function of local ties used

Ties used	TX	TY	TZ	Scale
Ties discrepancies <6 mm	2.5	2.5	2.5	1.4
Ties discrepancies up to 10 mm	1.4	1.1	1.2	1.2
All tie SINEX files (ITRF2008 results)	0.6	0.5	0.6	0.6

5 Conclusion

CATREF combination model was developed and designed for the purpose of the ITRF combinations and is well adapted for time series stacking and multi-technique combinations. It is based on the 7- or 14-parameter similarity transformation, a classical mathematical formula allowing transformation between distinct reference frames. Although these parameters are not all systematically estimated in all combination processes, we identified the advantages of including them in the ITRF combination procedure.

Using a sub-set of SLR time series incorporated in the ITRF2008 combination, we found that with or without including the weekly scales and origin translation components in the observation equations model, we obtain equivalent long-term solutions of station positions and velocities, at the level of their intrinsic uncertainties. Combining separately these two long-term solutions with GPS solution included in the ITRF2008, involving 21 local ties gave also equivalent results within the uncertainties of the estimated parameters.

One of the main lessons of the ITRF2008 results reported by Altamimi et al. (2011) is the large inconsistency between local ties and space geodesy estimates: 50 % of the available SLR and VLBI tie vectors to GPS exhibit residuals larger than 6 mm, and about 30 % have residuals larger than 10 mm. Thanks to the inclusion of the transformation parameters in the combination model, we were able to assess the level of uncertainty of the transfer of SLR origin and mean SLR and VLBI scale to GPS frame. We showed here that the more ties used (properly weighted) the better is the precision of the origin and scale determination.

References

Altamimi Z, Dermanis A (2011) The choice of reference system in ITRF formulation. In: Proceedings of Hotine-Marussi symposium, Rome, 2009 (in press)

Altamimi Z, Sillard P, Boucher C (2002) ITRF2000: a new release of the international terrestrial reference frame for Earth science applications. J Geophys Res 107(B10):2214. doi:10.1029/2001JB000561

Altamimi Z, Collilieux X, Legrand J, Garayt B, Boucher C (2007) ITRF2005: a new release of the international terrestrial reference frame based on time series of station positions and Earth orientation parameters. J Geophys Res 112(B09):401. doi:10.1029/2007JB004949

Altamimi Z, Collilieux X, Métivier L (2011) ITRF2008: an improved solution of the International Terrestrial Reference Frame. J Geodesy. doi:10.1007/s00190-011-0444-4

Blewitt G, Lavalée D, B7 (2002) Effect of annual signals on geodetic velocity. J Geophys Res 107. doi:10.1029/2001JB000570

Blewitt G, Altamimi Z, Davis J, Gross R, Kuo C-Y, Lemoine FG, Moore A, Neilan RE, Plag H-P, Rothacher M, Shum CK, Sideris MG, Schöne, Tregoning P, Zerbini S (2010) Geodetic observations and global reference frame contributions to understanding sea-level rise and variability. In: Church JA, Woodworth PH, Aarup T, Wilson WS (eds) Understanding sea-level rise and variability, 1st edn. Blackwell Publishing Ltd, Hoboken, NJ

Collilieux X (2008) Analyse des séries temporelles de positions des stations de géodésie spatiale: application au Repère International de Référence Terrestre (ITRF). Ph.D. dissertation, Paris Observatory

Collilieux, X. and Z. Altamimi (2012) External Evaluation of the Origin and the Scale of the International Terrestrial Reference Frame, Z. altamimi and X. Collilieux Ed., International Association of Geodesy Symposia 138, this issue, doi10.1007/978-3-642-32998-2_5

Collilieux X, Altamimi Z, Ray J, van Dam T, Wu X, B04402 (2009) Effect of the satellite laser ranging network distribution on geocenter motion estimation. J Geophys Res 114. doi:10.1029/2008JB005727

Collilieux X, Altamimi Z, Coulot D, van Dam T, Ray J (2010) Impact of loading effects on determination of the international terrestrial reference frame. Adv Space Res 45:144–154. doi:10.1016/j.asr.2009.08.024

National Research Council (2010) Precise geodetic infrastructure: national requirements for a shared resource. National Academy Press, Washington, DC. ISBN 978-0-309-15811-4

Pavlis E, Luceri C, Sciarretta C, Kelm R (2010) The ILRS contribution to ITRF2008. Available at. http://itrf.ensg.ign.fr/ITRF_solutions/2008/doc/ILRSSubmission4ITRF2008.pdf

Petit G, Luzum B (eds) (2010) IERS conventions 2010, IERS technical note 36, Verlag des Bundesamts für Kartographie und Geodäsie, Frankfurt am Main, Germany. http://tai.bipm.org/iers/conv2010/conv2010_c4.html

Plag HP, Pearlman M (eds) (2009) Global geodetic observing system. Meeting the requirements of a global society on a changing planet in 2020. Springer, Berlin. ISBN 978-3-642-02686-7

Distributed Processing for Large Geodetic Solutions 3

H. Boomkamp

Abstract

This paper reports on the activities of the IAG Working Group 1.1.1 on combination and comparison of precise orbits based on different space geodetic techniques. It will focus on the Dancer project which implements a distributed parameter estimation process that is scalable in the number of GPS receivers, so that an arbitrarily large number of receivers can be processed in a single reference frame realization. The background of this project will be summarized and its mathematical principles will be explained, as well as the essential aspects of the involved internet communication. It will show that the workload for data processing at a single participating receiver remains independent of the network size, while the data traffic only grows as a logarithmic function of the network size.

Keywords

Grid computing • Distributed process • Batch least squares • GPS • Orbit estimation

1 Introduction

Several current challenges in orbit estimation exceed the processing capacity of conventional Analysis Centres by three or four orders of magnitude. It is unlikely that an equivalent increase in capacity can be created at existing centres, or that hundreds of new centres might appear in the future. This means that alternative processing methods are of interest.

Two issues are of particular relevance to IAG sub-commission 1.1. The first is the simultaneous reprocessing of geodetic datasets over long solution arcs. At present, reprocessing occurs in the separate IAG services, but simultaneous analysis improves consistency and makes better use of local ties and ties at LEO satellites. The second issue is that the dominant dataset – GPS – has become too large and inaccessible to be analyzed at a single Analysis Centre. Only a small percentage of GPS sites have formal ITRF time series, which limits both the quality and the relevance of the ITRF.

To tackle these problems, IAG Working Group 1 has initiated three projects for grid computing on the internet. The *Digger* project is based on the Berkeley Open Infrastructure for Network Computing (Anderson 2004), and aims at fast reprocessing of all geodetic data of interest. The *Dancer* project builds a peer-to-peer network of the GPS receivers to allow least squares solutions for networks of unlimited size. Finally, *Dart* (Dancer Real-Time) offers an interface to the Dancer reference frame for real-time kinematic GPS users.

This paper reports on Dancer, which is the first of the three projects to approach operational status. Today there are around 20,000 permanent GPS receivers in national reference grids, networks for meteorology, earthquakes monitoring, wide-area augmentation, search and rescue, etc. Two problems prevent rigorous least-squares solutions in which all receivers are included. The first problem is that the processing capacity of conventional Analysis Centres is too small by several orders of magnitude. The second problem is that the majority of the GPS stations do not publish their data, or do this too late to be included in

H. Boomkamp (✉)
IAG sub-commission 1.1, Working Group 1, ESOC,
Robert-Bosch-Strasse 5, D-64293 Darmstadt, Germany
e-mail: henno.boomkamp@esa.int

Z. Altamimi and X. Collilieux (eds.), *Reference Frames for Applications in Geosciences*,
International Association of Geodesy Symposia 138, DOI 10.1007/978-3-642-32998-2_3,

routine analysis. Important receivers in that last category are most orbiting receivers on geodetic satellites.

The first problem can only be solved by using many more computers for routine analysis. Dancer aims at processing a hundred times more receivers than the current IGS, at a ten times higher data rate. The resulting dataset is three orders of magnitude larger than that of IGS, so that Dancer needs about a thousand times as many computers as the IGS.

The second problem is political, and would seem more difficult to solve. However, if the computer cluster is split in such a way that each data owner also owns part of the analysis capacity, the data itself does not have to be distributed in order for a receiver to be part of routine global analysis. This is also more practical than collocating thousands of computers at a single site.

Combining the above considerations, the natural way to implement the Dancer system was in the form of a distributed process on the internet. Many such systems exist, e.g. Andrzejak et al. (2010), Cappello et al. (2005), www.gridcomputing.com. Compared to the millions of computers in a project like SETI@home, Korpela et al. (2001), Dancer only has very modest requirements.

2 Hardware Infrastructure

The first task of the Dancer project was to find around 10,000 computers that might be collocated with GPS operators anywhere in the world, in such a way that a data owner may keep his observation data private while still taking part in the global estimation process. This problem can be solved easily, and at zero cost.

High-end GPS receivers output their observation data to a local or remote computer. At the majority of stations this computer is placed almost literally on top of the receiver. The network of permanent GPS sites is therefore primarily a network of 20,000 *computers*, to which GPS receivers are connected. Most of these computers are only used for sending RINEX files to Analysis Centres. A more efficient way of using the existing hardware will be to exploit these thousands of network computers for routine analysis, instead of trying to process the full workload with a handful of computers at Analysis Centres. This solves both problems from the previous section simultaneously. The combined processing capacity within the GPS network is three orders of magnitude larger than that of all Analysis Centres combined, while the computers are always perfectly collocated with the data owners. This processing capacity even grows proportionally to any future increase in network size.

Table 3.1 Target processing characteristics

Arc	48 h data batch, hourly solutions
Observations	Zero-differenced ionosphere-free
	Code and phase at 30 s epoch rate
Global param. shared by all receivers	8 for Earth rotation
	6 for pos/vel per satellite
	9 radiation pressure per satellite
	1 clock bias per satellite per epoch
Local param. for each receiver separately	3 station coordinates
	1 troposphere delay per 2 h
	1 ambiguity bias per pass
	1 clock bias per epoch
Process size with 20,000 receivers	750,000,000 observations
	75,000,000 local parameters
	150,000 global parameters

3 Distributed Least Squares Solutions

The second task of the Dancer project was to split a conventional batch least-squares solution in sub-tasks per receiver, keeping in mind that data traffic among the different processes should be minimized, so that all communication can take place via the public internet. The objective is to allow even the extremely large solution of Table 3.1, which would include all permanent receivers in the world today.

3.1 Pre-Elimination of Local Parameters

From the millions of estimated parameters in this process, only a small percentage consists of global parameters for Earth rotation, orbits and satellite clocks. All other parameters are local to a single receiver. The normal equation system can then be partitioned as in Eq. 3.1. The sub-vector $\vec{x}_A$ contains the global parameter corrections, while the sub-vectors $\vec{x}_j$ for $j = 1..N$ contain local parameter corrections for a single receiver. It will be clear how the dimensions of the other matrix partitions match those of the sub-vectors of $\vec{x}$.

$$\begin{pmatrix} A & B_1 & B_2 & \dots & B_N \\ {B_1}^t & M_1 & 0 & \dots & 0 \\ {B_2}^t & 0 & M_2 & \ddots & \vdots \\ \vdots & \vdots & \ddots & \ddots & 0 \\ {B_N}^t & 0 & \dots & 0 & M_N \end{pmatrix} \cdot \begin{pmatrix} \vec{x}_A \\ \vec{x}_1 \\ \vec{x}_2 \\ \vdots \\ \vec{x}_N \end{pmatrix} = \begin{pmatrix} \vec{y}_A \\ \vec{y}_1 \\ \vec{y}_2 \\ \vdots \\ \vec{y}_N \end{pmatrix} \tag{3.1}$$

The thousands of partitions B_j, M_j and $\vec{y}_j$ contain information that comes from just one receiver. Only partition A and its right-hand side sub-vector are accumulated from contributions of all receivers:

$$A = \sum_{j=1}^{N} A_j \tag{3.2}$$

$$\vec{y}_A = \sum_{j=1}^{N} \vec{y}_{Aj}$$

Using the second and later rows of Eq. 3.1, each local vector $\vec{x}_j$ can be expressed in terms of $\vec{x}_A$:

$$\begin{aligned} B_j^t \vec{x}_A + M_j \vec{x}_j &= \vec{y}_j \\ \Leftrightarrow \; \vec{x}_j &= M_j^{-1}\left(\vec{y}_j - B_j^t \vec{x}_A\right) \end{aligned} \tag{3.3}$$

Using Eqs. 3.2 and 3.3, the first row of Eq. 3.1 provides the pre-eliminated normal equation for $\vec{x}_A$ alone:

$$\sum_{j=1}^{N}\left(A_j - B_j M_j^{-1} B_j^t\right)\vec{x}_A = \sum_{j=1}^{N}\left(\vec{y}_{Aj} - B_j M_j^{-1}\vec{y}_j\right) \tag{3.4}$$

Each Dancer process can autonomously prepare its contribution j to this global normal equation. However, exchanging the large matrix contributions over the internet would lead to many Gigabytes of data traffic at all receivers, which is not realistic. Instead, a distributed solution method is used that only exchanges a few vectors.

3.2 Summary of the Solution Method

Each Dancer process divides its normal equation contribution j by the number of observations that were used to build it. This gives all contributions the same scale, of a single observation.

Dancer then computes the mean of the diagonals of all matrices j, using a network algorithm named *square dancing* that will be explained in a later section. This step gives all Dancer processes the same average diagonal, at the correct scale.

Each Dancer process now solves its own normal equation system j in which the original diagonal has been replaced by the global mean. This gives each process a different approximation for the vector $\vec{x}_A$.

The square dance method is now used once more to compute the average of all these solution vectors. It will be shown here below that this global mean is mathematically identical to $\vec{x}_A$ from Eq. 3.4.

With $\vec{x}_A$ solved, each Dancer process finds its local parameter corrections from Eq. 3.3 and can then update its orbits and observation residuals.

The solution process is iterated five times, so that network activity for a single run is dominated by the averaging of ten vectors. After the penultimate iteration, each process resolves integer ambiguities, which requires some additional internet traffic for the formation of suitable double differences.

3.3 Proof

A matrix contribution j in Eq. 3.4 is the sum of its diagonal matrix D_j and its off-diagonal matrix F_j. With $\vec{b}_j$ for the right-hand side, we can then write

$$\sum_{j=1}^{N}\left(D_j + F_j\right)\vec{x}_A = \sum_{j=1}^{N}\vec{b}_j \tag{3.5}$$

Dividing both sides by N shows that Eq. 3.5 is equivalent to the *mean* of all normal equation contributions j:

$$\frac{1}{N}\sum_{j=1}^{N}(D_j + F_j)\vec{x}_A = \frac{1}{N}\sum_{j=1}^{N}\vec{b}_j \tag{3.6}$$

Using the notation $\tilde{A}$ and $\tilde{a}$ for the global mean of a matrix or a vector respectively, this becomes

$$(\tilde{D} + \tilde{F})\vec{x}_A = \tilde{b} \tag{3.7}$$

Dancer replaces all local diagonals D_j by the global mean $\tilde{D}$. Each local process then solves a vector $\vec{x}_{Aj}$ from its modified equation system:

$$(\tilde{D} + F_j)\vec{x}_{Aj} = \vec{b}_j \qquad \Rightarrow \vec{x}_{Aj} \tag{3.8}$$

Rearranging this as in Eq. 3.3 gives a form for $\vec{x}_{Aj}$ known from the Jacobi method, e.g. Sameh (1971):

$$\vec{x}_{Aj} = \tilde{D}^{-1}\left(\vec{b}_j - F_j\vec{x}_{Aj}\right) \tag{3.9}$$

The Dancer solution $\tilde{x}$ is the mean of all $\vec{x}_{Aj}$, which is then also equal to the mean of the right-hand side of Eq. 3.9:

$$\tilde{x} = \frac{1}{N}\sum_{j=1}^{N}\vec{x}_{Aj} = \frac{1}{N}\sum_{j=1}^{N}\tilde{D}^{-1}\left(\vec{b}_j - F_j\vec{x}_{Aj}\right) \tag{3.10}$$

Because all processes used the same mean diagonal, it can be removed from the averaging operation:

$$\tilde{D}\tilde{x} = \frac{1}{N}\sum_{j=1}^{N}\vec{b}_j - \frac{1}{N}\sum_{j=1}^{N}F_j\vec{x}_{Aj} \tag{3.11}$$

We now use the statistical property that the mean of a product of two *independent* distributed variables is equal to the product of their mean values, e.g. Chung (1974):

$$\frac{1}{N}\sum_{j=1}^{N}F_j\vec{x}_{Aj} = \tilde{F}\tilde{x} \tag{3.12}$$

Equation 3.11 then becomes the same as Eq. 3.7:

$$\tilde{D}\tilde{x} = \tilde{b} - \tilde{F}\tilde{x} \quad \Leftrightarrow \quad (\tilde{D}+\tilde{F})\tilde{x} = \tilde{b} \tag{3.13}$$

so that $\tilde{x} \equiv \vec{x}_A$.

In summary: the off-diagonal matrix and right-hand side vector influence the solution Eq. 3.9 in linear combinations. Taking their average is another linear combination of the same numbers and can be done before or after the solution without changing the result. An Analysis Centre averages the entire normal equation, because it solves Eq. 3.5 which is equivalent to the mean Eq. 3.7. Dancer does not average the off-diagonal matrix or right-hand side, but averages the solution instead, which forms a compact linear combination of the same numbers.

This simple method for distributing a least squares solution over a computer grid reduces data traffic by several orders of magnitude, at the more affordable cost of solving an equation system Eq. 3.8 at each computer. It even reduces the overall process duration, because the local systems Eq. 3.8 are much sparser than the global Eq. 3.5, and are all accumulated and solved in parallel. The method can be scaled to any size N, just by adding computers.

3.4 Some Further Details

The method relies on independence of the distributions of the terms F_j and $\vec{x}_{Aj}$. A more accurate form of this equation would be

$$\frac{1}{N}\sum_{j=1}^{N}F_j\vec{x}_{Aj} = \tilde{F}\tilde{x} + \frac{1}{N}\sum_{j=1}^{N}\mathrm{cov}(F_j, \vec{x}_{Aj}) \tag{3.14}$$

The distribution of the matrices F_j is determined by errors in modelling partials and properties of the tracking geometry. The distribution of the solution vectors $\vec{x}_{Aj}$ depends on the modelling partials and on the noise in the observation residuals. Because this noise can be assumed independent from the errors in the partials, the covariance term in Eq. 3.14 is negligible. Minor defects in this assumption are compensated by the fact that the solution is iterated.

The contributions j in Eq. 3.5 might already be weighted sums over multiple receivers, without changing the validity of the method. This means that a solution for thousands of receivers can be produced on a cluster of a few dozen computers, in which each computer accumulates normal equation contributions from perhaps a hundred receivers, or whatever is feasible. This still imposes the condition that the observations are available to the centralized process. Nonetheless, there is no technical reason why the IGS Analysis Centres could not run cluster processes for all receivers that publish their data in near real-time. This would lead to a much denser formal ITRF solution than today.

By distributing the process to the level of a single task per receiver, Dancer instances can run on the local or remote computer that collects the receiver's data – or *anywhere* else on the internet. The exchanged information consists of anonymous vector sums, so that participation in the Dancer solution neither requires publication of the tracking data, nor of the estimation products. Of course, both may still be published voluntarily, like today.

Before pre-eliminating its local partition Eq. 3.3, each receiver applies minimum constraints as described in Altamimi et al. (2002). This requires knowledge of the a priori positions of (a subset of) the reference frame receivers, in order to build an a priori covariance matrix. Each Dancer process shares its formal position and velocity with the rest of the network on a voluntary basis. An adequate subset of Dancer receivers *must* do this in order to constrain the reference frame, but with 5,000 public receivers today this should not be a problem.

4 Square Dance Algorithm

The solution method must accumulate vector sums over a network in such a way that all computers end up with the same result. In a local area network this could be done by a central computer. However, Dancer is deliberately designed as a peer-to-peer process on the internet to avoid the need for central elements such as a server. The only available form of communication is then direct exchange of data between any two computers. This section explains the square dance method for accumulating data in such a decentralized network environment.

4.1 Basic Logic

The square dance method runs on a sub-network of a size that is an exact power of two. Any surplus computers pass their vector to a computer in this *core network* before the square dance process, and receive the sum of all vectors back afterwards.

The core nodes are numbered from *0* to 2^m-1 so that the binary representations of the numbers contain *m* bits. Each node maintains internet connections to the *m* other nodes with numbers in which just one bit is different from its own number.

The square dance accumulation method requires each node to upload its local vector to its *m* connections in succession, and download the vector from the other node in return. An incoming vector is added to the local vector before its next upload.

Before any of these pair-wise exchanges, the two nodes in a pair have different vectors and all their recursively included earlier vectors are different. This is certain, because one later bit in the involved node numbers was always different, otherwise a pair would not be connected to each other.

After an exchange, both nodes end up with the same sum of their exchanged vectors. This means that each successive cycle doubles the number of nodes with identical vectors. After *m* cycles, all 2^m nodes in the network have the same sum. Every cycle also doubles the number of local sums in which a given initial vector is included. The common global sum after *m* cycles therefore includes precisely all 2^m initial vectors.

4.2 Details of Network Organisation

The Dancer internet layer uses the JXTA protocol for peer-to-peer communication, Verstrynge (2008), Wilson (2002). This layer deals with low level tasks such as discovering the Dancer network on the internet, communicating through firewalls, or connecting to other nodes by their node number.

A Dancer network of arbitrary size *N* is split in a sub-network of *base nodes* and a sub-network of *folding nodes*. The size of the base node network is the largest power of two not larger than the network size *N*. The base node network in turn splits into equally sized networks of *core* and *spare* nodes.

Before the start of a square dance process, folding nodes send their vector to a base node found by toggling the most significant bit of their node number. The receiving nodes add the vector to their own. Spare nodes now send their vector to a core node, found by toggling the second bit. Receiving nodes add an incoming vector to their own.

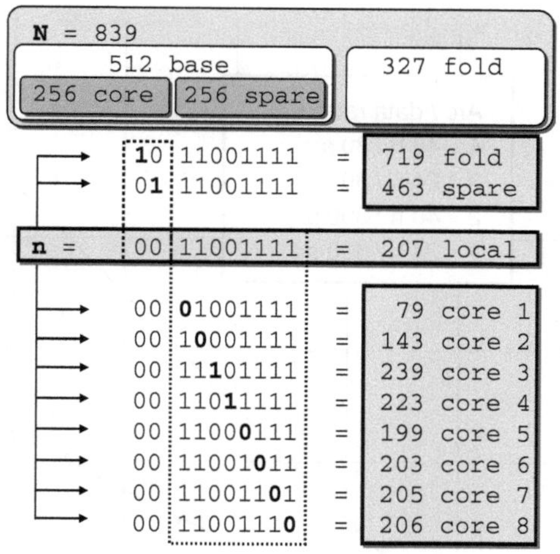

Fig. 3.1 Example for a network of size 839. Node number 207 is a core node with connections to 10 other nodes

The core nodes then perform a square dance process as described before, and send the result to their spare node. Finally, folding nodes receive the global result from their base node.

On the basis of its node number *n* and the network size *N*, each node immediately knows its own role within the global exchange process. It finds its exchange partners by toggling individual bits of its number *n*, as illustrated in Fig. 3.1.

All spare nodes and folding nodes – at least half the network – serve as backup computers in case that a core node goes off-line unexpectedly. A core computer is also replaced if it repeatedly lags behind with its exchange partners. This makes the network self-optimizing: the computers with the slowest connections will end up as folding nodes.

4.3 Data Budget

From the previous it follows that the data traffic at the core nodes grows as a logarithmic function of the network size *N* while data traffic at spare nodes and folding nodes is independent of the network size. The large solution of Table 3.1 would have 3,616 folding nodes, 8,192 spare nodes and 8,192 core nodes that run 13 square dance cycles. Dancer performs five least squares iterations, each of which requires two vectors to be averaged. Core nodes then transfer 150 vectors of a size of 1.2 MB, as well as some overhead data for housekeeping. Fig. 3.2 shows the data budget for various cases. The fastest half of the internet can easily handle such amounts of data in an hour (OECD 2009). Dancer should therefore be able to run hourly high-rate solutions for all GPS receivers in the world.

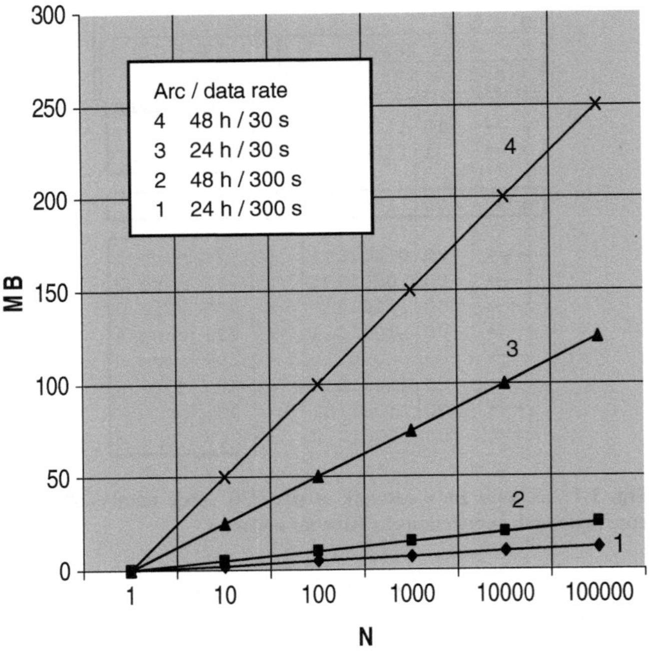

Fig. 3.2 Core node data budget for different solution characteristics, for a single dancer run

5 Summary and Conclusions

The conventional approach to geodetic analysis is to collect observation data at an Analysis Centre which then runs a central estimation process. A distributed process does the opposite: it disperses the analysis over the data sources while exchanging a minimum of process information via the internet. In that case, input data does not have to move at all while estimation products appear directly at the data source. This offers the advantages of exploiting many more computers in routine analysis, and including observation data that is not publicly available. It allows solution sizes that are three or four orders of magnitude larger than a centralized process, without needing any of the elaborate infra-structure of the latter.

This paper explained a simple method of distributing a conventional batch least squares solution over a network of computers. In particular, it demonstrated that it suffices to accumulate just two vectors per least squares iteration, rather than the complete normal matrix and right-hand side vector. If current Analysis Centres would use this approach on a local computer cluster, the formal ITRF solution for GPS could include all geodetic receivers that publish their data in near real-time.

The Dancer project pushes the concept further by implementing the distributed solution method in a JAVA peer-to-peer application on the internet. This allows rigorous solutions for *all* permanent GPS receivers in the world, at zero operational cost.

Future generations of GNSS receivers may be equipped with sufficient processing power to run embedded Dancer processes. This leads to an autonomous network of *smart receivers* that are connected to the internet – hence, to each other – and can then produce precise estimation products instead of, or in addition to, the observation data.

Further details on Dancer can be found in Boomkamp (2010), or at the project homepage www.GPSDancer.com, where the latest version of the software is also available for free download.

References

Altamimi Z, Boucher C, Sillard P (2002) New trends for the realization of the international terrestrial reference system. Adv Space Res 30 (2):175–184

Anderson DP (2004) BOINC: a system for public-resource computing and storage. In: Proceedings 5th IEEE/ACM international workshop on grid computing. Pittsburgh

Andrzejak A, Kondo D, Anderson DP (2010) Exploiting non-dedicated resources for cloud computing. 12th IEEE/IFIP network operations & management symposium Osaka

Boomkamp HJ (2010) Global GPS reference frame solutions of unlimited size. Adv Space Res 46(2):136–143

Cappello F, Djilali S, Fedak G, Herault T, Magniette F, Neri V, Lodygensky O (2005) Computing on large scale distributed systems. Future Gener Comp Syst, Elsevier ISSN 0167-739X, 21(3):417–437

Chung KL (1974) Elementary probability theory with stochastic processes. Springer, New york. ISBN 0-387-90159-0

Korpela E, Werthimer D, Anderson D, Cobb J, Leboffsky M (2001) SETI@home – massively distributed computing for SETI, computing in science and engineering 3 iss 1, Dubois ed (on-line)

OECD (2009) Indicators of broadband coverage. Directorate for science, technology and industry, report DSTI/ICCP/CISP(2009)3/FINAL (on-line)

Sameh AH (1971) On Jacobi and Jacobi-like algorithms for a parallel computer. Math Comput 25(115):579–590 (on-line)

Verstrynge J (2008) Practical JXTA. Lulu.com, Raleigh ISBN 978–1409215646

Wilson BD (2002) JXTA. New Riders Publishing, Indianapolis. ISBN 0-73571-234-4

Geocenter Variations from Analysis of SLR Data

4

M.K. Cheng, J.C. Ries, and B.D. Tapley

Abstract

The Earth's center of mass (CM) is defined in the satellite orbit dynamics as the center of mass of the entire Earth system, including the solid earth, oceans, cryosphere and atmosphere. Satellite Laser Ranging (SLR) provides accurate and unambiguous range measurements to geodetic satellites to determine variations in the vector from the origin of the ITRF to the CM. Estimates of the Global mass redistribution induced geocenter variations at seasonal scales from SLR are in good agreement with the results from the global inversion from the displacements of the dense network of GPS sites and from ocean bottom pressure model and GRACE-derived geoid changes.

Keywords

Geocenter • Satellite laser ranging (SLR) • Center of mass

1 Introduction

The Terrestrial Reference System (TRS) is a fundamental concept for all studies in the geosciences, and it is of critical importance for satellite navigation and other geo-information applications. The Earth's center of mass (CM or geocenter) is the center of mass of the entire Earth system, including the solid earth, oceans, cryosphere surface water and atmosphere. The CM is the point about which any Earth satellite will orbit and can be determined from observations of an Earth-orbiting satellite motion. The origin of the International Terrestrial Reference Frame (ITRF) is currently derived from long-term analysis of Satellite Laser Ranging (SLR) data, but its dynamic tie to the CM is not yet a component of the conventional model of ITRF. The variation of the CM with respect to the origin of ITRF is known as the geocenter motion and reflects the global scale mass redistribution and the interaction between the solid Earth and the mass loading. Determination of the Earth's center of mass is an important component for the realization of the Terrestrial Reference System, and has attracted considerable attention; see for example Trupin et al. (1992), Dong et al. (1997), Watkins and Eanes (1997), Pavlis (2002), Angermann and Müller (2008), Pavlis and Kusmicz-Cieslak (2009), and those papers that contributed to the *IERS Analysis Campaign to Investigate Motions of Geocenter* (Ray 1999).

In this study, the geocenter motion is simultaneously estimated along with the low-degree portion of the gravity field, providing a unified recovery of the signals in the SLR data. This paper discusses the analysis of estimates of the annual geocenter variations from a multi-satellite SLR data set and compares the results with the ILRS SLR network translation solutions (Collilieux et al. 2009; Altamimi et al. 2010) based on the International Laser Ranging Service (ILRS; Pearlman et al. 2002) and the global inversion mostly based on the 3-dimensional displacement of GPS stations (Wu et al. 2010a).

2 Theory

The Earth's gravitational potential is expressed by a spherical harmonic with coefficients (C_{nm} and S_{nm}), which are a function of the global mass distribution determined by a volume

M.K. Cheng (✉) • J.C. Ries • B.D. Tapley
Center for Space Research, University of Texas at Austin, Austin, TX 78759-5321, USA
e-mail: cheng@csr.utexas.edu

Z. Altamimi and X. Collilieux (eds.), *Reference Frames for Applications in Geosciences*, International Association of Geodesy Symposia 138, DOI 10.1007/978-3-642-32998-2_4,

integral enveloping the entire Earth system (see Heiskanen and Moritz 1967; Torge 1980; Lambeck 1988). By definition, the coordinates of the CM in the adopted reference system are determined by the degree-one normalized spherical harmonic coefficients C_{10}, C_{11} and S_{11}. The geocenter vector from origin of the reference system to the mass center is given by

$$\vec{r}_{cm} = a_e\sqrt{3}(C_{11}, S_{11}, C_{10}) \tag{4.1}$$

where a_e is the Earth's radius.

In a reference frame centered exactly at the CM, the degree-one harmonics are identically zero. The origin of the geocentric inertial reference frame, for high accuracy satellite orbit determination, is typically chosen to coincide with the CM, and the geocenter vector $\vec{r}_{cm}$ represents the offset between the ITRF origin and the instantaneous CM. As an alternate, a geocentric frame at the ITRF origin can be used, but in this case the degree-one terms are non-zero. However, it is also necessary for proper modeling to include a Coriolis-type force due to the fact that the origin is no longer an inertial reference point (Kar 1997).

The geocenter vector $\vec{r}_{cm}$ can be determined using space geodetic measurements, which link a satellite (which orbits about the center of mass of the Earth system) and the tracking sites (which realize the origin of the crust-fixed International Terrestrial Reference Frame (ITRF)) based on the geometric relation:

$$\vec{r}(t) = \vec{\rho}(t) + \vec{R}_{sj} - \vec{r}_{cm}(t) \tag{4.2}$$

where $\vec{r}$ is the position vector of the satellite related to the Earth's mass center and determined by integrating Newton's equations in a non-rotating geocentric reference frame, $\vec{\rho}$ is the range vector from the tracking site to the satellite. The vector $\vec{r}$ and $\vec{\rho}$ are expressed in the earth-body-fixed system. $\vec{R}_{sj}$ is the position vector of a tracking site with respect to the ITRF. $\vec{r}_{cm}$ is the vector from the origin of the ITRF to the CM (identical to the vector O_G described in the IERS conventions (McCarthy and Petit 2003; Petit and Luzum 2010)). Global mass redistribution alters Earth rotation, produces temporal variations of the gravitational field, and variations of the CM with respect to the origin of ITRF (as convention, the origin of ITRF is fixed to the crust of the Earth). This variation is referred to as geocenter motion.

The Earth consist of the solid and fluid layer, the volume integral over the entire Earth for the degree one spherical harmonic coefficients can be expressed mathematically as the sum of the volume integral for the solid earth and a shell integral for fluid thin layer (with thickness much less than a_e), which depends on the mass distribution and variable boundary as the deformable earth surface in response to mass loading. Let $\Delta\sigma(\phi, \lambda)$ be the change in the surface density (mass/area) as radial integral of the density distribution through the thin layer (Eq. (5) of Wahr et al. 1998). Thus, the mass loading of the thin layer induced normalized coefficients $C^f_{10}, C^f_{11}, S^f_{11}$ are determined by the surface integral of density changes $\Delta\sigma(\phi, \lambda)$ taken over the Earth's fluid surface thin layer with the surface element dS as follows (Chao et al. 1987, Eq. (6) of Wahr et al 1998):

$$C^f_{10} = \frac{a_e^2}{3M} \iint_{E_s} \Delta\sigma(\phi, \lambda) \sin\phi dS \tag{4.3}$$

$$C^f_{11} = \frac{a_e^2}{3M} \iint_{E_s} \Delta\sigma(\phi, \lambda) \cos\phi \cos\lambda dS \tag{4.4}$$

$$S^f_{11} = \frac{a_e^2}{3M} \iint_{E_s} \Delta\sigma(\phi, \lambda) \cos\phi \sin\lambda dS \tag{4.5}$$

Correspondingly, the geocenter vector can be expressed as $\vec{r}_{cm} = \vec{r}^{\,s}_{cm} + \vec{r}^{\,f}_{cm}$. The vector $\vec{r}^{\,s}_{cm}$ represents the coordinate of the center of mass of the solid Earth without the fluid load and deformation, evaluated by the volume integral for the solid earth assuming the mass of the solid earth approximates to the mass of whole Earth. The vector $\vec{r}^{\,f}_{cm}$ represents the coordinate of the center of the fluid thin surface layer, within which the mass is free to be redistributed (Blewitt 2003). $\vec{r}^{\,f}_{cm}$ is evaluated by the mass loading of the thin layer (described by $C^f_{10}, C^f_{11}, S^f_{11}$) and loading induced deformation of the earth surface (Eq. (7) of Wahr et al. 1998). Both vectors, $\vec{r}^{\,s}_{cm}$ and $\vec{r}^{\,f}_{cm}$, are with respect to the origin of the selected reference frame (ITRF). Unfortunately, $\vec{r}^{\,s}_{cm}$ is not directly observable from crust-based observations while the modeling of $\vec{r}^{\,f}_{cm}$ is uncertain. However, $\vec{r}_{cm}$ (the sum of $\vec{r}^{\,s}_{cm}$ and $\vec{r}^{\,f}_{cm}$) is observable from analysis of SLR data.

The evaluation of $\vec{r}^{\,f}_{cm}$ has been a principal concern in the past because of the complexity of the density redistribution and variable boundary surface, which covers the solid earth and separates from the fluid thin layer. In the earlier study, such as Trupin et al. (1992) and Dong et al. (1997), $\vec{r}^{\,f}_{cm}$ referred to the center of figure of the "outer surface of the solid Earth". This surface is assumed to be covered by a uniform infinitely dense array of points and the motions of these points are taken into account (Blewitt 2003), but this is very difficult to realize in practice (Dong et al. 2003). Even the current dense global GPS network represents only a portion of the surface of the solid Earth.

In recent developments, $\vec{r}^{\,f}_{cm}$ is evaluated by the 'thin-shell loading model' (Blewitt 2003). In this model, the boundary is defined by a sphere of radius a_e used for the approximate spherical harmonic expansion of the density variations (Eq. (9) of Wahr et al. 1998). The normalized spherical harmonic coefficients (denoted as C^σ_{lm} and S^σ_{lm}) can be inferred from a global inversion based on the GPS-determined three-dimensional displacements for a global dense network and the ocean bottom pressure (OBP) model along with the GRACE-derived geoid changes (Wu et al. 2006, 2010a,b). Thus, the notation of CF is still used for $\vec{r}^f_{cm}$ from the 'thin-shell loading model', which better represents $\vec{r}^f_{cm}$ than the earlier studies.

The variations of $\vec{r}_{cm}$ are determined by the time varying density distribution within the Earth system as the response of the Earth to the movements of the planetary fluid masses and the crustal movements due to tectonic motion and earthquakes. Among these variations, the changes of $\vec{r}^f_{cm}$ are mostly due to the climate change induced movements of the planetary fluid masses and surface displacement in response to the changes of the loading on the Earth's surface. In the frequency domain, the vector $\vec{r}^s_{cm}$ is relatively static with variations only on geologic time scales; $\vec{r}^f_{cm}$ describes the variations with time scales from sub-daily to seasonal and longer. In the past decades, considerable efforts were made to determine the changes of $\vec{r}^f_{cm}$ by estimating or modeling the distribution of the mass density changes, $\Delta\sigma(\phi, \lambda)$, which can be represented by the equivalent water height (Wahr et al. 1998).

Center of the surface loading. For an elastic Earth, the thin fluid layer produces a gravitational (load) potential at the Earth's surface, and resulting in surface displacements (deformation). The surface displacement further produces additional potential change (Lambeck 1988) based on the load Love number theory. Thus, the vector of the surface loading center $\vec{r}^L_{cm}$can be determined by the degree one terms in the surface density change induced potential (load plus additional potential from the loading induced deformation) based on the models of the redistributions of the atmosphere surface pressure, hydrosphere (soil moisture and water storage) and ocean mass (Dong et al. 1997; Chen et al. 1999) and ice melting.

Center of the fluid shell and global inversion. The coefficients $\vec{r}^L_{cm}$can be expressed in terms of the degree-one spherical harmonic coefficients of the surface density anomaly: C^σ_{1m} and S^σ_{1m} $(m = 0, 1)$ $(m = 0,\ 1)$ accounting for the surface mass loading induced potential (Eq. (12) of Wahr et al. 1998). Thus, $\vec{r}^f_{cm}$ is determined from

$$\vec{r}^f_{cm} = \frac{a_e \rho_w}{\rho_e}(1 + k'_1)\sqrt{3}(C^\sigma_{11}, S^\sigma_{11}, C^\sigma_{10}) \tag{4.6}$$

where ρ_e and ρ_w are the mean density of the solid earth and water, respectively. k'_1 is the degree-one load Love number for an elastic lithosphere (Farrell 1972). In the frame associated with the center of mass of just the solid earth (CE), $(1 + k'_1) = 1$. This relation can be applied to the estimates of geocenter variations from global inversion assuming the center of figure of the GPS network (CF) and CE are approximately equivalent (Blewitt 2003; Jansen et al. 2006).

The response of the solid Earth surface to loading changes caused by mass redistribution in the thin surface layer is the displacement in the local vertical and horizontal directions. Those displacements are proportional to the load potential and the gradient of the load potential characterized by the load Love number (Lambeck 1988). They can be expressed in terms of the spherical harmonic coefficients of the surface density changes (or anomalies) based on Farrell's (1972) loading theory.

Thus, the three dimensional displacement vectors of the GPS network (combined with the observed geoid changes from GRACE measurements or OBP models) were used to infer the degree-one terms of the loading density distribution to obtain the estimates of $\vec{r}^f_{cm}$ This method is described as a global inversion (Blewitt and Clarke 2003; Wu et al. 2003; Kusche and Schrama 2005; Jansen et al. 2009; Wu et al. 2010a). The surface density changes observed from GRACE were used to determine the induced geoid changes in the global inversion to improve separation of the information related to the degree-one terms in the observed displacements from GPS data analysis, and can be used to determine the loading displacement of tracking stations due to the loading potential except for the degree-one terms.

3 Determining Geocenter from SLR

Among the space geodetic techniques, the difficulty in modeling of the nongravitational forces acting on satellites limits the ability of the GPS and DORIS techniques to accurately measure the geocenter vector $\vec{r}_{cm}$. Satellite laser ranging (SLR) is currently the best means of obtaining precise and unambiguous range measurements for the various passive geodetic satellites, and this accurate range information allows the determination of even very small gravitational forces acting on them. The SLR data has, over the past three decades, provided long-term, stable determinations of the origin of the ITRF. The geocenter motion is usually not modeled in the precise orbit determination for SLR data analysis, and this signal will remain in the residuals.

The geocenter vector $\vec{r}_{cm}$ is defined geometrically in the observation equation assuming the coordinates of the tracking stations are described by the ITRF system. Conventional (pre-GRACE) gravity models were obtained assuming the CM coincides with the origin of the ITRF used for the tracking station coordinates. Consequently, the offset of the origin of the ITRF from the CM could be aliased into the geopotential coefficients estimated from space geodetic measurements. This will not occur with GRACE-based models, since the GRACE inter-satellite range data is insensitive to the geocenter motion. In the approach used in this study, the geocenter motion is estimated simultaneously with the orbit, force and measurement parameters from the SLR data.

In order to study the station displacements and the realization of the ITRF, an alternative approach is commonly used. Instead of directly estimating the geocenter motion $\vec{r}_{cm}$, the station displacements $\Delta\vec{R}_{sj}(j = 1, N)$ are estimated for the entire SLR network (with N = ~20–30 over 1 month intervals) along with the estimation of a number of (constant or once-per-rev) empirical acceleration parameters. The estimated network station coordinates $(\vec{R}_{sj} + \Delta\vec{R}_{sj})$ are then projected into the ITRF by a Helmert transformation defined by seven parameters: three translation parameters, one scale and three coordinate rotation angles. The translation reflects the origin difference between the estimated displaced station network and the ITRF used in the analysis, for which the origin is not necessarily the CM, at least at the seasonal and shorter time scales (Dong et al. 2003). This approach is used by the ILRS and earlier analyses from SLR (e.g., Eanes et al. 1997). A concern with this approach is that the orbit may have accommodated some of the geocenter motion, particularly in the Z direction, and thus may have attenuated the geocenter variation in that component.

As noted previously, a gravity field including the degree-one geopotential coefficients can be obtained using the space geodetic tracking data by fixing the tracking station coordinates to the ITRF. This is equivalent to the estimation of the geocenter motion ($\vec{r}_{cm}$) along with the geopotential coefficients with degree greater than one in this study, though a completely correct modeling requires the addition of Coriolis-type terms with the former approach. Simultaneously estimating the degree-one coefficients and the entire network displacements would be ill conditioned because they are equivalent and determined using the same observation information.

The satellites used in this study (Starlette, Ajisai, Stella, LAGEOS-1 and LAGEOS-2) all have spherical shapes, which simplifies the modeling of the non-gravitational forces. In addition, except for Ajisai, they have dense metal cores and very low area-to-mass ratios, which further reduces the impact of the non-gravitational force modeling errors. The orbit inclination ranges from 50° to 109°, and the altitudes range from 700 to 6,000 km. To accommodate residual dynamic modeling errors, 12-h drag coefficients (C_D) or empirical along-track acceleration parameters (C_T) were conventionally estimated as part of the precision orbit determination process. While these parameters have the potential to affect the determination of the gravitational forces, they are essential to accommodate the residual errors in the surface force modeling.

The background gravity models, such as the solid earth and ocean pole tides as well as the atmosphere and ocean de-aliasing (AOD) employed in this analysis are generally consistent with those for the GRACE RL04 products (Bettadpur 2007). The station displacement due to solid earth tides and ocean loading are modeled based on the IERS 2003 conventions, including the effects due to the diurnal and semidiurnal tidally-induced geocenter variations (Watkins and Eanes 1997). The station coordinates are based on LPOD2005 (Ries 2008), which is consistent with ITRF2005 but with some additional stations, coordinate updates and range bias modeling. To measure the geocenter signal at the mm level, this study used a short arc orbit analysis with an arc length of 3 days for the monthly solutions and 7 days for the weekly solutions.

Two SLR time series were obtained in this study. The first (SLR-w) consists of weekly solutions for the geopotential coefficients (to degree and order 5) and the geocenter parameters spanning 18 years from November 1992 to December 2010. The EGM08 gravity model (Pavlis et al. 2008) was used without the AOD model to determine the weekly solutions that is consistent with models used by the ILRS Analysis Working Group (AWG). The second SLR time series (SLR-m) consists of 106 monthly solutions from five geodetic satellites over the period from January 2002 to October 2010. The AOD model was used in this case for consistency with the GRACE RL04 processing. Three components of the geocenter motion, $\vec{r}_{cm}$, are directly estimated along with other dynamical parameters, including the satellite state vector (weekly or 3 days for monthly), 12-h C_D (or C_T) and the lower portion of the gravity field up to degree 5, over time intervals from weekly to monthly (Cheng et al. 2011). The station positions are fixed and range biases are based on LPOD2005. This approach provides a unified recovery of the gravity signals in the SLR data. The C_{20} estimates from this series are used to replace the GRACE estimates for supporting the GRACE applications in extracting the mass variation signal in the hydrological, ocean and polar ice sheets (Cheng and Ries 2009).

Figures 4.1, 4.2 and 4.3 show the monthly estimates for the three components of the geocenter variation, as well as the seasonal and linear trends. Figures 4.4, 4.5 and 4.6 show the weekly estimates, along with the seasonal, linear and long-period variations. The SLR estimates of the geocenter motion $\vec{r}_{cm}$ reflect the weekly or monthly averaged integrated effects of the crustal deformation due to global-scale mass exchange between the Earth components. The mean of the estimates for $\vec{r}_{cm}$ represents the mean position of the geocenter in the ITRF as determined by the constant parts of the mass distribution of the entire Earth, which is dominated by $\vec{r}^{\,s}_{cm}$. Seasonal changes of the $\vec{r}_{cm}$ about the mean correspond to the $\vec{r}^{\,f}_{cm}$, but could contain the information of the seasonal surface loading displacement of site positions due to the variations of the terms with degree greater than one in a spherical harmonic expansion of density variations. The effects for height displacement was discussed by van Dam et al. (2007). Further study is required

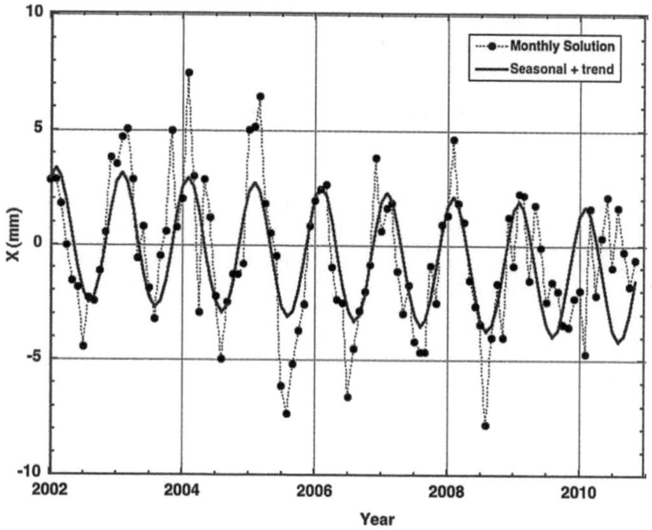

Fig. 4.1 Monthly geocenter variations in the X component. Linear + seasonal variation also shown

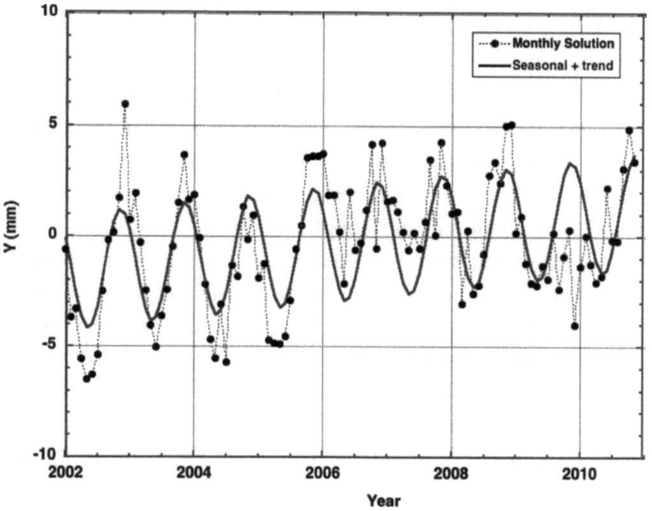

Fig. 4.2 Monthly geocenter variations in the Y component

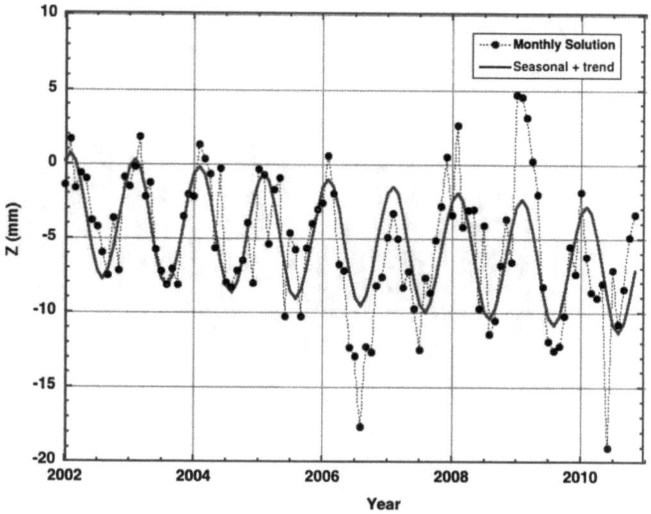

Fig. 4.3 Monthly geocenter variations in the Z component

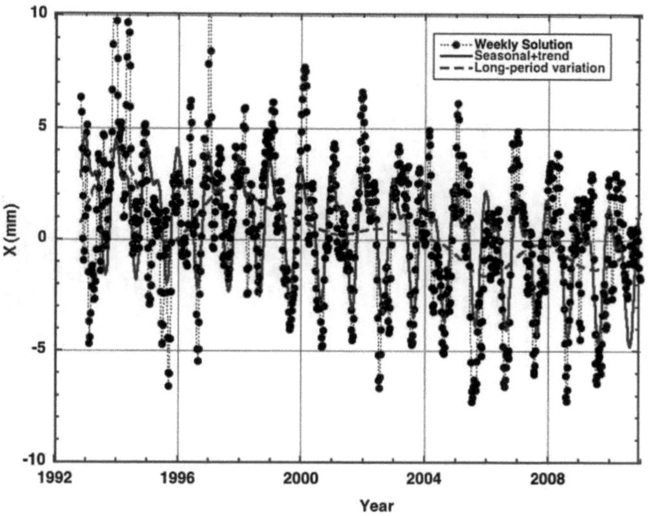

Fig. 4.4 Weekly geocenter variations in the X component. Seasonal + trend and long-period variations also shown

for evaluating these effects to correctly interpret the SLR-determined geocenter motion in comparison with the geophysical loading models and from the global inversion. While the loading surface mass change is expressed in a series spherical harmonics, the degree-one terms are directly related to the seasonal geocenter variations.

4 Comparison and Discussion

Table 4.1 compares the amplitude and phase for the annual variation of the geocenter motion from weekly (SLR-w) and monthly (SLR-m) solutions from this study, and the translation components estimated (with scale factor) from the 14-year weekly site position (denoted as ILRS-1) from the ILRS AWG reported by Collilieux et al., (2009), and that estimated from analysis of 26 years of weekly site positions (denoted as ILRS-2) in the development of ITRF2008 (Altamimi et al. 2010). The notation of 'Degree-one' refers to the geocenter variation converted from the degree-one terms estimated from the global inversion based on the GPS/OBP/GRACE (Wu et al. 2010), where the effects above degree-one are based on the time series of the GRACE derived-gravity fields and OBP model.

Table 4.1 shows that the weekly solution from this study is in good agreement with the ILRS–2 solution for the three translation components. The agreement with ILRS-1 is not quite as good for Y (ILRS-1 amplitude is larger) or Z (ILRS-1 is smaller). The monthly SLR solution is in good

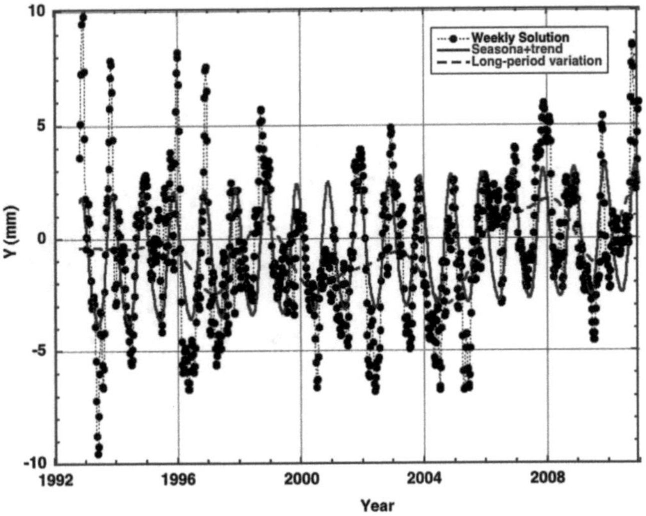

Fig. 4.5 Weekly geocenter variations in the Y component

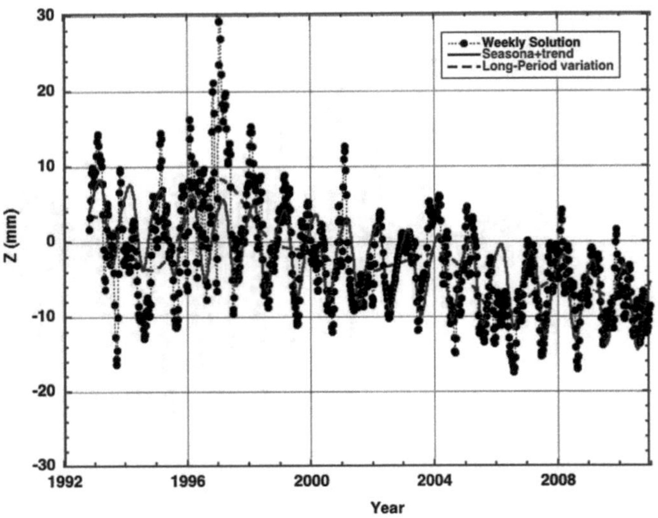

Fig. 4.6 Weekly geocenter variations in the Z component

agreement with the degree-one variations from the global inversion for the Y and Z component. The disagreement in X could be due to the network effect caused by an inadequate geographic distribution of SLR stations (Collilieux et al. 2009), or the global inversion results may be underestimating the variation in X.

The mean offset of the geocenter at epoch 2005.0 with respect to ITRF2005 is estimated to be −0.2, −0.7 and −4.8 mm for X, Y and Z components, respectively, from this analysis. It is reasonably consistent with the estimates (−0.5, 0.9, and−4.7 mm) from ITRF2008 (Altamimi et al. 2010). The translation rate difference between ITRF2008 and ITRF2005 is reported to be zero for Y and Z, and 0.3 mm/y for X (±0.2 mm/year for each) (Altamimi et al. 2010). However, drifts appear in the time series for all three

Table 4.1 Observed geocenter variations (amplitude in mm; phase in degrees). Error estimates are formal errors only. Convention is Amp cos (ω(t–t_0)-Phase) where t_0 is January 1

Case	X (Amp/Phase)	Y (Amp/Phase)	Z (Amp/Phase)
SLR-w	2.7 ± 0.2/ 40 ± 2	2.8 ± 0.2/ 323 ± 2	5.2 ± 0.2/30 ± 3
ILRS-1	2.7 ± 0.3/ 45 ± 4	3.8 ± 0.2/ 327 ± 4	3.6 ± 0.4/4 ± 7
ILRS-2	2.6 ± 0.1/ 40 ± 3	3.1 ± 0.1/ 315 ± 2	5.5 ± 0.3/ 22 ± 10
SLR-m	2.9 ± 0.4/ 35 ± 3	2.6 ± 0.2/ 306 ± 2	4.2 ± 0.3/33 ± 2
Degree-one	1.8 ± 0.2/ 49 ± 3	2.7 ± 0.2/ 325 ± 3	4.2 ± 0.2/31 ± 3

components from this analysis, as shown in Figs. 4.1, 4.2, 4.3, 4.4, 4.5 and 4.6. The rates are estimated to be −0.2, 0.3 and−0.5 mm/year for X, Y and Z, respectively, although the uncertainty in these trends is probably at the same level. It is interesting that there appear a good agreement with the rates for X and Y (−0.2 and 0.4 mm/year, respectively) estimated from the global inversion (Wu et al. 2010a, b), though the agreement is poor for Z (−0.9 mm/year). Ice melting and postglacial rebound could produce a secular geocenter velocity on the order of 1 mm/year (Greff-Lefftz 2000; Métivier et al. 2010), and further study is required to understand if these signals are geophysically meaningful.

A scale parameter is not estimated in this analysis since the length scale of the reference frame determined from SLR is only related to the velocity of light (for the ranging measurement reduction) and GM (the product of the gravitational constant G and the mass of the Earth), although accurate knowledge of the satellite center of mass offset is also critical (Ries 2007). The variations and trends in the translation and scale seen in other analyses may be due to position/velocity errors or unmodeled ranging biases (Coulot et al. 2009).

In summary, time series of geocenter motion were determined from analysis of SLR data from five geodetic satellites over a period of 18 years from Nov. 1992 to Dec. 2010 for weekly solutions, and a period of 9 years from 2002 through 2010 for monthly solutions. Results reveal significant seasonal and possible interannual variations in all three components (X, Y, Z) of the geocenter motion. The estimates of the annual variations are generally in good agreement from the various SLR analyses, as well as with the global inversion approach based on GPS/OBP/GRACE. Simultaneous estimation of geocenter motion and gravity field, the approach used here, can provide a unified recovery of the mass redistribution signals in the SLR data.

Acknowledgments This research was supported by NASA grants NNX08AE99E and JPL1368074.

References

Altamimi Z, Collilieux X, Métivier L (2010) ITRF2008: an improved solution of the international terrestrial reference frame. J Geodesy. doi:10.1007/s00190-011-04444-4

Angermann D, Müller H (2008) On the strengths of SLR observations to realize the scale and origin of the terrestrial reference system. In: Sideris MG (ed) Observing our changing earth: Proceedings of the 2007 IAG General Assembly, vol 133, pp 21–29. Springer, Perugia, 2–13 July 2007

Bettadpur S (2007) UTCSR level-2 processing standards document for level-2 product release 0004, GRACE 327-742, ftp://podaac-ftp.jpl.nasa.gov/allData/grace/

Blewitt G (2003) Self-consistency in reference frames, geocenter definition, and surface loading of the solid Earth. J Geophys Res 108 (B2):2103. doi:10.1029/2002JB002082

Blewitt G, Clarke P (2003) Inversion of Earth's changing shape to weigh sea level in static equilibrium with surface mass redistribution. J Geophys Res 108(B6):2311. doi:10.1029/2002JB002290

Chao BF, O'Connor WP, Change ATC, Hall DY, Foster J (1987) Sno load effects on the Earth's rotation and Gravitational field, 1979–1985. J Geophys Res 92(B9):9415–9422

Chen JL, Wilson CR, Eanes RJ, Nerem RS (1999) Geophysical interpretation of observed geocenter variations. J Geophys Res 104:2683–2690

Cheng MK, Ries J (2009) Monthly estimates of C20 from 5 SLR satellites, GRACE Technical Note 05, ftp://podaac.jpl.nasa.gov/allData/grace/docs/TN-05_C20_SLR.txt

Cheng M, Ries JC, Tapley BD (2011) Variations of the Earth's figure axis from satellite laser ranging and GRACE. J Geophys Res 116: B01409. doi:10.1029/2010JB000850

Collilieux X, Altamimi Z, Ray J, van Dam T, Wu X (2009) Effect of the satellite laser ranging network distribution on geocenter motion estimation. J Geophys Res 114:B04402. doi:10.1029/2008JB005727

Coulot D, Bério Ph, Bonnefond P, Exertier P, Féraudy D, Laurain O, Deleflie F (2009) Satellite laser ranging biases and terrestrial reference frame scale factor, observing our changing earth. Proceedings of the 2007 IAG general assembly, vol 133, pp 39–46. Springer, IAG, Perugia, 2–13 July 2007

Dong D, Dickey JO, Chao Y, Cheng MK (1997) Geocenter variations caused by atmosphere, ocean and surface ground water. Geophys Res Lett 24(15):1867–1870

Dong D, Yunck T, Heflin M (2003) Origin of the international terrestrial reference frame. J Geophys Res 108:B42200. doi:10.1029/2002JB002035

Eanes RJ, Kar S, Bettadpur SV, Watkins MM (1997) Low-frequency geocenter motion determined from SLR tracking data. Eos Trans AGU, 78(46), Fall meeting, Supplementary, F156

Farrell WE (1972) Deformation of the Earth by surface loading. Rev Geophys Space Phys 10:761–797

Greff-Lefftz M (2000) Secular variation of the geocenter. J Geophys Res 105(B11):25685–25692

Heiskanen WA, Moritz H (1967) Physical geodesy. W. H. Freeman and Co., San Francisco/London

Jansen MJF, Kusche J, Schrama EJO (2006) Low-degrees load harmonics coefficients from combining GRACE, GPS time series and a priori dynamics. In: Proceedings IAG symposium, 2006

Jansen MJF, Gunter BC, Kusche J (2009) The impact of GRACE, GPS and OBP data on estimates of global mass redistribution. Geophys J Int 177:1–13. doi:10.1111/j.1356-246X.2008.04031x

Kar S (1997) Long-period variations in the geocenter observed from laser tracking of multiple satellites, CSR report CSR-97-2, The University of Texas Center for Space Research, Mail stop R1000, Austin, TX

Kusche J, Schrama E (2005) Surface mass redistribution inversion from global GPS: a unified observation model. J Geophys Res 110: B09409. doi:10.1029/2004JB003556

Lambeck K (1988) Geophysical geodesy. Clarendon, Oxford

McCarthy DD, Petit G (2003) IERS conventions (2003) IERS technical note no. 32, International earth rotation and reference systems service. Frankfurt, Germany

Métivier L, Grefftz-Lefftz M, Atamimi Z (2010) On secular geocenter motion: the impact of climate changes. Earth Planet Sci Lett 296 (3–4):360–366. doi:10.1016/j.epsl.2010.05.021

Pavlis EC (2002) Dynamical determination of origin and scale in the earth system from satellite laser ranging. In: Adam J, Schwarz K-P (eds) Vistas for geodesy in the new millennium. Springer, New York, pp 36–41

Pavlis EC, Kuzmicz-Cieslak M (2009) Geocenter motion: causes and modeling approaches. In: Schillack S (ed) Proceedings of 16th international laser workshop, Poznan pp 16–26

Pavlis NK, Holmes SA, Kenyon SC, Factor JK (2008) An earth gravitational model to degree 2160: EGM2008, 2008 general assembly of the European geosciences union, Vienna, Austria

Pearlman MR, Degnan JJ, Bosworth JM (2002) The International laser ranging service. Adv Space Res 30(2):135–143. doi:10.1016/S0273-1177(02)00277-6

Petit G, Luzum B, IERS Conventions (2010) IERS technical note no. 36, international earth rotation and reference systems service. Frankfurt, Germany

Ray J (ed) (1999) IERS analysis campaign to investigate motions of the geocenter, IERS technical note 25, Observatoire de Paris, Paris

Ries J (2007) Satellite laser ranging and the terrestrial reference frame: principal sources of uncertainty in the determination of the scale, Geophys Res Abst 9:10809, EGU General Assembly, Vienna, 15–20 April 2007 [SRef-ID: 1607-7962/gra/EGU2007-A-10809]

Ries J (2008) LPOD2005: a practical realization of ITRF2005 for SLR-based POD, Ocean Surface topography science team meeting, 10–12 Nov 2008, Nice, France; available at ftp.csr.utexas.edu/pub/jason/models/coords

Torge W (1980) Geodesy, De Gruyter, New York

Trupin AS, Meier MF, Wahr J (1992) Effects of melting glaciers on the Earth's rotation and gravitational field: 1965–1984. Geophys J Int 108:1–15

van Dam T, Wahr JM, Lavallée D (2007) A comparison of annual vertical crustal displacements from GPS and gravity recovery and climate experiment (GRACE) over Europe. J Geophys Res 114: B04402. doi:10.1029/2006JB004335

Wahr J, Molenar M, Bryan F (1998) Time variability of the Earth's gravity field: hydrological and oceanic effects and their possible detection using GRACE. J Geophys Res 103(B12):30,205–30,229

Watkins MM, Eanes RJ (1997) Observations of tidally coherent diurnal and semi-diurnal variations in the geocenter. Geophys Res Lett 24:2231–2234

Wu X, Argus D, Heflin M, Ivins E, Webb F (2003) Site distribution and aliasing effects in the inversion for load coefficients and geocenter motion from GPS data. Geophys Res Lett 30(14):1742. doi:10.1029/2003GL017546

Wu X, Heflin M, Ivins E, Fukumori I (2006) Seasonal and interannual global surface mass variations from multisatellites geodetic data. J Geophys Res 111:B09401. doi:10.1029/2005JB004100

Wu X, Heflin M, Schotman H, Vermeersen B, Dong D, Gross R, Ivins E, Moore A, Owen S (2010a) Simultaneous estimation of global present-day water transport and glacial isostatic adjustment. Nat Geosci. doi:10.1038/NGEO938

Wu X, Collilieux X, Altamimi Z (2010b) Data sets and inverse strategies for global surface mass variations. Geophys Res Abstr 12, EGU2010-5484, http://meetingorganizer.copernicus.org/EGU2010/EGU2010-5484.pdf

External Evaluation of the Origin and Scale of the International Terrestrial Reference Frame

5

X. Collilieux and Z. Altamimi

Abstract

The International Terrestrial Reference Frame (ITRF) datum definition is of primary importance for many Earth Science Applications. While accurate origin information (Earth Center of Mass) is required for any precise satellite orbit determination, an accurate scale is indispensable for various calibrations (altimeter absolute bias, GNSS satellite antenna phase center offsets). Studies involving vertical motion determination, such as mean sea level and Glacial Isostatic Adjustment (GIA) are also affected by the choice of the underlying Terrestrial Reference Frame. ITRF datum accuracy evaluation has been traditionally performed by comparing independent space geodetic technique performances and successive ITRF solutions. While the ITRF2005 to ITRF2000 comparison may lead to pessimistic evaluation of the ITRF datum accuracy, the question is raised whether the error budget deduced from the ITRF2008 to ITRF2005 comparison would be optimistic, especially for the time evolution of the origin. It is fundamental to explore external ways to evaluate the ITRF frame parameters and especially their time evolution which impacts the results of many climatic studies. The state of art of available methods is reviewed by stressing their advantages and drawbacks. Most of them have been already implemented and show that the ITRF2005 origin rate is probably reliable at the millimeter per year level. However these methods have been applied on different velocity field with different models and required assumptions which make their mutual comparison difficult. Thus, new analyses are required in the future.

Keywords

Terrestrial reference system • Terrestrial reference frame

1 Introduction

A Terrestrial Reference System (TRS) is required from many Earth Science applications. It can be used to express positions at the Earth's surface and deformations of the Earth, to compare independent measurements of the same phenomena and to represent the Earth in its rotational motion in space. A TRS is a physical concept which is defined by the specifications of its origin, orientation and scale. As the axes of the system are inaccessible, a TRS is materialized by a Terrestrial Reference Frame (TRF), through the realization of its origin and scale, the orientation of its axes with respect to the crust, as well as their time evolution. This is usually achieved through the determination of a set of coordinates as a function of time of some points distributed at the Earth's surface as illustrated by Fig. 5.1.

The origin, orientation and scale information of the frame is inherent to its coordinates. But as these are the result of an estimation process, it is necessary to use a set of stations to mitigate as much as possible the random errors. A global

X. Collilieux (✉) • Z. Altamimi
IGN/LAREG and GRGS, Université Paris Diderot, 35 rue Hélène Brion, 75013 Paris, France
e-mail: xavier.collilieux@ign.fr

Z. Altamimi and X. Collilieux (eds.), *Reference Frames for Applications in Geosciences*, International Association of Geodesy Symposia 138, DOI 10.1007/978-3-642-32998-2_5,

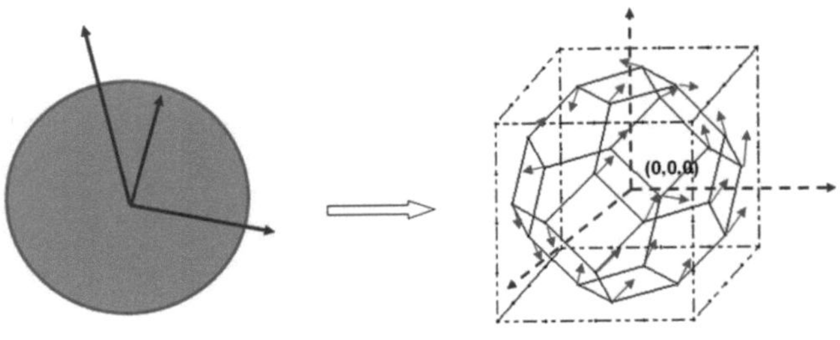

Fig. 5.1 From the terrestrial reference system to the terrestrial reference frame

Table 5.1 Transformation parameters between successive ITRFs

	TX	TY	TZ	Scale
At 2000.0	mm	mm	mm	ppb
ITRF2000- > ITRF2005	0.1	−0.8	−5.8	0.40
Rates (/year)	−0.2	0.1	−1.8	0.08
ITRF2005- > ITRF2008	−2.0	−0.9	−4.7	0.94
Rates (/year)	0.3	0.0	0.0	0.00

network of stations is also required to access to the TRF datum parameters for precise applications.

The availability of new stations and new geodetic data processing strategies make it necessary to update regularly TRFs. That is why the realizations of the International Terrestrial Reference System (ITRS), maintained by the International Earth Rotation and Reference Systems Service (IERS) are updated every few years. However, although the three latest releases ITRF2000, ITRF2005 and ITRF2008 are some realizations of the same TRS, they show systematic differences. The switch from ITRF2000 to ITRF2005 especially highlighted the importance of having an accurate reference frame since the estimation of the sea level rise by satellite altimetry was modified by −0.26 mm/year with larger regional differences (Beckley et al. 2007).

A traditional way to evaluate TRF origin and scale error is to compute a transformation using 14 parameters between the successive ITRFs. These parameters are composed by three translations, three rotations, a scale factor as well as their time derivatives. Table 5.1 extracted from Altamimi et al. (2007, 2011) gives the relationship between the three latest ITRFs. Some of the parameters are significantly different from zero, especially between ITRF2000 and ITRF2005 with a large z-translation rate. Differences between ITRF2005 and ITRF2008 velocities tend to be smaller but there are still rather large offsets at the reference epoch. It would probably be too optimistic to conclude that ITRF2008 origin and scale rates are free of any error. It is important to quantify the level of TRF datum uncertainty especially to build a conservative error budget for sea level rise estimation.

After recalling the standard way to evaluate the ITRF in Sect. 2, we discuss in Sect. 3 alternative methods that can be used to evaluate ITRF origin and scale rates. Finally the perspectives of this work are given in Sect. 4.

2 Internal Evaluation

The ITRF solutions are obtained by a combination of station positions and velocities estimated using the Doppler Orbitography and Radiopositioning integrated by Satellites (DORIS) orbitography technique, Global Navigation Satellite System (GNSS), Satellite Laser Ranging (SLR) and Very Long Baseline Interferometry (VLBI). All the satellite techniques are theoretically sensitive to the Earth Center of mass (CM) and all the techniques should give identical radial positioning. However, the agreement on origin and scale between the techniques is not sufficient at the current level of science requirement of a TRF, that is 1 mm at the reference epoch and 0.1 mm/year for velocities (Plag and Pearlman 2009).

A rigorous assessment of origin and scale differences between the techniques can be evaluated only by a multi-technique combination. Indeed, the estimation of origin and scale offsets is possible thanks to the availability of local ties, which provide the relative positions of the co-located instruments. The estimation of translation and scale rates between the specific technique frames mainly relies on the assumption of equal velocities at co-location sites but may also be influenced by local ties if several are available at different epochs for some sites.

Altamimi et al. (2011) have evaluated the scale agreement between SLR and VLBI at the level of 1.05 (±0.13) ppb at epoch 2005.0 and 0.049 (±0.010) ppb/year. DORIS TRF scale is consistent with the mean of SLR and VLBI at the level of 0.70 (±0.20) ppb at epoch 2005.0 and 0.002 (±0.030) ppb/year but the origin agreement is not satisfactory. For GPS, Collilieux et al. (2010) showed that GPS reprocessed TRF origin is still questionable, since the estimated z-translation rate with respect to SLR is between 0.3 and 0.7 mm/year. The scale of GPS reprocessed solutions that contributed to ITRF2008 are dependent on the Antenna Phase Center Offsets (APCOs) that have been adopted (Schmid et al. 2007). However, Collilieux and Schmid (2012) show that it is possible to infer inherent GPS scale rate information by evaluating APCO parameter drifts. They evaluated that the ITRF2008 scale rate accuracy may be at the level of 0.2 mm/year.

It seems that new reprocessed space geodesy solutions show a better agreement in terms of scale time evolution, but the origin, solely determined by SLR would benefit from evaluations performed with other datasets and models.

3 External Evaluation

The external evaluation of ITRF2008 origin and scale drifts can be summarized with those two following questions. Are ITRF2008 velocities related to the CM? Does the mean station vertical velocity have a real physical meaning?

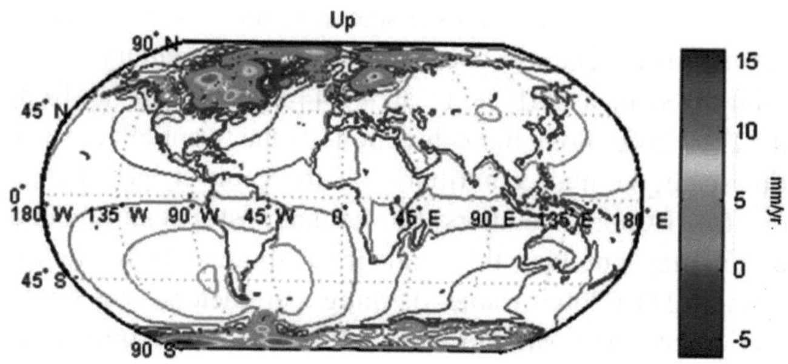

Fig. 5.2 Height vertical velocities predicted by the post-glacial rebound model ICE-5G VM2 (Peltier 2004) available at http://www.sbl.statkart.no/projects/pgs/

Changing the frame origin and scale directly impacts the coordinates. For example, if the CM moves downward with respect to the crust along the z-axis, the z-component of all the station coordinates will be modified positively. As a consequence, if the velocities of one point M of longitude λ and latitude φ and approximated position $\mathbf{X_0}$ in the CM is $\mathbf{V}_{\mathrm{CM}}$, the velocity in any frame could be derived according to

$$\mathbf{V} = \mathbf{V}_{\mathrm{CM}} + \dot{\mathrm{d}} \cdot \mathbf{X}_0 + \mathbf{G}(\lambda, \varphi) \cdot \dot{\mathbf{T}} \qquad (5.1)$$

where $\dot{\mathrm{d}}$ and $\dot{\mathbf{T}}$ are respectively the scale and translation rates. Assuming that $\mathbf{V}$ is the velocity of a station in ITRF2008, it is possible to evaluate ITRF2008 origin and scale rate errors by using a network of stations and estimating four parameters: three translations and one scale factor. This requires that some properties of $\mathbf{V}_{\mathrm{CM}}$ are known. Published studies that used this method or alternative approaches that could be used are discussed in the following.

3.1 Kinematic Models

There are many phenomena that could influence station velocities, including local effects such as water pumping or tectonic effects at regional scale. However, the main contributors to velocities at larger spatial wavelengths are the global tectonic plate motions and Global Isostatic Adjustment (GIA), please see Figure 5.2.

Argus (2007) fitted an origin rate error for the TRF using the following approach. He assumed that vertical velocity should be close to those delivered by GIA models and that horizontal velocities should be consistent with an Euler pole model. His approach, although including biases between various input technique frames, can be summarized by the following equations:

$$\begin{aligned} \mathbf{V_{up}} &= \mathbf{V}_{\mathrm{CE}} + \mathbf{G_{up}}(\lambda, \varphi) \cdot \dot{\mathbf{T}} \\ \mathbf{V_{2D}} &= \Omega \wedge \mathbf{X} + \mathbf{G_{2D}}(\lambda, \varphi) \cdot \dot{\mathbf{T}} \end{aligned} \qquad (5.2)$$

$\mathbf{V_{2D}}$ and $\mathbf{V_{up}}$ are the horizontal and vertical velocities, $\mathbf{V}_{\mathrm{CE}}$ is the vertical velocity given by the GIA model, $\boldsymbol{\Omega}$ is the rotation pole of the plate where the point is located, $\mathbf{T}$ is the origin rate error and $\mathbf{G_{2D}}$ and $\mathbf{G_{up}}$ are sub-matrices of the transformation between the global and the local frame. Using Eq. 5.2, Argus (2007) fitted the origin rate error as well as the rotation pole for each major tectonic plate. As the GIA model origin is the Centre of Mass of the solid Earth (CE), he found that the modeled CE is distinct from the ITRF2005 origin by 1.2 mm/year but also mentioned that CE should differ from CM by no more than 0.5 mm/year. He also noted that distinct GIA model may lead to differences up to 0.5 mm/year on the z-component. A similar study carried out by Kogan and Steblov (2008) found a translation of 2.5 ± 0.2 mm/year for quite a similar origin, that they designated as the Centre of Plate rotation (CP), using only the horizontal velocity field and Euler pole models. However, the same study carried out by Argus et al. (2010) confirms an origin which departed by 1.3 mm/year in norm to what was obtained by Kogan and Steblov (2008). The inclusion of a scale rate parameter in the method presented by Argus (2007) would be also worth investigating.

3.2 Gravimetric Data

Changes in gravity can be related either to vertical displacement or to redistribution of masses within the Earth (Vanicek and Krakiwsky 1986). As a consequence if absolute gravity trends are determined, ones should determine how much the contribution of vertical displacement is. Changes in absolute gravity can be converted into height velocities following:

$$\mathrm{dg}/\mathrm{dt} = \alpha(\lambda, \varphi) \mathrm{V}_{\mathrm{up}} \qquad (5.3)$$

where α(λ,φ) is the admittance factor which is a function of the station location. This factor depends on the physical process that causes the deformation. This ratio generally fluctuates between −0.35 and −0.15 μGal/mm (Richter et al. 2004). However, some processes such as volcanic filling of cavity generate no height changes so that this ratio may go to infinity (Vanicek, and Krakiwsky 1986). The range of variations of this admittance factor is quite important so it is fundamental to know the physical process responsible of the deformation. Moreover, although absolute gravimeter (AG) are very accurate, the mean uncertainty of determined trend is about 0.1 μGal/year using at least 14 years of observations (van Camp et al. 2005), which is equivalent to 0.5 mm/year using an admittance factor of −0.2 μGal/mm.

Several studies have monitored absolute gravity changes at GPS station locations. Without being exhaustive, we mention here the study of Lambert et al. (2006) who studied measurements over the North American mid-continent. They compared absolute gravimeter trends and GPS velocities determined in ITRF2000. They fitted a slope of −0.18 ± 0.03 μGal/mm, very close to the theoretical value of −0.16 μGal/mm. We compared their velocities with those of ITRF2008. More than the datum changes, velocities have been updated individually, probably leading to similar agreement although the frame has been changed. Mazzotti et al. 2007 studied the Northen Cascadia area which is a subduction zone. They fitted an admittance factor based on ITRF2000 results close to the theoretical value (−0.19 μgal/mm) but they reported a constant bias of 2.2 ± 1.3 mm/year. The velocity change between ITRF2000 and ITRF2008 at the site DRAO (Penticton, Canada) used for the reference frame definition is only 0.2 mm/year which show that reference frame scale error is unlikely to explain this bias. Moreover, tide gauge records in the area are consistent with GPS velocity determination (1.7 ± 0.5 mm/year) (Mazzotti et al. 2007). This shows how much it is complex to properly interpret the admittance factors. The work of Teferle et al. (2009) is also worth citing since they reported that TRF scale rate error can be a possible cause of discrepancy between the results from GPS and two absolute gravimeters in England.

It is necessary to use a network to investigate any reference frame error. The only study of that kind has been led by Plag et al. (2007). The authors fitted a translation rate bias with respect to ITRF2000 velocities at co-located AG/GPS sites using AG trend measurements. Their result shows a better agreement between the CM position determined by AG and the ITRF2005 origin than with the ITRF2000 origin.

3.3 Tide Gauges

GPS vertical velocities have been used to correct tide gauge (TG) records for ground motion. While they can be very precise, they are still potentially affected by reference frame origin and scale rate errors. As a consequence, geocentric sea level rise evaluated by this method at TGs should be analyzed with caution. Conversely, if the oceanic signal is known at a tide gauge location, it is possible to infer the vertical velocity at the TG from the TG records (Mitchum 2000). Bouin and Wöppelmann (2010) used that method to compute vertical velocities by assuming an overall constant oceanic signal of 1.8 mm/year. By comparing their GPS vertical velocities with those at 70 sites, they were able to infer a scale rate error for ITRF2005 of −0.13 ± 0.08 mm/year. Although these results are encouraging, they are fully dependent on the oceanic signal value used. Collilieux and Wöppelmann (2011) suggested an alternative method to evaluate the origin rate error using tide gauges and GPS. They found that the sea level spatial variations evaluated at 10 distinct regions over the twentieth century were minimized in a TRF lying at −0.44±0.22 mm/year from the ITRF2005 along the z-direction. At the twentieth century timescale, the main contribution is the geoid height change due to GIA but the authors showed that post-glacial rebound effect may not bias this statistic. Although these studies discuss ITRF2005 frame defining parameters, we could extrapolate the results to ITRF2008 since the z-component of the origin and scale rates are zero between these two frames. However, the two presented methods require strong assumptions. Altimetry supplies direct estimate of the oceanic signal over the ocean and should be a relevant alternative to these approaches.

3.4 Tide Gauges and Altimetry

Oceanic signal determined by satellite altimetry can be potentially computed at coastal sites by extrapolating determined signals along the track of the satellites. As reliable space altimetry results are available since 1993.0, vertical velocities at TG can be computed using more than 17 years of observations. The typical uncertainty of the method consisting in differentiating sea surface heights with TG records can be evaluated at the level of 0.5–3 mm/year (Cazenave et al. 1999; Nerem and Mitchum 2002; Kuo et al. 2004; Ray et al. 2010). The main limitations of the methods are the precision of space altimetry near coastal areas, and the required extrapolation of the altimetry signal as a function of the distance to the satellite track.

Ray et al. (2010) have compared such derived velocities with vertical velocities determined by DORIS technique in ITRF2005. As the DORIS technique is used to compute altimeter satellite orbits, the two datasets are not fully independent. However, as DORIS velocities are constrained to those of ITRF2005 in terms of scale rate and as satellite orbit is not constrained by the a priori TRF scale rate error (Morel and Willis 2005), the evaluation of ITRF2005 scale by this method should be reliable. By considering only 14 co-location sites separated by less than 6 km, they evaluated the median difference between these two velocity sets at the level of 1.2 mm/year. Although this value could be interpreted as TRF scale error, the number of sites is probably too small to achieve sufficient accuracy. The same study conducted by incorporating GPS sites could be worth doing.

4 Perspectives

The evaluation of the ITRF origin and scale accuracy is mandatory for many applications in geosciences. While the inter-comparison of space geodetic techniques supplies a first estimation of TRF possible error, the error budgets built with space geodesy also need to be completed with an external data set and physical considerations.

We described the studies that have already been done to assess space geodesy TRF using post-glacial rebound models, gravimetric data, tide gauges and altimetry. Due to the limited accuracy of each method and the necessary assumptions of each of them, all the results need to be merged. The comparison of the results we reported here shows a consistency of ITRF2005 better than 2 mm/year in origin and probably smaller than 1 mm/year in scale. These figures are preliminary since the various authors did not use the same input data and the same reference frame. These studies need to be updated for analyzing the same set of velocities, for example ITRF2008 since the improvement of ITRF2008 compared to previous version has been clearly demonstrated.

New studies are required, especially to confirm the potential of absolute gravity data. We also think that the inversion methods that allows estimating station displacements in the CM frame related to GIA, presented day ice melting and tectonic plate motion (Wu et al. 2010), would be worth investigating in the purpose of TRF evaluation. A working group of the IAG sub-commission 1.2 (Global Reference Frames) untitled: "External Evaluation of Terrestrial Reference Frames" has been recently created to coordinate this activity.

References

Altamimi Z, Collilieux X, Legrand J, Garayt B, Boucher C (2007) ITRF2005: a new release of the international terrestrial reference frame based on time series of station positions and Earth orientation parameters. J Geophys Res 112:B09401. doi:10.1029/2007JB004949

Altamimi Z, Collilieux X, Métivier L (2011) ITRF2008: an improved solution of the international terrestrial reference frame. J Geodesy 85(8):457–473. doi:10.1007/s00190-011-0444-4

Argus DF (2007) Defining the translational velocity of the reference frame of Earth. Geophys J Int 169:830–838. doi:10.1111/j.1365-246X.2007.03344.x

Argus DF, Gordon RG, Heflin MB, Ma C, Eanes RJ, Willis P, Peltier WR, Owen SE (2010) The angular velocities of the plates and the velocity of Earth's centre from space geodesy. Geophys J Int 180 (3):916–960. doi:10.1111/j.1365-246X.2009.04463.x

Beckley BD, Lemoine FG, Luthcke SB, Ray RD, Zelensky NP (2007) A reassessment of global and regional mean sea level trends from TOPEX and Jason-1 altimetry based on revised reference frame and orbits. Geophys Res Lett 34:L14608. doi:10.1029/2007GL030002

Bouin M-N, Wöppelmann G (2010) Land motion estimates from GPS at tide gauges: a geophysical evaluation. Geophys J Int 180. doi:10.1111/j.1365-246X.2009.04411.x

Cazenave A, Dominh K, Ponchaut F, Soudarin L, Crétaux JF, Le Provost C (1999) Sea level changes from Topex-Poseidon altimetry and tide gauges, and vertical crustal motions from DORIS. Geophys Res Lett 26(14):2077–2080. doi:10.1029/1999GL900472

Collilieux X, Schmid R (2012) Evaluation of the ITRF2008 GPS vertical velocities using satellite antenna z-offsets. GPS solutions, online first. doi:10.1007/s10291-012-0274-8

Collilieux X, Wöppelmann G (2011) Global sea-level rise and its relation to the terrestrial reference frame. J Geodesy 85(1):9–22. doi:10.1007/s00190-010-0412-4

Collilieux X, Métivier L, Altamimi Z, van Dam T, Ray J (2010) Quality assessment of GPS reprocessed terrestrial reference frame. GPS Solut 15(3):219–231. doi:10.1007/s10291-010-0184-6

Kogan MG, Steblov GM (2008) Current global plate kinematics from GPS (1995–2007) with the plate-consistent reference frame. J Geophys Res 113:B12. doi:10.1029/2007JB005353

Kuo C, Shum CK, Braun A, Mitrovica JX (2004) Vertical crustal motion determined by satellite altimetry and tide gauge data in Fennoscandia. Geophys Res Lett 31:L01608. doi:10.1029/2003GL019106

Lambert A, Courtier N, James TS (2006) Long-term monitoring by absolute gravimetry: tides to postglacial rebound. J Geodyn 41:307–317. doi:10.1016/j.jog.2005.08.032

Mazzotti S, Lambert A, Courtier N, Nykolaishen L, Dragert H (2007) Crustal uplift and sea level rise in northern Cascadia from GPS, absolute gravity and tide gauge data. Geophys Res Lett 34. doi:10.1029/2007GL030283

Mitchum GT (2000) An improved calibration of satellite altimetric heights using tide Gauge Sea levels with adjustment for land motion. Mar Geod 23(3):145–166. doi:10.1080/01490410050128591

Morel L, Willis P (2005) Terrestrial reference frame effects on global sea level rise determination from TOPEX/Poseidon altimetric data. Adv Space Res 36(3):358–368. doi:10.1016/j.asr.2005.05.113

Nerem RS, Mitchum GT (2002) Estimates of vertical crustal motion derived from differences of TOPEX/POSEIDON and tide gauge sea level measurements. Geophys Res Lett 29(19):40–41. doi:10.1029/2002GL015037

Peltier WR (2004) Global glacial isostasy and the surface of the ice-age Earth: the ICE-5G (VM2) Model and GRACE. Annu Rev Earth Planet Sci 32:111–149. doi:10.1146/annurev.earth.32.082503.144359

Plag HP, Pearlman M (eds) (2009) Global geodetic observing system. Meeting the requirements of a global society on a changing planet in 2020. Springer, Berlin (ISBN: 978-3-642-02686-7)

Plag HP, Hammond W, Kreemer C (2007) Combination of GPS-derived vertical motion with absolute gravity changes constrain the tie between reference frame origin and Earth center of mass. EarthScope National Meeting, Monterey

Ray RD, Beckley BD, Lemoine FG (2010) Vertical crustal motion derived from satellite altimetry and tide gauges, and comparisons with DORIS measurements. Adv Space Res. doi:10.1016/j.asr.2010.02.020

Richter B, Zerbini S, Matonti F, Simon D (2004) Long-term crustal deformation monitored by gravity and space techniques at Medicina, Italy and Wettzell, Germany. J Geodyn 38:281–292. doi:10.1016/j.jog.2004.07.013

Schmid R, Steigenberger P, Gendt G, Ge M, Rothacher M (2007) Generation of a consistent absolute phase-center correction model for GPS receiver and satellite antennas. J Geod 81:781–798. doi:10.1007/s00190-007-0148-y

Teferle FN, Bingley RM, Orliac EJ, Williams S, Woodworth PL, McLaughlin D, Baker T, Shennan I, Milne GA, Bradley SL, Hansen DN (2009) Crustal motions in Great Britain: evidence from continuous GPS, absolute gravity and Holocene sea level data. Geophys J Int 178:23–46. doi:10.1111/j.1365246X.2009.04185.x

van Camp M, Williams S, Francis O (2005) Uncertainty of absolute gravity measurements. J Geophys Res 110:B9. doi:10.1029/2004JB003497

Vanicek P, Krakiwsky E (1986) Geodesy: the concepts, 2nd edn. Elsevier, New York

Wu X, Heflin MB, Schotman H, Vermeersen B, Dong D, Gross RS, Ivins E, Moore A (2010) Simultaneous estimation of global present-day water transport and glacial isostatic adjustment. Nat Geosci. doi:10.1038/NGEO938

GRGS Evaluation of ITRF2008, from SLR Data

6

F. Deleflie, D. Coulot, B. de Saint-Jean, O. Laurain, and P. Exertier

Abstract

As an ILRS Analysis Center (AC), we report further the official (final) ITRF2008 solution delivered by the ITRS product center. Following the operational analysis scheme of SLR data, that we perform over the period 1995–2010, we compute for the Lageos-1 and Lageos-2 satellites weekly arcs with ITRF2005 and ITRF2008. Then, we evaluate the sets of orbital parameters, of Earth Orientation Parameters (EOPs), and of Station Coordinates (SSCs). We also compare our results to those obtained by other ACs, in terms of SSCs, EOPs, translations and scale factors.

Keywords

ITRF • SLR • ILRS

1 Introduction

The International Terrestrial Reference System (ITRS) realizations are established and maintained with the global space geodetic networks. The network measurements must be precise, continuous, worldwide, and interconnected by co-locations of different observing techniques. The requirements to be followed in the framework of the Global Geodetic Observing System (GGOS) are to perform a global Terrestrial Reference Frame (TRF) with an accuracy of 1.0 mm, and a stability of 0.1 mm/year, ensuring a sea level rise measurement coherent with the altimetric data precision (Plag and Pearlman 2009).

F. Deleflie (✉)
IMCCE-Observatoire de Paris/GRGS, 77 Avenue Denfert Rochereau, Paris F-75014, France
e-mail: Florent.Deleflie@obs-azur.fr

D. Coulot
IGN/LAREG and GRGS, Université Paris Diderot, 35 rue Hélène Brion, 75013 Paris, France

B. de Saint-Jean • O. Laurain • P. Exertier
Observatoire de la Côte d'Azur, Geoazur/GRGS, Av. N. Copernic, Grasse F-06130, France

An enormous effort has been achieved by the space geodesy technique services of IAG, namely IVS, ILRS, IGS, IDS and their Analysis and Combination Centers (ACs and CCs), to provide reprocessed solutions in view of ITRF2008. The Satellite Laser Ranging (SLR) technique is one of these techniques, organized through the International Laser Ranging Service (ILRS) (Pearlman et al. 2002). The Analysis Working Group (AWG) of the ILRS worked on the ITRF2008 submission during the first part of 2009. The combined solution was based on the contribution of seven ACs (ASI, DGFI, GA, GFZ, our own contribution for GRGS, JCET, and NSGF). Each AC solution contains Sets of Station Coordinates (SSCs) and daily EOPs, using Lageos and Etalon data, according to ILRS/AWG guidelines. After a few dedicated AWG meetings on spring 2009, the final contribution of ILRS for ITRF2008 was sent to IERS on August 2009. The ITRS product center has then released ITRF2008 on spring 2010.

Following Deleflie and Coulot (2009) and Deleflie and Coulot (2010), we analyse in this paper the relevance of ITRF2008. We compare the quality of the products that we regularly provide as an official ILRS AC with the ones obtained using ITRF2008 instead of SLRF2005.

Z. Altamimi and X. Collilieux (eds.), *Reference Frames for Applications in Geosciences*, International Association of Geodesy Symposia 138, DOI 10.1007/978-3-642-32998-2_6,

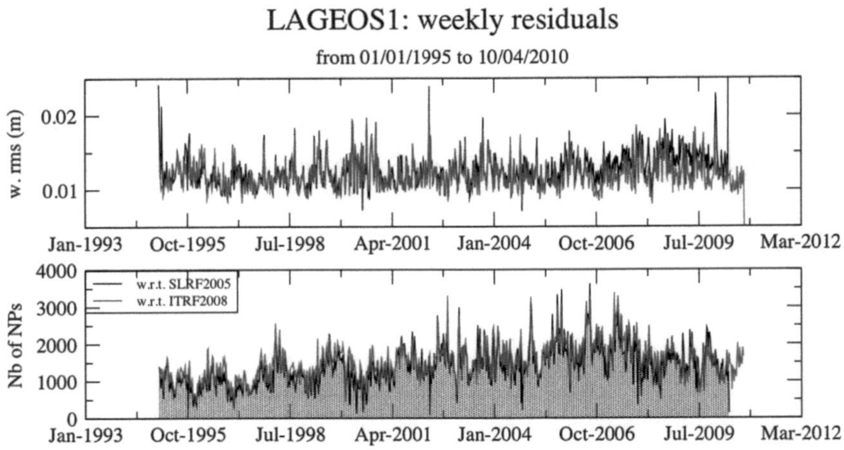

Fig. 6.1 Residuals of weekly arcs of Lageos-1, from 01/01/1995 to 10/04/2010, and numbers of normal points per week

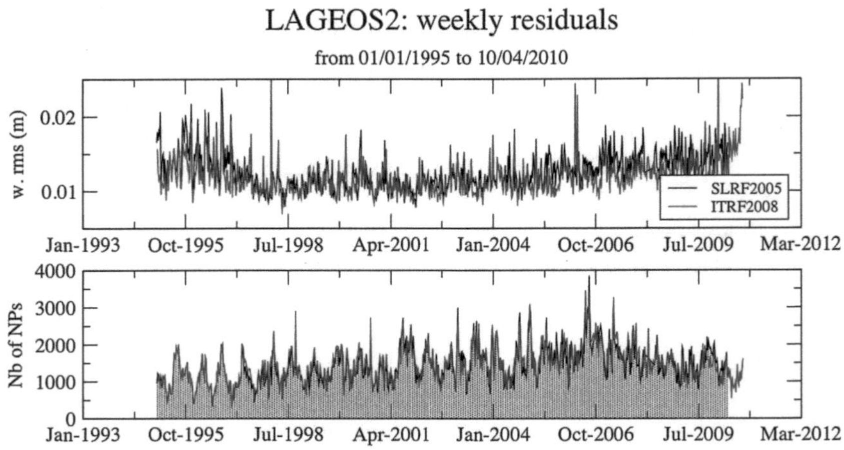

Fig. 6.2 Residuals of weekly arcs of Lageos-2, from 01/01/1995 to 10/04/2010, and numbers of normal points per week

2 Orbit Post-Fit Residual Analysis

Two geodetic satellites, Lageos-1 and Lageos-2, were used in this study.

2.1 Orbit Computation

The orbital modelling that we used follows the AWG guidelines. In particular, we accounted for the last release of the file containing all the data corrections to be applied to SLR data. These data came from about 30 tracking stations (most of them located in the northern hemisphere, due to a well-known heterogeneity of the ILRS network), gathering up a total of 2,000–3,000 normal points per week and per satellite.

Two computations were carried out, the first one using SLRF2005 (AWG 2007) for a priori SSC, the second using ITRF2008 (Altamimi et al. 2011). The levels of magnitude of weekly residuals are very similar (for Lageos-1, a mean of 1.27 cm $\pm$ 0.25 cm for SLRF2005, and 1.20 cm $\pm$ 0.20 cm for ITRF2008 ; for Lageos-2, a mean of 1.26 cm $\pm$ 0.25 cm for SLRF2005, and 1.19 cm $\pm$ 0.24 cm for ITRF2008), even if a slight difference can be seen over the period 2007–2009. Figure 6.1 shows the level of weekly residuals for Lageos-1 whereas Fig. 6.2 shows the level of weekly residuals for Lageos-2. There are no significant differences between the time series of orbital parameters deduced from the post-fit analysis with SLRF2005 or ITRF2008, except over the period, as expected, 2006–2010. Let us mention, additionally, that the 3D-rms (Fig. 6.3) of the overlap differences (on the basis of two common days per week) is of the order of 5.73 cm $\pm$ 21.89 cm (deduced from time series) for Lageos-1 and of 3.64 cm $\pm$ 1.75 cm for Lageos-2 with SLRF2005 (of 5.80 cm $\pm$ 21.91 cm for Lageos-1 and of 3.59 cm $\pm$ 1.74 cm for Lageos-2 with ITRF2008, respectively). From these values, it seems reasonable to assume that no difference between the two TRF realizations is absorbed through usual traditional parameters used to compensate the lack of non gravitational forces modelling.

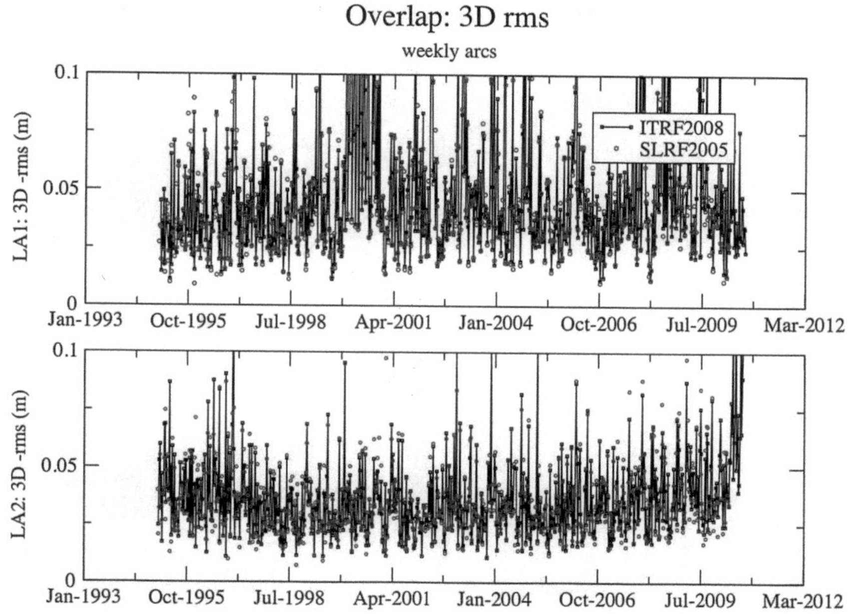

Fig. 6.3 3D-Overlaps for Lageos-1 and Lageos-2 orbits, deduced from 9-day weekly orbital arcs, two consecutive ones having 2 days in common

2.2 Analysis of a Priori Residuals

We had a look at the differences between the observations (range) and the theoretical corresponding values (distances between the tracking stations and the satellites), computed from (1) the orbit and (2) the SSCs provided in SLRF2005 or ITRF2008. These differences were reported before adjusting any "geometrical" parameter linked to the TRF. When using ITRF2008, we show that there is an improvement of the RMS of the a priori residuals for the great majority of the stations (for Lageos-1, improvement for 84 % of the stations, with a median improvement of 1.5 mm ; for Lageos-2, improvement for 88 % of the stations, with a median improvement of 2.2 mm) ; the core station sub-network, supposed to be merely composed by stations of "good quality" is improved for 80 % of the stations for Lageos-1, (with a median improvement of 0.9 mm) and for 85 % of the stations for Lageos-2 (with a median improvement of 1.1 mm). The a priori residuals of the station number 7810 (Zimmerwald, Switzerland) have ever been improved at the level of 5.9 mm for Lageos-1, and 7.9 mm for Lageos-2. We can as well mention the improvement for 7403 (Arequipa, Peru), at the levels of 53.7 mm and 32.3 mm respectively. The different geometrical configurations from the two satellites (Fig. 6.4) should be investigated in detail to explain these orders of magnitude depending on the satellite. Let us mention, furthermore, that in the case of 7403, the velocity of the station expressed in SLRF2005 was supposed to be the same before and after the 2001-earthquake, inducing consequently a quite high level of a priori residuals (Fig. 6.5), and for most of weekly orbital arcs, an elimination of most of the normal points coming from this station ; when using ITRF2008, different velocities depending on the epoch are used for 7403, and the adjusted orbits can account for most of SLR measurements tracked by the Arequipa ILRS station 7403.

3 Results and Comparisons Between SLRF2005, ITRF2005 and ITRF2008

To compare various terrestrial frames, realized as SSCs, a 7-parameter transformation (Altamimi et al. 2002) is estimated, and described by translations (T_x, T_y, T_z), rotations (R_x, R_y, R_z), and a scale factor (D). We carried out weekly transformations between ILRS combined solution and official TRF realizations (SLRF2005 over the period 1982.9–1993.0, ITRF2008 over the period 1993.0–2008.9). For each computation, we (1) project the variance-covariance matrix, (2) we reject raw residuals at the level of 10 cm. For the statistics after transformation, all the position residuals are considered.

Figure 6.6, Tables 6.1 and 6.2 show the comparison, for the main parameters defining a TRF and Earth's rotation, between the ilrsa-v24 ILRS combined solution[1] and using ITRF2005 and ITRF2008 for the period 1993.0–2008.9 (SLRF2005 and ITRF2008 for the period 1982.9–1993.0). It appears that for all parameters, the differences are lower when using ITRF2008, and that there is a better stability of time series achieved with ITRF2008. Let us notice, moreover, that the difference for the scale of the ilrsa-v24 solution is much lower with ITRF2008 than with ITRF2005, as shown Fig. 6.6. Let us mention, furthermore, that this figure can be seen as the difference between the two ILRS solutions.

[1] Described in: http://itrf.ign.fr/ITRF_solutions/2008/doc/ILRSSubmission4ITRF2008.pdf

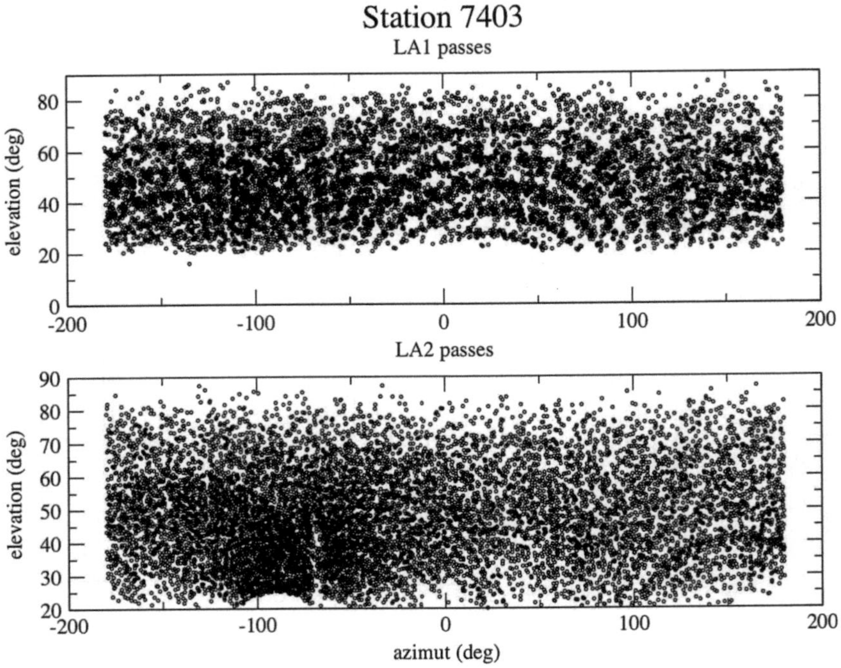

Fig. 6.4 Geometrical configurations of Lageos-1 and Lageos-2 passes, for the Arequipa station (7403). Let us note that many NPs for Lageos-2 seem to be gathered up around an azimut of $-100°$

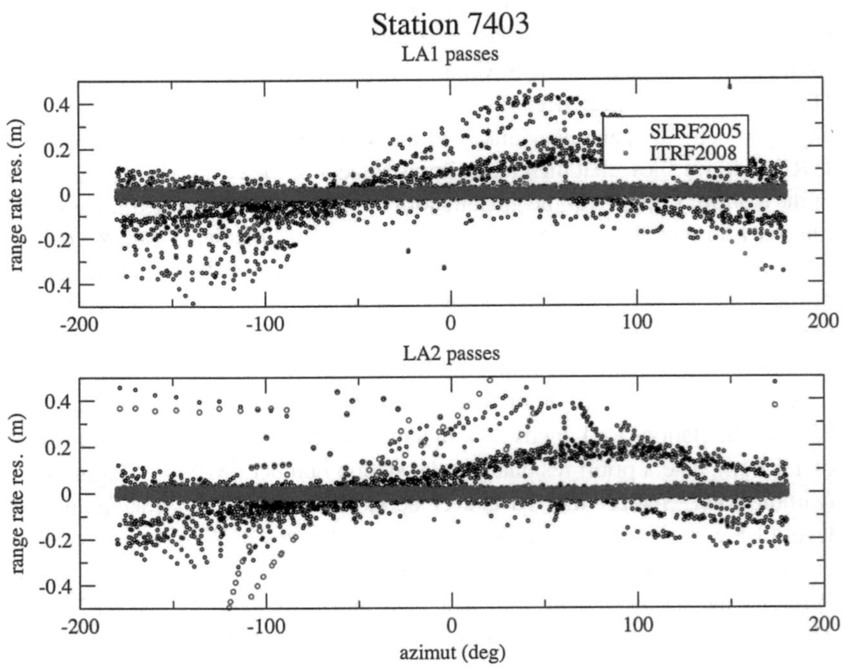

Fig. 6.5 A priori range rate residuals w.r.t. the azimut, for Lageos-1 and Lageos-2 passes, for the Arequipa station (7403), using SLRF2005 or ITRF2008 as an a priori

4 Conclusion

We analysed the SLR part of ITRF2008. Based on an evaluation in terms of Helmert parameters and 3D WRMS of the coordinate residuals, it seems that the new ITRF version is performing better. Station position series WRMS have a better stability in the three components, and we noticed big improvements for SLR stations 7810 and 7403. Let us note that since ITRF retrieves coordinates and velocities from coordinate time series, under the assumption of linear station motion, all the realizations, including ITRF2008, are potentially affected by earthquakes, as the one that occurred near Concepcion, Chile, on spring 2010. It should be kept in mind that the big differences

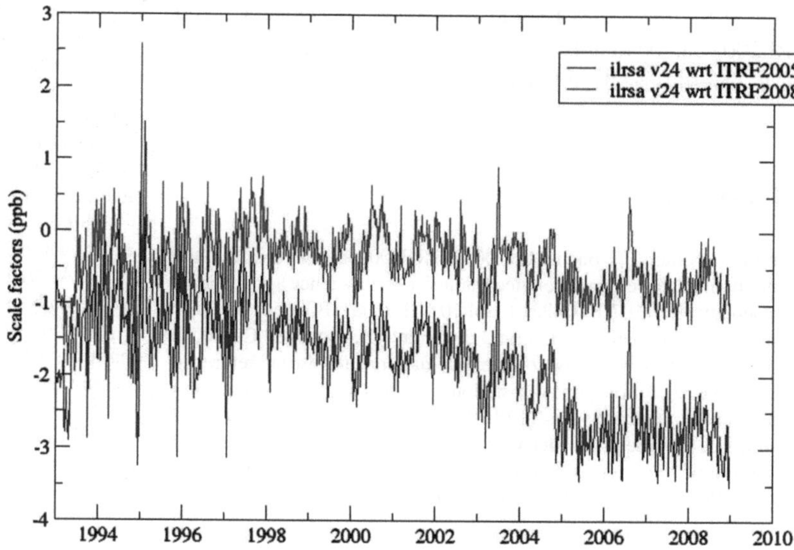

Fig. 6.6 Scale factors between the ilrsa-v24 solution and (i) ITRF2005 and (ii) ITRF2008

Table 6.1 Differences between the pole coordinates from ilrsa-v24 solution consistent with (1) ITRF2005, (2) ITRF2008, and the IERS 05 C04 reference time series. Let us note in particular a better WRMS and a reduction of all biases and (especially the T_z bias) with ITRF2008.

		ITRF2005	ITRF2008	SLRF2005	ITRF2008
		period:1993.0–2008.9		period:1982.9–1993.0	
T_x	Weighted mean	−0.83	−0.01	4.92	3.99
(mm)	Weighted std. dev.	3.97	3.53	8.65	6.60
	WRMS	4.05	3.53	9.95	7.72
T_y	Weighted mean	−0.13	0.06	0.31	2.36
(mm)	Weighted std. dev.	3.76	3.36	7.79	6.51
	WRMS	3.76	3.36	7.80	6.93
T_z	Weighted mean	1.29	0.76	−14.55	−4.06
(mm)	Weighted std. dev.	7.35	7.02	21.39	19.93
	WRMS	7.46	7.06	25.88	20.34
scale	Weighted mean	−1.91	−0.47	−0.15	0.06
(ppb)	Weighted std. dev.	0.69	0.41	1.23	1.11
	WRMS	2.03	0.63	1.24	1.11
X_p	Weighted mean	40	6	348	304
(μas)	Weighted std. dev.	228	203	700	679
	WRMS	232	203	782	744
Y_p	Weighted mean	43	−35	84	−223
(μas)	Weighted std. dev.	222	203	738	689
	WRMS	226	206	743	724

Table 6.2 Station positions residuals between ilrsa-v24 and the solution consistent with (1) ITRF2005, (2) ITRF2008. Let us note in particular a better stability of the series achieved with ITRF2008, through the three components

TRF	E (mm)	N (mm)	U (mm)
(Period: 1993.0–2008.9)	Median WRMS of residuals		
	All stations (20 core stations)		
ITRF2005	11.87 (7.23)	14.33 (8.09)	13.93 (7.61)
ITRF2008	11.14 (6.56)	10.77 (7.53)	8.74 (5.65)
(Period: 1982.9–1993.0)	Median WRMS of residuals		
	All stations (20 core stations)		
SLRF2005	19.44 (15.22)	21.00 (18.23)	21.29 (13.47)
ITRF2008	13.59 (14.02)	18.98 (16.75)	12.91 (11.82)

in the SLR time series for the realization of ITRF2005 and ITRF2008 can be mainly explained by the treatment of the site biases, carefully analyzed by the ILRS AWG.

References

Altamimi Z, Sillard P, Boucher C (2002) ITRF2000: a new release of the international terrestrial reference frame for earth science applications. J Geophys Res 107(B10):2214. doi:10.1029/2001JB000561

Altamimi Z, Collilieux X, Legrand J, Garayt B, Boucher C (2007) ITRF2005: a new release of the international terrestrial reference frame based on time series of station positions and earth orientation parameters. J Geophys Res 112:B-09401. doi:10.1029/2007jb004949

Altamimi Z, Collilieux X, Métivier L (2011) ITRF2008: an improved solution of the international terrestrial reference frame. J Geodesy 85(8):457–473. doi:10.1007/s00190-011-0444-4

Bizouard C, Gambis D (2009) The combined solution C04 for earth orientation parameters consistent with international terrestrial reference frame 2005. In: Drewes H (ed) Geodetic reference frames, IAG symposia 134. doi:10.1007/978-3-642-00860-3_41

Deleflie F, Coulot D (2009) GRGS ILRS analysis center contribution for the ITRF2008 realization. In: Heydari-Malayeri M, Reylé C, Samadi R (eds) Proceedings des Journées SF2A 2009, Besançon, pp 97–101

Deleflie F, Coulot D (2010) GRGS evaluation of the ITRF2008p solution, from SLR data. In: Boissier S, Heydari-Malayeri M, Samadi et R, Valls-Gabaud D (eds) Proceedings des Journées SF2A 2010, Marseille, pp 117–121

ILRS Analysis Working Group (2007) SLRF2005. ftp://cddis.gsfc.nasa.gov/slr/products/resource/reanalysis_2007/slrf20005.pdf

Pearlman MR, Degnan JJ, Bosworth JM (2002) The international laser ranging service. Adv Space Res 30(2):135–143. doi:10.1016/S0273-1177(02)00277-6

Plag HP, Pearlman M (eds) (2009) Global geodetic observing system. Meeting the requirements of a global society on a changing planet in 2020. Springer, Berlin

Alternative Definitions of the Terrestrial Reference System and Its Realization in Reference Frames

7

H. Drewes, D. Angermann, and M. Seitz

Abstract

The International Terrestrial Reference System (ITRS) is defined by the IERS Conventions as a geocentric system with the origin in the Earth's centre of mass. It is realized by a crust-fixed frame of reference stations (ITRF). The paper deals with alternative realizations of these specifications with the high accuracy needed in geosciences research.

A geocentric frame fixes the origin permanently in the Earth's centre of mass, while a crust-fixed frame moves with the Earth's crust, and the origin of the coordinate system may depart from the geocentre ("geocentre motion"). The characteristics and realizations of both definitions are discussed along with their advantages and shortcomings.

The computation of the reference frame is highly correlated with the observed network. In a global reference frame, the network stations should be distributed homogeneously over the Earth. Clusters of stations affect the frame by possible systematic (e.g. climatic) effects, in particular when applying similarity (Helmert) transformations. Densifications of the global frame in sparsely occupied regions of the network suffer from eventual distortions created by inhomogeneous station distributions.

The time evolution of the reference frames is at present done by linear station coordinate changes (constant velocities) over long time intervals only. Seasonal variations are not considered. Experiences with the Chile 2010 earthquake demonstrate the necessity of successive reference frames with short time lag. Alternatives are discussed in the paper.

Keywords

International Terrestrial Reference System (ITRS) • International Terrestrial Reference Frame (ITRF) • Geocentric origin • Crust-fixed TRF

1 Introduction

Three topics are discussed in the following:

- Alternative definitions of global and regional terrestrial reference systems and its realizations in reference frames (geocentric versus crust-fixed);
- Effects of the geographical distribution of the stations forming the terrestrial reference frame (clusters versus homogeneously distributed);
- Time evolution of the terrestrial reference frame (multi-year solutions with constant station velocities versus such with non-linear velocities, or frequent epoch solutions, respectively).

2 Definition of Global and Regional Terrestrial Reference Systems and Its Realization in Reference Frames

The International Terrestrial Reference System (ITRS) is defined by the IERS Conventions as follows (Petit and Luzum 2010, Sect. 4.1.4):

H. Drewes (✉) • D. Angermann • M. Seitz
Deutsches Geodätisches Forschungsinstitut, Alfons-Goppel-Str. 11, D-80539 Munich, Germany

Z. Altamimi and X. Collilieux (eds.), *Reference Frames for Applications in Geosciences*, International Association of Geodesy Symposia 138, DOI 10.1007/978-3-642-32998-2_7,

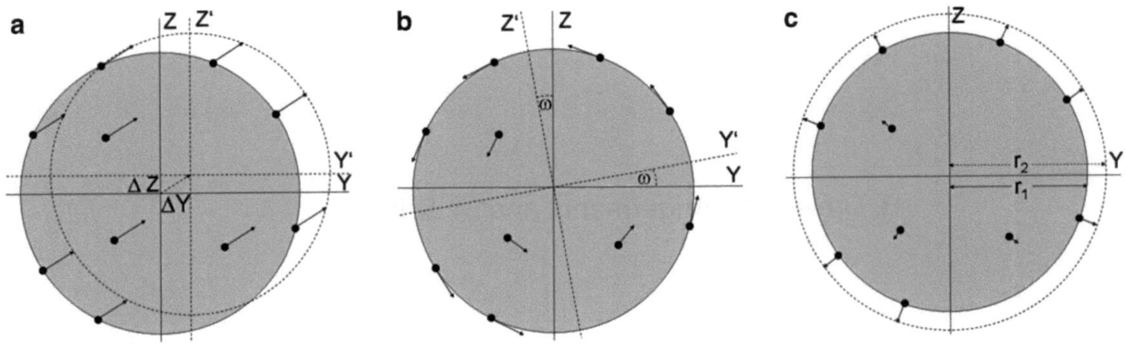

Fig. 7.1 Crust-fixed reference frame: datum parameters change with systematic motions of the reference stations. (**a**) Common shifts are seen as motions of the origin (ΔX, ΔY, ΔZ), (**b**) common rotations as changes in Earth orientation (ω_X, ω_Y, ω_Z), and (**c**) dilatation as change of scale (r_2/r_1)

1. It is geocentric, its origin being the centre of mass for the whole Earth, including oceans and atmosphere;
2. The unit of length is the meter (SI). The scale is consistent with the TCG time coordinate for a geocentric local frame, in agreement with IAG (1992) and IUGG (1992) resolutions. This is obtained by appropriate relativistic modelling;
3. Its orientation was initially given by the BIH orientation at 1984.0;
4. The time evolution of the orientation is ensured by a no-net-rotation condition with regards to horizontal tectonic motions over the whole Earth.

The International Terrestrial Reference Frame (ITRF) is the global realization of the ITRS by a set of physical points (observation stations) at the Earth's surface and their given coordinates (positions and constant velocities). The coordinates are computed by combining solutions of the analysis centres of the individual techniques (DORIS, GPS, SLR, VLBI) given as crust-based reference frames (ibid, Sects. 4.1.3 and 4.2.5).

At regional scale we may mention exemplarily:

- The European Terrestrial Reference System (ETRS89) is defined coincident with the ITRS at the epoch 1989.0 and moving with the stable part of the Eurasian Plate, i.e., it is crust-fixed (e.g. Adam et al. 2000).
- The Reference System of the Americas (SIRGAS) is defined geocentric and realized as a densification of the ITRF (e.g. Hoyer et al. 1998).

We have thus, in principle, two different definitions and realizations: geocentric and crust-fixed reference systems and frames. They may be characterized by their principal distinction:

In a *crust-fixed reference system*, the position coordinates of the reference stations are kept fixed according to the definition (in ETRS89 they were fixed to the Eurasian Plate in its position in 1989). If the stations move (e.g. with the tectonic plates or due to crustal deformation), the coordinates are not changed, but the coordinate system moves with the frame, i.e., the datum (origin, orientation and scale unit) changes w.r.t. the originally defined quantities (geocentre, BIH orientation, metric length). The datum parameters depend thus on the behaviour of the reference station network on the Earth crust.

We get implicitly crust-fixed reference frames, if we estimate the station coordinates using similarity transformations from one epoch to another. We fix coordinates at one epoch and transform those of the other epoch upon them. If the entire network moves collectively w.r.t. the defined datum (e.g. in global scale through opposite seasonal loading on the northern and southern hemispheres), the common displacements of the whole network do not go into the station coordinates of the second epoch, but to the transformation parameters, i.e. they change the datum (Fig. 7.1). If station velocities are included (e.g. in 14 parameter transformations), the average deviation of the station velocities from the reference velocity model (e.g. NNR NUVEL-1A) goes into the transformation parameters and not into the individual station velocities (Drewes 2009a).

The transformation parameters strongly depend on the selected stations of the network. We may demonstrate this by the transformation of weekly coordinates of the global IGS network (available as files iglyyPwwww at ftp://cddis.gsfc.nasa.gov) to the reference frames IGS05 and ITRF2008, respectively. Both frames are based on station positions at a reference epoch and linear velocities. It is, in general, anticipated that seasonal station variations average out at zero in the global frame and do not affect the datum realization. We see however that there are clear seasonal effects in the transformation parameters (in Fig. 7.2 shown for Z-translation), i.e. the complete IGS network moves seasonally with respect to the reference frames. This effect can also be seen in the ITRF computations (e.g. Altamimi et al. 2007).

The seasonal effect is significantly larger when transforming to ITRF2008 instead of IGS05. This is obviously a network effect demonstrating that the seasonal variations are compensated more by the IGS05 than by the ITRF2008 (cf. Chap. 3).

A *geocentric reference frame* keeps the origin (X_0, Y_0, Z_0) always in the Earth's centre of mass. This is achieved by

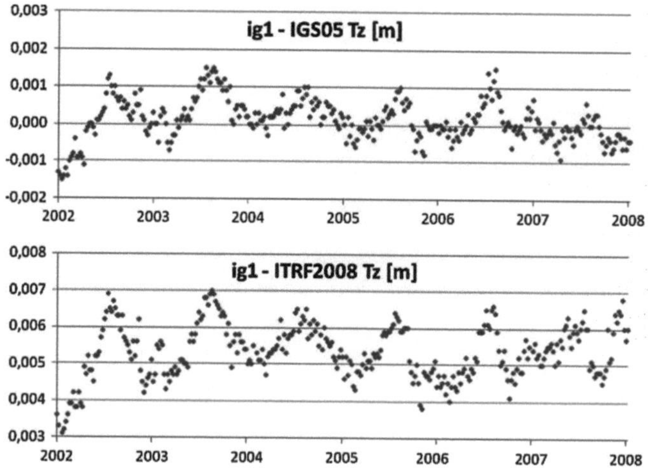

Fig. 7.2 Transformation parameters (Tz) of IGS weekly coordinates (ig1yyPwwww) transformed to the reference frames IGS05 (*above*) and ITRF2008 (*below*)

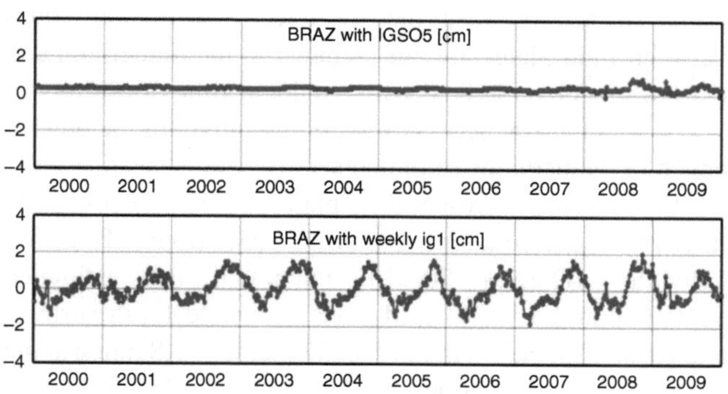

Fig. 7.3 Time series of heights of station Brasilia in SIRGAS obtained from network transformation to the global frame IGS05 (*above*) and by a common adjustment of the regional network with weekly reference coordinates (ig1yyPwwww) from a global network (*below*)

coordinate adjustment with data from precise satellite techniques, e.g. Satellite Laser Ranging (SLR), without network transformations. The geocentric origin is given externally by the gravity field parameters. If the first degree and order spherical harmonic coefficients (C_{11}, S_{11}, C_{10}) of the gravity model applied in orbit determination are zero and no other constraints are introduced, then the orbit coordinates (ephemeris) are always geocentric (formula 1). They are transferred in the common orbit and network adjustment to the station coordinates which also become geocentric.

$$\begin{aligned} C_{11} &= \iiint X\,dm / a\,M; & X_0 &= \iiint X\,dm / M \\ S_{11} &= \iiint Y\,dm / a\,M; & Y_0 &= \iiint Y\,dm / M \qquad (1) \\ C_{10} &= \iiint Z\,dm / a\,M; & Z_0 &= \iiint Z\,dm / M. \end{aligned}$$

In regional scale, a geocentric reference frame may be computed by introducing the coordinates of stations of the global reference frame with their standard deviations as quasi-observations in the network adjustment. We demonstrate the difference between this procedure and a transformation of the regional network to a global reference frame in Fig. 7.3. Above we see the time series of heights obtained by introducing the coordinates of the IGS05 reference frame with constant velocities as a NNR/NNT condition, i.e. similarity transformation. Below we have the equivalent time series when introducing the weekly coordinates of the global IGS network in a common adjustment. We clearly see that seasonal variations in the order of ±1 cm, which are known to be caused by atmospheric and hydrospheric loading, appear only in the latter case.

The conclusion is that the computation of the reference frame through direct adjustment of the observations in a geocentric system provides results closer to reality (depicting actual seasonal effects) than the network transformation to a crust-fixed reference frame with constant velocities.

3 Geographical Distribution of Stations Forming the Reference Frame

Above we saw that the network transformation parameters strongly depend on the distribution of stations in the network. In the ITRF we have dense clusters in Europe and North America: 35 % of the sites are in located in North America and 22 % in Europe (see Fig. 7.4). These clusters dominate the transformation parameter determination. Physical effects in these regions, e.g. seasonal variations due to atmospheric and hydrospheric loading, are over-represented in the global network. As both regions are in the northern hemisphere, the seasons are synchronous and multiply the effect.

In order to demonstrate the effect, we repeated the transformation of the IGS weekly station coordinates to the IGS05 reference frame (see Chap. 2, Fig. 7.2) by stations with a more homogeneous distribution. We selected each one station in a global equal area grid coming up with 71 stations instead of the total IGS05 set of 132 stations. The difference of transformation parameters is shown in Fig. 7.5 exemplarily for the scale factor.

We see a significantly larger seasonal signal when transforming to the homogeneous network than to the original IGS05 frame. This can be explained by the smoothing effect of the clusters in Europe and North America which produce analogical seasonal coordinate variations due to similar climate effects. The opposite climatic behaviour of the northern and southern hemispheres, which enters in particular into the scale factor, is more visible if the southern sites are better represented in the complete network.

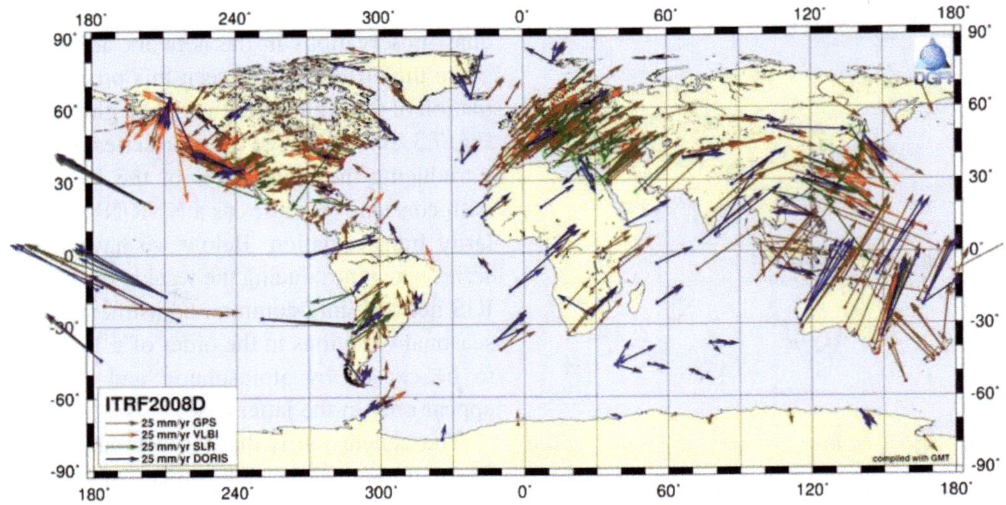

Fig. 7.4 ITRF2008 velocity field showing clusters in Europe and North America

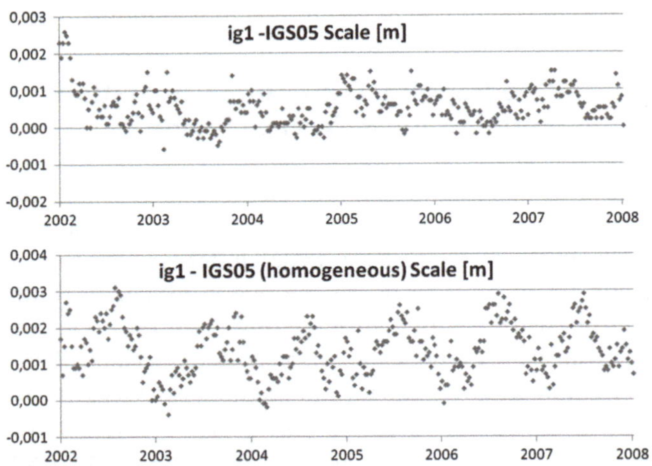

Fig. 7.5 Scale parameter of IGS weekly coordinates (ig1yyPwwww) transformed to the original reference frame IGS05 (*above*) and IGS05 stations reduced to homogeneous distribution (*below*)

4 Time Evolution of Reference Frames

According to IERS conventions, the time variation of station coordinates is today estimated by constant velocities. Non-linear coordinate variations (e.g. seasonal or episodic) are not included in the model.

The datum of the velocities is defined such that no net rotation is produced with regard to horizontal motions over the whole Earth surface.

The realization of the kinematic datum is done by the geologic-geophysical model NNR NUVEL-1A (DeMets et al. 1994) which is assumed to have no net rotation. There are two principal shortcomings in this assumption: (1) the NNR condition cannot be fulfilled over the whole Earth surface because the extended deformation zones between rigid plates are not included in the modelling, and (2) the model is derived over geologic times (millions of years) which is not necessarily representative for the present epoch of geodetic measurements.

We estimated present-day plate rotation vectors and inter-plate deformations (APKIM2005) from ITRF2005 velocities (Drewes 2009b), integrated the estimated velocities in a $1° \times 1°$ grid over the whole Earth surface, and found a global rotation of 0.057 mas/a (=1.8 mm/a at the equator of rotation) around a pole at $\varphi = -23°$, $\lambda = 143°$ (see Fig. 7.6).

Besides the global rotation there is a translation of $\Delta X = -11$ mm/a, $\Delta Y = 13$ mm/a, $\Delta Z = 15$ mm/a. This is due to the configuration of plates, i.e. there is an average motion of all the plates w.r.t. the geocentre towards the Northwest Pacific where we have the extended subduction zones. These translations are considered in the estimated station velocities, but they affect the realization of the datum parameters indirectly in the transformation procedure: If the centre of the transformed network (average of station coordinates) is far off the coordinate origin, as it is in the space-geodetic networks, then the translation, rotation and scale parameters become highly correlated.

As a consequence, we have to use a present-day plate rotation model derived from space geodetic observations for the time evolution of the reference frame. Using NNR NUVEL-1A instead does not only change rotation, but the correlations between the transformation parameters estimated from the unequally distributed stations (see Chap. 3) and the presence of global translation (see above) affect translation, rotation and scale parameters.

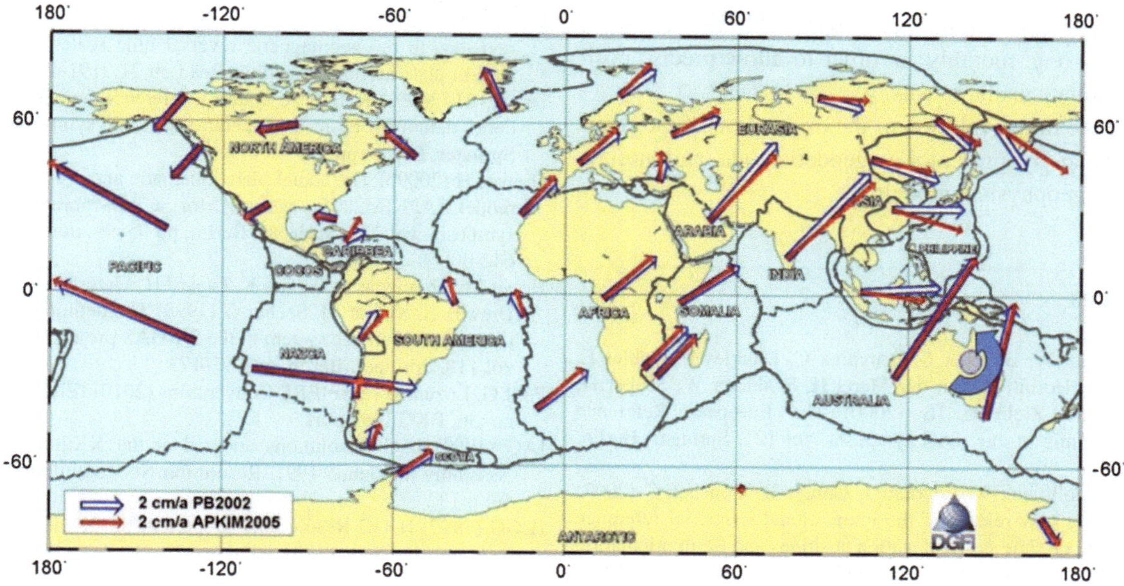

Fig. 7.6 APKIM2005 in NNR NUVEL-1A (PB2002) reference frame with rotation around $\varphi = -23°$, $\lambda = 143°$

5 Conclusions

The use of the terrestrial reference frame for global change and geodynamics research requires ultimate accuracy of station positions and velocities. We want to see processes like global sea level rise and tectonic plate motions. Sea level rise is quite well known from satellite altimetry with an average present-day rate of ± 3.4 mm/a. But we want to see in particular possible acceleration, which requires an accuracy in the order of 0.1 mm/a for altimetry tracking stations in a stable global reference frame. Plate motions are today known with precision better than ± 1 mm/a. But to detect abnormal movements, e.g. as precursors of seismic events, we need higher accuracy in global and regional scale. We therefore have to exploit all methods to achieve a highly accurate frame referring to a unique, long-term stable geocentric datum not affected by the actually employed station network.

The fundamental requirement when applying satellite techniques is the conformity of reference frames of station coordinates and gravity field parameters. If coordinates of tracking stations are not geocentric, one must not use a geocentric gravity field model with the first degree and order spherical harmonic coefficients equal zero. The parameters must be corrected correspondingly. Motions of the origin of the coordinate system with respect to the geocentre (dX/dt, dY/dt, dZ/dt), often called geocentre motions, require this correction for each epoch of terrestrial coordinate estimation using geodetic satellite techniques (e.g. weekly). If we do not correct it, the motion of the origin is transferred to the satellite orbits and affects the estimated coordinates (station positions and/or sea level heights). Such motions are for instance generated by non-modelled seasonal station displacements, i.e. coordinates are no longer geocentric, and in general when applying transformations of networks with non-geocentric datum.

The more reasonable approach is leaving the origin permanently in the geocentre. Station coordinates must change if the station is moving and must not be fixed to the moving crust. This can be achieved by realizing the reference frame through simultaneous adjustment of station positions, station velocities and satellite orbits in the geocentric reference frame of the gravity field ($C_{11} = S_{11} = C_{10} = 0$) using satellite observations, if applicable in combination with other observations (e.g. VLBI), without other datum constraints.

We conclude the following recommendations for the computation of the ITRF

- Realize the ITRF by combination of original geodetic data on the observation level (or datum free normal equations generated with identical standards and conventions) without network transformations.
- Realize the geocentric datum by satellite methods (SLR) using geocentric gravity field parameters without additional constraints.
- Realize regional densifications of the ITRF with identical definition and procedure (geocentric, common adjustment in a global reference frame). This requires the inclusion of SLR observations or geocentric reference stations.
- Realize in multi-year solutions non-linear station motions (e.g. seasonal variations) by estimating them within the time-dependent reference frame.

- Realize successive reference frames with high temporal resolution (e.g. monthly) in order to allow precise positioning after seismic events.
- Realize the time evolution of the orientation of the reference frame by present-day geodetic data and not by geologic-geophysical models.

References

Adam J, Augath W, Boucher C, Bruyninx C, Dunkley P, Gubler E, Gurtner W, Hornik H, van der Marel H, Schlueter W, Seeger H, Vermeer M, Zielinski JB (2000) The European Reference System coming of age. IAG symposia, vol 121. Springer, Berlin, pp 47–54

Altamimi Z, Collilieux X, Legrand J, Garayt B, Boucher C (2007) ITRF2005: a new release of the international terrestrial reference frame based on time series of station positions and Earth orientation parameters. J Geophys Res 112(B09401):19. doi:10.1029/2007JB004949

DeMets C, Gordon R, Argus DF, Stein S (1994) Effect of recent revisions to the geomagnetic reversal time scale on estimates of current plate motions. Geophys Res Lett 21:2191–2194

Drewes H (2009a) Reference systems, reference frames, and the geodetic datum – basic considerations. IAG symposia, vol 133. Springer, Berlin, pp 3–9

Drewes H (2009b) The actual plate kinematic and crustal deformation model (APKIM2005) as basis for a non-rotating ITRF. IAG symposia, vol 134. Springer, Berlin, pp 95–99, doi:10.1007/978-3-642-00860-3_15

Hoyer M, Arciniegas S, Pereira K, Fagard H, Maturana R, Torchetti R, Drewes H, Kumar M, Seeber G (1998) The definition and realization of the reference system in the SIRGAS project. IAG symposia, vol 118. Springer, Berlin, pp 167–173

Petit G, Luzum B (eds) IERS Conventions (2010) IERS technical note no. 36, BKG Frankfurt

IAG (1992) IAG Resolutions adopted at the XXth IUGG General Assembly in Vienna 1991, Resolution No. 1. Bulletin Géodésique 66: p 132

IUGG (1992) IUGG Resolutions adopted at the XXth IUGG General Assembly in Vienna 1991 and related to Geodesy, Resolution No. 2. Bulletin Géodésique 66: p 128

Evaluation of GNSS Monument Stability

8

R. Haas, S. Bergstrand, and W. Lehner

Abstract

We report on an evaluation of the stability of four different GNSS monuments that was conducted in the summer of 2010. The monuments were monitored by forward intersections using a survey system consisting of two robotic total stations and a set of retro reflecting prisms. The system was operated for almost 3 months, performing observations in two faces with a repetition cycle of 5 min. Movements in excess of 6 mm were detected. The results show clear evidence that the detected deformations are related to variations in temperature and solar radiation and can be suppressed by simple shielding of the monument. Furthermore, our project is a step towards the realization of continuous cartesian connections at geodetic fundamental stations.

Keywords

Geodetic monitoring • GNSS monuments • Local-tie monitoring • Fundamental geodetic stations • Global geodetic observing system (GGOS) • Co-located space geodetic techniques • Continuous cartesian connections (CCC)

1 Introduction

Space geodetic techniques and their applications are today dominated by Global Navigation Satellite Systems (GNSS). One of the main objectives of static GNSS is to maintain local, regional, and global reference frames and networks. In Sweden, the National Land Survey (Lantmäteriet—LM) operates the SWEPOS network since the early 1990s. During the last years, LM has augmented the original network of 21 SWEPOS sites that are used for geophysical research, e.g. the BIFROST project (Scherneck et al. 2003), and the whole network now also comprises about 150 additional RTK-stations (SWEPOS 2011). The original SWEPOS monuments consist of 3 m tall heated circular concrete pillars that are firmly connected to crystalline bedrock (Scherneck et al. 2002). The monuments were designed this high in order to guarantee full satellite visibility above 10 ° of elevation, to prevent vandalism and disturbances due to people and animals, and to mitigate snow effects.

In the near future, the original 21 SWEPOS sites are to be equipped with additional GNSS monuments for redundancy purposes, and a favorable ratio of monument stability versus financial expense is sought for. The Onsala Space Observatory (OSO) is one of these 21 sites and the decision on the monument design requires a reliable stability evaluation.

On their homepage, UNAVCO (2011) suggests a number of monument designs utilized in different regional networks, and lists important aspects when choosing monuments. When external factors (multipath, elevation cut-off, ground stability etc.) are excluded, two important criteria for a good monument design remain: stability, and a narrow antenna

R. Haas (✉)
Department of Earth and Space Sciences, Chalmers University of Technology, Onsala Space Observatory, SE-439 92 Onsala, Sweden
e-mail: rudiger.haas@chalmers.se

S. Bergstrand
SP Technical Research Institute of Sweden, Measurement Technology, SE-501 15 Borås, Sweden

W. Lehner
Institute of Geodesy and Geophysics, Vienna University of Technology, AT-1040 Wien, Austria

Z. Altamimi and X. Collilieux (eds.), *Reference Frames for Applications in Geosciences*, International Association of Geodesy Symposia 138, DOI 10.1007/978-3-642-32998-2_8,

mount within a wavelength from the antenna base; the latter to not interfere with satellite signal reception.

However, the basis for assessments of monument stability is usually derived directly from GNSS results, e.g. Williams et al. (2004); Beavan (2005); Bergstrand et al. (2005); Langbein (2008). As an alternative, independent measurements can be performed with classical geodetic survey techniques. These techniques allow high precision in the local frame and high temporal resolution. In the summer of 2010 we evaluated the stability of four GNSS monument designs at OSO.

2 GNSS-Monuments

At OSO, Precambrian crystalline bedrock is exposed at large parts of the observatory. Four GNSS monuments were erected within an area of 10 m × 10 m.

The first monument was a 3.20 m high truss mast with a triangular cross section used in temporary real-time-kinematic (RTK) networks and was provided by LM for the investigation. Haas and Bergstrand (2010) indicated a potential bending effect of truss masts with quadratic cross section due to differential solar heating, and LM has pondered upon a solution to prevent potential vandals from climbing the antenna. We therefore agreed to evaluate the effect of a shield made of a plastic sewage pipe around the truss mast, and compare the shielded and unshielded mast over periods long enough to achieve responses that are representative of the ambient conditions.

The other three monuments were manufactured at the workshop at OSO. The hexagonal mast of 3.20 m height is an extended version of the design used for the mounting of small so-called Salsa radio telescopes at OSO. The Earlconic mast follows the design used by the Michigan Department of Transportation (2011). It is a steel pipe with reinforcement gussets and was also constructed to be 3.2 m high. The fourth mast is a 1.25 m high welded reinforced tripod constructed to resemble the shallow drilled braced monument (SDBM) as described by UNAVCO (2011). The four monuments are described in more detail by Lehner (2011).

The narrow base of the Earlconic and Salsa monuments allowed for horizontal fixations, whereas the considerably wider base of the SDBM monument was adjusted to the rock surface by subsequent telescopic extension of the three outwards pointing legs and fastening. Figure 8.1 shows the four monuments set up at OSO.

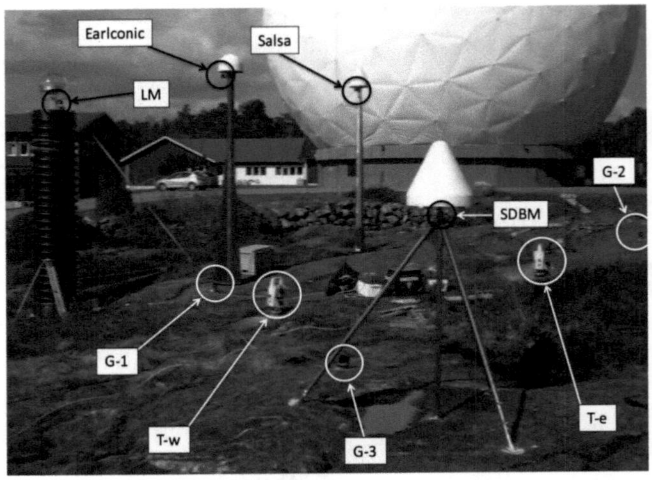

Fig. 8.1 Picture of the small survey network at the Onsala Space Observatory. Shown are the two total stations *T-w* (total station west) and *T-e* (total station east), the three survey prisms *G-1, G-2, G-3* mounted directly in bedrock, and the four prisms mounted on the *LM*, Earlconic, Salsa, and *SDBM* monument

3 Survey Method

The developed survey method is based on the most accurate method for classical surveying with theodolites: the standard method of forward intersection observing a target with double instruments in two faces. The advantages of the method are measurement redundancy and that the weakest part of the measurement system—the electronic distance meter (EDM)—is largely omitted. A detailed description of the method is provided by Lehner (2011).

We deployed two Leica TS30 robotic total stations (Zogg et al. 2009) to monitor Leica GMP 104 prisms and Leica RFI reflectors. The two total stations were attached to tribrachs that were mounted directly in the bedrock and formed a baseline of about 3.5 m length with east–west orientation. We refer to an eastern and western total station in the following. The utilized coordinate system is based on the direction between these two stations and may deviate slightly from true east and north. Around the two total stations, three survey prisms were mounted directly in bedrock. The target reflectors were mounted at the monument tops adjacent to the antenna mounts (Fig. 8.1), in order to represent potential movements of the antennas. The two total stations were computer controlled with the Leica GeoMoS 5.0 software, which handled the measurement cycle and logging of data. Observations were done almost continuously from mid of May to mid of August 2010. Additionally, air and material temperature observations were made throughout the campaign.

As the hair cross of modern total stations cannot be observed backwards through the optics as in classic forward intersection, a traditional alignment of the two total stations is not possible. Instead, the orientation and the distance between the two total stations were determined from redundant observations of a calibrated scale bar that was set up in four geometrically well distributed positions with respect to the baseline between the total stations. A Newton iteration method was then used to determine the relative orientation of the two total stations, and the distance between them, thus defining the scale. The uncertainty of the relative orientation was estimated to be 0.15 mgon, based on a variation analysis. The distance estimate between the two total stations was verified by interferometric laser tracker measurements to be in agreement on the 0.1 mm level.

4 Data Analysis and Results

The raw data taken with the pair of robotic total stations were analyzed in post-processing. The measurements to the reference survey prims mounted directly in solid bedrock were used to check the stability of the survey system. The time series of horizontal angles for these prisms show small diurnal sinusoidal signatures with amplitudes of up to 4 mgon, and a drift of 10 mgon during the whole campaign. Based on these measurements, corrections for the horizontal angles to all other targets were applied on an epoch by epoch basis. Forward intersection was then applied to determine time series of local xyz-coordinates for the prisms. These time series reveal movements on the order of up to 6 mm, in particular for the horizontal components. As an example, Fig. 8.2 presents time series for xyz-coordinates during 6 days in the end of May and beginning of June 2010, together with the recorded ambient temperature. Temperature and coordinate variations show a similar pattern, with temperature ahead in the time domain.

The coordinate uncertainties were derived by error propagation, using the instrumental precision of the total stations as provided by the manufacturer and the geometrical situation. For the four investigated monuments the expected uncertainties for the coordinate determination were between 0.01 mm and 0.15 mm for the X-, 0.02 mm and 0.15 mm for the Y-, and 0.03 mm and 0.09 mm for the Z-component.

Figure 8.3 shows the coordinate variations in the xy-plane. The two designs that exhibit larger movements are the gusset enforced Earlconic and the Salsa monument. For the Salsa monument, which in all essence is symmetrical, the elliptical pattern reflects the difference between the large thermal gradients when the sun rises and sets and the smaller gradients at noon. The Earlconic monument exhibits movements of the same magnitude as the Salsa monument, but the movement directions are confined to those of the gussets. Considerably smaller are the movements of the truss mast and the SBDM. When comparing the smaller movements of these two monuments, one should remember that the height ratio of the two monuments are almost 3:1, which inhibits a direct comparison. However, both movements have a strong directional signature. This is comprehensible for the SBDM which have tripod-like features similar to the Earlconic monument, but is not as obvious for the truss mast.

Investigating the truss mast a bit further, we compare the truss mast when being shielded from the sun with a plastic pipe to being fully exposed. The effect of the shield is shown when comparing the graphs in Fig. 8.4, which were recorded under similar conditions. In the top graph of Fig. 8.4, the shield was fitted around the monument and the movements are orientated between the second and fourth quadrant, which is quite contrary to the other monuments. When taking a closer look at the figure, there appears to be two populations.

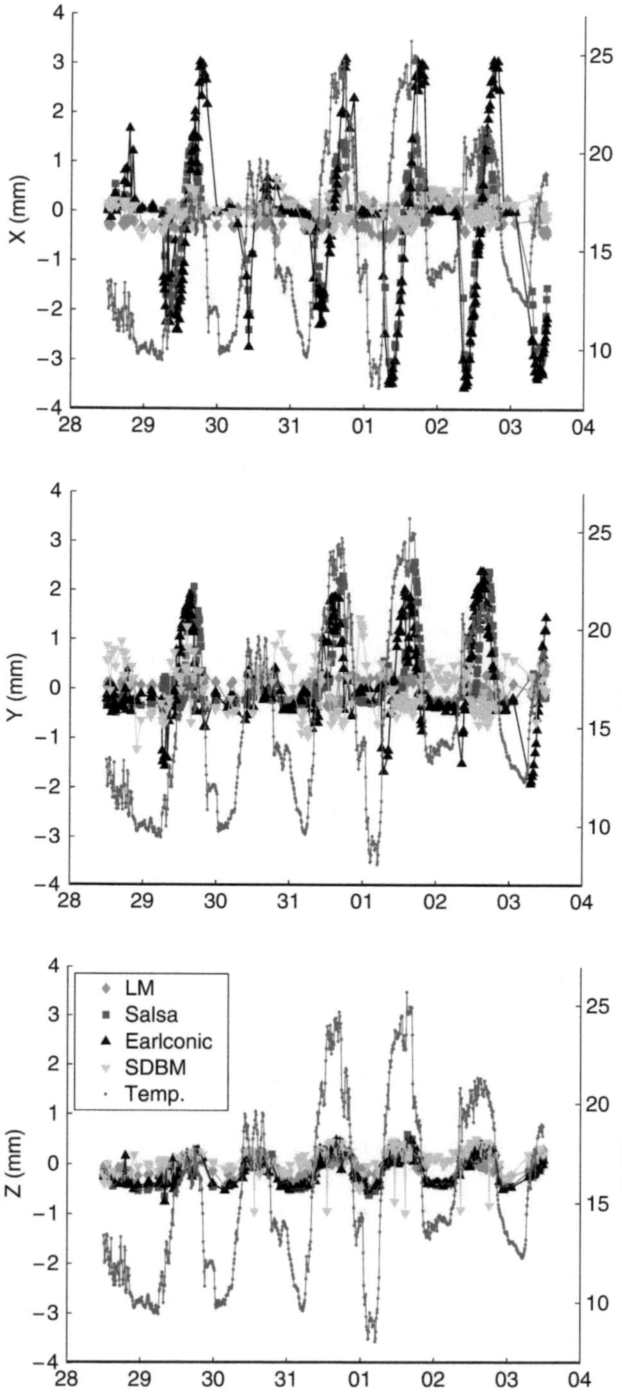

Fig. 8.2 X-, Y- and Z-coordinates (*top, middle and bottom plot*) of the survey prisms on four monuments (*left scale*), together with the recorded ambient temperature (*right scale*) during six days in the end of May/beginning of June

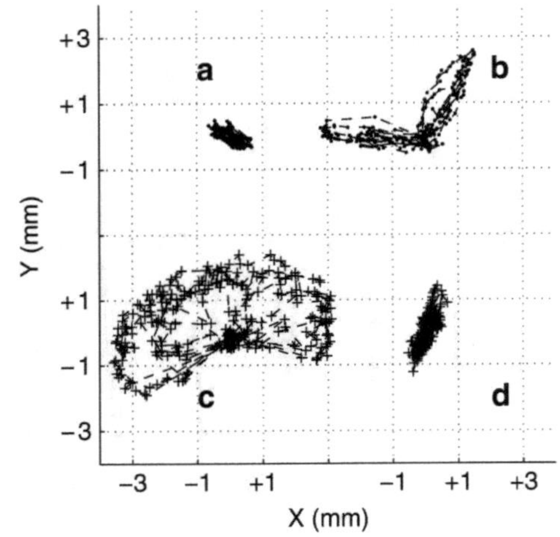

Fig. 8.3 Horizontal positions of the survey prisms mounted at the top of the four monuments during 6 sunny days in the end of May and beginning of June 2010. Shown are (**a**) the truss mast (LM), (**b**) the Earlconic, (**c**) the Salsa, and (**d**) the SDBM monument. The shown values refer to a mean position for each of the prisms. The values for the individual prisms are offset from each other by 3 mm in X and Y for improve visibility and allow easy comparison. The size of the crosses represent the uncertainty of each individual measurement

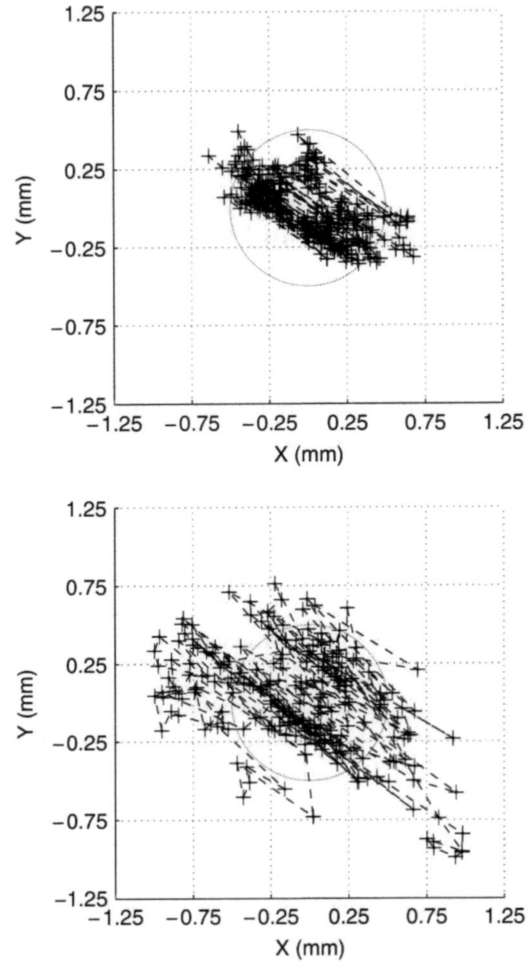

Fig. 8.4 Horizontal positions of the survey prism mounted at the top of the truss mast (M-T) during several days when the mast was covered by a protective pipe (*top graph*), and when the monument was not covered by a protective pipe (*bottom graph*). The shown values refer to a mean position and the dotted circle has a diameter of 1 mm. The size of the crosses represent the uncertainty of each individual measurement

It turns out that the direction coincides with that of a flat bar iron mounted on top of the mast to move the antenna ground plane away from metal objects and in order to improve signal reception. Even so, the observations can almost be circumscribed by a circle of 1 mm diameter. When the shield has been removed, the noise level increases beyond the 1 mm diameter, see the bottom graph in Fig. 8.4, but the pattern is not as articulate as for the solid Salsa and Earlconic monuments. This probably reflects that the solar heating through the truss is constantly changing due to the irregular shape. Nevertheless, the directivity remains and it appears that the flat bar still has considerable impact on the movements.

Figure 8.5 presents the relation of ambient temperature and Z-component for the four monuments. Vertical movements on the range of 1 mm are detected, corresponding to temperature changes of about 15 °C. The Earlconic and Salsa monuments show a larger deformation than the other monuments, which can be explained by that the LM monument was shielded from direct solar radiation, and that the SDBM monument is shorter than the others.

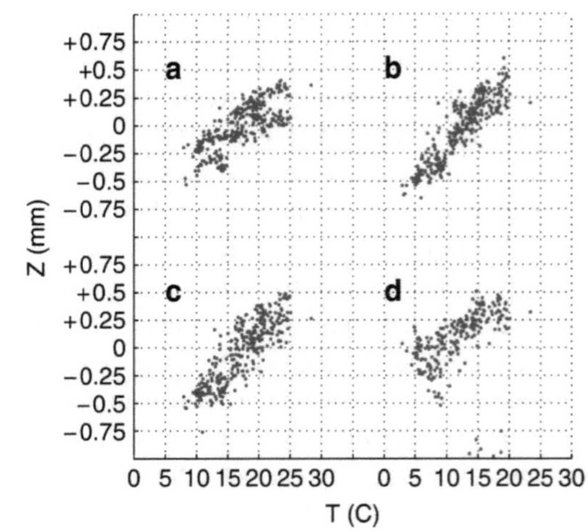

Fig. 8.5 Scatter plots of Z-coordinate versus ambient temperature. Shown are (**a**) the truss mast (LM), (**b**) the Earlconic, (**c**) the Salsa, and (**d**) the SDBM monument

5 Conclusions and Outlook

We developed a survey method to continuously observe local movements of GNSS monuments and employed it for 3 months. This method uses two computer controlled robotic total stations in a forward intersection approach and provides time series of monument positions with high accuracy and high temporal resolution.

Our results reveal a clear connection between temperature and deformation of GNNS monuments, both in the horizontal and the vertical domain. Effects like these are of course expected and the vertical movements can probably be modeled with reasonable performance based on temperature measurements and standard material properties. However, Figs. 8.3 and 8.4 prove that the movement patterns reflect solar interaction with the monument geometry, and show that thermal expansion due to solar heating is an important factor when space geodetic observations need to be fine-tuned. As the observed patterns are asymmetrical, the solar induced movements will cause biases of the monument position estimates. Nevertheless, the patterns seem to motivate the use of lighter monument constructions, e.g. truss masts, probably due to improved heat dissipation and reduced thermal expansion. Shielding the monument proved to significantly reduce the observed movements and should be applicable to all designs. It goes without saying that, had the antenna support atop the truss mast been symmetrical and shielded, the movements could probably have been suppressed further. GNSS signal reception must not be compromised, but complementing the current shield design with a thinner plastic pipe outside the support could be realized without disturbing the antenna pattern. Note that the bedrock at OSO is very stable and that the presented evaluation addresses monument movements above ground only. Other aspects that might be more important for some applications (e.g. swaying during rapid earth movements) have not been addressed.

The presented survey method appears to be well applicable to local tie surveys as well as the proposal for continuous cartesian connections (CCC) at fundamental geodetic stations (Bergstrand and Haas 2011, Haas and Bergstrand 2012). A continuous monitoring with high accuracy and high temporal resolution is relevant for Global Geodetic Observing System (GGOS) (Rummel et al. 2005). The GGOS intends to combine and integrate different geodetic techniques in order to exploit the individual strength of the techniques and thus depends on that local ties between co-located geodetic techniques at fundamental geodetic stations are known with an accuracy on the order of 0.1 mm (Rothacher et al. 2009; Ray and Altamimi 2005; Altamimi et al. 2007). Our method is a first step to realize this ambitious goal.

The next step in the system development is to observe targets on a moving platform (e.g. a VLBI telescope) during operation and to monitor several reference points at a co-location station with high temporal resolution.

Acknowledgement We thank Christer Thunell from Leica Geosystems Sweden, Magnus Herbertsson from SP Technical Research Institute of Sweden, Lasse Wennerbäck, Christer Hermansson, and Håkan Millqvist from the OSO workshop for their support.

References

Altamimi Z, Collilieux X, Legrand J, Garayt B, Boucher C (2007) ITRF2005: a new release of the international terrestrial reference frame based on time series of station positions and Earth orientation parameters. J Geophys Res 112:B09401. doi:10.1029/2007JB004949

Beavan J (2005) Noise properties of continuous GPS data from concrete pillar geodetic monuments in New Zealand and comparison with data from U.S. deep drilled braced monuments. J Geophys Res 110:B08410. doi:10.1029/2005JB003642

Bergstrand S, Scherneck H-G, Lidberg M, Johansson JM (2005) BIFROST: noise properties of GPS time series. Dyn Planet Int Assoc Geodesy Symp 130(II):123–130. doi:10.1007/978-3-540-49350-1_20, 2007

Bergstrand S, Haas R (2011) Can Continuous Cartesian Connections realize local ties at 0.1 mm level? Proc. of the 17th International Workshop on Laser Ranging – Extending the Range –, edited by U. Schreiber, M. Pearlman and G. Appleby, Mitteilungen des Bundesamts für Kartographie und Geodäsie, http://cddis.gsfc.nasa.gov/lw17/docs/papers/session8/05-Bergstrand_Haas_ILRS2011.pdf

Haas R, Bergstrand S (2010) COLD MAGICS—continuous local deformation monitoring of an arctic geodetic fundamental station. In: Behrend D, Baver KD (eds) International VLBI service for geodesy and astrometry 2010 general meeting proceedings, NASA Conference Publication, NASA/CP-2010Ð215864, pp 118–120

Haas R, Bergstrand S (2012) Continuous Cartesian connections at geodetic collocation sites. in preparation

Langbein J (2008) Noise in GPS displacement measurements from Southern California and Southern Nevada. J Geophys Res 113: B05405. doi:10.1029/2007JB005247

Lehner W (2011) Evaluation of environmental stress on GNSS-monuments. M.Sc. thesis, Chalmers University of Technology and Technical University of Vienna, http://publications.lib.chalmers.se/records/fulltext/165054.pdf

Michigan Department of Transportation (2011). http://www.mdotcors.org/

Ray J, Altamimi Z (2005) Evaluation of co-location ties relating the VLBI and GPS reference frames. J Geodesy 79:189–195. doi:10.1007/s00190-005-0456-z

UNAVCO Resources: GNSS Station Monumentation (2011). http://facility.unavco.org/kb/questions/104/UNAVCO+Resources

Rothacher M, Beutler G, Bosch W, Donnellan A, Gross R, Hinderer J, Ma C, Pearlman M, Plag H-P, Richter B, Ries J, Schuh H, Seitz F, Shum CK, Smith D, Thomas M, Velacognia E, Wahr J, Willis P, Woodworth P (2009) The future global geodetic observing system. In: Plag H-P, Pearlman M (eds) Global geodetic observing system. Springer, Heidelberg/Berlin, pp 237–272. doi:10.1007/978-3-642-02687-49

Rummel R, Rothacher M, Beutler G (2005) Integrated global geodetic observing system (IGGOS)—science rationale. J Geodyn 40:357–362

Scherneck H-G, Johansson JM, Elgered G, Davis JL, Jonsson B, Hedling G, Koivula H, Ollikainen M, Poutanen M, Vermeer M, Mitrovica JX, Milne GA (2002) BIFROST: observing the three-dimensional deformation of fennoscandia. In: Mitrovica JX, Vermeersen BLA (eds) Glacial isostatic adjustment and the Earth system, Geodynamics series. American Geophysical Union, Washington, DC, pp 69–93

Scherneck H-G, Johansson JM, Koivula H, van Dam T, Davis JL (2003) Vertical crustal motion observed in the BIFROST project. J Geodyn 35:425–441. doi:10.1016/S0264-3707(03)00005-X

SWEPOS (2011) A national network of reference stations for GPS. http://swepos.lmv.lm.se/english/index.htm

Williams SDP, Bock Y, Fang P, Jamason P, NikolaidisR M, Prawirodirdjo L, Miller M, Johnson DJ (2004) Error analysis of continuous GPS position time series. J Geophys Res 109:B03412. doi:10.1029/2003JB002741

Zogg H-M, Lienhart W, Nindl D (2009) Leica TS30 white paper. Leica Geosystems AG, Heerbrugg, Switzerland. http://www.leica-geosystems.com/downloads123/zz/tps/general/white-tech-paper/WhitePaper_TS30_en.pdf

Comparison of Realizations of the Terrestrial Reference Frame

9

C. Ma, D. MacMillan, S. Bolotin, K. Le Bail, D. Gordon, and J. Gipson

Abstract

IGN and DGFI both generated realizations of the terrestrial reference frame under the auspices of the IERS from combination of the same space geodetic data. We compared the IGN and DGFI TRFs with a GSFC CALC/SOLVE TRF. WRMS position and velocity differences for the 40 most frequently observed sites were 2–3 mm and 0.3–0.4 mm/year. There was a scale difference of −0.39/−0.09 ppb between the IGN/DGFI realizations and the GSFC solution. When we fixed positions and velocities to either the IGN or DGFI values in CALC/SOLVE solutions, the resulting EOP estimates were not significantly different from the estimates from a standard TRF solution.

Keywords

Radio wave technique • Terrestrial reference frame • Space geodesy

1 Introduction

Recently IGN generated ITRF2008, the latest version of the IERS international terrestrial reference frame. This frame is a combination solution based on input from the VLBI, SLR, GPS, and DORIS technique combination analysis centers. For this realization of the TRF (Terrestrial Reference Frame), the input consisted of the technique EOP (Earth orientation parameters) and station position time series as well as the available site tie vectors. To provide an alternative to cross-check the IGN solution, DGFI also generated a combination TRF, DTRF2008, from the same input data available to IGN.

In this paper we evaluate ITRF2008 and DTRF2008, by comparing them with CALC/SOLVE (Ma et al. 1990) VLBI TRF solutions and by investigating the effects of applying the two ITRFs in VLBI solutions. Essentially, we investigate how well the VLBI information provided to the combination is recovered. In Sect. 2, we directly compare the site positions and velocities from the IGN and DGFI solutions with those from a standard VLBI solution. In Sect. 3, we examine the EOP series estimated from solutions in which the TRF positions and velocities are not estimated but are instead fixed to ITRF2008 or DTRF2008. In Sect. 4, we summarize our conclusions.

C. Ma (✉)
NASA Goddard Space Flight Center, Greenbelt, MD 20771, USA
e-mail: chopo.ma@nasa.gov

D. MacMillan • S. Bolotin • K. Le Bail • D. Gordon • J. Gipson
NVI, Inc. and NASA Goddard Space Flight Center, Greenbelt, MD 20771, USA

2 Comparisons of TRFs

For our comparisons, we ran an operational-type VLBI TRF solution with the CALC/SOLVE software using data from 1979 until February 2010. The solution estimated global positions and velocities and source positions from the entire time period as well as EOP. We then compared this VLBI TRF with the positions and velocities extracted from the ITRF2008 and DTRF2008 SINEX files. Tables 9.1 and 9.2 show the position and velocity Helmert 7-parameter transformation values between the VLBI TRF and the IGN or DGFI TRFs. One significant difference is the scale difference of −0.39 ± 0.15 ppb for the IGN solution. For DGFI, the scale difference was only −0.09 ± 0.10 ppb. Translating

Z. Altamimi and X. Collilieux (eds.), *Reference Frames for Applications in Geosciences*,
International Association of Geodesy Symposia 138, DOI 10.1007/978-3-642-32998-2_9,

Table 9.1 Seven-parameter position transformation at epoch 2005

	IGN–VLBI	DGFI–VLBI
T_x (mm)	−0.04 ± 1.1	+0.8 ± 0.7
T_y (mm)	−1.7 ± 1.0	−1.4 ± 0.7
T_z (mm)	+0.8 ± 0.9	−0.5 ± 0.6
R_x (mm)	−5.7 ± 1.2	−5.9 ± 0.8
R_y (mm)	+1.6 ± 1.2	+0.3 ± 0.8
R_z (mm)	−1.9 ± 1.2	+2.5 ± 0.8
Scale (ppb)	−0.39 ± 0.15	−0.09 ± 0.10
Scale		

Table 9.2 7-Parameter velocity transformation

	IGN–VLBI	DGFI–VLBI
$\dot{T}_x$ (mm/year)	−0.39 ± 0.11	+0.07 ± −0.07
$\dot{T}_y$ (mm/year)	−0.56 ± 0.10	−0.13 ± 0.07
$\dot{T}_z$ (mm/year)	−0.25 ± 0.10	−0.30 ± 0.08
$\dot{R}_x$ (mm/year)	−0.27 ± 0.13	−0.30 ± 0.08
$\dot{R}_y$ (mm/year)	+0.00 ± 0.12	+0.14 ± 0.08
$\dot{R}_z$ (mm/year)	+0.15 ± 0.10	−0.06 ± 0.08
Scale (ppb/year)	+0.023 ± 0.016	−0.005 ± 0.010

to site vertical, the IGN discrepancy corresponds to about 2.5 mm. The scale differences must arise from the treatment of scale in the respective combinations. IGN found an SLR–VLBI scale difference of −1.05 ± 0.13 ppb and for the combination, they weighted the VLBI and SLR scales equally (Altamimi et al. 2010; this issue). In contrast, DGFI found essentially no difference in VLBI and SLR scale so that the difference in scale between DTRF2008 and VLBI (SLR) was 0.01 ± 0.03 ppb (0.02 ± 0.03 ppb) respectively (Seitz et al. 2010, this issue). Since VLBI is insensitive to geocenter, the translation differences are due to differences between ITRF2008, DTRF2008, and the a priori coordinates used in the TRF solution. The IGN (DGFI) velocity transformation parameters are all less than 0.6 (0.3) mm/year, where the formal uncertainties are ~0.1 mm/year.

After removing the effect of the seven-parameter transformations, there are some significant residual differences between the VLBI TRF and the IGN and DGFI solutions. In Figs. 9.1, 9.2, 9.3 and 9.4, we show the horizontal and vertical residual differences for sites that observed in at least 20 observing sessions. Differences less than 5 mm are indicated by open circles. The displacement vector differences for larger residuals are plotted. The largest differences for both IGN and DGFI solutions are mainly for Japanese network sites and for US mobile VLBI sites, which observed in the 1980s and early 1990s. The residual 3D differences between the IGN/DGFI solution and the VLBI TRF solution were less than 10 mm for 49/45 sites and greater than 10 mm for 66/64 sites. The number of sessions for the 40 most frequently observed sites ranged from 123 to 2,386 sessions. For these sites, the WRMS of the residual

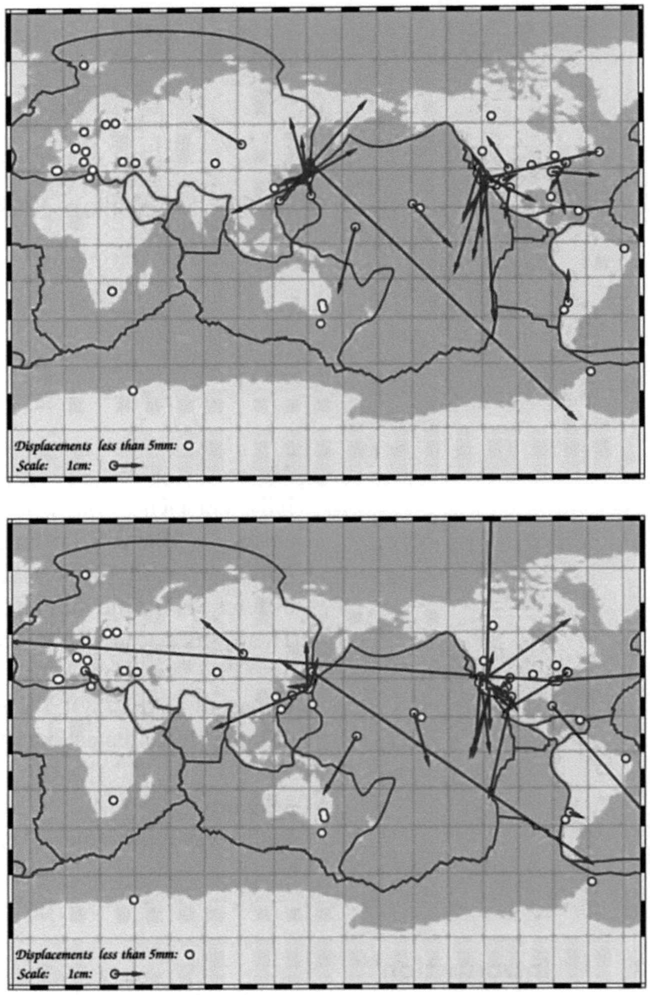

Fig. 9.1 IGN (*upper*) and DGFI (*lower*) horizontal residual vectors (mm) relative to VLBI TRF solution

differences in NEU positions (velocities) for the IGN solution were 1.6 mm (0.3 mm/year), 2.4 mm (0.3 mm/year), and 2.8 mm (0.4 mm/year). For the DGFI solution the NEU residual WRMS values were 2.2 mm (0.3 mm/year), 1.8 mm (0.3 mm/year), and 3.1 mm (0.4 mm/year). For comparison, the VLBI analysis center solution height estimates differed from the IVS ITRF2008 combination solution by 1–2 mm in WRMS (Böckmann et al. 2010).

3 Effects of IGN and DGFI TRFs in VLBI Solutions

In our standard VLBI TRF solutions, we estimate a TRF along with EOP and a CRF (celestial reference frame). In this way, EOP connect the estimated TRF and CRF in a self-consistent way. To evaluate the ITRF2008 solutions, we ran two additional solutions in which we fixed the positions and velocities to those in either the IGN or DGFI solution TRF and then estimated EOP and the CRF. For sites where there

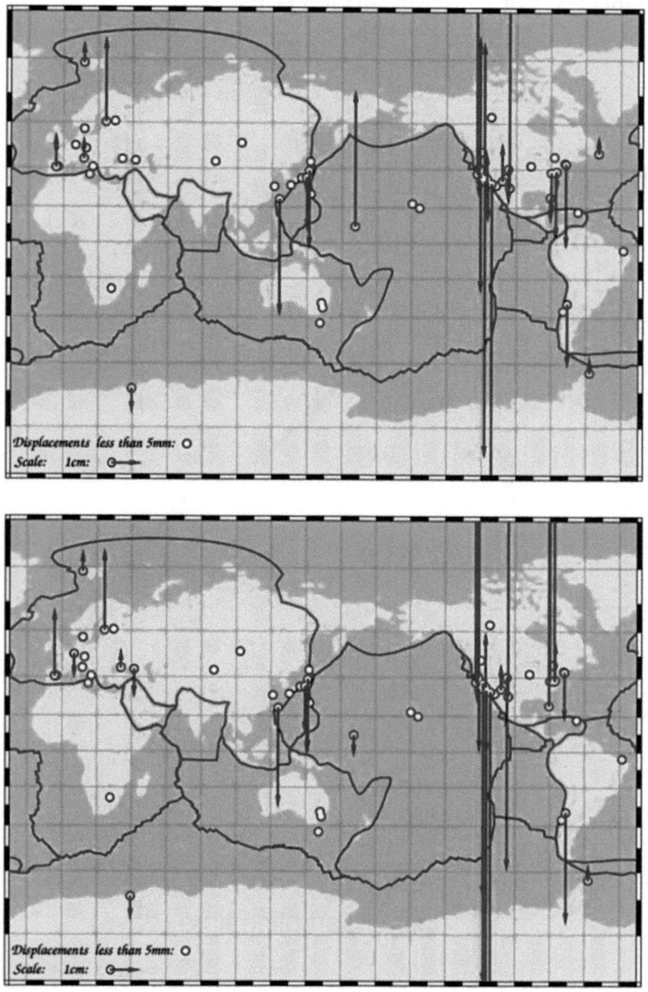

Fig. 9.2 IGN (*upper*) and DGFI (*lower*) vertical position (mm) relative to VLBI TRF solution

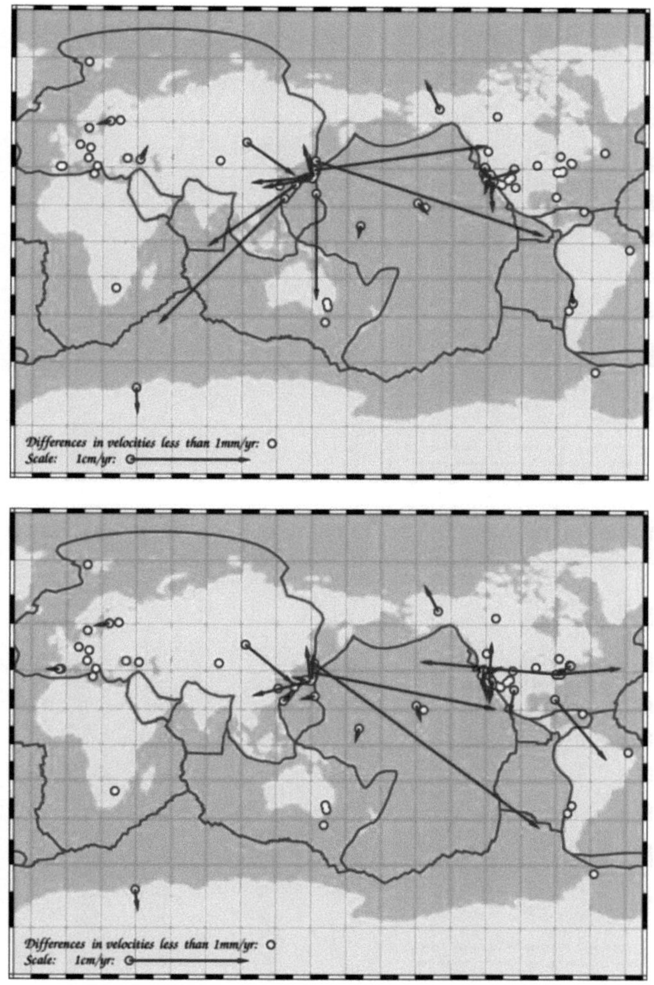

Fig. 9.3 IGN (*upper*) and DGFI (*lower*) horizontal velocity vectors (mm/year) relative to the VLBI TRF solution

were episodic jumps, we applied the jumps given in the DGFI and IGN SINEX files. In the case of Fairbanks, where there was nonlinear postseismic motion after the Denali Earthquake in 2002, we applied the models determined by DGFI or IGN that each consisted of series of XYZ offsets and rates.

As expected, the overall solution fit was best for the standard VLBI solution (22.500 ps) since the TRF was estimated. Fixing the TRF to the DGFI a priori gave a solution residual WRMS fit of 22.650 ps, which was somewhat better than fixing to the IGN a priori which had a fit of 22.733 ps. The IGN 24-h session fits were especially bad for a number of Japanese network sessions. Generally the IGN and DGFI solution daily session WRMS residual fits were similar. Figure 9.5 shows the distribution of the (DGFI-IGN) differences in session fits. Solution fits were slightly better using the DGFI positions and velocities. Typically, 24-h-session solution fits are 20–40 ps. However, the solution fits for many domestic Japanese network sessions were significantly worse for the IGN solution because some of the Japanese station positions (for example, AIRA, GIFU3, SINTOTU2, CHICHI10) were much different from the GSFC VLBI TRF positions as seen in Fig. 9.1.

We compared EOP estimates from the three solutions (IGN, DGFI, VLBI TRF) with the IGS EOP time series. As summarized in Table 9.3, the agreement between the IGS series and the IGN and DGFI series are not significantly different. The χ^2 are greater than 1 mainly because the combined formal EOP uncertainties are too small. VLBI sigmas are small by a factor of 1.5–1.7 and IGS sigmas of 15–30 μas are also too small. IGS agreement with the GSFC VLBI TRF EOP series is slightly better. For X-pole and Y-pole, we also computed as a function of sampling time the Allan variance (Allan 1966, 1987; Le Bail 2006) of the differences between each of the three EOP series and the IGS series. The results shown in Fig. 9.6 indicate that there is no significant difference (much less than one-sigma) between the three solutions and the IGS series.

The Allan variance of the differences between polar motion estimates from the IGN or DGFI solutions and the GSFC VLBI TRF solution are shown in Fig. 9.7. Given the formal uncertainties of the Allan variance, there is no

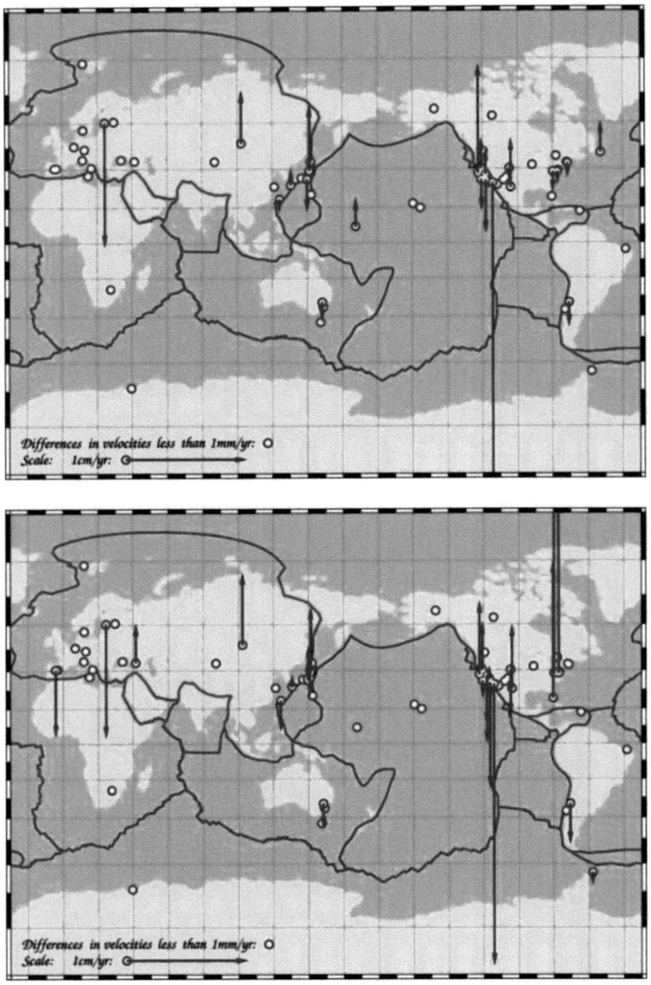

Fig. 9.4 IGN (*upper*) and DGFI (*lower*) vertical rates (mm/year) relative to the VLBI TRF solution

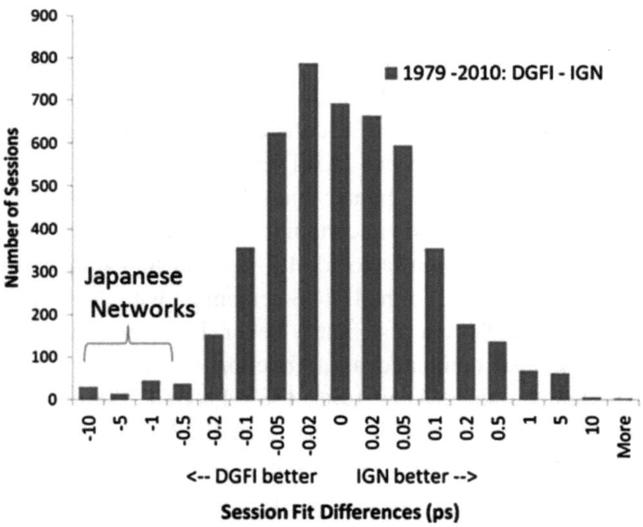

Fig. 9.5 Distribution of differences between 24-h session fits for solutions in which the TRF was fixed to either the IGN or DGFI positions and velocities

Table 9.3 EOP differences with IGS EOP series

	VLBI TRF		DGFI		IGN	
	WRMS	χ^2	WRMS	χ^2	WRMS	χ^2
X, μas	115	3.1	118	3.9	117	3.8
Y, μas	116	3.4	118	4.3	118	4.3
LOD, μs/d	19.7	3.7	19.7	3.7	19.8	3.7

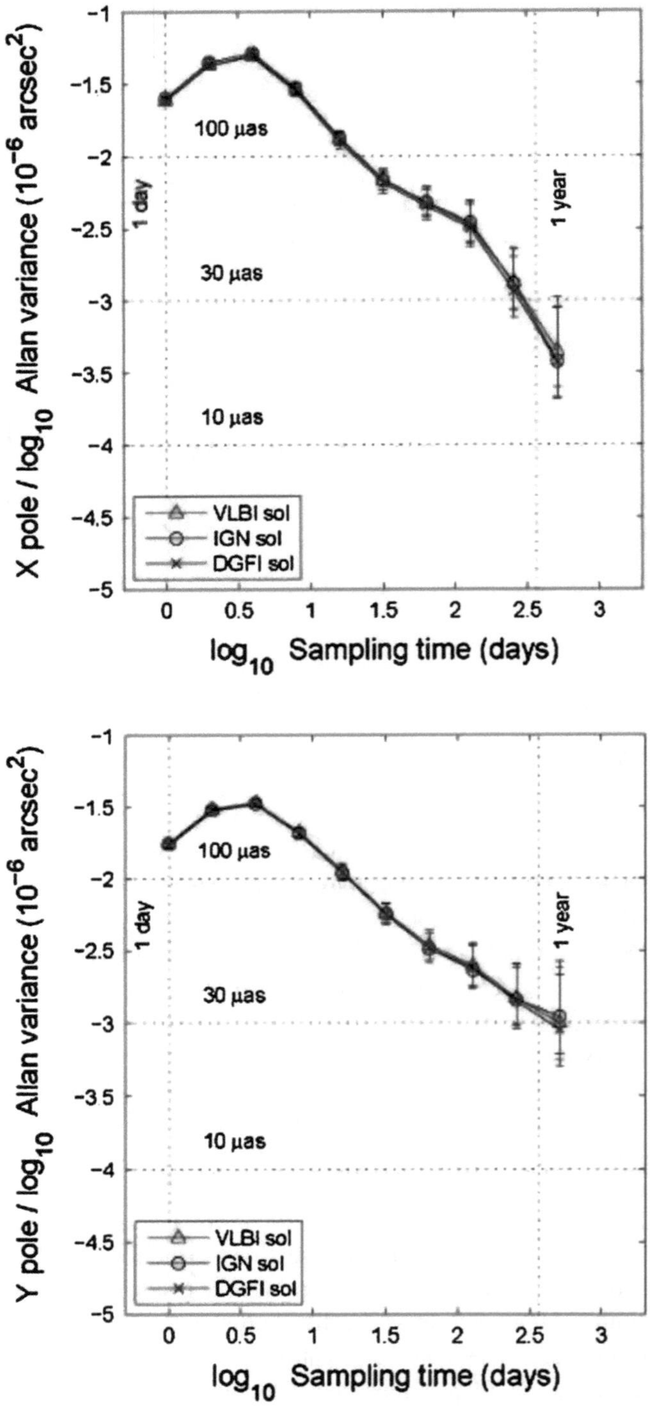

Fig. 9.6 Allan variances of X-pole and Y-pole differences between each of the solutions (VLBI TRF, IGN, DGFI) and the IGS series

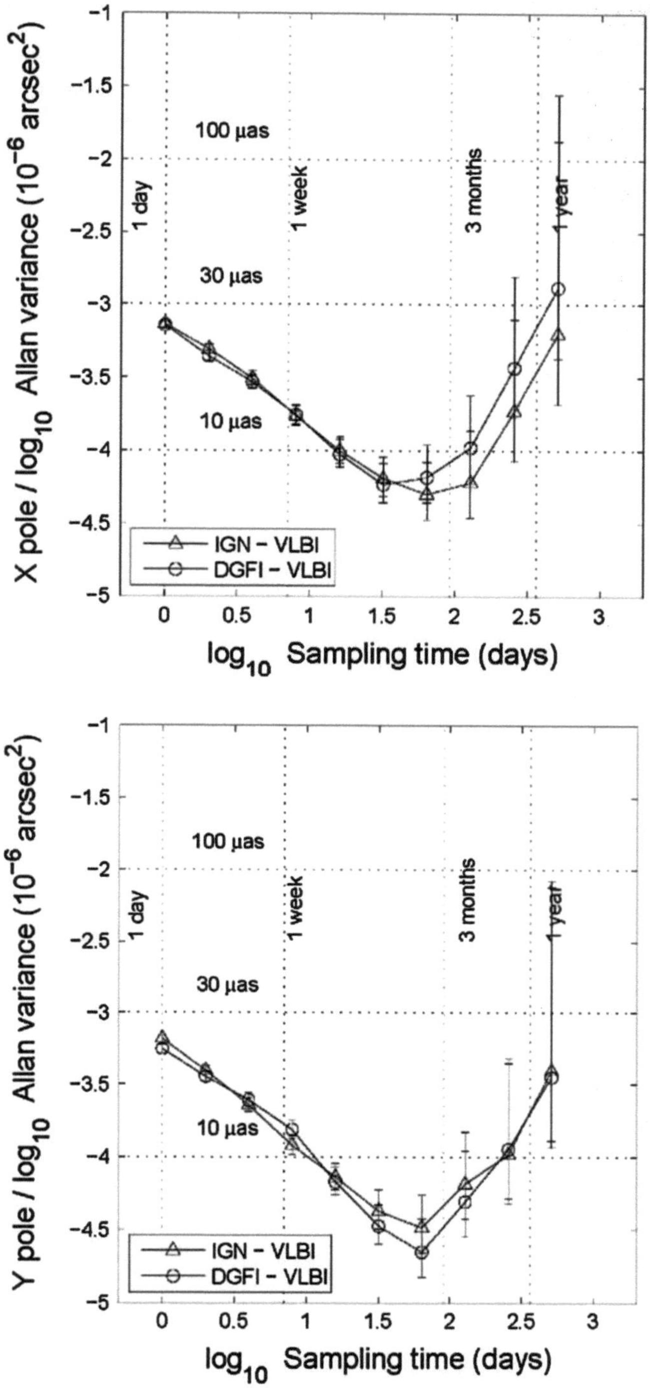

Fig. 9.7 Allan variance between IGN and DGFI solution EOP and VLBI TRF EOP

significant difference in the agreement of the DGFI and IGN solutions with the TRF solution. The WRMS differences between EOP (X, Y, UT1) estimates from the VLBI TRF solution and the IGN (DGFI) solution are 52 µas (46 µas), 46 µas (42 µas), and 3.3 µs (3.3 µs) respectively. Agreement is at the 1-sigma level and is slightly better for DGFI.

The effect on the estimated CRF using the IGN or DGFI a priori TRFs is not significant. The CRF XYZ frame rotation angles (in mas) between the IGN (DGFI) solutions and the full VLBI TRF solution are only 0.026 ± 0.050 (0.050 ± 0.050), -0.020 ± 0.060 (-0.023 ± 0.060), and 0.000 ± 0.020 (0.003 ± 0.020). WRMS differences (in mas) between source positions after taking out the rotation angle differences are only 0.011 (0.009) in declination and 0.005 (0.003) in right ascension.

4 Conclusions

We have compared the IGN and DGFI TRFs with the TRF estimated from a GSFC operational-type CALC-SOLVE solution. For the 40 most frequently participating stations, the WRMS differences are 2–3 mm in position components and 0.3–0.4 mm/year for velocity components. The scale difference between the GSFC TRF solution and the IGN/DGFI solution is −0.39/−0.09 ppb. This IGN scale difference occurs because IGN found a scale difference of −1.05 ppb between the SLR and VLBI and input solutions and then weighted the two solutions equally in their combination. Possible explanations for the difference between the scales of the IGN and DGFI solutions are differences in their treatment of site ties and more generally differences between their strategies used to define the TRFs. But these are questions that would need to be addressed by IGN and DGFI.

There are large differences between GSFC and DGFI and/or IGN positions/velocities for a large number of sites including Japanese domestic sites, mobile VLBI sites in North America that observed in the 1980s and early 1990s, and other VLBI stations that have not observed recently.

We also evaluated the effect of using the IGN or DGFI TRFs in CALC/SOLVE solutions. Fixing the positions and velocities to either IGN or DGFI solutions yield EOP series that agree with the IGS combined series equally well within formal uncertainties. The DGFI solution EOP estimates agrees slightly better than the IGN solution with a GSFC VLBI TRF solution where positions and velocities are estimated. Both solutions agree with the TRF solution within formal uncertainties.

References

Allan DW (1966) Statistics of atomic frequency standards. Proc IEEE 54:221–231

Allan DW (1987) Time and frequency characterization, estimation and prediction of precision clocks and oscillators. IEEE Trans UFCC 34:647–654

Altamimi Z, Collilieux X, Métivier L (2010) ITRF combination: theoretical and practical considerations and lessons from ITRF2008. REFAG2010, IAG commission 1 symposium, Marne la Vallée

Böckmann S, Artz T, Nothnagel A, Tesmer V (2010) VLBI terrestrial reference frame contributions to ITRF2008. J Geod 84:201–219. doi:10.1007/s00190-009-0357-7

Le Bail K (2006) Estimating the noise in space-geodetic positioning: the case of DORIS. J Geod 80:541–565. doi:10.1007/s00190-006-0088-y

Ma C, Sauber J, Bell LJ, Clark TA, Gordon D, Himwich WE, Ryan JW (1990) Measurement of horizontal motions in Alaska using very long baseline interferometry. J Geophys Res 95(B13):21991–22011

Seitz M, Angermann D, Drewes H (2010) Accuracy assessment of ITRF2008D. REFAG2010, IAG commission 1 symposium, Marne la Vallée

The 2010 Reference Edition of the IERS Conventions

10

G. Petit and B. Luzum

Abstract

This paper presents the IERS Conventions (2010) the new reference edition replacing the Conventions (2003) and describes their most significant features: new realizations of the celestial and terrestrial reference systems; a new conventional geopotential model along with updated model and implementation for the ocean tides; updated models for several components of the station displacement; new models for all aspects of atmospheric propagation.

Keywords

Celestial reference system • Terrestrial reference system • Standards • Astrometry • Space geodesy

1 Introduction

The principal products of the International Earth Rotation and Reference Systems Service (IERS) are the international celestial and terrestrial reference frames, ICRF and ITRF, and the set of Earth orientation parameters required to transform between them. These products are generated using several independent techniques, therefore their quality strongly depends on using a consistent and complete set of models and procedures, which are formalized in the IERS Conventions. Following the publication of the reference edition IERS Conventions (McCarthy and Petit 2004), the Conventions have been continually updated in an electronic version available at http://tai.bipm.org/iers/convupdt/convupdt.html. Updates were discussed through an Advisory Board, established in 2005, where all components of the IERS are represented and the most significant changes were explicitly endorsed by the IERS Directing Board. At a workshop held in September 2007, the IERS then decided to produce a new reference edition of the Conventions, which is now completed as the Conventions (2010) (Petit and Luzum 2010). In the next sections the main features of the new reference edition, highlighting the most important changes with respect to the Conventions (2003), are presented.

G. Petit (✉)
Bureau International des Poids et Mesures, 92312 Sèvres, France
e-mail: gpetit@bipm.org

B. Luzum
US Naval Observatory, Washington, DC, USA

2 General Features of the IERS Conventions (2010)

The reference systems and procedures of the IERS are based on the resolutions of the International Astronomical Union (IAU) and the International Union of Geodesy and Geophysics (IUGG), as discussed in Sect. 2.2. It follows that the IERS reference systems are defined in the framework of the General Relativity Theory (GRT) as outlined in Sect. 2.3. A new feature in the Conventions (2010) is an attempt to formalize rules to classify models and their usage in generating IERS products, see Sect. 2.1. For a more extensive presentation of these concepts, see the introduction of the Conventions (2010).

Z. Altamimi and X. Collilieux (eds.), *Reference Frames for Applications in Geosciences*,
International Association of Geodesy Symposia 138, DOI 10.1007/978-3-642-32998-2_10,

2.1 Guiding Principles for Models

Models and effects to be considered in the scope of the Conventions are classified into three categories: Class 1 ("reduction") models are those recommended to be used a priori in the reduction of raw space geodetic data in order to determine parameter estimates. These models are to be accepted as known a priori, therefore it is intended that their accuracy is at least as good as the expected accuracy of the geodetic data. Class 2 ("conventional") models are those that eliminate an observational singularity and are purely conventional in nature. Class 3 ("useful") models are those that are beneficial but are not required as either Class 1 or 2.

The IERS Conventions intend to present a complete and consistent set of the necessary models of the Class 1 and Class 2 types, including implementation software. Where conventional choices must be made (Class 2), the Conventions provide a unique set of selections to avoid ambiguities among users. The resolutions of the international scientific unions and historical geodetic practice provide guidance when equally valid choices are available. Class 3 models are included when their use is likely to be sufficiently common, as guidance to users. In the next sections, information on the model class will be provided as required.

2.2 The Framework Set by the Resolutions of the Scientific Unions

The reference systems and procedures of the IERS are based on the resolutions of international scientific unions. The celestial system is based on IAU (International Astronomical Union) Resolution A4 (1991). It was officially initiated and named International Celestial Reference System (ICRS) by IAU Resolution B2 (1997) and its definition was further refined by IAU Resolution B1 (2000) and by IAU Resolution B3 (2009). The terrestrial system is based on IUGG (International Union of Geodesy and Geophysics) Resolution 2 (1991). It was officially endorsed as the International Terrestrial Reference System (ITRS) by IUGG Resolution 2 (2007). The transformation between celestial and terrestrial systems is based on IAU Resolution B1 (2000) and was complemented by IAU Resolutions B1 and B2 (2006). The definition of time coordinates and time transformations, the models for light propagation and the motion of massive bodies are based on IAU Resolution A4 (1991), further refined by IAU Resolution B1 (2000) and IAU Resolution B3 (2006). In some cases, the procedures used by the IERS, and the resulting conventional frames produced by the IERS, do not completely follow these resolutions. These cases are identified in the Conventions and procedures to obtain results consistent with the resolutions are indicated.

The text of all IAU Resolutions may be found at http://www.iau.org/administration/resolutions/general_assemblies/.

In addition the IERS Conventions take advantage of the work of several working groups and commissions:

- The IAU Division 1 working group "Numerical Standards in Fundamental Astronomy" (NSFA; see Luzum et al. (2011)) recommendations are implemented in the IERS Conventions in many ways, primarily in its Chaps. 1 and 3.
- The IAU Division 1 working group "Nomenclature in Fundamental Astronomy" (NFA; see Capitaine (2009)) which acted over the period 2003–2006. Nomenclature used in the Conventions follows the NFA and all relevant terms are reproduced *verbatim* in the Conventions glossary.
- The IAU Commission 52 "Relativity in Fundamental Astronomy" has issued a series of recommendations, see Klioner et al. (2010).

For a more complete account on the issues presented in this section, see Petit (2010).

2.3 Relativistic Models for Space-Time Coordinates and Signal Propagation

The relativistic models for space time coordinates and time transformations are treated in Chap. 10 of the Conventions, where the presentation of coordinate time scales (TCB and TDB for the barycentric system, TCG and TT for the geocentric system) accounts for all IAU Resolutions. Coordinate times TT (since 2000) and TDB (since 2006) now have conventional definitions. In the Conventions (2003), several options were given to realize transformations between TCB and TCG, none of which provided a complete realization. Instead, one conventional transformation, XHF2002_IERS, is proposed in the Conventions (2010). The transformation between proper and coordinate time in the vicinity of the Earth (typically up to geosynchronous orbit or slightly above) is also covered in the Conventions (2010).

The relativistic models for signal propagation are treated in Chap. 11 of the Conventions, with no significant change with respect to the Conventions (2003).

3 The Reference Systems and Frames

While the basic principles for the definition of the reference systems have not changed, significant modifications have occurred in their realization since the edition of the IERS Conventions (2003). For what concerns the realization of the celestial frame and the conventional transformation between the celestial and terrestrial systems, these modifications were endorsed by IAU Resolutions. On the other hand, successive realizations of the ITRF resulted from the work of the various IERS components under the responsibility of the ITRF product center.

3.1 The Celestial Reference System and Frame

The second realization of the International Celestial Reference Frame (ICRF2) was selected as the conventional celestial reference frame. The ICRF2, which was adopted with Resolution B3 of the XXVII IAU General Assembly, consists of 295 defining and 3,119 additional compact extragalactic radio sources (Fey et al. (2009)). The noise floor for the positions of the sources is estimated to be approximately 40 μas and the axis stability is 10 μas. The celestial reference system is defined by the J2000.0 equator and the origin of right ascension, the Celestial Intermediate Origin (CIO). The conventional positions of this fiducial plane and the direction of the CIO are realized by Very Long Baseline Interferometry (VLBI) and Lunar Laser Ranging (LLR) observations. The optical realization of the ICRF is the Hipparcos Celestial Reference Frame (HCRF) as defined by IAU Resolution B1.2 (2000) while the conventional dynamical realization of the ICRS is provided by a modern ephemeris such as the DE421 (Folkner et al. (2009)).

3.2 The Terrestrial Reference System and Frame

Since the publication of ITRF2000 (Altamimi et al. 2002) which was the conventional realization of the terrestrial frame in the Conventions (2003), a new strategy has been used where time series of station positions (weekly from satellite techniques and 24-h sessions from VLBI) and daily EOPs are used as input data. On this basis, ITRF2005 (Altamimi et al. 2007) was issued in 2007 then ITRF2008 (Altamimi et al. 2011) was specified in 2010 and became the conventional realization of the Conventions (2010). The ITRF2008 is based on reprocessed solutions of four space geodesy techniques: VLBI, SLR, GNSS and DORIS, spanning 29, 26, 12.5 and 16 years of observations, respectively. Solutions are provided by the respective International Services of the International Association of Geodesy, namely the IVS (Schlüter and Behrend 2007), ILRS (Pearlman et al. 2002), IGS (Dow et al. 2009) and IDS (Willis et al. 2010). The ITRF2008 is composed of 934 stations located at 580 sites, of which 105 have co-located techniques. The ITRF2008 is specified by the following frame parameters:

- Origin: The ITRF2008 origin is defined in such a way that there are zero translation parameters at epoch 2005.0 and zero translation rates with respect to the ILRS SLR time series.
- Scale: The scale of the ITRF2008 is defined in such a way that there is a zero scale factor at epoch 2005.0 and a zero scale rate with respect to the mean scale and scale rate of VLBI and SLR time series.
- Orientation: The ITRF2008 orientation is defined in such a way that there are zero rotation parameters at epoch 2005.0 and zero rotation rates between ITRF2008 and ITRF2005. These two conditions are applied over a set of 179 reference stations located at 131 sites.

3.3 The Transformation Between Reference Systems

The transformation between reference systems is accomplished through the application of a sequence of rotation matrices. These rotation matrices are consistent with the latest IAU resolutions and are comprised of rotations due to precession/nutation, Earth rotation and polar motion. The rotations can be performed using the CIO-based algorithm or an equinox-based algorithm. The conventional models used in the transformation are the IAU 2006 precession, IAU 2000 nutation model, the diurnal and semidiurnal variations due to ocean tides (Ray et al. 1994), and the libration effects (Brzezinski and Capitaine 2010).

4 The Geopotential Model

Following the availability of GRACE data in the 2000s, new gravitational models have been developed (see e.g. Parts I and II of Flechtner et al. (2010)) that represent a significant improvement with respect to EGM96, the past conventional model of the IERS Conventions (2003). The IERS Conventions (2010) conventional model is based on the EGM2008 model but for some of the low-degree coefficients, the conventional geopotential model uses values which are different from the original EGM2008 values. The static field also assumes values for the secular rates of low-degree coefficients, up to degree 4. In order to use the static field properly and project it in time, the secular rates should be accounted for. The models for the low degree terms are generally consistent with the past long term trends, but these trends are not strictly linear in reality as there may be decadal and other variations that cannot be captured by linear models.

In addition to the static field, other time varying effects (Class 1) must be taken into account by variations in the Stokes coefficients of the geopotential and are described in the Conventions (2010):

- The model for solid Earth tides is unchanged from the Conventions (2003);
- The effect of ocean tides is presented in some detail and implementation is provided for the tide model FES2004 from Lyard et al. (2006);
- The solid Earth pole tide is unchanged from the Conventions (2003), except for a typo correction with a marginal effect of order 2 % of the total effect;
- The effect of the ocean pole tide is introduced in a model following Desai (2002).

5 Conventional Models for Station Displacements

The IERS Conventions distinguish three categories of station displacements:

1. Conventional displacements of reference markers on the crust relate the regularized positions of the reference points to their conventional instantaneous positions. The regularized positions are provided by the ITRF in the form of linear models and represent the long-term variations due to e.g. plate motion. Generally the conventional instantaneous positions are used in data analyses as a priori coordinates for subsequent adjustment of observational data. They include tidal motions (mostly near diurnal and semidiurnal frequencies) and other accurately modeled displacements of reference markers (mostly at longer periods);
2. Other displacements of reference markers include nontidal motions associated with changing environmental loads and have a very broad spectral content;
3. Displacements that affect the internal reference points within the observing instruments, which are generally technique-dependent.

Models in the first category should be applied at the observation level and must be used in all analyses, except possibly when the magnitude of the effect is completely negligible with respect to the expected accuracy of the observations. The Conventions (2010) provides conventional models (Class 1) for the displacement due to:

- The body tides arising from the direct effect of the external tide generating potential, unchanged from the Conventions (2003);
- The ocean tidal loading, with a new formulation and the provision of a conventional routine;
- The diurnal and semidiurnal atmospheric pressure loading, based on a new model by Ray and Ponte (2003);
- The centrifugal perturbations caused by Earth rotation variations, including the pole tide which is to be computed with respect to a new mean pole model IERS (2010);
- The loading caused by the ocean pole tide, based on a new model by Desai (2002) consistent with the treatment for the geopotential.

Effects in the second category vary in general more slowly and models are usually less accurate than in the first category, therefore it is generally recommended that they not be included in computing conventional instantaneous positions. This is an active scientific field and the situation is likely to evolve in the future. At this time, no such model is described in the Conventions (2010).

Models in the third category are to be applied in the analysis for each individual technique. They are listed in the Conventions with a short description and a link to provide more complete information.

6 Models for Atmospheric Propagation

To make better use of improved observational capability, the new mapping function of Mendes et al. (2002) and the zenith delay model of Mendes and Pavlis (2004) have been adopted for optical techniques. For radio techniques, the hydrostatic delay is modeled using the Saastamoinen (1972) formula as provided by Davis et al. (1985). Since no adequate physical model exists to account for the wet tropospheric delay, for precise applications the wet delay is estimated using a standard mapping function. The Vienna Mapping Function (VMF1) from Boehm et al. (2006), which uses numerical weather model inputs, is the conventional radio mapping function. A model is provided to compensate for the ionospheric dispersive effects.

7 Supporting Material

The conventional models are not the only things that have been improved with the new Conventions. Significant effort has been spent to improve the usability of the Conventions. An example of this is the revision of the associated conventional software. All Conventions software now uses a standard template that provides better readability of the code and more extensive commenting. In addition to the augmented documentation, the code was modified to improve robustness. To ensure that the code is working as expected, test cases are now provided with each routine. Finally, since it is expected that this code will be used in a wide variety of applications, a software license has been added to provide the conditions under which the code may be used.

8 Conclusion

The IERS Conventions (2010) are now available. They implement the framework set by the Resolutions adopted by the scientific unions and describe the standard reference systems realized by the IERS and the models and procedures used for this purpose. As any reference edition, the Conventions (2010) are designed to represent the situation at a given moment. The Conventions will continue to evolve with the progresses in the field and we expect important changes in some of the directions e.g. the treatment of non-tidal motions associated with changing environmental loads or the link with vertical reference systems. This is a work in progress.

Acknowledgments We thank all the individuals who contributed to the IERS Conventions (2010), a number of whom are mentioned in the introduction chapter of the Conventions (2010). Special thanks to Jim Ray who led the Advisory Board for the IERS Conventions updates and provided advice for many years.

References

Altamimi Z, Sillard P, Boucher C (2002) ITRF2000: a new release of the international terrestrial reference frame for Earth science applications. J Geophys Res 107(B10):19. doi:10.1029/2001JB000561

Altamimi Z, Collilieux X, Legrand J, Garayt B, Boucher C (2007) ITRF2005: a new release of the international terrestrial reference frame based on time series of station positions and Earth orientation parameters. J Geophys Res 112(B9):B09401, 19. doi: 10.1029/2007JB004949

Altamimi Z, Collilieux X, Métivier L (2011) ITRF2008: an improved solution of the international terrestrial reference frame. J Geod. doi:10.1007/s00190-011-0444-4

Boehm J, Werl B, Schuh H (2006) Troposphere mapping functions for GPS and very long baseline interferometry from European centre for medium-range weather forecasts operational analysis data. J Geophys Res 111:B02406. doi:10.1029/2005JB003629

Brzezinski A, Capitaine N (2010) Semi-diurnal signal in UT1 due to the influence of tidal gravitation on the triaxial structure of the Earth. In: Corbett IF (ed) Highlights of astronomy, vol 15. Cambridge University Press, Cambridge UK

Capitaine N (2009) Nomenclature and numerical standards for IAU models and IERS conventions for earth rotation. In: Soffel M, Capitaine N (eds) Proceedings Journées Pub. 2008, Lohrmann-Observatorium and Observatoire de Paris, pp 46–49

Davis JL, Herring TA, Shapiro II, Rogers AEE, Elgered G (1985) Geodesy by radio interferometry: effects of atmospheric modeling errors on estimates of baseline length. Radio Sci 20(6):1593–1607

Desai SD (2002) Observing the pole tide with satellite altimetry. J Geophys Res 107(C11):3186. doi:10.1029/2001JC001224

Dow JM, Neilan RE, Rizos C (2009) The international GNSS service in a changing landscape of global navigational satellite systems. J Geod 83:191–198. doi:10.1007/s00190-008-0300-3

Fey AL, Gordon D, Jacobs CS (eds) (2009) The second realization of the international celestial reference frame by very long baseline interferometry, IERS Technical Note 35, Frankfurt am Main: Verlag des Bundesamts für Kartographie und Geodäsie, 204pp

Flechtner F, Gruber T, Güntner A, Mandea M, Rothacher M, Schöne T, Wickert J (eds) (2010) System Earth via geodetic-geophysical space techniques. Springer (Publisher), Berlin, doi: 10.1007/978-3-642-10228-8

Folkner W, Williams JG, Boggs DH (2009) The planetary and lunar ephemeris DE 421, IPN progress report 42–178, 15 Aug 2009, 31pp. See http://ipnpr.jpl.nasa.gov/progress_report/42-178/178C.pdf

Klioner SA, Capitaine N, Folkner W, Guinot B, Huang T-Y, Kopeikin S, Pitjeva E, Seidelmann PK, Soffel M (2010) Units of relativistic time scales and associated quantities. In: S. Klioner, Seidelmann PK, Soffel MH (eds) Proceedings of IAU symposium, vol 261. Cambridge University Press, Cambridge, UK pp 79–84

Luzum BJ, Capitaine N, Fienga A, Folkner W, Fukushima T, Hilton J, Hohenkerk C, Krasinsky G, Petit G, Pitjeva E, Soffel M, Wallace P, (2011) The IAU 2009 system of astronomical constants: the report of the IAU working group on numerical standards for fundamental astronomy. Celest Mech Dyn Astr (2011) 110:293–304. doi: 10.1007/s10569-011-9352-4

Lyard F, Lefevre F, Letellier T, Francis O (2006) Modelling the global ocean tides: modern insights from FES2004. Ocean Dyn 56 (5–6):394–415. doi:10.1007/s10236-006-0086-x

McCarthy DD, Petit G (eds) (2004) IERS conventions (2003), IERS technical note 32, Frankfurt am Main: Verlag des Bundesamts für Kartographie und Geodäsie, 127pp

Mendes VB, Pavlis EC (2004) High-accuracy zenith delay prediction at optical wavelengths. Geophys Res Lett 31:L14602. doi:10.1029/2004GL020308

Mendes VB, Prates G, Pavlis EC, Pavlis DE, Langley RB (2002) Improved mapping functions for atmospheric refraction correction in SLR. Geophys Res Lett 29(10):1414. doi:10.1029/2001GL014394

Pearlman MR, Degnan JJ, Bosworth JM (2002) The international laser ranging service. Adv Space Res 30(2):135–143. doi:10.1016/S0273-1177(02)00277-6

Petit G (2010) The new edition of the IERS conventions: conventional reference systems and constants. In: Proceedings Journées Systèmes de reference spatio-temporels 2010, in print

Petit G, Luzum BJ (eds) (2010) IERS Conventions (2010). IERS technical note 36, Frankfurt am Main: Verlag des Bundesamts für Kartographie und Geodäsie, 179pp. See also http://tai.bipm.org/iers/conv2010 or http://maia.usno.navy.mil/conv2010

Ray RD, Ponte RM (2003) Barometric tides from ECMWF operational analyses. Ann Geophys 21(8):1897–1910. doi:10.5194/angeo-21-1897-2003

Ray RD, Steinberg DJ, Chao BF, Cartwright DE (1994) Diurnal and semidiurnal variations in the Earth's rotation rate induced by oceanic tides. Science 264(5160):830–832. doi:10.1126/science.264.5160.830

Saastamoinen J (1972) Atmospheric correction of the troposphere and stratosphere in radio ranging of satellites. In: Henrikson SW, Mancini A, Chovitz BH (eds) The use of artificial satellites for geodesy, Geophysical monograph series, vol 15, pp 247–251. AGU, Washington, D. C., doi:10.1029/GM015

Schlüter W, Behrend D (2007) The International VLBI Service for Geodesy and Astrometry (IVS): current capabilities and future prospects. J Geod 81(6–8):379–387. doi:10.1007/s00190-006-0131-z

Willis P, Fagard H, Ferrange P, Lemoine FG, Noll CE, Noomen R, Otten M, Ries JC, Rothacher M, Soudarin L, Tavermier G, Valette JJ (2010) The International DORIS Service, toward maturity. In: Willis P (ed) DORIS: scientific applications, Geodesy Geodyn Adv Space Res 45(12):1408–1420. doi 10.1016/j.asr.2009.11.018

Dependence of IGS Products on the ITRF Datum 11

J.R. Ray, P. Rebischung, and R. Schmid

Abstract

Throughout its nearly two decades, the International GNSS (Global Navigation Satellite Systems) Service (IGS) has sought to align its products closely to successive realizations of the International Terrestrial Reference Frame (ITRF). This has been disruptive for IGS users at times, especially during the 1990s when some radical ITRF datum choices were adopted. During the past decade, IGS impacts due to ITRF updates have been smaller and mostly caused by errors in the results from the contributing space geodetic techniques.

Frame orientations (rotations) are purely conventional, so the IGS relies on the ITRF via a subset of reliable, globally distributed stations. Except for the period when ITRF93 was used, this procedure has worked well. The IGS origin in principle could be self-reliant or contributory to ITRF by direct observation of a frame origin aligned to the long-term center of mass of the entire Earth system. In practice, however, GNSS-based results have been less reliable than those from satellite laser ranging (SLR). So the ITRF origin, based on SLR only, has been adopted historically. Until the transition from ITRF2005 to ITRF2008, there have sometimes been significant origin shifts as SLR results have evolved. However, the present stability of the ITRF origin may finally have reached the few-mm level.

In many respects, the IGS dependence on the ITRF scale is most subtle and problematic. In addition to an overall Helmert alignment of the IGS frame to match the ITRF scale (and other datum parameters), since 2006 the IGS calibration values for the GNSS satellite antenna z-offsets depend directly on the same ITRF scale (due to high correlations if the IGS frame scale is not fixed). We therefore face a non-linear situation to maintain full consistency between all IGS products and the ITRF scale: each IGS frame contribution to ITRF based on one set of antenna calibrations must be used, together with frames from other techniques, to determine an updated ITRF and new antenna calibrations, which are then no longer strictly consistent with the starting IGS frame. One can hope that the process will iteratively converge eventually. But large shifts in the ITRF scale, such as the -1 ppb change from ITRF2005 to ITRF2008, are highly disturbing, much more so than the associated rotational or translational shifts.

J.R. Ray (✉)
NOAA National Geodetic Survey, 1315 East–west Highway, Silver Spring, MD 20910, USA
e-mail: jim.ray@noaa.gov

P. Rebischung
IGN/LAREG, Université Paris Diderot, 35 rue Hélène Brion, 75013 Paris, France

R. Schmid
Institut für Astronomische und Physikalische Geodäsie, Technische Universität München, München, Germany

Z. Altamimi and X. Collilieux (eds.), *Reference Frames for Applications in Geosciences*, International Association of Geodesy Symposia 138, DOI 10.1007/978-3-642-32998-2_11,

Only SLR and very long baseline interferometry (VLBI) have been considered reliable and accurate enough to be used for the ITRF scale. But experience and theoretical studies have shown that neither is accurate to better than about 1 ppb. Note in particular that a 1 ppb uncertainty in the GM constant fundamentally limits the possible scale agreement between SLR and VLBI to no better. Consequently, the authors strongly urge that the ITRF scale hereafter be fixed conventionally to the ITRF2008 scale indefinitely until it is convincingly shown that VLBI and/or SLR can determine the ITRF scale within 0.5 ppb. If this is not done, the IGS might maintain its own ITRF2008 scaled frame to minimize future operational dislocations.

Keywords
Reference frames • International GNSS service (IGS) • International terrestrial reference frame (ITRF) • Terrestrial scale • Antenna calibration

1 Introduction

Over its 17-year history, the IGS has tried to adopt the references of the International Earth Rotation and Reference Systems Service (IERS) as closely as possible. The IERS Conventions are largely implemented, UT1 reference values are taken from the IERS (but propagated to data epochs via integration of IGS length-of-day observations), and successive ITRFxx datums have been used for IGS products. Starting in 2000, however, IGS internal frame realizations have been preferred in order to maintain the highest level of self-consistency, but aligned to the ITRF datum.

Some difficulties have been encountered. The IERS Earth orientation parameters (EOPs) were found to be too inaccurate for direct use, so the IGS adopted its own observed pole starting in 1995. In addition, the datum shifts between ITRFxx updates have sometimes been disruptive for users. The large rotations applied to ITRF93 to reduce EOP inconsistencies were a particularly serious problem. Lately, the leading problem for the IGS has been change in the ITRF scale, because this affects estimates of GNSS satellite antenna *z*-offsets.

2 ITRF Rotations

Conventionally, each ITRF realization has been rotationally aligned to its predecessor, except for ITRF93, which was offset to restore consistency with IERS published EOPs. The rotations applied to ITRF93, compared to ITRF2008 at epoch 2000.0, were −1.71, −1.48, and −0.30 mas about *X*, *Y*, and *Z*, respectively (see Table 4.1 of Petit and Luzum 2010). The rotation rates were non-zero also. After ITRF93, rotational consistency was restored.

The ITRF93 frame was used by the IGS from 1 Jan. 1995 (GPS week 782) till 29 Jun. 1996 (week 859). The original IGS Final orbits from that period were later compared to the homogeneously reprocessed orbits (IGS ACC 2010) that used the IGS05 frame (aligned to ITRF2005). Gendt et al. (2010) found that large rotational offsets, up to nearly 1 mas, were evident in the ITRF93 orbits, closely matching the expected shifts due to the ITRF rotations. Figure 11.1 (from Gendt et al. 2010) illustrates results for the *X* axis.

3 ITRF Translations

Since the first use of ITRF by the IGS (ITRF92 in 1994) the origin of successive frame realizations has moved by up to −24.0, +4.6, and −41.2 mm in *X*, *Y*, and *Z*, respectively, compared to ITRF2008 at epoch 2000.0 (Petit and Luzum 2010). The translation rates have reached up to −3.2 mm/year in *Z*. From ITRF88 onward, shifts in the *Z* direction have increased steadily from −125.2 mm to the latest change of −4.7 mm from ITRF2005 to ITRF2008. These are entirely due to the evolution of SLR geocenter results since no other technique contributes to the ITRF origin.

Contrary to the situation with ITRF rotations, the origin stability of the IGS orbit frame is not sensitive to the terrestrial frame origin. The results presented by Gendt et al. (2010) demonstrate this. The explanation is that the origin of IGS orbits should follow the actual Earth geocenter in the same way as for SLR, provided that no over-constraints are applied in the data analysis. However, the fidelity of IGS geocenter offsets is less reliable from GNSS measurements due to the greater importance of empirical orbit parameters. The original IGS orbit modeling caused a large drift in the *Y* direction of more than 50 mm from 1994 till 1998, reflecting refined IGS orbit modeling as more once-per-revolution satellite parameters were added later. Orbit differences in the *X* direction were minor but noisier before 1999, whereas there was a strong annual variation in the *Z* direction with an

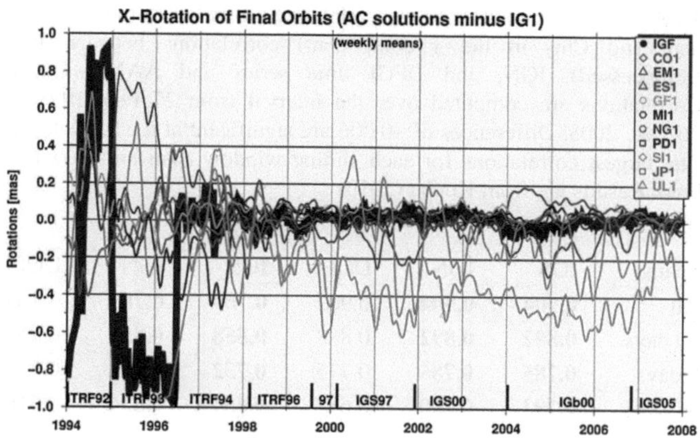

Fig. 11.1 The bold line shows the *X* rotations of the original IGS Final orbits compared to consistently reprocessed orbits. Frames used by the IGS are marked along the bottom of the plot, which comes from Gendt et al. (2010)

amplitude of >10 mm before 2002, decreasing since then. The strong dependence of GPS geocenter *Z* shifts on orbit modeling has been illustrated by Hugentobler (2005), who also found signals with periods equal to the GPS draconitic year (about 350 days). Such orbit-related effects dominate the IGS results over any impacts due to the ITRF origin.

4 ITRF Scale and Satellite Antenna Offsets

In 2006, the IGS adopted satellite antenna phase center offsets (PCOs, together with matching nadir-dependent phase center variations, PCVs) that were determined from the GNSS data to be consistent with "absolute" calibrations for the ground tracking antennas (Schmid et al. 2007). Even using long spans of tracking data, the PCO estimation problem is not robust because the data are only weakly sensitive to PCO errors (Cardellach et al. 2007):

$$\Delta\rho = -\Delta\text{PCO} \times \left(0.94 + 0.06\ \sin^2 e\right)^{1/2} \qquad (11.1)$$

where $\Delta\rho$ is the geometric range change between a satellite and a ground station due to a PCO error, ΔPCO, in the radial (z) direction toward the Earth and e is the elevation angle of the satellite viewed from the station. In addition, the ΔPCO parameters are highly correlated with station height and station zenith troposphere delay parameters, whose partial derivatives are proportional to sin(e) and 1/sin(e), respectively. In order to solve this system sensibly, the IGS has chosen to fix the terrestrial frame scale (i.e., the net height of a global set of stations) to the ITRF scale (determined by VLBI and SLR data) and use absolute robot calibrations for the tracking antennas. Doing this, the expected error (in mm) in ΔPCO estimates averaged over the satellite constellation is about $-(20\times\Delta s)$ where Δs is the net scale frame error (in mm).

Since 1994, ITRFxx scales have differed from ITRF2008 by up to 3.41 ppb or 21.7 mm. The scale accuracy of the present ITRF2008 is thought to be about 1.2 ppb or 8 mm (Altamimi et al. 2011), implying a frame-related error in IGS estimates of satellite PCOs of about 16 cm. Because there are no near-term prospects for much better VLBI or SLR scale accuracies (noting the estimated error in GM alone, which scales all satellite frames, is about 1 ppb; see Petit and Luzum 2010), we strongly recommend that future ITRF realizations use scales fixed to ITRF2008 as a conventional standard. This practice would greatly simplify ITRF usage and eliminate needless variations in applications that have no basis in physical reality.

5 Fixing ITRF Scale Would Simplify PCO Maintenance

The present IGS methodology to maintain and update satellite antenna PCOs is an iterative process. A particular GNSS-based frame (IGSxx, closely aligned to a given ITRFxx) is adopted for an extended operational period and for occasional reprocessing of the historic raw data set. Accompanying the frame must be a set of absolute antenna calibrations for ground and satellite antennas, igsxx.atx in the ANTEX format, wherein the satellite PCOs need to be consistent with the IGSxx frame scale. Normally, major changes in IGSxx and igsxx.atx are not allowed during their periods of usage in order to avoid instabilities in user results. The long-term solutions resulting from this system are the IGS inputs to the next ITRFyy realization.

If the ITRFyy scale differs from its predecessor, it is then necessary to generate updated PCOs and a new set of igsyy.atx calibrations, adding at the same time any major calibration changes for existing ground antennas. Computing the new PCOs can be done by defining a new IGSyy frame derived from ITRFyy and adopting its datum. Then, prior IGS solution files are back-solved to estimate consistent PCOs by fixing the IGSyy scale. If the ITRF scale were fixed conventionally, at least for an extended period, then the PCO maintenance could be greatly simplified and any adjustments that might be needed would be much smaller. However, if calibration values for existing ground antennas are changed (e.g., due to more recent measurements) then it is still necessary to compute station position corrections for any stations affected by the antenna calibration changes.

This process should converge if major calibration changes for existing antennas decline, as expected, but only if the ITRF scale also becomes much more stable over time. Within the current accuracy of VLBI and SLR, this is most easily assured by fixing the ITRF scale conventionally.

6 Competing Strategies for ITRF2008 Combination

In preparing ITRF2008, independent combination solutions were formed by Institut Géographique National (IGN) and by Deutsches Geodätisches Forschungsinstitut (DGFI). The same technique inputs were used by both groups, but their internal strategies differed in ways that have been difficult to compare independently. One of the few objective ways to evaluate the overall performance of the two approaches is to compare their combined polar motion (PM) estimates to independent excitation measures from atmosphere and ocean angular momentum (AAM and OAM, respectively). Such comparisons of geodetic and geophysical excitations have been made for many years relying on the products of general circulation models for the atmosphere and the ocean, mostly following the development by Barnes et al. (1983).

Kouba (2010) has kindly performed such a comparison between test IGN and DGFI PM time series and AAM + OAM excitations. AAM values, four times daily, are from the NCEP reanalysis model (Salstein and Rosen 1997) where the inverted barometer assumption has been applied. Daily averaged values are formed around noon epochs to match the respective geodetic series epochs. OAM values come from the ECCO_kf080 model (Gross et al. 2005). Both series are from the IERS Global Geophysical Fluids Center at www.iers.org. The PM and (AAM + OAM) time series were compared over the period 27 Feb. 1997 to 26 Dec. 2008 using methods described by Kouba (2005). Cross-correlations between the series are shown in Table 11.1 and residuals in Table 11.2. Also included is the IGS reprocessed PM time series, since this contribution dominates the combinations. The most recent 4.5-year period was also considered by Kouba but the conclusions are unchanged.

Kouba's results show that the DGFI combination has the lowest correlations and highest residuals over all spans compared to the geophysical excitations. Only the differences over spans shorter than 30 days are significant at the 95 % confidence level but the margin grows steadily for shorter intervals. The IGN and IGS series are nearly indistinguishable and always agree with (AAM + OAM) better than the DGFI PM. These independent comparisons strongly suggest that the DGFI procedures have introduced some measure of high-frequency noise into the combination, relative to the IGS and IGN combination, although the exact mechanism for that cannot be identified from these results alone.

Table 11.1 PM excitation (Chi_1 related to variations in the *y* component and Chi_2 in the *x* component) correlations between IGS (reprocessed), IGN, and DFGI time series and AAM + OAM. Correlations are computed over the interval from 27 Feb. 1997 to 26 Dec. 2008. Differences of ~0.006 are significant at the 95 % level. The largest correlations for each sliding window span are shown in bold. Results are from Kouba (2010)

	Chi_2			Chi_1		
Spans	IGS	IGN	DGFI	IGS	IGN	DGFI
All	**0.904**	**0.904**	0.902	**0.769**	**0.769**	0.765
30 days	**0.892**	**0.892**	0.888	**0.858**	**0.858**	0.852
5 days	**0.785**	**0.785**	0.775	**0.732**	**0.732**	0.719
3 days	**0.703**	0.700	0.687	**0.634**	**0.634**	0.616

Table 11.2 RMS of residuals (units are mas/d) between PM and AAM + OAM excitations for IGS (reprocessed), IGN, and DFGI time series computed over three different spans. The smallest residuals for each sliding window span are shown in bold. Results are from Kouba (2010)

	Chi_2			Chi_1		
Spans	IGS	IGN	DGFI	IGS	IGN	DGFI
All	**0.270**	**0.270**	0.273	0.255	**0.254**	0.257
<6 days	**0.162**	**0.162**	0.173	**0.139**	**0.139**	0.148
<3 days	**0.111**	**0.111**	0.122	**0.106**	**0.106**	0.112

7 Conclusions

The stability of the ITRF datum is critical for the general usefulness and continuity of the IGS products, especially the orientation and scale components. To improve from the present level of stability, considering the intrinsic scale uncertainty of ~1 ppb, the authors urge that future ITRF realizations maintain the ITRF2008 scale as a convention. There is no benefit to users to experience scale jumps with each ITRF update within that range (i.e., height changes within about ±6 mm). The most direct impact on the IGS is for its satellite antenna *z*-offset estimates, which rely on constraining the terrestrial scale to a specific datum. Meanwhile, the geodetic observing techniques need to focus their research efforts toward improving the stability and accuracy of their products, including the datum aspects.

It would also benefit the IGS if IERS procedures for handling future ITRF updates were improved. Clear and respected schedules for each new realization should be agreed and the criteria used to evaluate test combinations should be well defined and objective to ensure the highest possible quality. Considering the various factors, including the time needed for the next grand reprocessing, the IGS suggests that the next ITRF realization after 2008 aims for a delivery in about 2013.

Acknowledgements The private and very helpful contributions of Jan Kouba are greatly appreciated. Xavier Collilieux has been instrumental in developing the framework to determine PCO estimates using prior analysis solutions with a fixed terrestrial scale. All components of the IGS (Dow et al. 2009) have been indispensible in carrying out this work.

References

Altamimi Z, Collilieux X, Métivier L (2011) ITRF2008 : an improved solution of the International Terrestrial Reference Frame. J Geod 85(8):457–473. doi: 10.1007/s00190-011-0444-4

Barnes RTH, Hide R, White AA, Wilson CA (1983) Atmospheric angular momentum fluctuations, length-of-day changes and polar motion. Proc R Soc Lond A 387:31–73. doi:10.1098/rspa.1983.0050

Cardellach E, Elósegui P, Davis JL (2007) Global distortion of GPS networks associated with satellite antenna model errors. J Geophys Res 112:B07405. doi:10.1029/2006JB004675

Dow JM, Neilan RE, Rizos C (2009) The International GNSS Service in a changing landscape of Global Navigation Satellite Systems. J Geod 83(3–4):191–198. doi:10.1007/s00190-008-0300-3

Gendt G, Griffiths J, Nischan T, Ray J (2010) IGS reprocessing – summary of orbit/clock combination and first quality assessment. Presentation at IGS Workshop, Newcastle upon Tyne, available at acc.igs.org/repro1/repro1_IGSW10.pdf

Gross RS, Fukumori I, Menemenlis D (2005) Atmospheric and oceanic excitation of decadal-scale Earth orientation variations. J Geophys Res 110:B09405. doi:10.1029/2004JB003565

Hugentobler U (2005) Models in GNSS data analysis. Presentation at COMET – Advances in GPS data processing and modeling for geodynamics. University College London, London

IGS ACC (2010) IGS Analysis Center Coordinator. Website at acc.igs.org

Kouba J (2005) Comparison of polar motion with oceanic and atmospheric angular momentum time series for 2-day to Chandler periods. J Geod 79(1-3):33–42. doi:10.1007/s00190-005-0440-7

Kouba J (2010) ITRF2008 and IGS repro1 polar motion comparisons with AAM + OAM. Private communication, available electronically at acc.igs.org/trf/ITRF08ERPcomp.pdf

Petit G, Luzum B (eds) (2010) IERS Conventions (2010) IERS Technical Note 36, Verlag des Bundesamts für Kartographie und Geodäsie, Frankfurt am Main, 179 pp

Salstein DA, Rosen RD (1997) Global momentum and energy signals from reanalysis systems. In: 7th Conference on climate variations. American Meteorological Society, Boston, pp 344–348

Schmid R, Steigenberger P, Gendt G, Ge M, Rothacher M (2007) Generation of a consistent absolute phase center correction model for GPS receiver and satellite antennas. J Geod 81(12):781–798. doi:10.1007/s00190-007-0148-y

Recent Results from the IGS Terrestrial Frame Combinations

12

P. Rebischung and B. Garayt

Abstract

Since February 2010, the Institut Géographique National (IGN) has replaced Natural Resources Canada as the terrestrial frame coordinator of the International GNSS Service (IGS). One important task of this coordination consists in weekly combinations of the solutions provided by nine IGS Analysis Centres into weekly IGS solutions which include station positions, Earth rotation parameters and coordinates of the geocenter.

These combinations enable inter-comparisons of the AC solutions. We show that such comparisons reveal systematic distortions between the AC solutions and that relating them to analysis specificities can be a way to improve the quality of both the AC and combined solutions.

Because the geocenter determination by GNSS still suffers from mismodeling issues, a rigorous combination of the AC geocenter estimates is not feasible yet. The comparison of recently reprocessed geocenter time series from GNSS and SLR is however encouraging.

Keywords

IGS • Terrestrial frame • Combination • GNSS

1 Introduction

The International GNSS Service (IGS) routinely generates ultra-rapid, rapid and final products for the GNSS community, in support of Earth science research, multidisciplinary applications and education. The IGS final products, generated weekly, are the most accurate. They include:

- GNSS satellite orbits,
- Satellite and tracking stations clock offsets,
- Earth rotation parameters (ERPs),
- Coordinates of tracking stations,
- Coordinates of the geocenter,
- Ionospheric and tropospheric delay parameters.

P. Rebischung (✉)
IGN/LAREG and GRGS, Université Paris Diderot, 35 rue Hélène Brion, 75013 Paris, France
e-mail: paul.rebischung@ign.fr

B. Garayt
IGN/SGN, 2-4 Avenue Pasteur, 94165 Saint-Mandé Cedex, France

The final orbit, clock, ERP and station position products are obtained by combining the solutions from different IGS Analysis Centers (ACs). To keep the combination at a manageable size, it is performed in several steps. First, the AC SINEX (Software INdependent EXchange format) solutions, which include station positions and ERPs, are combined. The AC orbits and clocks are then combined in a subsequent step. Special attention is paid to the consistency between the combined SINEX solutions and the combined orbits and clocks. Details of this two-step combination process are described in Kouba et al. (1998).

The combination of the AC SINEX solutions has been performed at Natural Resources Canada (NRCan) from its beginning in 1999 until January 2010 (Ferland 2000; Ferland 2009). After an extensive test phase, the Institut Géographique National (IGN) then took this task on. Some changes were brought to the combination process but its principle remains the same: it consists in a standard least-squares adjustment of the combined station positions and ERPs using as input the AC solutions with their

Z. Altamimi and X. Collilieux (eds.), *Reference Frames for Applications in Geosciences*,
International Association of Geodesy Symposia 138, DOI 10.1007/978-3-642-32998-2_12,

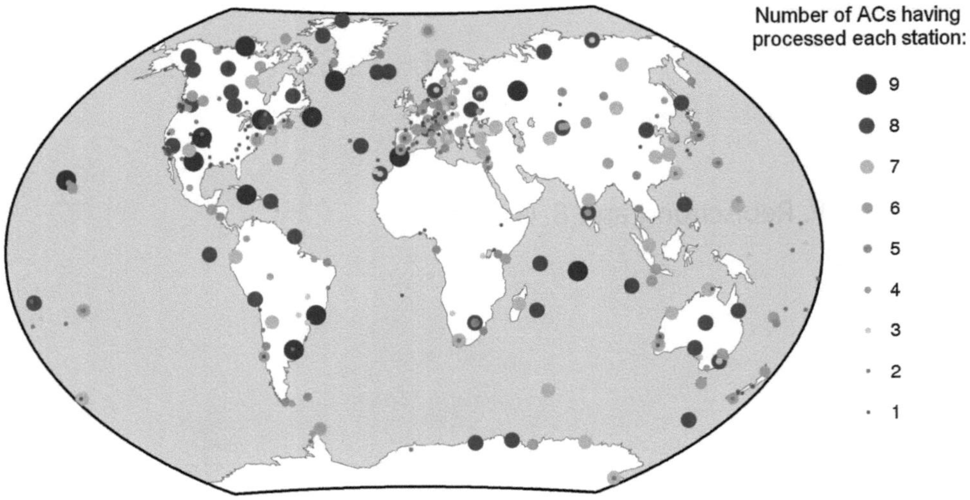

Fig. 12.1 Station network in the combined solution of GPS week 1,598. Symbols depend on the number of ACs having processed each station

Table 12.1 IGS Analysis Centres (ACs) and global network associate analysis centres (GNAACs)

AC	Description
cod	Centre for Orbit Determination in Europe, Switzerland
emr	Natural Resources Canada, Canada
esa	European Space Operation Centre, Germany
gfz	GeoForschungsZentrum, Germany
grg	Groupe de Recherche en Géodésie Spatiale, France
jpl	Jet Propulsion Laboratory, USA
mit	Massachusetts Institute of Technology, USA
ngs	National Geodetic Survey, USA
sio	Scripps Institution of Oceanography, USA
GNAAC	
mig	Massachusetts Institute of Technology, USA
ncl	University of Newcastle-upon-Tyne, UK

Table 12.2 Official products of the weekly IGS terrestrial frame combinations

File	Description
igsyyPwwww.snx	Weekly combined solution
igsyyPwwww.ssc	Weekly combined solution without covariance matrix
igsyyPwwww.erp	Weekly combined ERPs
igsyyPwwww.res	Residuals between AC solutions and weekly combined solution
igsyyPwwww.itr	Residuals between AC solutions and IGS reference frame
igsyyPwwww.sum	Summary
IGSyyPww.snx	Cumulative solution
IGSyyPww.ssc	Cumulative solution without covariance matrix
IGSyyPww.res	Residuals between AC solutions and cumulative solution
igs00p03.erp	Long-term series of combined ERPs

complete covariance matrices. The CATREF software (Altamimi et al. 2004) is at the heart of the IGN combination process.

Independent combinations of the AC SINEX solutions are also performed by two Global Network Associate Analysis Centres (GNAACs). They are used for comparison and validation of the IGN combination. Table 12.1 lists the nine IGS ACs currently contributing to the SINEX combination and the two IGS GNAACs.

Figure 12.1 shows the network of stations in a particular combined solution. It usually includes about 400 stations in total. All are not processed by the nine ACs. (Each AC processes its own network, ranging from 80 to 250 stations.) But the AC networks overlap enough to ensure that they can be well tied to each other. (The formal errors of the transformation parameters estimated during the combination have recently all been below 1 mm.)

Together with the combined SINEX solution, a long-term cumulative solution is also updated and released each week. It contains station positions at reference epoch 2005.0 and station velocities which are obtained by stacking the weekly combined solutions in a standard least-squares adjustment.

Table 12.2 lists the official products of the weekly IGS terrestrial frame combinations available at the IGS data centres. Note that many by-products such as time series and plots of station coordinates, combination residuals and combination statistics are available at ftp://igs-rf.ign.fr and ftp://igs-rf.ensg.eu.

2 Combination Methodology

2.1 Preparation of AC Solutions

The AC SINEX solutions are obtained by fitting parameters to GNSS observations through least-squares adjustments. Such an adjustment results in a normal equation which can be written as:

$$(N + N_c).(x - x_0) = K,$$

where x_0 is the vector of a priori parameter values, x is the vector of estimated parameters, N is the unconstrained normal matrix, N_c is the normal matrix of the constraints and K is the second member of the constraint-free normal equation. (The constraints applied by the ACs should not contribute to the right-hand side of the normal equation.)

In their SINEX files, ACs provide x, x_0, the covariance matrix of the constraints $Q_c = N_c^{-1}$ and either the constraint-free normal equation (N, K) or the total covariance matrix $Q_{tot} = (N + N_c)^{-1}$. In the latter case, the first operation is to recover the original constraint-free normal equation by computing $N = Q_{tot}{}^{-1} - Q_c{}^{-1}$ and $K = Q_{tot}{}^{-1}.(X - X_0)$. The next step consists in fixing parameters that are not combined to their estimated values. This mainly includes UT1-UTC and the satellite phase centre offsets which are estimated by some ACs.

At this step, the AC normal matrices normally have three singularities corresponding to three rotations of the whole system. Three constraints defining the orientation of the terrestrial frame are therefore necessary and sufficient to invert them. But two different solutions are in fact generated from each AC constraint-free normal equation:

- A "R-solution" obtained by adding the strictly necessary orientation constraints,
- Another "A-solution" obtained by adding supplementary constraints in origin and scale.

Because of the applied constraints, the "A-solutions" have their origins, scales and orientations statistically determined at the same level. On the other hand, the origins of the "R-solutions" still reflect the uncertainty of the geocenter determination by GNSS.

2.2 Iterative Combinations of A-Solutions

Once the AC solutions have been prepared, they are iteratively combined. Each combination consists in a least-squares adjustment where the observations are the AC estimates for station positions and ERPs and the estimated parameters are the combined station positions, the combined daily ERPs and a set of seven transformation parameters θ_s (three translations, one scale factor and three rotations) between each solution *s* and the combined solution.

The employed weight matrix is block-diagonal, its blocks being the inverts of the AC covariance matrices scaled by the AC variance factors: $Q_s{}^{-1}/\sigma_s{}^2$. The AC variance factors used for the first iteration are those from the previous weekly combination. They are then updated after each iteration using the "degree of freedom estimator" method (Sillard 1999).

The estimation of transformation parameters between each AC solution and the combined solution introduces seven singularities in the system. To define the origin, scale and orientation of the combined solution, internal constraints are applied. They can simply be written as $\Sigma\ \theta_s = 0$ and define the frame of the combined solution as a mean of the AC frames.

After each iteration, observations with residuals exceeding 5 cm or normalized residuals exceeding 5 are looked for, and the corresponding stations are deleted from the incriminated AC solutions. Iterations are performed until no new outliers are found. A last combination is then made in order to estimate the variance factors of the outlier-free AC solutions.

Experience showed that combining "R-solutions" while estimating transformation parameters results in loosely determined translations which can cause residuals of the combination to be not centred on zero. That is why all these first iterations make use of the more balanced "A-solutions".

However, because origin constraints were applied to the "A-solutions", they do not reflect anymore the uncertainty of the geocenter determination by GNSS. In order to keep this uncertainty in the combined solution, a last combination of the "R-solutions" is thus necessary.

2.3 Last Combination of R-Solutions

When no outliers remain in the AC solutions and their final variance factors are computed, a last combination is performed using the "R-solutions". As just mentioned, the goal of this last combination is to keep, in the combined solution, the uncertainty of the geocenter determination by GNSS. In this last combination, transformation parameters between the AC solutions are fixed to those previously estimated, so that no constraints are needed to define the frame of the combined solution.

Once this last combination has been performed, the origin of the combined frame is made explicit: its three coordinates (0, 0, 0) are added as explicit parameters (the geocenter coordinates) to the combined solution with the appropriate covariance information.

Finally, the seven transformation parameters from the combined solution to the current IGS Reference Frame, IGS05 (Ferland 2006), are estimated and applied to the combined solution. In that way, the final combined solution is expressed in IGS05. In particular, the geocenter coordinates are not (0, 0, 0) anymore. They can now be considered as the instantaneous geocenter coordinates in IGS05.

2.4 Paradox of the Combined Geocenter

In the previous section, we showed how the combined geocenter was obtained. It is in fact nothing but the origin of the combined solution, which is defined, by the internal constraints, as the mean of the origins of the AC solutions.

But this method is not fully satisfying. For a rigorous geocenter combination, we should consider the origins of all AC solutions as observations of the same point, the geocenter. In concrete terms, this means that no translation parameters should be estimated between the AC solutions.

In practice however, these translation parameters are mandatory. The origins of the AC solutions can indeed be distant of several centimetres from each other so that, if no translations were estimated, the residuals of the combination would be highly biased. The estimated translations can be interpreted as corrections to the individual AC geocenters, estimated during the first iterations and applied in the last combination.

The discrepancies between the AC geocenters illustrate the fact that the geocenter determination by GNSS still suffers from mismodeling issues and correlations with other parameters such as empirical solar radiation pressure parameters (Hugentobler et al. 2006). Work will be necessary to understand and overcome these discrepancies before a rigorous geocenter combination can be performed.

3 Results

3.1 Earth Rotation Parameters

Several studies have already been made on the accuracies, strengths and weaknesses of the IGS combined ERPs (Ray 2008, 2009b; Ray and Ferland 2009). This paper will thus rather focus on the terrestrial frame aspects. Some essential facts about the IGS combined ERPs are however reminded.

Ray and Ferland (2009) assessed the accuracy levels of the combined IGS ERPs at 30 μas, 150 μas/days and 10 μs for the pole coordinates, their rates and the length of day respectively. These accuracies make the GNSS technique the main contributor to combined ERP series such as the IERS Bulletin A or the ITRF2008 ERP series (Altamimi et al. 2011).

Some anomalies can however be detected when analyzing the polar motion discontinuities at day boundaries (Ray 2008, 2009b). Spurious spectral peaks from weekly to fortnightly periods can be observed and probably explained by deficiencies in the a priori models used for the ERP subdaily variations. Harmonics of the GPS draconitic year are also present, certainly induced by orbit mismodelings.

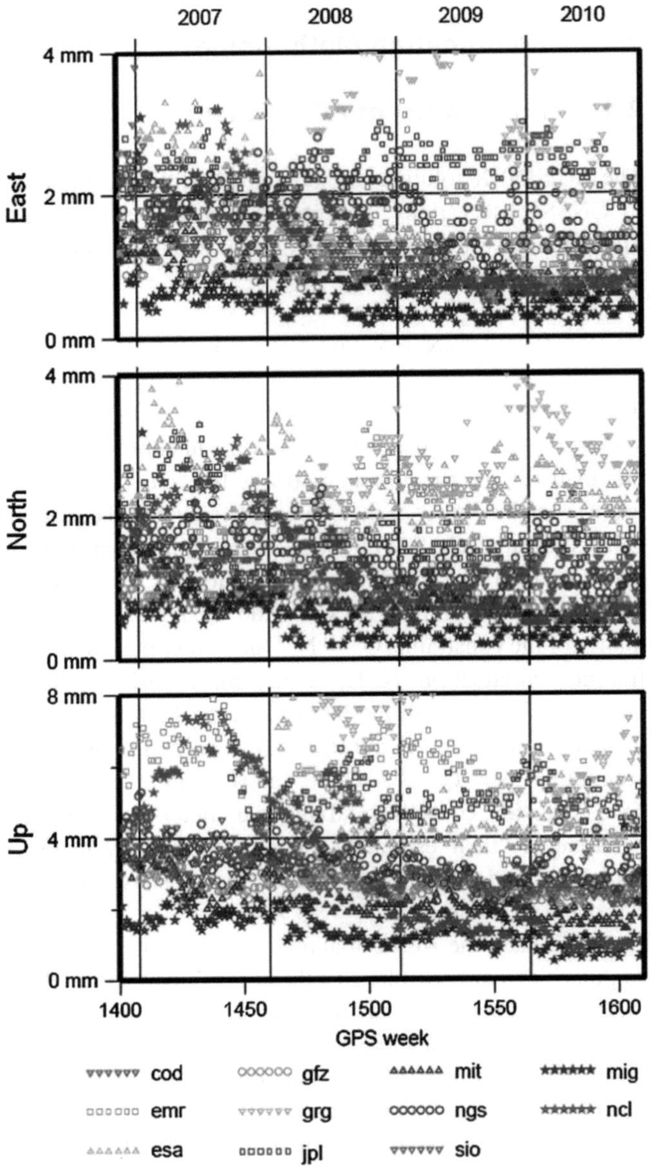

Fig. 12.2 RMS of the AC and GNAAC station position residuals in the East (*top*), North (*middle*) and Up (*bottom*) components

3.2 Terrestrial Frame

3.2.1 Level of Agreement Between Solutions

The weekly SINEX combinations make possible comparisons of AC solutions with each other. Each combination indeed gives a set of station position residuals for each AC. And these residuals reflect the differences in shape between the combined terrestrial frame and the AC estimated terrestrial frame.

To assess the global level of agreement between AC solutions, RMS of the AC residual sets are computed in the East, North and Up components. The temporal evolutions of these RMS since GPS week 1,400 are shown on Fig. 12.2.

The AC solutions currently agree at the levels of 1–3 mm in the horizontal components and 2–6 mm in vertical. These values have significantly improved since the beginning of the IGS SINEX combination in 1999 because of continuous refinements in the AC analysis procedures. A slowdown of this progress can however be noticed since 2007.

One should also note on Fig. 12.2 the high level of agreement between the combined solution from IGN and those from the two GNAACs (currently below 0.5 mm in horizontal and 1 mm in vertical between the IGN and MIG combined solutions). This can be interpreted as an independent validation of the weekly combinations performed at IGN.

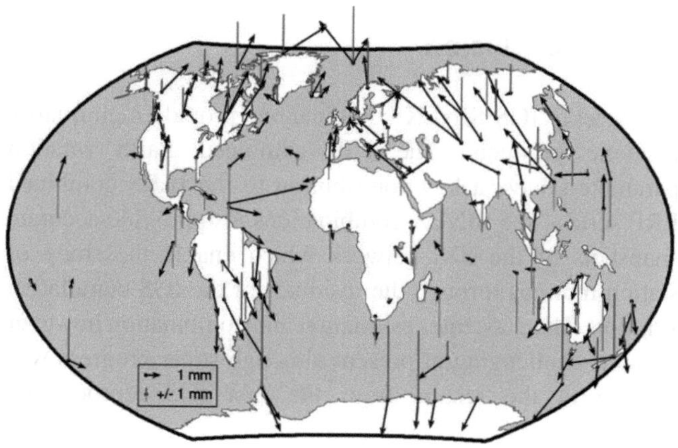

Fig. 12.3 Station position residuals of the ESA solution from week 1,596 combination. Horizontal residuals are shown by the *black arrows*. Vertical residuals are shown in red when positive and in green when negative

3.2.2 Distortions Between Solutions

The study of the AC station position residuals is of high interest. Systematic distortions between the terrestrial frame of one AC (or a group of ACs) and the combined frame are indeed potentially related to deficiencies or improvements in the analysis strategy of this AC. If such deficiencies or improvements can be identified, this can be a way to improve the overall quality of the IGS products.

In that scope, a first basic idea was to generate maps of the AC station position residuals. An example of such a map is given in Fig. 12.3. To make potential global distortions even clearer, spherical harmonics can be fitted to the AC residual vector fields and represented, like on Fig. 12.4. Such maps are now routinely generated and available at ftp://igs-rf.ign.fr/pub/AC_res.

In order to detect potential constant distortions between the individual AC frames, spherical harmonics fits of the AC residuals were computed and averaged over 1 year. This confirmed for example that the strong zonal pattern of the ESA North residuals visible on Fig. 12.4 is not proper to week 1,596, but a real systematic effect. A probable explanation for this pattern is that ESA is currently one of the two ACs which does not estimate horizontal tropospheric gradients. This defect should be sorted out very soon (Springer, personal communication).

Constant distortions of some other AC frames could also be observed. But they are still under investigation at this moment.

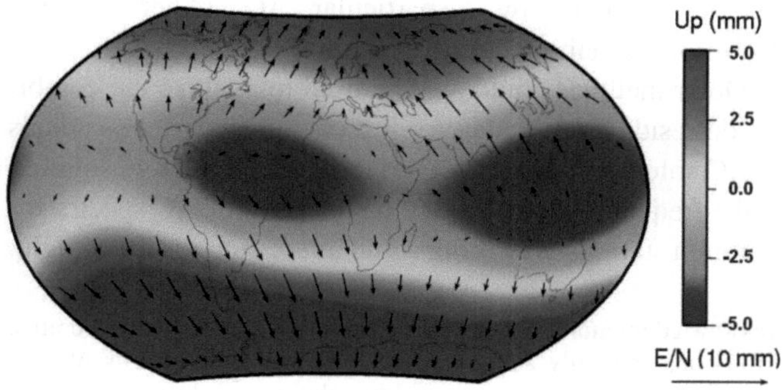

Fig. 12.4 Spherical harmonics fit (up to degree and order 2) of the residual vector field shown on Fig. 12.3

Table 12.3 Annual amplitudes and phases of the geocenter series from the SLR contribution to ITRF2008 (SLR) and the IGS reprocessed products (IG1), according to the model A·cos(2πt - φ) with t in decimal year

	TX	TY	TZ
	A(mm) φ(deg)	A(mm) φ(deg)	A(mm) φ(deg)
SLR	2.9 225	3.2 141	5.5 205
IG1	2.3 178	3.2 134	3.0 346

3.3 Geocenter

As mentioned in Sect. 2.4, the current performance of the geocenter estimation by the ACs does not allow a rigorous geocenter combination. The IGS combined geocenter product should thus be used with utmost care.

One should however point out that the geocenter determination by GNSS has already strongly benefited from recent advances in GNSS analysis. The IGS recently reprocessed its full history of GPS data using the latest models and methodology (Ray 2009a). And the resulting geocenter series shows interesting correlations with geocenter estimates from the SLR technique.

Table 12.3 compares the annual signals observed in the geocenter series from the IGS reprocessed products (Ferland 2010) and the SLR contribution to ITRF2008 (Altamimi et al. 2011) over the period 1997.0–2009.0. A good agreement in amplitude and phase is observed in Y; the agreement is slightly worse in X; no correlations are observed between the two series in the Z component.

4 Conclusion

The weekly IGS SINEX combinations provide high quality geodetic products. The IGS combined Earth rotation parameters have a key contribution to the IERS combined ERP series. The SINEX combinations also provide accurate snapshots of the IGS network which enable the study of station motions through the residuals of the IGS cumulative solution. The geocenter estimation and combination however remains challenging at present although some progress was noticed in the results from the first IGS reprocessing campaign.

Comparing the AC solutions with each other and understanding their differences can lead to further improvements of the combination results. For example, a direct examination of the AC station position residuals revealed constant distortions of some AC frames. One particular pattern could be explained by the non-estimation of horizontal tropospheric gradients by a particular AC. Other constant distortions remain under investigation.

Other methods could also be used to analyze the combination residuals. We plan for instance to study the residuals of AC solutions with respect to the IGS cumulative solution in the frequency domain. The comparison of AC spectra for common stations may bring some interesting results to light.

Acknowledgments All the participants to the IGS collaborative efforts are gratefully acknowledged. The significant efforts of ACs and GNAACs for submitting quality solutions week after week are particularly appreciated.

References

Altamimi Z, Sillard P, Boucher C (2004) CATREF software: combination and analysis of terrestrial reference frames. LAREG technical note SP08, Institut Géographique National, France

Altamimi Z, Collilieux X, Métivier L (2011) ITRF2008: an improved solution of the international terrestrial reference frame. J Geod. doi:10.1007/s00190-011-0444-4 (in press)

Ferland R (2006) Proposed IGS05 realization. IGS mail 5447, http://igscb.jpl.nasa.gov/mail/igsmail/2006/msg001 70.html

Ferland R (2010) Description of IGS submission to ITRF2008. Available at http://itrf.ign.fr/ITRF_solutions/2008/doc/IGSsubmission4ITRF2008.txt

Ferland R, Piraszewski M (2009) The IGS-combined station coordinates, earth rotation parameters and apparent geocenter. J Geod 83:385–392. doi:10.1007/s00190-008-0295-9

Ferland R, Kouba J, Hutchison D (2000) Analysis methodology and recent results of the IGS network combination. Earth Planets Space 52:953–957

Hugentobler U, van der Marel H, Springer T (2006) Identification and mitigation of GNSS errors. Position paper, IGS 2006 workshop proceedings

Kouba J, Ray J, Watkins MM (1998) IGS reference frame realization. IGS analysis center workshop proceedings, European Space Operations Centre, Darmstadt, pp 139–172

Ray J (2008) Analysis effects in IGS polar motion estimates. Available at http://acc.igs.org/erp-index.html

Ray J (2009a) IGS data reprocessing campaign. Available at http://acc.igs.org/reprocess.html

Ray J (2009b) IGS reprocessed polar motion estimates. Available at http://acc.igs.org/erp-repro1.html

Ray J, Ferland R (2009) Status and prospects for IGS polar motion measurements. Presentation at IERS workshop on EOP combination and prediction, Warsaw. Available at http://acc.igs.org/erp/eopp09-pole.pdf

Sillard P (1999) Modélisation des systèmes de référence terrestres. Contribution théorique et méthodologique. PhD dissertation, Observatoire de Paris, France

Local Ties and Co-Location Sites: Some Considerations After the Release of ITRF2008

13

P. Sarti, C. Abbondanza, and Z. Altamimi

Abstract

Tie vectors (TVs) measured at co-location sites carry fundamental information for the computation of the International Terrestrial Reference Frame (ITRF). The combination of the different frames stemming from each space geodetic (SG) technique relies on the availability and accuracy of the relative positions between reference points of co-located SG instruments, i.e. TV. If, on the one hand, TVs accurate at 1 mm level are sought to preserve the accuracy of the global frame and fulfill the requirements of the global geodetic observing system (GGOS), on the other hand, the assessment of TVs accuracy is not easy. Their accuracies are often questioned on the base of their agreement within the combination of SG solutions and the combination residuals. Though, the final discrepancies highlighted by the combination residuals do not depend uniquely on the accuracy of the TVs but are influenced by several factors of different origin. In this paper, we identify some of these factors and investigate their possible origin adopting different perspectives: local ties and terrestrial surveying, SG techniques and frames combination. Our purpose is to highlight some of the possible systematic errors in terrestrial and SG data analysis as well as to identify actions to be taken in the near future to mitigate the biases highlighted by the residuals of the combination. In contrast to what is commonly assumed, we show that the residuals are potentially influenced by a combination of biases affecting the TVs, their alignment and the SG solutions. Therefore, an objective evaluation of the error sources is necessary for each SG technique in order to improve their results as well as the combined SG products.

Keywords

Tie vector • Local tie • Co-location site • ITRF • Combination residuals

1 Introduction

The most recent computation of the combined global frame, the International Terrestrial Reference Frame 2008 (ITRF2008) (Altamimi et al. 2011), represents the highest level of accuracy that can be achieved in the realization of frames with space geodesy. The computation is under the responsibility of the International Earth Rotation and Reference Systems Service (IERS) through the IERS ITRS Centre hosted by the Institut Géographique National (IGN) based in Marne la Vallée (France): it is in charge of providing regular ITRF solutions. Two ITRS Combination Centres (CCs),

P. Sarti (✉) • C. Abbondanza
Istituto di Radioastronomia (IRA)- Istituto Nazionale di Astrofisica (INAF), Via P. Gobetti, 101, Bologna 40129, Italy
e-mail: p.sarti@ira.inaf.it

Z. Altamimi
Institut Géographique National, LAREG, 6-8 Avenue Blaise Pascal, Marne-la-Vallée 77455, France

Z. Altamimi and X. Collilieux (eds.), *Reference Frames for Applications in Geosciences*, International Association of Geodesy Symposia 138, DOI 10.1007/978-3-642-32998-2_13,

the Deutsches Geodätisches Forschungsinstitut (DGFI) and the Laboratoire de Recherche en Géodésie (LAREG) of the IGN, are nowadays in charge of computing ITRF solutions based on different combination strategies. It is worth highlighting that TVs are dealt with differently in the two combination approaches (see e.g. Altamimi et al. 2007; Krügel and Angermann 2007). TVs are fundamental to combine global frames defined by each SG technique whose description can be found at the following web page: http://itrf.ign.fr/ITRF_solutions/2008/input_data.php. The post-fit residuals of the combinations carry along important information about the overall agreement of SG techniques and TVs. The discrepancies highlighted by the residuals are often much larger than the formal errors associated to TVs estimates or SG solutions. This is a clear indication of the presence of potentially large systematic errors that, in principle, can reside in both estimation processes. In order to highlight the various stages where the biases can originate, we discuss shortly the whole estimation process of terrestrial TVs and some specific aspects of SG data analysis that, as already shown by recent studies, can cause large systematic errors (Schmid et al. 2007; Sarti et al. 2011). In Sect. 2, we discuss the potential sources of biases within the TV estimation process based on terrestrial observations. We deal with the contribution of several factors which may affect the magnitude of the discrepancies within the different computation steps. In particular, the different nature of SG instruments reference points (RPs) is recalled and acknowledged as an important cause of disagreement between SG and terrestrial observations. In Sect. 3, we focus on the biases that the different nature of RPs can cause in SG data analysis. In Sect. 4, we shortly discuss the magnitude of the ITRF2008 tie residuals, showing some interesting patterns that can be found correlating, e.g. the age and length of the TVs with the magnitude of the residuals.

2 Steps Towards Tie Vector Estimation

The estimation process of a TV usually comprises four sequential steps: (1) surveying of the SG instrument RP with terrestrial or Global Positioning System (GPS) technique, (2) reduction of the observations and data analysis, (3) data conditioning and estimation of the TV and, when terrestrial observations are used, (4) alignment of the TV from the native local topocentric frame into the global frame. In principle, each step can introduce systematic errors and bias the final TV estimate, thus affecting the discrepancies between the TVs and the SG solutions at the co-location sites. The standard deviations of the terrestrial TVs provided to the ITRF CCs can be as small as a fraction of millimeter. Nevertheless, this can only be regarded as a formal precision of the TV; the accuracy of the TV is much more difficult to assess and certainly affects the combination and the magnitude of the residuals. In the following subsections we summarize and quantify the sources of biases that have been investigated so far.

2.1 Phase 1: Surveying

We disregard here all the systematic errors which can potentially arise during a terrestrial survey, e.g. uncalibrated electronic diastimeters and theodolites, atmospheric effects etc. We only focus on a potential source of systematic errors which is intrinsically embedded in the definition of the instrument RP during the survey. It is strictly connected to the different kind of RP that SG techniques and local ties are linked and refer to. In order to make this difference clear, it is sufficient to note that SG observations define the Electronic RP, where the SG observable is acquired, whereas local ties aim at estimating the position of the Conventional RP which is defined according to the geometric properties of the SG instrument (see e.g. Sarti et al. 2009). There is no possibility to measure the location of the Electronic RP with terrestrial observations. As a consequence, local tie results are always and can only be referred to the Conventional RP. It is usually not difficult to survey the position of Conventional RP in Doppler Orbitography and Radiopositioning Integrated by Satellites (DORIS) and GPS antennas. The Conventional RP is defined as the Antenna Reference Point (ARP) and it can be easily measured with triangulation, without any intervention on the permanent observing station (see e.g. Sarti et al. 2004). The survey of Conventional RP in Satellite Laser Ranging (SLR) and Very Long Baseline Interferometry (VLBI) instruments is more subtle and it is based on a varying degree of conditioning of the terrestrial measurements performed on targets attached to the structure of the instrument. A detailed study on terrestrial observations of VLBI telescopes showed that gravitational flexure may bias the position of the Conventional RP by a non-negligible amount (Sarti et al. 2009). Particularly, the authors showed that the location of the targets placed on different parts of the Medicina (Northern Italy) VLBI telescope, a 32 m AZ-EL antenna, may influence the accuracy of the estimated Conventional RP Up component up to 1 cm, depending on the flexure experienced by the different parts of the structure where the targets are located. Similarly, unmodelled thermal deformations on VLBI telescopes can bias the Conventional RP estimate derived by terrestrial observations. Although, quantitative studies on this latter aspect are still missing. Gravitational and thermal deformations only affect VLBI telescopes, while SLR, GPS and DORIS instruments do not suffer similar biases. Other sources of errors that can potentially affect the TV estimate may be related to the geometry of the local ground control network and to the different observation schemes adopted during the survey of the Conventional RP. Particularly, indirect methods relying on terrestrial measurements (Sarti et al. 2004; Dawson et al. 2007; Leinen et al. 2007; Lösler 2009) adopt different observation strategies (e.g. redundant forward or backward intersection, sideshots) whose effects on the accuracy of the estimated Conventional RPs (i.e. TV) have not been quantified, yet. Indeed, detailed quantitative investigations on these topics are required but are still missing.

2.2 Phases 2 and 3: Data Reduction and Conditioning

Phase 2 concerns the reduction of terrestrial observations prior to the conditioning applied for the TV estimation. This is usually accomplished with commercial software whose performances are believed to be similar and not to affect the following Phase 3. This latter regards the introduction of proper geometric conditions that lead to the estimation of the TV. Inter-comparison of software must be regarded as a crucial asset to investigate potential sources of biases of the TV estimate. It has been continuously encouraged as one of the crucial activities of the IERS working group on site surveys and co-location sites. Despite these efforts, the investigation remains scarce and only one complete investigation has been carried out by Dawson et al. (2007). The authors clearly show that the mitigation of biases up to 3 mm on the estimate of e.g. a VLBI antenna RP depends on the geometric conditions imposed on the observations in the post-processing Phase 3. They also show that the use of different software for the terrestrial data reduction pertaining Phase 2 does not impact the accuracy of the RP estimation. The statistical information associated to TVs has to be as complete as possible: a full variance covariance matrix has to be considered a mandatory result of the local tie and must therefore be provided along with the TV estimate. Although, not all methods and software nowadays applied are capable of fulfilling this duty. An inter-comparison of software and methods must be regarded as a crucial task by all groups which are involved in local tie surveying.

2.3 Phase 4: TV Alignment

The alignment of a TV measured with terrestrial observations concerns the transformation of the estimated vector from a local topocentric frame (whose Up axis is oriented according to the local vertical) into a global frame. It is a very important stochastic phase that can straightforwardly influence the discrepancies between the TV and the SG solutions.

A remarkable example is highlighted by Ray and Altamimi (2005) where the authors used an old VLBI-GPS tie for Medicina and found a large tie residual in the East component of 7.3 mm. Using a new estimated TV where the difference with the old one resides uniquely in the alignment of the TV into the global frame, reduced the discrepancy to 2.7 mm. This is a significant error reduction, while the information concerning the tie between the two instruments being the same in the native local topocentric frame. In other words, the same TV estimated with terrestrial observations in 2001 was aligned with two different approaches into the ITRF. The accuracy of the local tie can therefore be well regarded as being identical but this example shows that the residuals on the vector components differ by several millimetres, depending on the approach adopted to align the TV.

An investigation on the impact of a mis-alignment of the TV on the residuals was carried out by Abbondanza et al. (2009). The authors progressively rotate the TV of Medicina (modulus $\approx$ 63 m) and show how the orientation of the TV changes the values of the residuals up to 1 cm, depending on the extent of the rotation. These studies clearly highlight the dependency of the combination residuals on this fourth phase and suggest that further quantitative studies concerning the identification of a proper alignment procedure of terrestrial TV are urgently needed.

3 Space Geodetic Techniques: Electronic RP Variation Models

SG data analysis uses observations acquired at the Electronic RP and estimates positions referred to the Conventional RP. Therefore, in principle both SG and terrestrial observations refer to the same RP. The ability to mitigate potential biases embedded in the dichotomy Electronic/Conventional RP in SG data processing relies on the capability of providing accurate models to connect the Electronic and the Conventional RP. These models can be regarded as some sort of intra-technique eccentricity between the two RPs and they are crucial for achieving a high level of accuracy in SG data analysis. They must be introduced in the SG data processing since they counterbalance the variations of the Electronic RP with respect to the Conventional RP.

The most remarkable example is given by the variation of the Electronic RP of GPS antennas with respect to the Conventional RP (i.e. the ARP). It is described using Phase Center Variation (PCV) models to ensure a high level of accuracy in GPS data analysis (Schmid et al. 2007). Uncalibrated radomes are still severely limiting the accuracy of GPS observations since their effect is not quantified in PCV models. They may cause discrepancies with respect to terrestrial TV as large as 16 mm, as shown by the investigation of Ray et al. (2007) for site Fortaleza (Brasil).

In VLBI technique, according to investigations performed more than 30 years ago (Carter et al. 1980), the variation of the signal path induced by gravitational flexure of the VLBI telescope structure can bias the height component of the estimated telescope's Conventional RP. Recent investigations on the Medicina and Noto (Southern Italy) 32-m AZ-EL antenna have proved that the extent of the signal path variation can be as large as 1 cm and the Up component of the Conventional RP can be biased by ≈ 7 mm (Sarti et al. 2011). Therefore, signal path variations must be investigated and, if present, corrected using antenna-specific models in VLBI data processing.

Satellite Laser Ranging (SLR) technique could be affected by timing and range biases but detailed investigations on the instabilities of the Electronic RP (the photodetector) are missing. Finally, DORIS technique might present instabilities analogous with GPS but, again, no specific investigations have been carried out. In fact, as discussed in Sect. 4, the magnitude of the TV residuals involving DORIS are the largest among the four SG techniques. They might be related to biases caused by an unmodelled variation of the relative position between the Electronic and the Conventional RPs of the beacons.

4 Combination Residuals

As mentioned in Sect. 1, the ITRF CCs apply a different approach in the combination of SG results and TVs. In the ITRF2008 combination, all available local ties provided in SINEX files were used as observations, with proper weighting as a function of their consistency with space geodesy solutions. For more details, the reader may refer to Altamimi et al. (2011). The DGFI group uses a selection of local ties satisfying certain conditions based on the best agreement between the technique solutions and the combined one, both on station positions and polar motion, but with a uniform weighting of 1 mm on all ties and components (Krügel and Angermann 2007).

The residuals represent an interesting source of information about the extent of agreement between SG observations and terrestrial connections realized at co-location sites and may be used to spotlight the presence of potential biases.

The ITRF2008 combination residuals of the TVs involving GPS and the other SG techniques are available at ITRF2008 web site http://itrf.ign.fr/ITRF_solutions/2008/. We have investigated some simple relations between the magnitude of the ties residuals and a few characteristics of the tie vectors, namely their age and their modulus. Older TVs might show higher residuals due to three different

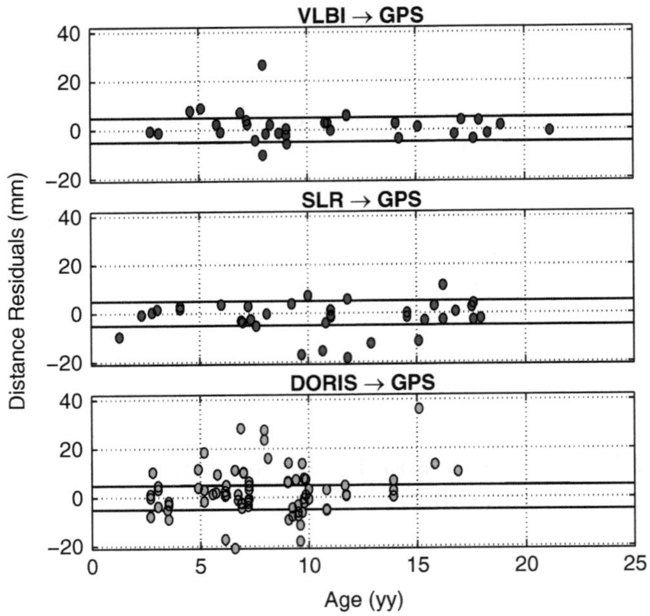

Fig. 13.1 TV distance residuals stemming from ITRF2008 combination (Altamimi et al. 2011) as function of the TV age. *Black lines* identify a ±5 mm boundary

reasons, at least. First, a lower accuracy might be associated to the surveys of older tie vectors, due to the use of lower class instruments and a reduced capacity of acquiring large number of observation sets. Second, the TV estimation process described in Sect. 2 has been noticeably refined in the last decade. Finally, the alignment of terrestrial TV into the global frame was realized mostly with GPS equipment whose precision has considerably increased from the 1990s on. These reasonable statements suggest that, to some extent, a dependence between TV age and residuals might be expected. We have investigated the residuals of the TV length since they are independent of any possible misalignment effect. In Fig. 13.1, the VLBI-GPS, SLR-GPS and DORIS-GPS TV residuals are shown as a function of the TV age. It is worth highlighting that in the combination no down-weighting is applied as a function of age. No clear relation between the age of the TV and the magnitude of the residuals can be identified for any of the co-locations. Taking into account a 5 mm boundary, the outliers are rather randomly distributed above this limit over the whole time span. At least one VLBI-GPS TV and one SLR-GPS TV are older than 20 years though their residuals are comparable with those of more recent TVs. Analogous results can be drawn for DORIS-GPS TVs, whose residuals show a rather uniform distribution with respect to age but are remarkably higher than those mentioned above. Almost half of the TVs involving DORIS exceed the 5 mm boundary, with residuals that can be as large as 30 mm or more. Indeed, this behaviour is rather surprising since the surveying procedures required to measure the Conventional RP of DORIS (and GPS)

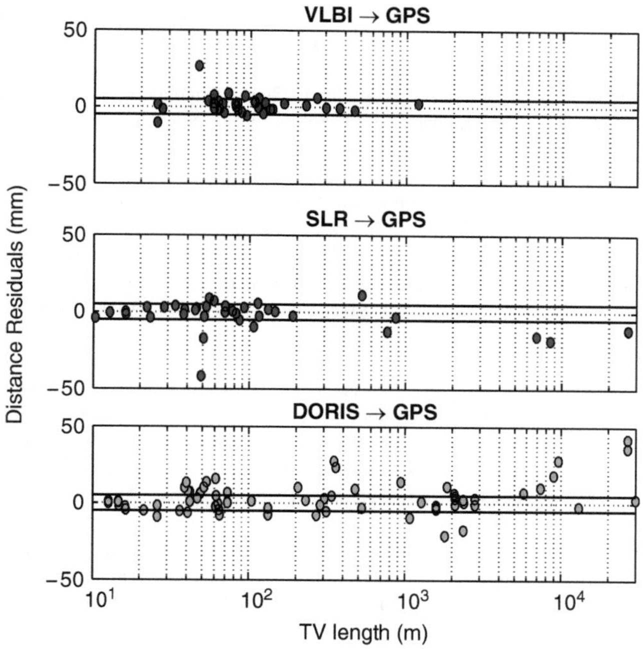

Fig. 13.2 TV distance residuals stemming from ITRF2008 combination (Altamimi et al. 2011) as a function of the TV length. *Black lines* identify a ±5 mm boundary

antennas can be regarded as being easier and effortless than the procedures applied to SLR and VLBI Conventional RP surveying. It basically concerns the survey of the antenna reference point and requires a lower number of observations and a reduced degree of conditioning.

Likewise, we have investigated the relation between the magnitude of the residuals and the length of the TV. Here, a dependence of the residual on the separation between the two instruments (i.e. the TV modulus) might be present. In fact, the precision of terrestrial measurements depends on the extension of the network to be surveyed and it is expected to decrease as the length of the TV increases.

Almost all VLBI-GPS and SLR-GPS TVs have lengths not exceeding 1 km whereas DORIS has a large number of co-located beacons that are separated from GPS antennas by more than 1 km, with few DORIS-GPS TVs exceeding 10 km. When such distances are involved, it is indeed legitimate to question whether such geodetic sites can be considered as co-location sites. The graphs of Fig. 13.2 show the magnitude of the residuals as a function of the TV length; the residuals do not appear to depend on the value of the TV modulus over the whole range of distances.

Finally, it is interesting to focus on each single TV component in the local geodetic frame, taking into consideration the value of the associated residuals (Table 13.1). It is rather surprising to see that the best overall agreement is related to VLBI-GPS co-locations. On the contrary, the TVs involving DORIS have impressively large residuals, more than half

Table 13.1 Percentages of TVs whose residuals (in East, North, Up and Distance) exceed 5 mm

	East	North	Up	Distances
GPS-VLBI	31	26	43	23
GPS-SLR	26	26	74	32
GPS-DORIS	53	53	62	49

being larger than 5 mm on all three components. As mentioned before, this is particularly surprising given the relative simplicity of the surveying approach that, in principle, must be adopted when measuring a DORIS beacon or a GPS antenna Conventional RP. The overall agreement of the TVs involving DORIS suggests the presence of biases that must be investigated. This is a clear example of the utility of combination residuals in spotting out problems that need further investigation and attention.

Details on the insertion and handling of the statistical information carried by TVs and their treatment in the combination of frames for ITRF computation can be found in (Altamimi et al. 2011).

5 Conclusions

The extent of the discrepancies between SG solutions and TVs stemming from the combination is much larger than their formal errors, the residuals being approximately one order of magnitude larger and at the centimeter level. These discrepancies may originate from errors in local surveys, systematic-errors in space geodesy estimates, or in both.

On the local tie side, the diversities in the estimation procedure recalled in Sect. 2 can play an important role. Site-dependent methods are often applied: different surveying approaches, data reduction and computation strategies as well as not standardized alignment procedures of the terrestrial TVs may combine and originate biased estimates whose effects on the combination are not easy to trace and to mitigate. The inter-comparison of the methods remains poor and it must be considered as a priority to homogenize the local tie procedures at co-location sites. A comparison has been successfully carried out uniquely on two indirect methods (Dawson et al. 2007). When performed on other approaches, it can represent an important step towards the identification of local tie biases and towards the standardization of the TV estimation process. Also, the estimated terrestrial TVs in the topocentric frame are not routinely stored and saved. As explained in Sect. 3, where the effects of the alignment of the TV are recalled with respect to the combination, the pre-alignment estimate is, indeed, useful being straightforwardly related to the accuracy of the TV in its native frame.

On the SG technique side, technique and site-dependent systematic errors may introduce biases and may contribute to the extent of the observed discrepancies (see e.g. Ray et al. 2007). One of the main issue is related to a proper and accurate connection of the Electronic and Conventional RP of geodetic instruments. The technique Services must take into account possible errors in the realization of this connection. The Services must promote investigations apt at mitigating the technique-related biases and must introduce all necessary corrections. Examples are given by the PCV files in GPS (Schmid et al. 2007) and SPV models in VLBI (Sarti et al. 2011). This task cannot be, under any circumstance, be regarded as a problem to be solved within the local tie operations.

Finally, on the combination side, it must be spotlighted that the alignment process of a terrestrial TV may alias the accuracy of the TV itself (Ray and Altamimi 2005). This very important aspect requires serious consideration and further investigation by the groups involved in TV estimation, since it can affect the global frame computation. The residuals of the combination are a by-product of the computation. They must be regarded as an important source of information and, as such, be made easily available to the whole geodetic community. Geodesists involved in SG data analysis and/or in local tie operations should be aware of the information embedded in the combination residuals and properly use this information to better understand the causes of the large, empirically determined discrepancies observed in the combination and expressed by its residuals.

References

Abbondanza C, Altamimi Z, Sarti P, Negusini M, Vittuari L (2009) Local effects of redundant terrestrial and GPS-based tie vectors in ITRF-like combinations. J Geodesy 83(11):1031–1040. doi:10.1007/s00190-009-0321-6

Altamimi Z, Collilieux X, Legrand J, Garayt B, Boucher C (2007) ITRF2005: a new release of the international terrestrial reference frame based on time series of station positions and Earth orientation parameters. J Geophys Res 112(B09401). doi:10.1029/2007JB004949

Altamimi Z, Collilieux X, Métivier L (2011) ITRF2008: an improved solution of the international terrestrial reference frame. J Geodesy. doi:10.1007/s00190-011-0444-4

Carter E, Rogers AEE, Counselman CC, Shapiro II (1980) Comparison of geodetic and radio interferometric measurements of the Haystack-Westford base line vector. J Geophys Res 85:2685–2687

Dawson J, Sarti P, Johnston G, Vittuari L (2007) Indirect approach to invariant point determination for SLR and VLBI systems: an assessment. J Geodesy 81(6–8):433–441. doi:10.1007/s00190-006-0125-x

Krügel M, Angermann D (2007) Frontiers in the combination of space geodetic techniques. In: Tregonin P, Rizos C (eds) Dynamic planet: monitoring and understanding a dynamic planet with geodetic and oceanographic tools. Springer, Berlin/Heidelberg, International Association of Geodesy Symposia, vol 130, pp 158–165, IAG Symposium on Dynamic Planet, Cairns, Australia, Aug 22–26, 2005

Leinen S, Becker M, Dow J, Feltens J, Sauermann K (2007) Geodetic determination of radio telescope antenna reference point and rotation axis parameters. J Surv Eng 133(2):41–51. doi:10.1061/(ASCE)0733-9453 (2007)133:2(41)

Lösler M (2009) A new mathematical model for reference point determination of an azimuth-elevation type radio telescope. J Surv Eng 135(4):131–135. doi:10.1061/(ASCE)SU.1943-5428.0000010

Ray J, Altamimi Z (2005) Evaluation of co-locations ties relating the VLBI and GPS reference frames. J Geodesy 79(4–5):189–195. doi:10.1007/s00190-005-0456-z

Ray J, Crump D, Chin M (2007) New global positioning system reference station in Brazil. GPS Solut 11(1):1–10. doi:10.1007/s10291-006-0032-x

Sarti P, Sillard P, Vittuari L (2004) Surveying co-located space geodetic instruments for ITRF computation. J Geodesy 78(3):210–222. doi:10.1007/s00190-004-0387-0

Sarti P, Abbondanza C, Vittuari L (2009) Gravity dependent signal path variation in a large VLBI telescope modelled with a combination of surveying methods. J Geodesy 83(11):1115–1126. doi:10.1007/s00190-009-0331-4

Sarti P, Abbondanza C, Petrov L, Negusini M (2011) Height bias and scale effect induced by antenna gravitational deformations in geodetic VLBI data analysis. J Geodesy 85(1):1–8. doi:10.1007/s00190-010-0410-6

Schmid R, Steigenberger P, Gendt G, Ge M, Rothacher M (2007) Generation of a consistent absolute phase center correction model for GPS receiver and satellite antennas. J Geodesy 81(12):781–798. doi:10.1007/s00190-007-0148-y

Small Trends and Oscillations in the 25 Year ILRS Translations and Scale Time Series

14

C. Sciarretta, V. Luceri, and G. Bianco

Abstract

The new realization of the International Terrestrial Reference System, ITRF2008, follows the same strategy of ITRF2005 and is based on an inter-technique combination of geodetic solutions, in turn obtained from an intra-technique combination strategy performed at the Technique Centre (ILRS, IGS, IVS, IDS) level. In Summer 2009, ILRS has provided IERS with its official contribution to the ITRF2008: a homogeneous set of coordinate time series for the period 1983–2008, obtained by ASI/CGS ILRS Primary Combination Center combining time series sets of solutions provided by seven official ILRS Analysis Centers.

A detailed analysis of the ILRS solution origin and scale parameters is presented looking at the time series of the translations and scale with respect to newly released ITRF2008. The analysis is made both for combined and individual solutions, to highlight especially their non-linear behavior; these tiny, non-linear, residual effects must be investigated and possibly minimized.

Keywords

SLR • Helmert parameters • Geocenter • Scale • ILRS

1 Introduction

The role of the Satellite Laser Ranging technique for the ITRF maintenance is quite critical: due to the spatial and temporal features of the tracking network and its specific sensitivity, it is expected to realize, alone, the ITRF origin and, in conjunction with VLBI, its scale. ITRF2008 is the new realization of the International Terrestrial Reference System (Altamimi et al. 2011) and is expected to be an improved solution compared to ITF2005; ITRF2008 uses as input data time series of station positions and Earth Orientation Parameters (EOPs) provided by the Technique Centers of the four space geodetic techniques (GPS, VLBI, SLR, DORIS).

Thus, the Combination Centers' role implies the responsibility of generating an official mono-technique solution merging in an optimal way all the available Analysis Center (AC) solutions. The official ILRS (Pearlman et al., 2007) ILRSA, generated at ASI/CGS, has been obtained by a direct combination of the loosely constrained solutions provided in the late Spring 2009 by seven official ILRS ACs (ASI, DGFI, GA, GFZ, GRGS, JCET, NSGF), each one following strict standards agreed upon within the ILRS Analysis Working Group to provide qualified products of the highest possible quality.

The ILRSA solution covers a long period, more than 25 years and allows a detailed analysis of the translation and scale Helmert parameters with a weekly (bi-monthly in 1983–1992) granularity. The remarkable coherence of the contributing ILRS AC series makes the final combined estimates very precise and the main components of the derived Helmert

C. Sciarretta (✉) • V. Luceri
e-GEOS, ASI-CGS, Matera, Italy
e-mail: cecilia.sciarretta@e-geos.it; cinzia.luceri@e-geos.it

G. Bianco
Agenzia Spaziale Italiana, CGS, Matera, Italy
e-mail: giuseppe.bianco@asi.it

Z. Altamimi and X. Collilieux (eds.), *Reference Frames for Applications in Geosciences*, International Association of Geodesy Symposia 138, DOI 10.1007/978-3-642-32998-2_14,

parameters (origin and scale) time series very clean: a comparison of the ILRSA solution with SLRF2005 (Luceri and Bianco 2007) and ITRF2008 shows very neat linear trend (with very low slope uncertainty and WRMS or the residuals of a few mm) and clear periodic terms of less than 2 mm amplitude at the main once-per-year frequency.

2 ILRSA Long Term TRF Solution

2.1 ILRSA Solution Basic Facts

Seven ILRS ACs submitted their SLR SSC/EOP 15-day arc solutions covering the period 1983–1992 and 7-day arc solutions covering the period 1993–2008 to contribute to the ITRF2008 construction. Each AC performs data analysis following the ILRS recommendations but under independent processing strategy, involving the estimation of many parameters (site coordinates, EOPs, satellites State Vectors, biases). The estimated parameters used in the combination are the SLR site coordinates, a set once per arc, and daily EOP estimates (3-day estimates for the 1983–1992 period); station range biases, provided by the ACs as well, are presently removed before the combination (Table 14.1).

The coverage percentage is intended as the number of qualified arc solutions by an AC (i.e. those formally perfect and with a suitable looseness degree not causing distortions in the final combined solution), divided by the number of the expected arc solutions for the period. The 1993–2008 period has a remarkable level of coverage for all the ACs.

The included AC solutions, generated as a time extension of the current weekly solutions, produced with different data analysis SW, have undergone a testing procedure to become ILRS official products and concur to the official ILRSA combined solution; moreover, for the ITRF2008 contribution, due to the large number of system updates and network variations taking place during the huge time span considered, all the AC solutions have been checked (and if necessary corrected) in terms of estimated/applied system bias and/or used dataset according to the ILRS AWG recommendations.

The contributing solutions are expected to be loose constrained: even this aspect has been checked and, when necessary, amended.

2.2 ILRSA Solution Overall Quality

The ILRSA solution, produced at the ASI/CGS ILRS primary combination center, is obtained following a combination procedure based on the direct combination of loose constrained solutions (Davies and Blewitt 2000), performed along the lines of the iterative Weighted Least Square technique including a rigorous outlier editing.

Table 14.1 Input solutions features

AC	Version	Issue date	Coverage 83–92 %	Coverage 93–08 %
ASI	V23	19/05/09	88.84	95.92
DGFI	V23/ V24	27/05/09	88.84	91.49
GA	V22	19/05/09	69.42	90.89
GFZ	V23	19/05/09	86.36	95.56
GRGS	V24	03/06/09	–	96.04
JCET	V23	24/05/09	88.02	92.69
NSGF	V23	13/07/09	85.54	95.08

Table 14.2 Comparison with SLRF2005

Analysis center	All sites 3d WRMS (mm)	Core sites 3d WRMS (mm)
ASI	13±8	10±6
DGFI	8±6	7±4
GA	15±9	8±4
GFZ	16±9	11±7
GRGS	26±30	16±20
JCET	11±7	8±5
NSGF	22±12	17±10

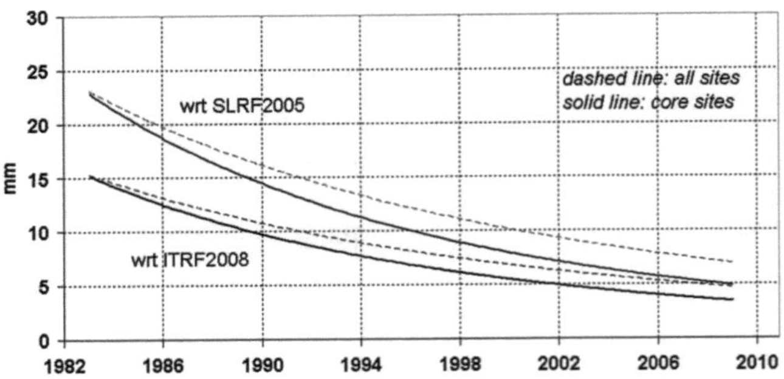

Fig. 14.1 Trend of site residual WRMS w.r.t. TRFs

The overall coherence of all the AC contributing solutions is a strong prerequisite to derive a high quality combination product. When a solution is "far" (5σ criterion) from the combination solution and/or is far from the reference EOPs and "Core Sites" coordinates, its covariance is properly rescaled to down-weight the solution in the combination. The table below shows the 3D WRMS of the residuals measured comparing the site coordinate estimates with respect to SLRF2005, the ILRS internal reference frame tied to ITRF2005; the "Core Sites" column indicates roughly, for the whole period, the adherence of each single solution to the given reference frame measured on the reference (Core) sites, selected by ILRS for their stability, data history and current ITRF model agreement (Table 14.2).

The figure below plots the time trends, modeled by power functions, of the coordinate WRMS residuals of the combined ILRSA solution, with respect to the SLRF2005 and the

Table 14.3 Comparison of ILRSA solution with TRFs

	SLRF2005		ITRF2008	
	Mean	σ	Mean	σ
ILRSA solution	WRMS (mm)	WRMS (mm)	WRMS (mm)	WRMS (mm)
All sites	13.21	*18.80*	10.93	*18.43*
Core sites	8.44	*5.79*	7.13	*4.99*

Table 14.4 ILRSA Helmert Tx versus SLRF2005

Tx	Tx_slope (mm/year)	σTx_slope (mm/year)	WRMS (res) (mm)
ASI	−0.35	0.02	5.37
DGFI	−0.57	0.03	6.27
GA	0.05	0.02	4.18
GFZ	−0.49	0.03	5.46
GRGS	−0.32	0.03	4.50
JCET	−0.18	0.02	4.19
NSGF	−0.41	0.03	6.70
ILRSA	−0.29	0.02	4.16

new ITRF2008. The WRMS value decreases with time in all the cases, while a better fit to the ITRF2008 is evident, with a decrease of 20 % in the mean WRMS value. In the ITRF2008 frame, the overall coordinate WRMS decreases below the 5 mm level for the Core Sites in the recent years and approaches the same level including all the sites, indicating a significant upgrading in time of the TRF models caused also by a significant upgrading of the ILRS contributions in time: e.g. ILRS contribution to ITRF2005 included only the 1993 onwards data, causing an imperfect modeling of the oldest SLR stations having tracked shortly after 1993 (Fig. 14.1 and Table 14.3).

3 ILRSA Reference Frame Long Term Stability

3.1 Assessment w.r.t. SLRF2005

To fulfill the science and ITRF requirements, the construction of a “stable” TRF is pursued, i.e. showing a dominant linear behavior in time of the origin and scale components, without discontinuities.

The initial evaluation of the ILRSA combined solution is made with respect to the SLRF2005, the current terrestrial reference at epoch; it showed a clear superposition of linear trends and small amplitude periodic oscillations in the Helmert (origin and scale) parameters time series, derived from combined and individual solutions. A linear fit, rigorously estimated on each translation and scale parameter time series, shows neat and small slope values (see Tables 14.4, 14.5, 14.6 and 14.7 below), in spite of weak and noisy estimates for the 1983–1992 period, depending on the data strength and geographical distribution.

Table 14.5 ILRSA Helmert Ty versus SLRF2005

Ty	Ty_slope (mm/year)	σTy_slope (mm/year)	WRMS (res) (mm)
ASI	−0.12	0.02	4.50
DGFI	0.09	0.03	5.78
GA	0.17	0.02	4.29
GFZ	0.11	0.02	4.98
GRGS	0.04	0.03	3.71
JCET	0.10	0.02	3.99
NSGF	−0.08	0.03	7.26
ILRSA	0.06	0.02	3.82

Table 14.6 ILRSA Helmert Tz versus SLRF2005

Tz	Tz_slope (mm/year)	σTz_slope (mm/year)	WRMS (res) (mm)
ASI	0.24	0.06	10.38
DGFI	0.88	0.08	13.07
GA	0.83	0.04	8.58
GFZ	0.36	0.06	10.89
GRGS	0.06	0.02	7.11
JCET	0.25	0.04	8.32
NSGF	0.11	0.08	14.06
ILRSA	0.38	0.03	7.45

Table 14.7 ILRSA Helmert scale versus SLRF2005

Scale	Scale_slope (mm/year)	σScale_slope (mm/year)	WRMS (res) (mm)
ASI	−0.31	0.02	4.26
DGFI	−0.48	0.03	4.98
GA	−0.22	0.01	3.64
GFZ	−0.08	0.03	4.71
GRGS	−0.46	0.02	3.34
JCET	−0.23	0.01	2.88
NSGF	−0.62	0.03	6.00
ILRSA	−0.30	0.01	3.15

The coherence of the different solutions concurring to the ILRSA combined solution is evident from the tables below, containing slopes estimated in a linear fit (T_slope) with formal uncertainty (σT_slope) and WRMS w.r.t. the linear fit: Tx and Ty Helmert translations are very stable, with comparable slope values and with low WRMS of residuals, Tz is noisier, as expected due to the hardly separable coupling of the orbit errors with the unbalanced north–south distribution of the SLR systems, the same direction of the Earth's rotation. The estimated long term scale is more stable than the previous ILRS contribution to the ITRF2005, due to a careful handling of the measurement station biases by ACs and CCs, as prescribed by ILRS AWG, influencing the height determination and hence the scale; the linear fit on the scale time series presents a low residual WRMS (3 mm for the ILRSA combined solution) and a clear negative slope (−0.3 mm/year).

Under the adopted stability concept, the key parameters to be focused are the formal uncertainty of the slope, indicating the precision of the estimated Helmert parameters, and the post-fit WRMS, indicating the agreement of the estimated Helmert parameters with the linear (i.e. "stable") model.

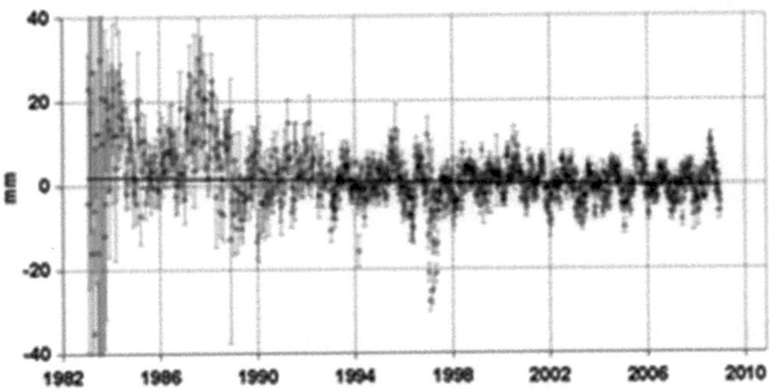

Fig. 14.2 ILRSA Tx versus ITRF2008

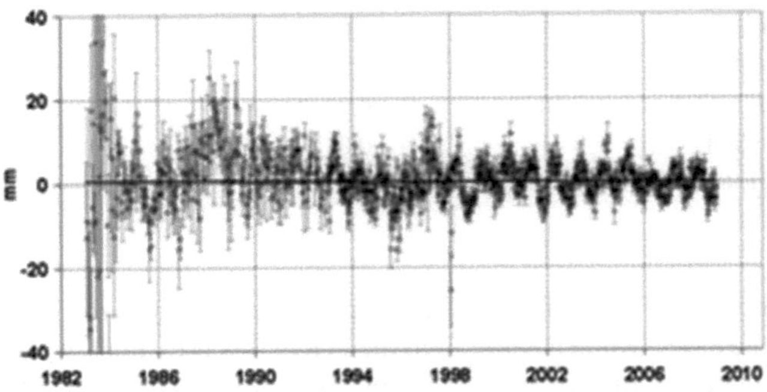

Fig. 14.3 ILRSA Ty versus ITRF2008

3.2 Comparison w.r.t. ITRF2008

The new international terrestrial reference frame, ITRF2008 (Altamimi et al. 2011), which uses in the combination the ILRSA solution as official ILRS product, has been used to re-evaluate again the origin and scale determinations from ILRSA. The translations and scale value time series obtained by a comparison of all the solution arcs to the ITRF2008 model have been rigorously fitted by a linear trend (slope and constant term). As expected, the estimated slope values for the parameters with respect to ITRF2008 are smaller than those with respect to SLRF2005; they are even more neat, with a very low uncertainty (Figs. 14.2, 14.3, 14.4, 14.5; Table 14.8). The WRMS of the parameter residuals shows a significant decrease (around 10 %) with respect to SLRF2005 for the whole 1983–2008 period. The slope of the translation parameters is also very low (almost "0-slope" values); the scale slope value maintains a clear negative value (−0.22±0.01 mm/year). All the estimated constant terms are around +1 mm.

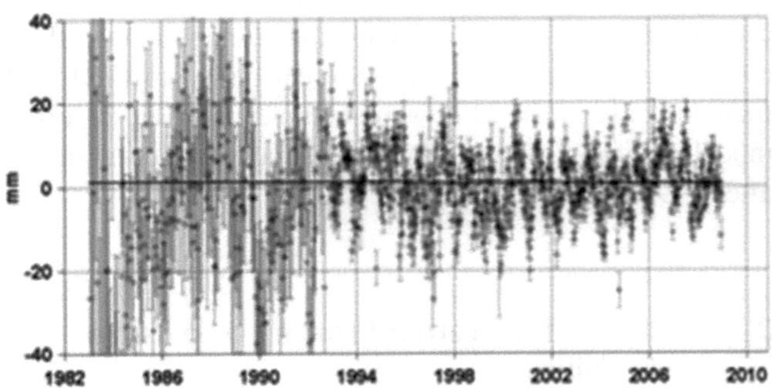

Fig. 14.4 ILRSA Tz versus ITRF2008

4 Spectral Analysis of ILRSA *Detrended* Translations and Scale Parameters

Once the Tx, Ty, Tz and Scale time series with respect to ITRF2008 have been *detrended* (i.e. after a removal of slope and constant term), the residuals show more clearly the non-linear terms, in particular the periodic terms in the translation parameters (Fig. 14.6). A Discrete Fourier Transform (Frigo and Johnson 1998) has been applied to the series in their clean part (after 1993.0), where the sampling frequency is homogeneous and equal to 0.142857 day^{-1} (one point per

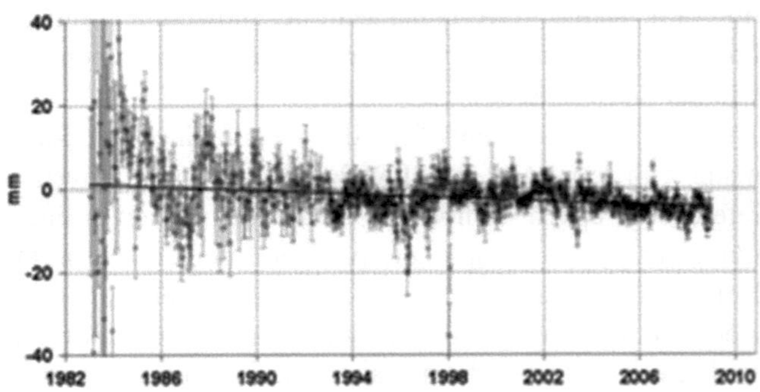

Fig. 14.5 ILRSA scale versus ITRF2008

Table 14.8 ILRSA Helmert parameter versus ITRF2008

Helmert parameter	Slope (mm/year)	σSlope (mm/year)	WRMS (res) (mm)
Tx	−0.10	0.01	3.71
Ty	−0.02	0.01	3.63
Tz	−0.03	0.01	6.80
Scale	−0.22	0.01	2.80

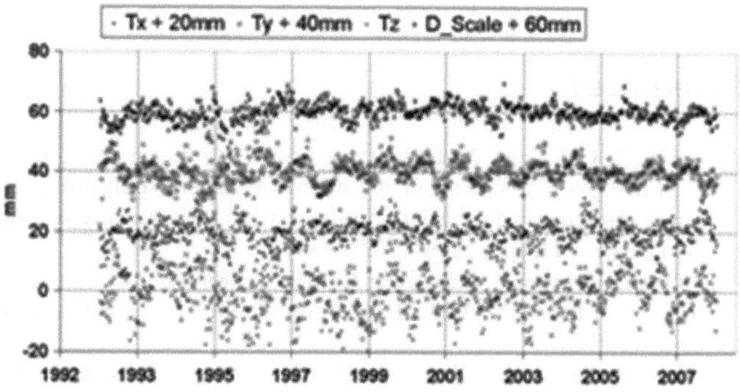

Fig. 14.6 Detrended Tx, Ty, Tz, scale

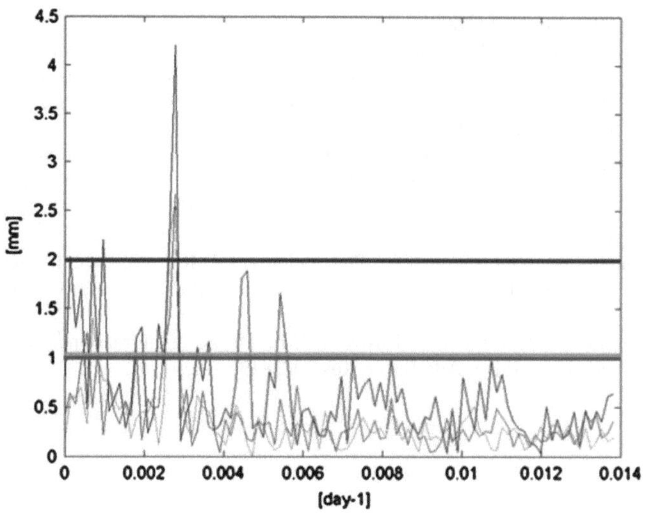

Fig. 14.7 Single-sided amplitude spectrum of Tx, Ty, Tz

week). The total number of points, for each series, is 834 over 16 years.

The resulting spectral plots (FFT amplitude vs. frequency) indicate significant periodic terms in the low frequency portion of the spectra.

For the translation parameters, the plot (Fig. 14.7) shows all frequencies lower than 0.014 day^{-1} (roughly, once per two-and-half months); empirical thresholds (red line: Tx; green line: Ty; blue line: Tz), selected roughly as the half of the highest peak, have been applied to discriminate the most significant frequencies in the Helmert translation signal.

It is evident that the dominant terms are concentrated around the annual frequency and that they appear in all the

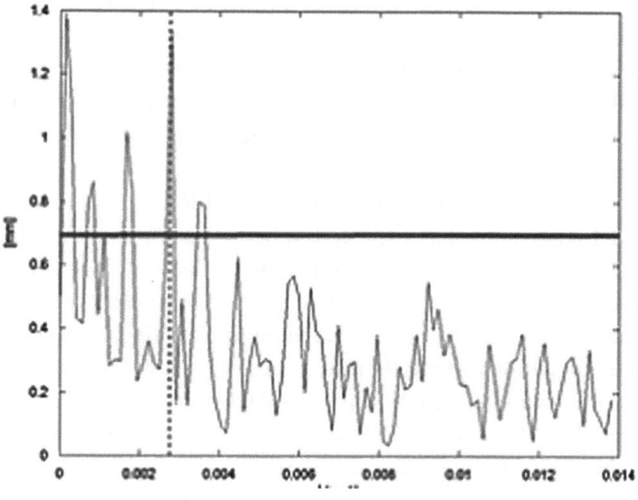

Fig. 14.8 Single-sided amplitude spectrum of scale

translation series (with rough amplitude around 2.5 mm for Tx and Ty, around 4 mm for Tz). The SLR translations series should reflect significantly the variations from atmosphere, ocean and surface ground water at the annual and semiannual scales (Dong et al. 1998): the geocenter motion (intended as the displacement between Earth's Center of Mass and Earth's Center of Figure) from geophysical loading models (Collilieux et al. 2009; Table 14.2), predicts amplitude values, for the dominant annual term, very close to those estimated from the ILRSA solution, indicating also that the "network effect" (depending on the SLR tracking geometrical configuration and availability), embedded in the detrended translation estimations, is relevant especially for the Tx component, reflecting the clustering of SLR European stations.

The spectral behavior of the scale (Fig. 14.8) shows less significant seasonal terms, partly related to the unmodeled surface loading effects and partly to residual mismodeled SLR instrumental range biases (Altamimi et al. 2007; Collilieux et al. 2009): no amplitude is greater than 1.4 mm; the majority of terms exceeding 1 mm amplitude are clustered at the very low frequencies, indicating the presence of a long-time-scale variation; however, an annual term of 1.3 mm amplitude is detectable, close to the prediction from geophysical models (Collilieux et al. 2009; Table 14.2), indicating a better handling of the SLR instrumental range biases in the ILRS contribution to ITRF2008 than in the previous solution for ITRF2005.

5 Conclusions

The ILRSA combined solution, the official ILRS contribution to ITRF2008, when issued, has been assessed with respect to the SLRF2005, the internal ILRS reference frame, derived from ITRF2005. This preliminary comparison showed good performance of the quality parameters (site coordinates WRMS, Helmert parameters time series) for the final combined solution and a remarkable coherence for the individual, contributing solutions. In particular, translation and scale parameters time series showed clean linear dominant trends.

The a-posteriori comparison of ILRSA solution to ITRF2008, shows as expected, a better performance of the 3D WRMS of the coordinate residuals than with respect to the SLRF2005 throughout the data span. The closer adherence to ITRF2008 reflects also in more clear linear trends in the translations and scale time series, with low formal error for the estimated slopes (order of 0.01 mm/year) and a few mm WRMS of the linear fit residuals (<4 mm for Tx, Ty; <7 mm for Tz; <3 mm for scale).

The translations show an almost "0-slope" value, accordingly to the fact that the SLR solution defines the ITRF2008 origin (Altamimi et al. 2011); small differences are due, presumably, to processing choices in the results assessment (e.g. weighting, reference sites selection) and disappear if the 1993–2008 period only is considered. The scale slope appears to be negative of about −0.2 mm/year, even removing the 1983–1992 estimates, weak in terms of quality, amount and geographical distribution of the data.

Once the estimated linear trends are removed, Helmert translations and scale have been analysed to detect residual non-linear terms from equally sampled 1993 onwards estimates. The use of a Discrete Fourier Transform method indicates a dominant annual term of few mm amplitude (2.5 mm for Tx, Ty; 4 mm for Tz) for the translations, very close to the prediction of the Geocenter motion from the variable surface loading models, indicating the influence of the "network effect" mainly for Tx, due to the clustering of the European SLR network.

The scale time series exhibits a lower amplitude annual term (1.3 mm), closer, than in the past ITRF contribution, to the geophysical prediction; that is due to a rigorous handling of the SLR instrumental biases by all the ILRS ACs following the ILRS AWG recommendations; nevertheless, scale time series shows a clear evidence (cluster of significant very low frequencies) of a non-linear long-time-scale variation, to be further investigated together with the negative slope of the linear trend.

References

Altamimi Z, Collilieux X, Legrand J, Garayt B, Boucher C (2007) ITRF2005: a new release of the international terrestrial reference frame based on time series of station positions and Earth orientation parameters. J Geophys Res 112:B09401

Altamimi Z, Collilieux X, Métivier L (2011) ITRF2008: an improved solution of the international terrestrial reference frame. J Geodesy 85(8):457–473

Collilieux X, Altamimi Z, Ray J, van Dam T, Wu X (2009) Effect of the satellite laser ranging network distribution on geocenter motion estimation. J Geophys Res 114:B04402

Davies P, Blewitt G (2000) Methodology for global geodetic time series estimation: a new tool for geodynamics. J Geophys Res 105 (B5):11083–11100

Dong D, Dickey JO, Chao Y, Cheng MK (1998) Geocenter variations caused by mass redistribution of surface geophysical processes. In: IERS technical note no. 25, Paris: Central Bureau of IERS - Observatoire de Paris, 1999. iv, 122 p.

Frigo M, Johnson GS (1998) FFTW: an adaptive SW architecture for the FFT. In: Proceedings of the 1998 IEEE international conference on acoustics, speech and signal processing. IEEE Service Center, Piscataway

Luceri V, Bianco G (2007) The temporary ILRS reference frame: SLR2005. Lecture notes, Fall 2007 ILRS workshop, Grasse Observatory, France

Pearlman MR, Degnan JJ, Bosworth JM (2007) The international laser ranging service. Adv Space Res 30(2):135–143

15 Accuracy Assessment of the ITRS 2008 Realization of DGFI: DTRF2008

Manuela Seitz, Detlef Angermann, and Hermann Drewes

Abstract

The DTRF2008 is a realization of the ITRS computed by the ITRS Combination Centre at DGFI. It is based on the same input data as the ITRF2008. In order to assess the internal and external accuracy of DTRF2008 several validation procedures are applied which are based on comparisons with technique-only multi-year solutions (for assessing the internal accuracy) and comparisons with ITRF2008 and ITRF2005 (for assessing the external accuracy). The analysis is done separately for the four space-geodetic techniques GPS, VLBI, SLR and DORIS. The internal accuracy for station positions is between 0.6 and 3.3 mm and the external accuracy between 7 and 10 mm depending on the space technique. For the velocities the internal accuracy is between 0.25 and 1.0 mm/a and the external between 0.2 and 2.0 mm/a.

Keywords

International terrestrial reference frame • ITRF2008 • DTRF2008 • Combination • GPS • SLR • VLBI • DORIS • Datum parameters

1 Introduction

Within the IERS two ITRS Combination Centres were in charge of the computation of a new ITRF solution, the ITRF2008. These Combination Centres are the Institut Géographique National (IGN) in Paris and the Deutsches Geodätisches Forschungsinstitut in Munich.

The computation is based on long-term input data time series of the four space-geodetic techniques Very Long Baseline Interferometry (VLBI), Satellite Laser Ranging (SLR), Global Positioning System (GPS) and the Doppler Orbitography and Radiopositioning Integrated by Satellite (DORIS) provided by the corresponding technique centres (IVS, ILRS, IGS and IDS). The data include solutions of station positions and the Earth Rotation Parameters (EOP). Both Combination Centres computed an ITRS realization released to the IERS community in Spring 2010. After an elementary comparison of the two frames which showed a good agreement, the IGN solution was released as the official ITRF2008 solution of the IERS at the end of May 2010. In order to avoid confusion, the DGFI solution is published as DTRF2008.

While the standard deviations estimated from the TRF adjustment provide the precision of the parameters, the accuracy has to be assessed by validation procedures. The availability of two Combination Centres and their ITRS realizations provides a great potential with respect to the validation and quality assessment. As the two Combination Centres followed different combination strategies and the computations are performed using different combination software packages, the solutions are formally independent.

The internal accuracy of the DTRF2008 which includes the agreement of the techniques and the adequacy of the combination strategy is assessed from comparisons with technique-only solutions. The external accuracy is derived from comparisons with the ITRF2008 and ITRF2005.

M. Seitz (✉) • D. Angermann • H. Drewes
DGFI, Alfons-Goppel-Straße 11, D-80539 Munich, Germany
e-mail: seitz@dgfi.badw.de

Z. Altamimi and X. Collilieux (eds.), *Reference Frames for Applications in Geosciences*,
International Association of Geodesy Symposia 138, DOI 10.1007/978-3-642-32998-2_15,

The paper concentrates on the reference frames. For the Earth Orientation Parameters (EOP), which also provide information on the accuracy of the solution, we refer to Seitz et al. (2012).

2 Combination Strategy at DGFI

The computation of DTRF2008 is based on the combination of normal equations of the four space-geodetic techniques VLBI, SLR, GPS and DORIS (Angermann et al. 2009), which is performed using the combinations software DOGS-CS (Gerstl et al. 2000). The input data are provided as weekly (SLR, GPS, DORIS) or session-wise (VLBI) data by the technique centres of the corresponding IAG services. The time series cover time spans of 25 years in case of VLBI and SLR, 16 years for DORIS and 12 years for GPS.

Figure 15.1 shows a simplified flowchart of the computation of DTRF2008. After the reconstruction of the normal equations from SINEX format by removing the specified constraints, solution time series per technique are computed and the time series of all the included parameters (station positions, EOP and datum parameters) are analysed in order to identify discontinuities, non-linear station motions and outliers. Then the normal equations are accumulated to one normal equation per technique by introducing station velocities as new parameters and considering the detected discontinuities and outliers. After this, the normal equations of the different techniques are combined. The selection of local ties, the relative weighting of the techniques and the datum realization of the terrestrial reference frame are the most important steps in this process. A detailed description of the processing is given by Seitz et al. (2012).

The strategy followed by IGN is based on the combination of input solutions. This implies the estimation of parameters of a similarity transformation within the combination process in order to consider datum differences between the input solution and the combined frame (Altamimi et al. 2011).

3 Internal Accuracy

The internal accuracy of DTRF2008, which comprises the agreement of the techniques and the accuracy of the combination strategy itself, can be assessed by internal validation procedures. Assessing the accuracy of the station positions the datum parameters and the network geometry are analysed separately.

3.1 Datum Parameters

The datum of DTRF2008 (and also ITRF2008) is realized as recommended by the IERS Conventions 2003 (IERS 2004):

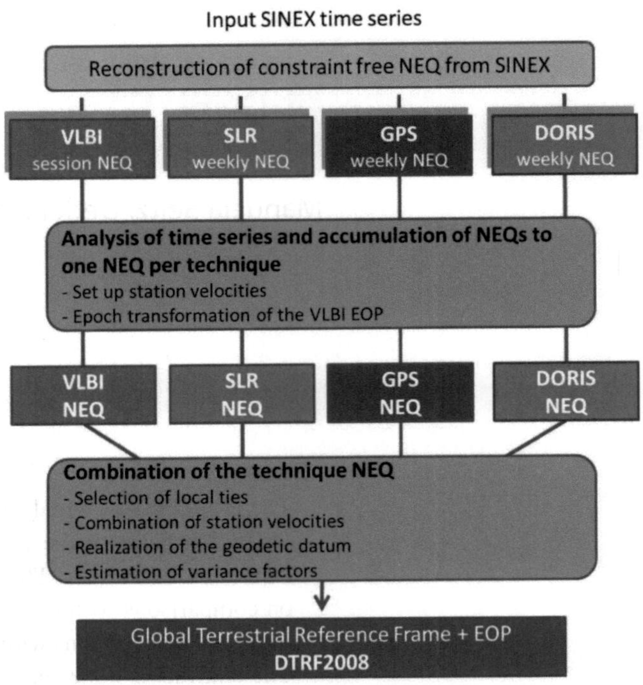

Fig. 15.1 Flowchart of computation process of DTRF2008

the origin is derived from SLR observations only, the scale is realized by SLR and VLBI observations and the orientation is realized by applying no-net-rotation conditions with respect to ITRF2005 (Altamimi et al. 2011). Figure 15.2 shows the translation time series for the SLR network derived from the transformation of weekly SLR solutions to the multi-year SLR solution computed by accumulating the weekly SLR normal equations.

Before LAGEOS-2 was launched, the time series is clearly affected by a long-periodic sinusoidal-like signal. But, as the standard deviations are large by comparison to those estimated for a combined LAGEOS-1/2 series (exemplarily shown for the z-component), the effect on the mean origin and origin rate is not significant. Hence, the complete SLR time series are used for realizing the origin of DTRF2008.

Figure 15.3 shows the time series for the SLR and the VLBI scale. Like for the translation, the SLR series shows a periodic signature during the time when only LAGEOS-1 was observed. The VLBI series shows a small trend at the early years. Both effects do not have a significant effect on the mean scale and both series are used for the scale realization of the DTRF2008.

Combining the space-geodetic techniques local tie misfits (disagreement between local ties and the coordinate differences derived from the space techniques) can lead to a deformation of the station networks but can also affect the geodetic datum parameters. In order to analyse to what extent the datum can be conserved in the combination, the

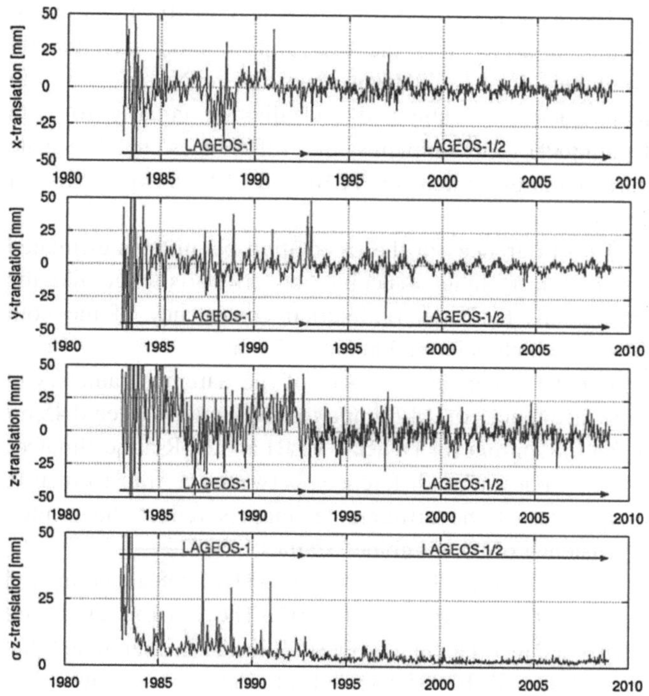

Fig. 15.2 SLR translation parameters and standard deviations of the z-translation. The corresponding WRMS values are: $WRMS_x = 3.8$ mm, $WRMS_y = 3.9$ mm and $WRMS_z = 8.3$ mm

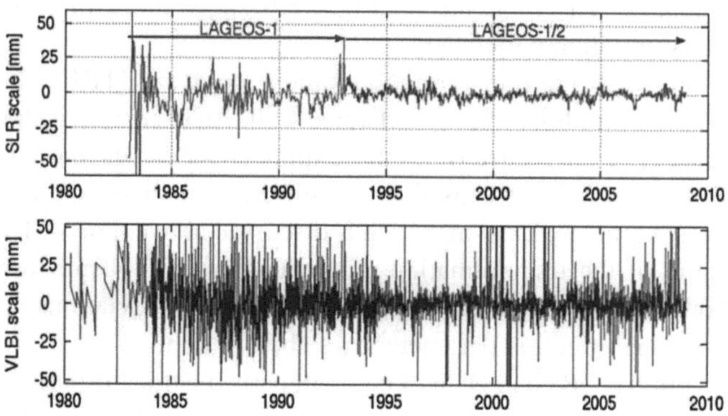

Fig. 15.3 SLR (*upper plot*) and VLBI (*lower plot*) scale parameters. The corresponding WRMS values are 3.6 and 8.2 mm

SLR and VLBI multi-year technique-only solutions are compared to the DTRF2008. Table 15.1 gives the translation parameters. They show a change of the origin due to the combination of maximum 0.3 mm and 0.1 mm/a per component.

For the scale differences of 0.0 mm and 0.0 mm/a (expressed at the Earth's surface) are estimated. A scale factor of 0.0 for both, SLR and VLBI, is an indication that no significant scale difference between both techniques exist (Table 15.2).

In total the change of the datum parameters due to the combination is a shift of <0.4 mm and a drift of <0.2 mm/a.

Table 15.1 Translation parameters of DTRF2008 w.r.t. SLR-only multi-year solution

SLR	tx	ty	tz
Bias (mm)	0.1 ± 0.2	−0.3 ± 0.2	0.2 ± 0.2
Drift (mm/a)	0.1 ± 0.1	0.2 ± 0.1	0.1 ± 0.1

Table 15.2 Scale parameters of DTRF2008 w.r.t. SLR- and VLBI-only multi-year solution

SLR	Scale	VLBI	Scale
Bias (mm)	0.0 ± 0.0	Bias	0.0 ± 0.0
Drift (mm/a)	0.0 ± 0.0	Drift	0.0 ± 0.0

The scale is expressed as a radial distance (mm) at the Earth's surface

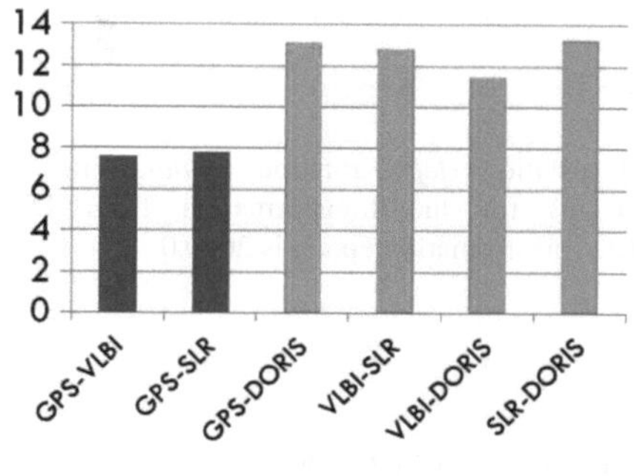

Fig. 15.4 Mean a posteriori local tie residuals (mm) for the different co-location types of DTRF2008. Number of co-location sites included/available: G-V: 33/41, G-S: 30/44, G-D: 34/42, V-S: 10/22, V-D: 9/12, S-D: 8/14

3.2 Network Geometry

Discrepancies between the space-geodetic techniques and the local tie measurements at co-location sites can lead to a deformation of the networks in the combination. Figure 15.4 shows the mean a posteriori local tie residuals of the introduced local ties for the different co-location types. They reach up to 13 mm for some co-location types. Even, if one of the criteria applied for the implementation of local ties is to minimize the network deformation, a deformation cannot be fully avoided, as the combined network should be as consistent as possible. Thus, local ties have to be implemented using adequate standard deviations in order to obtain a consistent network with a small deformation (Seitz et al. 2012).

A measurement for the change of the network geometry are the mean RMS values derived from the transformations of the single-technique multi-year solutions to DTRF2008. Table 15.3 summarizes the values for the four techniques. Whereas the deformation of GPS and VLBI network are very small, the SLR and DORIS networks are deformed by 2–3 mm and 0.5–1 mm/a.

Table 15.3 RMS values of the similarity transformation between DTRF2008 and the technique specific multi-year solutions

	Position (mm)	Velocity (mm/a)
GPS	0.6	0.09
VLBI	0.3	0.05
SLR	1.9	0.42
DORIS	3.3	0.83

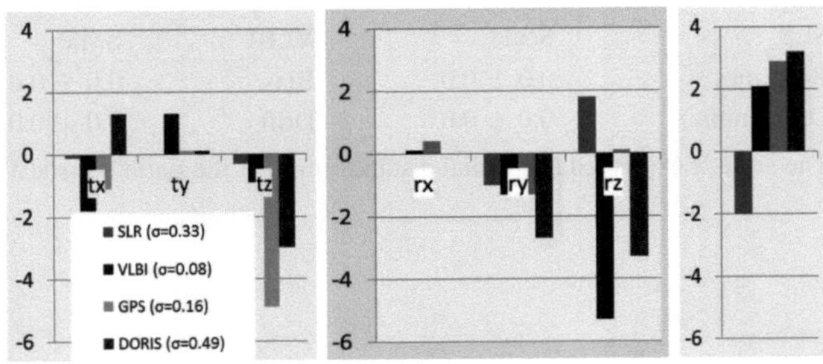

Fig. 15.5 Translation (*left*), rotation (*middle*) and scale (*right*) parameters (mm) for the transformation from DTRF2008 to ITRF2008. The transformation epoch is 2000.0

4 External Accuracy

The external accuracy of the DTRF2008 can be assessed from comparisons with the ITRF2008 and the ITRF2005 solutions. Differences between the two frames show the impact of the computation strategies. But they also provide the accuracy which can be reached for the ITRF today. The determination of external accuracy is also done separately for datum parameters and network geometry.

4.1 Datum Parameters

The accuracy of the datum parameters is derived by 14 parameter similarity transformations between DTRF2008 and ITRF2008. The transformations are done separately for each technique. Hence, not only the datum differences between the frames but also systematics between the techniques can be analysed. Figure 15.5 shows the transformation parameters for the station positions.

The translation parameters between the two frames are small for the SLR network part and reach up to 5.2 mm for the other techniques. The reason for the differences between the techniques is, that the origin is realized from the SLR observations only. How accurately the information about the origin can be transferred into the other network parts (GPS, VLBI and DORIS part) depends on how well the networks of the techniques are linked. This again depends very strongly on the handling of the local ties.

The scale is derived from SLR and VLBI observations. While in DTRF2008 the scale information was directly combined, for ITRF2008 a scale difference between both techniques was estimated and divided equally to both techniques. This explains the differences in the scale parameters estimated for VLBI and SLR network parts of the two frames.

The differences in the orientation parameters estimated for the four technique-specific network parts reflect like the translations the level of internal consistency of the both frames, which is in total about 5–7 mm.

For the linear development of the datum parameters in time differences of 0.17 mm/a for GPS and between 0.45 m/a and 0.85 mm/a for DORIS, VLBI and SLR are estimated.

The rather large differences between the transformation parameters of the different techniques reflect the limited consistency of the combined frames.

In a second step DTRF2008 and ITRF2008 are compared to ITRF2005 (Altamimi et al. 2007). The transformations are performed again technique-wise. It has to be mentioned that for SLR the ITRF2005 rescaled solution was used. In case of GPS the transformation was performed using the IGS05 (Ferland 2006) solution instead of ITRF2005. IGS05 is aligned to ITRF2005 but considers absolute antenna phase centre corrections (PCC) for the GPS satellite and ground antennas. The diagrams in Fig. 15.6 show the estimated transformation parameters.

For SLR the transformation parameters w.r.t. ITRF2005 are small for both frames and agree well. An exception is the scale. The difference of 2 mm (expressed as distance at the Earth's surface) is caused by the different methods applied for the scale realization, which is discussed above. For VLBI some of the transformation parameters reach 3–4 mm for both DTRF2008 and ITRF2008. The standard deviations of the transformation parameters are about 1 mm.

For the GPS and DORIS network parts the transformation parameters reach maximal 2 mm for DTRF2008, with the exception of the scale which shows a difference of about 5 mm. For ITRF2008 and ITRF2005 transformation parameters are partly larger. For the z-translation of GPS 6 mm and for the z-rotation of DORIS 4.5 mm are estimated. The scale differences reach 7 mm for both, GPS and DORIS. The parameter rates (not shown in the diagram) reach up to 0.2 mm/a for GPS, 0.4 mm/a for VLBI, 0.7 mm/a for SLR and 1.2 mm/a for DORIS.

The comparison with ITRF2005 highlights two facts: (1) The transformation parameters between DTRF2008 and ITRF2008, which are computed from the same input data sets are at the same level as the parameters derived from an individual comparison of the two frames with ITRF2005. The reason might be that the realization of a homogeneous and consistent combined network is one of the most challenging tasks in the TRF computation. (2)

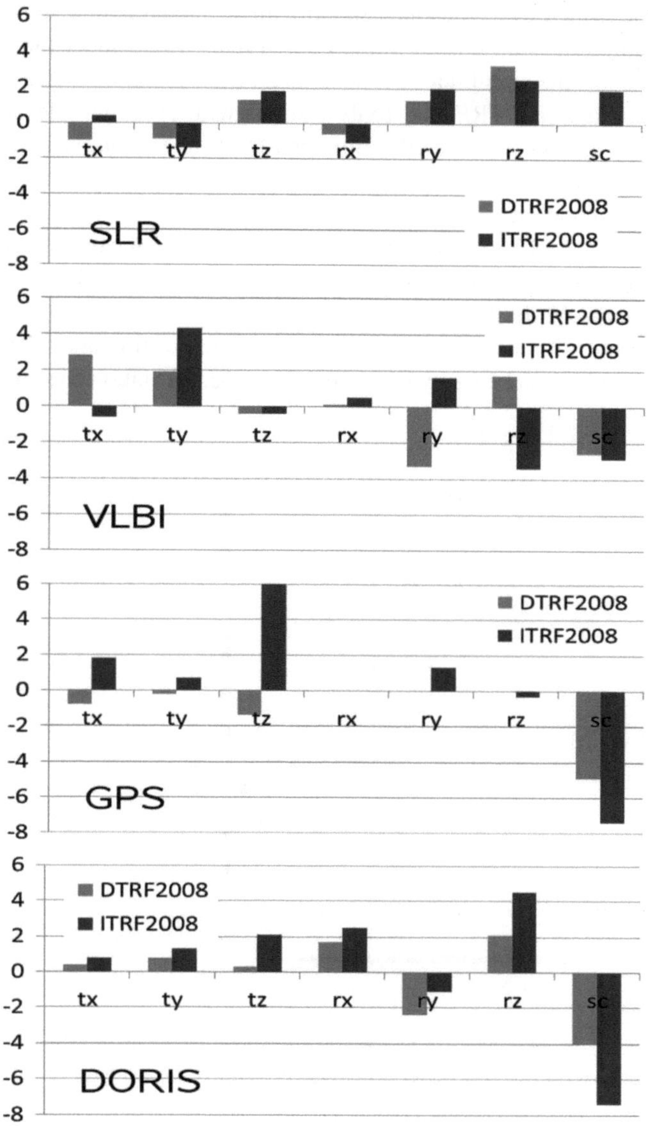

Fig. 15.6 Translation, rotation and scale parameters derived from the transformation between ITRF2005 and DTRF2008 and ITRF2008, respectively. The transformation epoch is 2000.0

The transformation parameters differ strongly for the different techniques. Thus, no unique set of transformation parameters can be applied for the transformation between different terrestrial reference frames.

4.2 Change of Network Geometry

The external accuracy of the network consistency is quantified by the RMS of station position and velocity residuals after removing the parameters of a similarity transformation between DTRF2008 and ITRF2008. In addition the RMS values of the transformation between DTRF2008 and ITRF2005 are analysed. Tables 15.4 and 15.5 give an overview about the RMS values.

Table 15.4 RMS values of the 14 parameter similarity transformation between DTRF2008 and ITRF2008

Tec./# core stat.	Position (mm)	Velocity (mm/a)
GPS (68)	1.33	0.19
VLBI (25)	0.38	0.09
SLR (39)	2.00	0.82
DORIS (46)	3.20	0.98

A set of stable stations are used for the transformation

Table 15.5 RMS values of the 14 parameter similarity transformation between DTRF2008 and ITRF2005

Tec./# core stat.	Position (mm)	Velocity (mm/a)
GPS (62)	3.04	0.34
VLBI (25)	1.80	0.33
SLR (11)	3.11	0.82
DORIS (44)	6.02	1.63

Core stations are used for the transformation

The agreement of DTRF2008 and ITRF2008 is very good for GPS. The mean differences reach values of only 0.4 mm for the positions and 0.2 mm/a for velocities. For VLBI, SLR and DORIS the differences are between 1.3 and 3.2 mm as well as 0.1 and 1.0 mm/a.

The comparison with ITRF2005 reveals larger differences. For the positions mean differences between 1.8 and 6.0 mm are derived, for the velocities differences of 0.3 up to 1.6 mm/a. The reason for the larger discrepancies w.r.t. ITRF2005 are the extension of observation time series by three more years of data and various model improvements which were implemented in the technique analysis software since ITRF2005 was released. For example, in case of GPS, the complete input data set was reprocessed by applying absolute PCC (Schmid et al. 2007). For VLBI the implementation of the pole tide model was corrected and thermal deformation of the antennas was considered (Nothnagel 2009). In case of SLR the station related range biases were improved and for DORIS the solar radiation pressure and air drag modelling was updated (Gobinddass et al. 2009, 2010).

The RMS values given in Tables 15.4 and 15.5 are computed considering only the core stations. Figure 15.7 shows histograms of the residuals of all stations obtained from the comparison of DTRF2008 and ITRF2008. Whereas for GPS the residuals of all stations are small, for VLBI and SLR a quite considerable number of stations show large residuals of 50 mm and more. Most of these stations provide only a few observations, have short observation time spans and were observing in the 1980s. A critical review of the list of stations contributing to the ITRF might be worth of being discussed if the accuracy of ITRF should be improved in future.

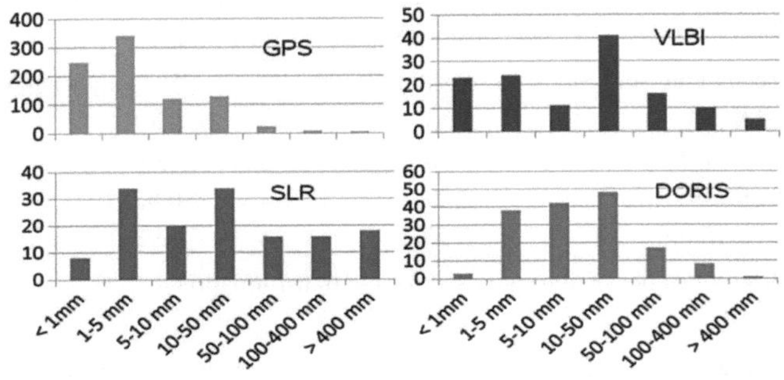

Fig. 15.7 Histograms for station position residuals resulting from the transformation between DTRF2008 and ITRF2008. Please note the different scales of the vertical axes

described by a sinusoidal signal with an annual period show amplitudes of several millimetres. Figure 15.8 (upper plot) shows the amplitudes of the annual signal in height of GPS stations in DTRF2008 (only stations with an amplitude of ≧5 mm are considered), estimated from the station position residual time series resulting from the DTRF2008 computation.

Figure 15.8 shows that the appearance of significant annual signals is not limited to certain regions but is a global phenomenon. While the maximum mean amplitude in Fig. 15.8 is about 13 mm, the residuals between true position and TRF position can be partly larger. For example, for station Brasilia the discrepancies in station height reach up to 20 mm during the last years (Fig. 15.8, lower plot). It can be concluded, that the agreement of the TRF position and the "true" position at a certain epoch can be limited to a few centimetres.

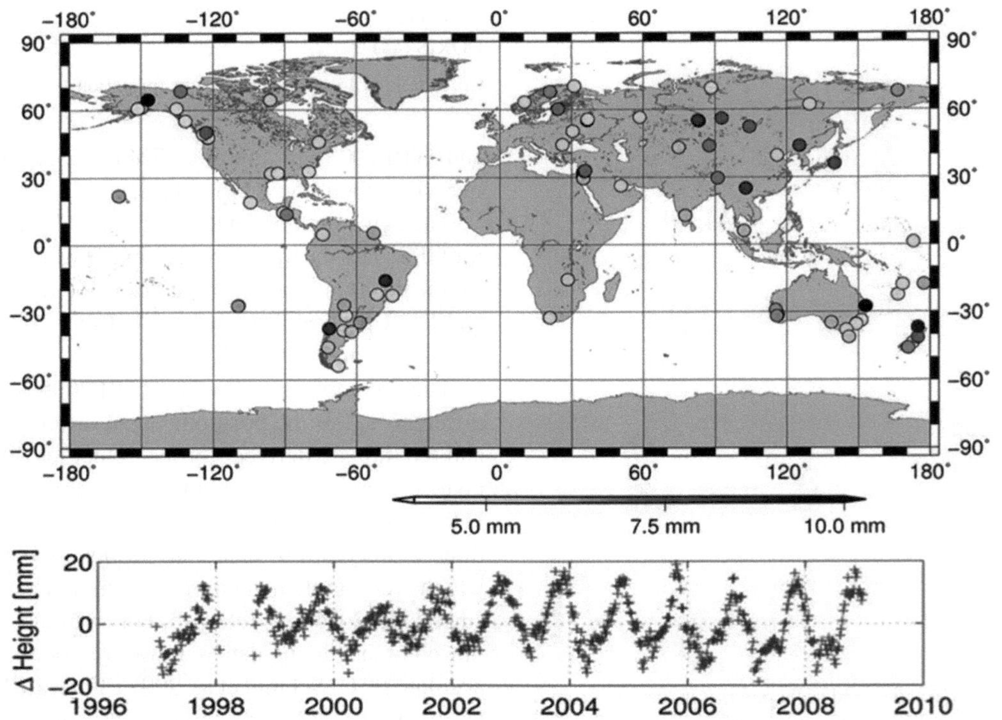

Fig. 15.8 Amplitudes of annual signals of station heights for GPS stations of DTRF2008 (*upper plot*). Only stations with amplitudes of 5 mm and more are considered. Station height residual time series of GPS station Brasilia (*lower plot*)

4.3 Application of TRF at Epoch

In the two sections above the external accuracy of the mean station positions and velocities are discussed. In view of the different applications of the reference frames (e.g., the alignment of epoch solutions of regional networks to ITRF) also the agreement between the TRF and the "true" station position at a certain epoch must be investigated.

Seasonal variations of station positions – especially, the station heights – caused mainly by atmospherical and hydrological loading are not considered in TRF computation at this time as both, the geophysical modelling and an extended parametrization, do not allow for a satisfying approximation of the variations. The annual variations, which can be nearly

5 Summary

DTRF2008 is the 2008 realization of the ITRS computed at the ITRS Combination Centre at DGFI. An assessment of internal and external accuracy of the frame was performed by using comparisons with the technique-only solutions and comparisons with ITRF2008 and ITRF2005. The internal accuracy in total is between 0.6 mm for GPS and 3.3 mm for DORIS, the internal long-term stability is between 0.25 and 1.0 mm/a. The external accuracy is between 7 and 10 mm and between 0.2 and 2.0 mm/a for the velocities, respectively, if only core stations are considered. The separate analysis of datum parameters and network geometry and the individual analysis of the techniques showed, that the lack of consistency of the combined network, is one of the factors, which limit the accuracy of TRF solutions. The consistency is directly related to the implementation of the local tie vectors. They are essential for the computation of TRF solutions, but because of local tie residuals of a few centimetres they also limit the accuracy of the TRF.

References

Altamimi Z, Collilieux X, Legrand J, Garayt B, Boucher C (2007) ITRF2005: a new release of the international terrestrial reference frame based on time series of station positions and Earth orientation parameters. J Geophys Res 112(B09401)

Altamimi Z, Collilieux X, Métivier L (2011) ITRF2008: an improved solution of the international terrestrial reference frame. J Geodesy 85(8):457–473

Angermann D, Drewes H, Gerstl M, Krügel M, Meisel B (2009) DGFI combination methodology for ITRF2005 computation. In: Drewes H (ed) Geodetic reference frames, IAG symposia, vol 134. Springer-Verlag, Berlin, pp 11–16

Ferland R (2006) Proposed IGS05 realization, IGS Mail 5447

Gerstl M, Kelm R, Müller H, Ehrnsperger W (2000) DOGSCS Kombination und Lösung großer Gleichungs-systeme. Manual VIII für DOGS version 4.05. Internal report. DGFI, Munich

Gobinddass M, Willis P, de Viron O, Sibthorpe A, Zelensky N, Ries J, Ferland R, Bar-Sever Y, Diament M, Lemoine F (2009) Improving DORIS geocentre time series using an empirical rescaling of solar radiation pressure models. Adv Space Res 44:1279–1287

Gobinddass M, Willis P, Menvielle M, Diament M (2010) Refining DORIS atmospheric drag estimation in preparation of ITRF2008. Adv Space Res 46:1566–1577

IERS (2004) IERS conventions 2003. In: Mc Carthy D, Petit G (eds). *IERS TN* 32, Frankfurt/Main

Nothnagel A (2009) Short note: conventions on thermal expansion modelling of radio telescopes for geodetic and astrometric VLBI. J Geodesy 83(9):787–792

Schmid R, Steigenberger P, Gendt G, Ge M, Rothacher M (2007) Generation of a consistent absolute phase centre correction model for GPS receiver and satellite antennas. J Geodesy 81(12): 781–798

Seitz M, Angermann D, Bloßfeld M, Drewes H, Gerstl M (2012) The 2008 DGFI realization of the ITRS: DTRF2008. J Geod DOI:10.1007/s00190-012-0567-2

Monitoring Site Stability at the Space Geodesy Facility, Herstmonceux, UK

16

M. Wilkinson, G. Appleby, R. Sherwood, and V. Smith

Abstract

Deformations at the local site level will directly introduce errors both into site coordinates that are determined from individual geodetic techniques and the measured site ties that link those techniques at co-located sites. Such errors will be present in individual-technique solutions and affect their combination in the formation of an International Terrestrial Reference Frame (ITRF). The NERC Space Geodesy Facility at Herstmonceux, UK, operates a highly precise and prolific International Laser Ranging Service satellite laser ranging (SLR) station, two International Global Navigational Satellite Systems Service (IGS) receivers and a permanently-installed absolute gravimeter. As a result, the site remains an important contributor to the maintenance of the ITRF. Site deformation is monitored using a recently-instigated programme of precise digital levelling and by short-baseline GPS analyses using data from on-site and regional GNSS receivers. The stability of the Herstmonceux site will impact on the validity of future comparisons of site height changes observed using the three independent techniques, SLR, GPS and absolute gravimetry.

Keywords

Site stability • Co-location • Levelling • GPS • Short baseline

1 Introduction

The International Terrestrial Reference Frame, e.g. ITRF2008, Altamimi et al. (2011), is defined by the combination of official products from the International GNSS Service (IGS), Dow et al. (2009), the International Laser Ranging Service (ILRS), Pearlman et al. (2002), the International DORIS Service (IDS), Willis et al. (2010) and the International VLBI Service for Geometry and Astrometry (IVS), Schlüter and Behrend (2007). Sites on the Earth's surface that employ more than one of these techniques in co-location are crucially important to this combination through the use of surveyed site ties (Ray and Altamimi 2005). Responsibility for the quality-control of the data provided is with each operator of the technique and the aim is to produce high quality, consistent and bias-free data. The Space Geodesy Facility (SGF), situated near to the village of Herstmonceux in East Sussex, UK operates an SLR station, two IGS GNSS sites, HERS and HERT and a permanently installed absolute gravimeter (AG). This paper describes results of ongoing work at the SGF to determine and monitor vertical and horizontal site stability using short-baseline GPS analyses as well as high-quality digital levelling across the site.

2 Satellite Laser Ranging

The UK SLR station (HERL) is in continuous operation, weather permitting and tracks over 40 satellites, including the geodetic passive sphere pairs Lageos and Etalon which orbit the Earth at altitudes of approximately 6,000 and 19,000 km

M. Wilkinson (✉) • G. Appleby • R. Sherwood • V. Smith
NERC Space Geodesy Facility, Herstmonceux, Hailsham, East Sussex BN27 1RN, UK
e-mail: matwi@nerc.ac.uk

Z. Altamimi and X. Collilieux (eds.), *Reference Frames for Applications in Geosciences*, International Association of Geodesy Symposia 138, DOI 10.1007/978-3-642-32998-2_16,

respectively. Using this range data, and data collected from a network of approximately 40 ILRS stations around the world, weekly station coordinates and velocities are calculated by the SGF ILRS Analysis Centre (AC) for combination with solutions from other ACs to form daily ILRS reference frame products. SLR, along with VLBI, determines the scale of the ITRF and SLR alone refers the frame to the centre-of-mass of the Earth (Altamimi et al. 2011).

The SLR system is closely monitored for errors that would impact on the range measurements. Calibrations are made frequently to a nearby target, the distance to which has been precisely surveyed. Regular calibration removes from the observations systematic effects of order 1–2 cm due for example to ambient temperature changes. Clearly, it is crucial that the site itself remains stable over time.

3 GNSS

3.1 SGF GNSS Sites

HERS is the primary UK IGS station and so ties the national GPS network to the global and regional networks. The site was installed in September 1991 in close proximity to the SLR telescope and is situated on top of a 7 m high latticed mast made from hot rolled mark steel with a concrete monument base. The site has a near-unobstructed full horizon view. The closest structure for potential multipath reflections is the SLR dome, which is approximately 10 m away with its top at an estimated 0.5 m above the antenna base. The site originally used a Rogue SNR-8C receiver, which was replaced in 1998 with an ASHTECH Z-XII3, which was then replaced in August 2010 with a Septentrio GPS/GLONASS PolaRx3 receiver and a Leica AR25 antenna (Fig. 16.1).

To the west of HERS at a distance of 136.5 m is the HERT GPS/GLONASS site, mounted on a small brick monument on the corner of a brick-built "Water Tower", in close proximity to the primary SLR calibration target. HERT was installed in March 2003 using an ASHTECH Z18 receiver, which was replaced in December 2007 with a LEICA GRX1200GGPRO. HERT is part of the EUREF network and is a EUREF and IGS real-time data-provider, streaming data for general real-time navigational applications. A third GNSS receiver (HERO) owned and managed by the UK Ordnance Survey (OS) was installed on a braced-monument midway between HERS and HERT in April 2009 as one of 12 OS GeoNet "zero-order" GNSS reference stations for the UK.

3.2 GPS Analysis

The GAMIT/GLOBK GPS analysis software package, version 10.4, documented by Herring et al. (2010), is

Fig. 16.1 HERS (*top of tower*) and HERT GNSS sites

used to process the satellite pseudo-range and phase data and produce daily site coordinates. A global network of approximately 40 sites was selected and is constrained with IGS reprocessed satellite orbits (reprol) and IERS Earth orientation parameters in GLOBK to provide a stable, well-defined frame of reference, leaving the HERS site unconstrained. The GAMIT default of automatically resolving ambiguity parameters in the processing was used with IGS absolute antenna offsets and an NMF mapping function. This analysis gives a daily height value with a formal error of 4–5 mm for comparison with the independent height measurements from SLR.

3.3 Hydrogen Maser Driving Stability

In August 2010 the HERS receiver was connected to a newly commissioned T4Science iMaser Hydrogen Maser frequency source. This made the local maser time (unofficially UTC-SGF) available to the timing community via the RINEX data for IGS timing-group analysis. The new receiver was calibrated in collaboration with the UK National Physical Laboratory. The maser has consistently performed to specification and, along with other similar external-clock-driven IGS sites, is used in IGS clock analyses. In May 2010 the maser one-second tick and 10 MHz frequency also became the source for driving all laser ranging systems so that the SLR measurements also benefit from the very stable frequency source.

4 Absolute Gravimetry

An FG-5 absolute gravimeter was installed in the SGF basement in 2006 and is run for 24 h mid-week to measure local gravity, which can be determined with an accuracy of 2 μGal. Variations in the local acceleration of gravity may be converted into equivalent height changes, by assuming an appropriate value for a Bouguer conversion factor as done for example by Appleby et al. (2010), and thus provide a

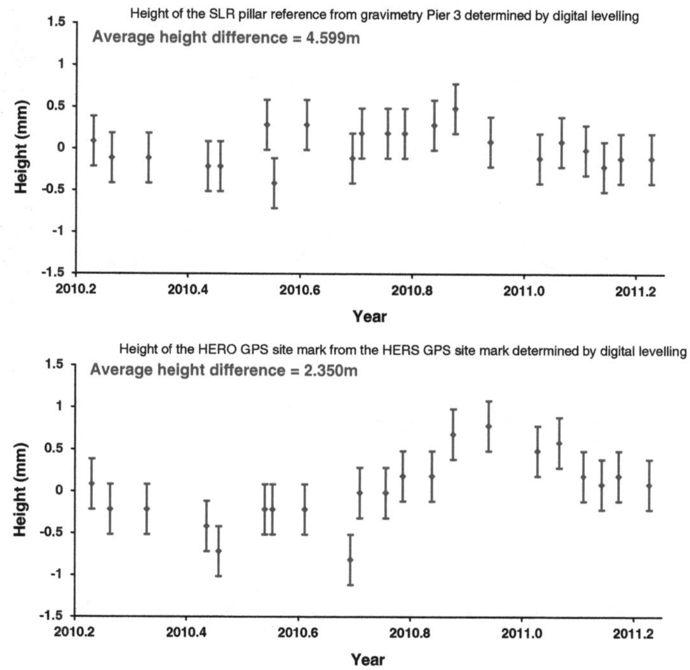

Fig. 16.2 Height differences measured by digital levelling: SLR pillar – gravimeter (*top*); HERS-HERO monuments (*bottom*). Error-bars are 0.3 mm instrumental error

measure of site stability that is independent of the SLR and GPS solutions. The hypothesis of this ongoing work is that height variation detected by the satellite techniques but which may in fact be due to local site instability will either appear as confirmation in the gravity dataset or, if not, suggest that relative height instability exists between the sites of the various techniques.

5 Site Stability and Surveying

The co-located techniques offer independent measurements of station height change. To understand fully and interpret any differences between these results it is necessary to monitor closely all local site changes that could have an impact. This is of particular interest at the SGF since the site is situated on a clay bed, which may expand and contract as forced by seasonal hydrological changes. A 15 year record of the depth of the water table below the site exists from a borehole situated a few metres from the SLR pillar, and shows a good empirical correlation with height variations as determined from analysis of SLR and GPS observations from the site (Appleby et al. 2010).

The site ties that link the independent geodetic techniques are re-surveyed every 2–3 years. The most recent survey was carried out by the Institut Géographique National (IGN) in June 2008 (http://itrf.ign.fr/doc_ITRF/Herstmonceux_Colocation_Survey_aug08.pdf), and also included a redetermination of the SLR calibration target distance. The reference point for all SLR measurements is the intersection of the telescope azimuth and elevation rotation axes and the HERS and HERT measurements are referred to the antenna reference points, with a small vertical offset for HERS. The IGN results showed a significant difference of about 5 mm in the magnitude of the vector between the SLR and HERS compared to the adopted vector that was determined by the OS in 1993. This new value has led to a much better agreement of the site tie with that independently determined during the technique combinations carried out for the production of ITRF2008 (Altamimi et al. 2011). In addition, the reported IGN-surveyed length of this tie agrees to better than 1 mm to that calculated as discussed below using short baseline GPS analysis for the 2 days of the survey.

In addition to the site survey, a program of digital levelling to monuments around the site every 1–2 weeks was begun in 2010 using a Leica DNA03. The levelling run includes one of the gravimeter piers, the monuments of GNSS sites HERS, HERT and HERO, and the SLR pillar. Figure 16.2 shows two plots of height differences obtained during 1 year of observations; one is between the gravimeter pier and a marker on the SLR pillar and the other is between the HERS and HERO monuments. There is a suggestion in both series of an annual variation in relative heights of perhaps 0.5 mm. As this dataset builds up it will provide an independent method to monitor and constrain the magnitude and time-scale of any changes at the site that may help to explain changes in the derived heights of the other techniques.

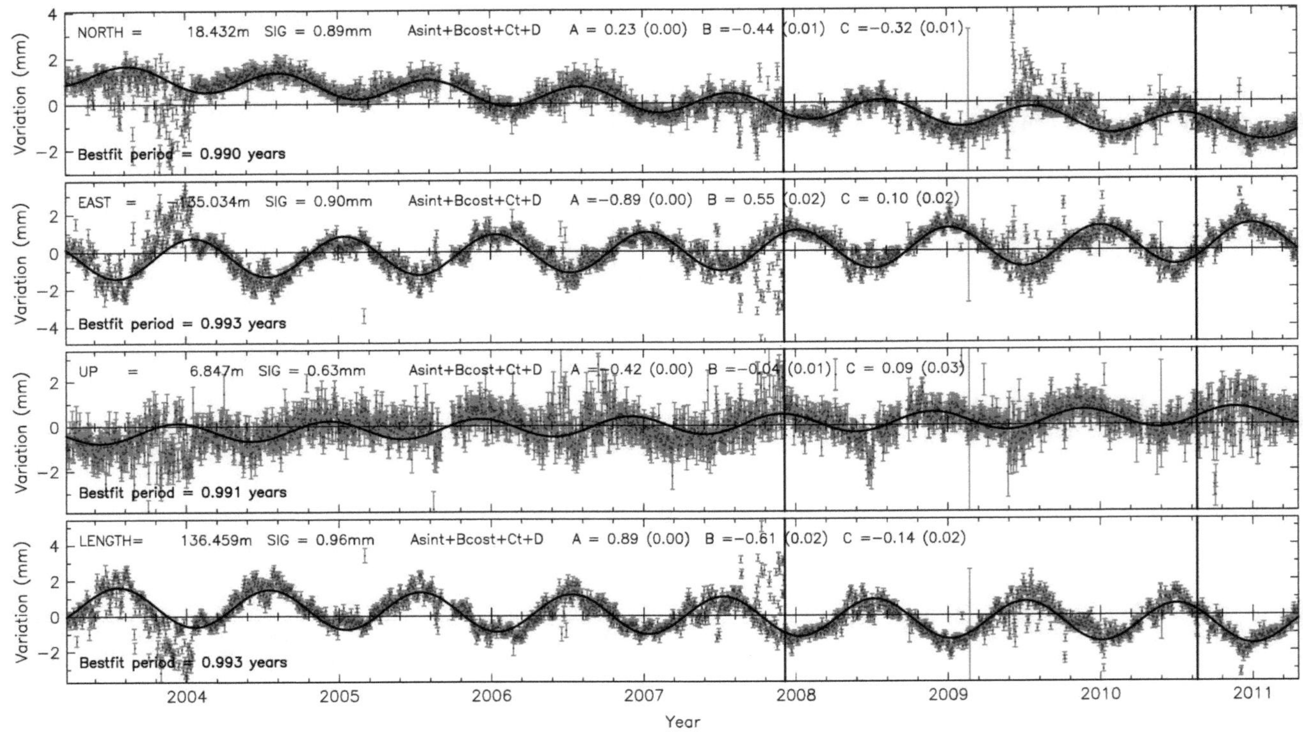

Fig. 16.3 The north, east, up and length components of the HERS-HERT baseline based on GAMIT analysis of the L1 frequency

6 Short Baseline GPS Analysis

6.1 SGF Herstmonceux

A powerful and precise technique to monitor short baselines from analysis of GPS pseudo-range and phase data is in use at the SGF using the GAMIT software, IGS absolute antenna offsets, IERS Earth orientation parameters and IGS orbits.

The HERS-HERT baseline, of magnitude about 136.5 m, is considered along with the other on-site baselines to be "short", with all receivers experiencing very similar atmospheric delays and impact of orbital errors, which largely cancel out in this relative comparison. Ambiguities were fixed in the GAMIT processing, and, using only a single GPS frequency, L1 or L2, the baselines can be calculated with a precision of better than 1 mm. A daily HERS-HERT baseline solution is produced soon after the RINEX data, IGS orbits and IERS EOPs become available, thus providing a timely and powerful data quality check.

Figures 16.3 and 16.4 contain the HERS-HERT baseline components north, east, up and length calculated using the L1 and L2 frequencies respectively. The baseline components contain near-annual signals, with amplitude of approximately 1 mm in the baseline length from the L1 solution. Small discontinuities in the baseline time series due to the upgrade of the HERT site in December 2008 and of the HERS site in August 2010 were solved-for simultaneously along with constant, linear and periodic components through least-squares fits to the data; the epochs are indicated on the plots by vertical lines. There is good agreement between the periods determined from the L1 and L2 data sets. Significant differences are, however, apparent in the results, including the baseline-length linear terms which are -0.14 ± 0.02 mm/year for the L1 dataset and -0.29 ± 0.02 mm/year for L2.

A physical variation in the HERS-HERT baseline due to movement in one, or both, of the sites would be significant for the output of the SGF. The HERS vector to the SLR reference point (HERL) is the station's primary site tie and a physical movement of the HERS monument would impact on the accuracy to which this tie can be defined. Using the coefficient of expansion for steel as $12 \times 10^{-6}\,K^{-1}$, the 7 m HERS tower would expand vertically by 1.68 mm over a 20 °C temperature range, which is of the order of the observed annual variation. Thermal expansion was considered in detail by King and Williams (2009) and identified as a factor, but probably not the only one to contribute to an annual signal, and is unlikely to explain horizontal (E, N) motion. Additionally, relative movement of the HERT site would imply a movement of the monument that holds the primary SLR calibration target. Any such changes in this surveyed "known" distance would directly appear as a variable bias in laser range measurements.

To assess the potential for multipath effects being another possible cause of the observed annual variations, the

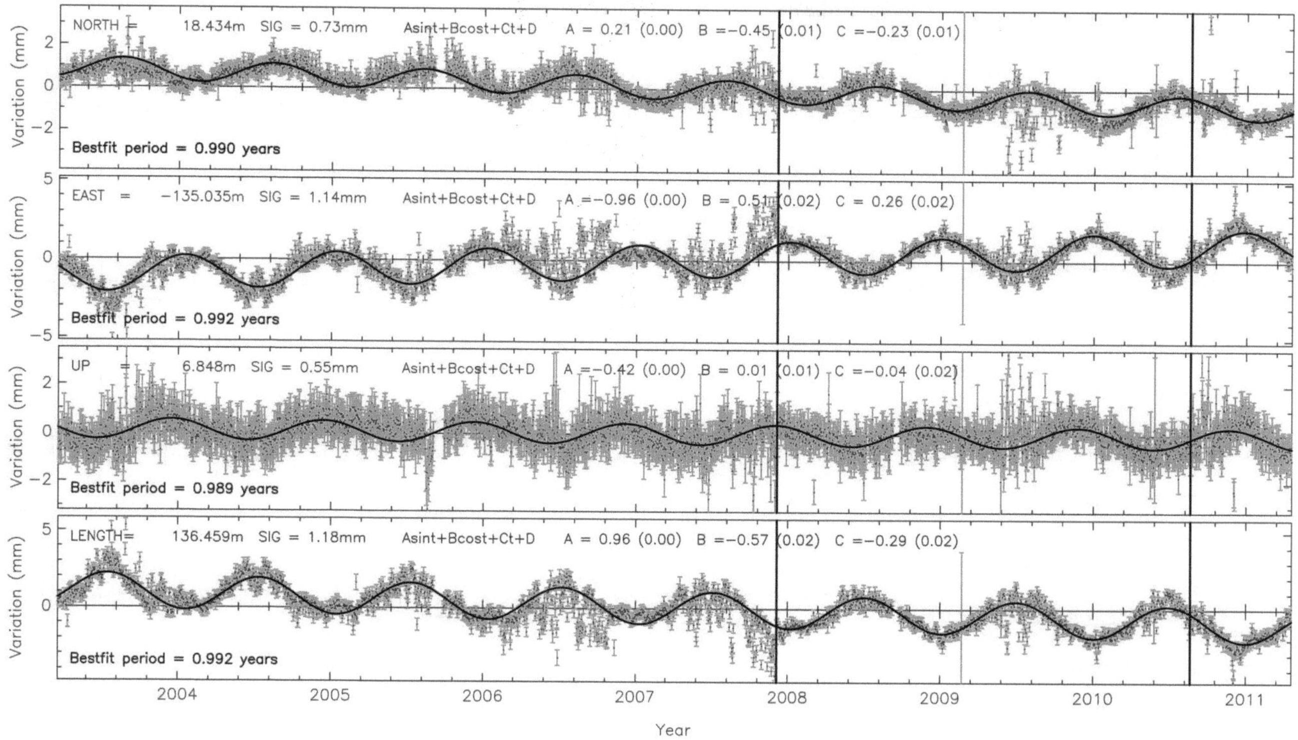

Fig. 16.4 The north, east, up and length components of the HERS-HERT baseline, using only the L2 frequency

HERS-HERT L1 baseline was reprocessed using elevation cut-off values of 0°, 10°, 20°, 30° and 40°, an approach similar to that of King and Williams (2009) and Hill et al. (2009). HERS was set to use an elevation cut-off of 4° with the ASHTECH Z-XII3 receiver and of 0° with the Septentrio PolaRx3. HERT has always tracked only above 5° elevation. As expected, the precision of the baseline components reduced with increasing elevation cut-off, but the annual signal persisted in each component, suggesting that multi-path reflection within the HERS site environment is not causing the signal. This conclusion agrees with Hill et al. (2009) who observed small close-to-annual signals in short baselines in a network of braced GPS monuments in Yucca Mountain, Nevada. The authors suggested that thermal effects, including bedrock expansion, were a more likely cause, rather than the multipath environment. In the paper by King and Williams (2009), the authors analyse sub-daily signals to avoid any propagation into long period signals, yet they still see similar near-annual signals in series of their daily solutions.

To investigate further the variations revealed in the HERS-HERT GPS baselines, a third GPS site, locally-named SOLA, was installed in October 2007 for 2 years on an OS pillar situated between the two sites and re-established in December 2010. If the annual signal in the HERS-HERT baseline is due to a physical movement of one site only then the introduction of a third site, and consequently two more baselines, should identify which site is responsible. In addition to this, from April 2009, data from the OS HERO site has also been analysed. Signals, possibly annual, are present in each of the four baselines, HERS-SOLA, HERT-SOLA, HERS-HERO and HERT-HERO, Figs. 16.5 and 16.6 showing the baseline components for HERS-SOLA and HERT-HERO. Having more than two sites at the SGF for baseline analysis is of course an advantage over other studies, e.g., that of King and Williams (2009), and as more data is accumulated the characterisation of the signals will be strengthened. However, at present these time series do suggest that one site alone cannot be identified as solely responsible for the periodic variations.

6.2 Further Afield

It is also of value to look at some baselines to sites outside the immediate SGF site. The OS runs a network of GNSS sites across the UK and the data is managed, archived and distributed by the British Isles GNSS Facility (BIGF, www.bigf.ac.uk). The nearest sites to the SGF are EAST (10.22 km away), WEIR (34.77 km) and DUNG (45.52 km). Baselines from HERS and HERT to these nearby locations were calculated using L1 and L2 separately, but the results were noticeably noisier. Therefore, for these longer baselines, both frequencies were used in an ionosphere-free combination, LC. Figure 16.7 shows the derived components for the HERS-EAST baseline. Periodic signals are visible in the north (amplitude 0.90 mm), east (0.68 mm) and length

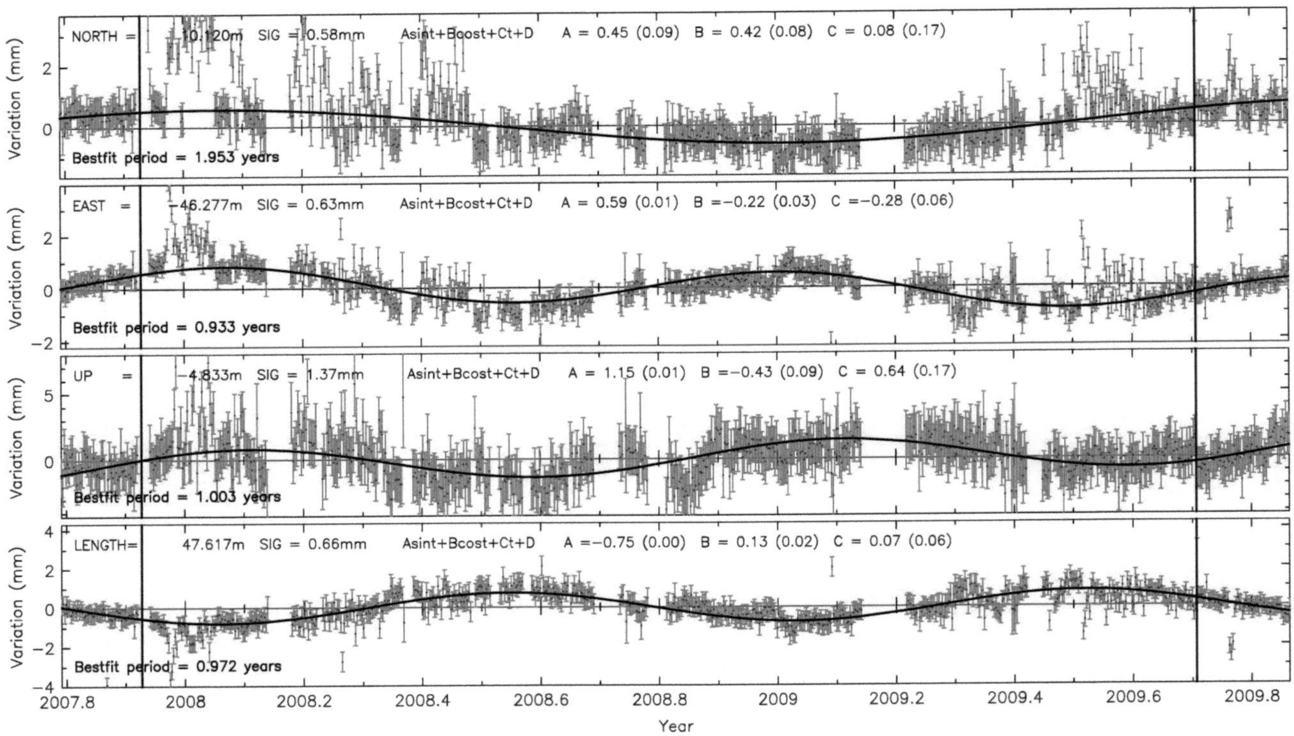

Fig. 16.5 The HERS-SOLA baseline. The amplitudes of the near-annual signals are poorly determined

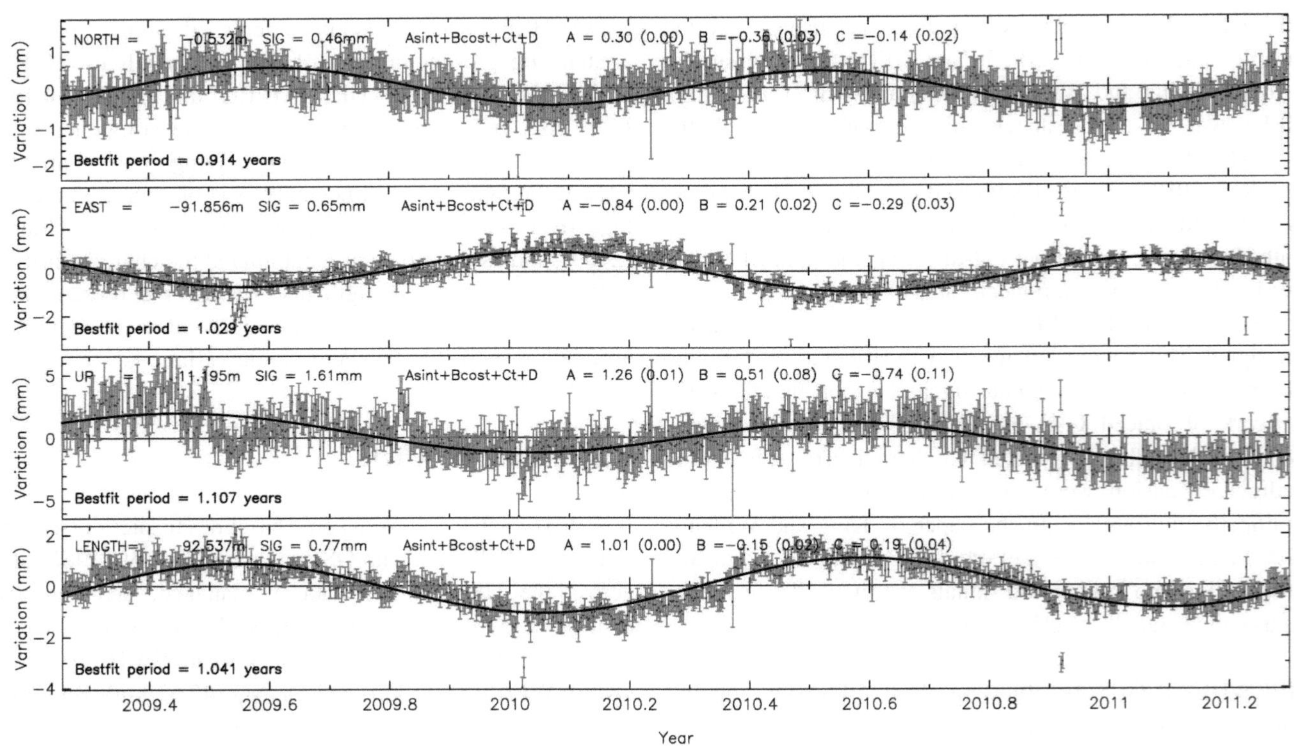

Fig. 16.6 The HERT-HERO baseline. A close-to-annual signal is visible in each component

(0.85 mm) components. The up time-series is more scattered, but a signal of amplitude 0.78 mm is determined. The HERT-EAST baseline shows similar small variations to those in the HERS-EAST baseline. Higher-amplitude signals are present in the components of HERS and HERT to the WEIR site. The most distant baselines analysed, to DUNG, show very similar characteristics to those of the close sites. Another close pair of OS sites, distant from the SGF, is POOL

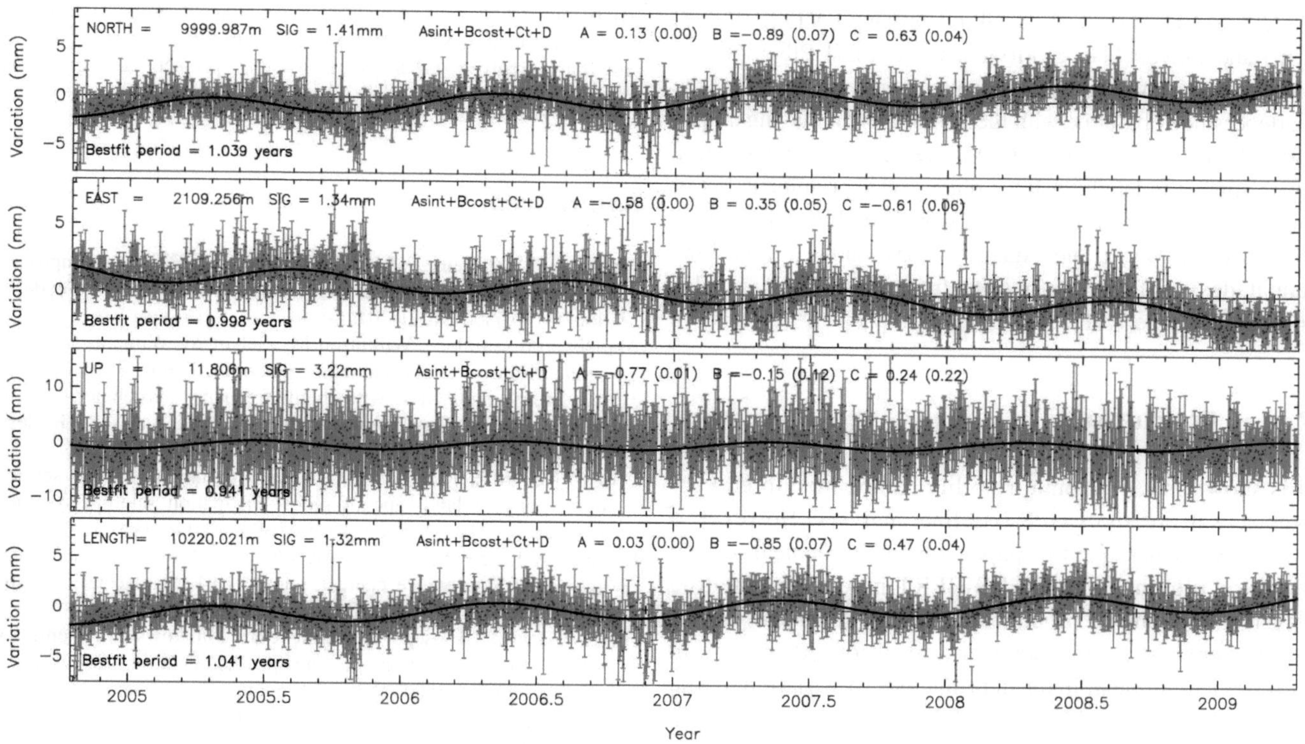

Fig. 16.7 The HERS-EAST baseline. An annual signal is still present for this longer baseline, using the LC frequency combination

Table 16.1 Summary of short-baseline results

Baseline	Length Distance	Period (years)	Slope (mm/year)	Amplitude (mm) North	East	Up	Length
HERS-HERT (L1)	136.5 m	0.993	−0.14	0.49	1.05	0.42	1.08
HER-HERT (L2)	136.5 m	0.992	−0.29	0.50	1.09	0.42	1.11
HERS-SOLA	47.6 m	0.972	0.07	0.62	0.63	1.23	0.76
HERT-SOLA	89.9 m	1.643	1.39	0.54	1.33	1.16	1.32
HERS-HERO	47.4 m	0.920	−0.40	0.42	0.41	1.86	0.43
HERT-HERO	92.5 m	1.041	0.19	0.47	0.87	1.36	1.02
HERS-EAST	10.2 km	1.041	0.47	0.90	0.68	0.78	0.85
HERT-EAST	10.2 km	1.034	−0.19	1.06	0.66	0.68	0.95
HERS-WEIR	34.8 km	0.945	−0.52	1.05	1.63	2.00	0.37
HERT-WEIR	34.7 km	1.020	0.03	1.61	0.58	1.84	0.81
HERS-DUNG	45.5 km	0.944	0.67	0.60	0.59	0.66	0.54
HERT-DUNG	45.7 km	1.036	0.61	0.17	0.63	0.52	0.62
HURN-POOL	5.3 km	0.947	−0.09	0.78	0.52	0.49	0.49
HARB-HRAO	2.1 km	1.004	0.27	0.37	0.45	6.91	0.86
MAT1-MATE	10.7	1.001	0.14	0.60	0.46	0.19	0.77
WTZR-WTZZ	1.6	0.999	0.08	0.21	0.08	0.11	0.22
PIN1-PIN2	50.4	0.99	−0.01	0.09	0.09	0.08	0.09

The quoted period refers to that of the signal in the values of baseline length

and HURN in Dorset, which are 5.27 km apart. The baseline between them likewise exhibits clear signals in some components, but not all. Table 16.1 summarises the results of the baseline solutions, including the amplitude and period of the near-annual terms as well as the linear terms.

6.3 International Short Baselines

Further analysis was undertaken on other short baselines at international sites in the IGS archive, namely at Hartebeesthoek, South Africa (HARB-HARO, 2,066 m), Matera, Italy

(MAT1-MATE, 11 m), Wettzell in Germany (WTZR-WTZZ, 2 m) and Pinyon Flat, California, USA (PIN1-PIN2, 50 m). Very strong near-annual signals were present in the Hartebeesthoek baseline with the up-component having amplitude 6.9 mm. The signals in the Matera and Wettzell baselines, much smaller in length than the others, were prominent, but of small amplitude. Similarly the Pinyon Flat short baseline, as discussed by Wyatt and Agnew (2005), showed very low-amplitude variation. The results from these sites are also included in Table 16.1.

7 Conclusions

It is shown that horizontal stability can be monitored very precisely using short GPS baseline analyses. However, a difficulty discussed in this paper is whether the near-annual variation present in the HERS-HERT baseline is due to a physical movement of one or both of the sites, due to another environmental factor or a result of either noise in the GPS technique or in the data analysis. A variation in true distance between the HERS and HERT sites at the level of a mm or more would be of great importance in particular for the interpretation of the SLR data. However, the appearance of similar variations in the SOLA and HERO baselines, and also in other short baselines both in the UK and internationally, suggests that the signals may originate within the GPS technique itself. The results to date of a new inter-technique digital levelling programme appear promising, with the potential to monitor vertical site motion at sub-mm levels. The site tie measurement between HERS and HERL made by IGN in 2008 may indicate a significant change in this vector. However, the good numerical agreement with that reported as a by-product of the ITRF2008 combination process, suggests that the IGN measurement represents an improved value compared to the previous OS one, rather than a real change in this important vector.

Acknowledgements The services of the Natural Environment Research Council (NERC) British Isles continuous GNSS Facility (BIGF, www.bigf.ac.uk), the OS (www.ordnancesurvey.co.uk), and the International GNSS Service (IGS, http://igscb.jpl.nasa.gov), in providing archived GNSS data to this project, are gratefully acknowledged.

References

Altamimi Z, Collilieux X, Métivier L (2011) ITRF2008: an improved solution of the international terrestrial reference frame. J Geodesy. doi:10.1007/s00190-011-0444-4

Appleby GM, Smith V, Wilkinson M, Ziebart M, Williams S (2010) Comparison of height anomalies determined from SLR, absolute gravimetry and GPS with high frequency borehole data at Herstmonceux. Grav Geoid Earth Observ Int Assoc Geodesy Symp 135(1):107–113. doi:10.1007/978-3-642-10634-7_15

Dow J, Neilan R, Rizos C (2009) The international GNSS service in a changing landscape of global navigation satellite systems. J Geodesy 83:191–198. doi:10.1007/s00190-008-0300-3

Herring TA, King RW, McClusky SC (2010) GAMIT reference manual and GLOBK reference manual, release 10.4. Massachusetts Institute of Technology, Cambridge

Hill EM, Davis JL, Elosegui P, Wernicke BP, Malikowski E, Niemi NA (2009) Characterization of site-specific GPS errors using a short-baseline network of braced monuments at Yucca Mountain, southern Nevada. J Geophys Res 114:B11402. doi:10.1029/2008JB006027

King MA, Williams SD (2009) Apparent stability of GPS monumentation from short-baseline time series. J Geophys Res 114:B10403. doi:10.1029/2009JB006319

Pearlman MR, Degnan JJ, Bosworth JM (2002) The international laser ranging service. Adv Space Res 30(2):135–143. doi:10.1016/S0273-1177(02)00277-6

Ray J, Altamimi Z (2005) Evaluation of co-location ties relating the VLBI and GPS reference frames. J Geodesy 79:189–195. doi:10.1007/s00190-005-0456-z

Schlüter W, Behrend D (2007) The international VLBI service for geodesy and astrometry (IVS): current capabilities and future prospects. J Geodesy 81(6–8):379–387. doi:10.1007/s00190-006-0131-z

Willis P, Fagard H, Ferrage P, Lemoine FG, Noll CE, Noomen R, Otten M, Ries JC, Rothacher M, Soudarin L, Tavernier G, Valette J-J (2010) The international DORIS service (IDS): toward maturity. DORIS Sci Appl Geodesy Geodyn Adv Space Res 45 (12):1408–1420. doi:10.1016/j.asr.2009.11.018

Wyatt FK, Agnew DC (2005) The PIN1 and PIN2 GPS sites at Pinon flat observatory. UC San Diego: Scripps Institution of Oceanography Retrieved from: http://escholarship.org/uc/item/3z16h0zm

Part II

Strengths, Weaknesses, Modelling Standards and Processing Strategies of Space Geodetic Techniques

A Priori Gradients in the Analysis of Space Geodetic Observations

17

J. Böhm, L. Urquhart, P. Steigenberger, R. Heinkelmann, V. Nafisi, and H. Schuh

Abstract

We introduce a static a priori gradient model (APG) based on a spherical harmonic expansion up to degree and order nine to describe the azimuthal asymmetry of tropospheric delays. APG is determined from climatology data of the European Centre for Medium-Range Weather Forecasts (ECMWF), and the refined model can be used in the analysis of observations from Global Navigation Satellite Systems (GNSS) and Very Long Baseline Interferometry (VLBI). Comparisons reveal that gradients estimated in GNSS analysis are mostly smaller than those provided by APG. This difference is also confirmed by station and source coordinate changes if APG is used in GNSS and VLBI analysis.

Keywords

Tropospheric delay • Tropospheric gradients • VLBI • GNSS • APG

1 Introduction

Azimuthal asymmetry of troposphere delays has to be taken into account in the analysis of space geodetic observations, e.g. from Global Navigation Satellite Systems (GNSS) or Very Long Baseline Interferometry (VLBI). This asymmetry can be due to systematic effects, e.g., a north–south oriented gradient (see Fig. 17.1) is due to the larger extension of the troposphere above the equator, or due to random effects, like any change of weather patterns. Typically, north and east gradients, G_n and G_e (Eq. 17.1), are estimated with a resolution of 2–24 h to account for this asymmetry. Based on Davis et al. (1993) who set up a gradient model for the wet refractivity, MacMillan (1995) proposed to use

$$\Delta L(a,e) = \Delta L_0(e) + \\ + mf(e) \cdot \cot(e) \cdot [G_n \cos(a) + G_e \sin(a)] \quad (17.1)$$

to describe the troposphere delay $\Delta L(a,e)$ at azimuth a and elevation e, where ΔL_0 denotes the symmetric part of the delay (Petit and Luzum 2010), and *mf* the mapping function. Usually, the hydrostatic mapping function is applied in the analysis of space geodetic observations. Chen and Herring (1997) proposed the model

$$\Delta L(a,e) = \Delta L_0(e) + \\ + \frac{1}{\sin(e)\tan(e) + C} \cdot [G_n \cos(a) + G_e \sin(a)] \quad (17.2)$$

J. Böhm • H. Schuh (✉)
Vienna University of Technology, Gußhausstraße 27-29, 1040 Vienna, Austria
e-mail: johannes.boehm@tuwien.ac.at; harald.schuh@tuwien.ac.at

L. Urquhart
University of New Brunswick, Fredericton, Canada

P. Steigenberger
Technische Universität München, Munich, Germany

R. Heinkelmann
Deutsches Geodätisches Forschungsinstitut, Munich, Germany

V. Nafisi
Vienna University of Technology, Gußhausstraße 27-29, 1040 Vienna, Austria

University of Tehran, Tehran, Iran

University of Isfahan, Isfahan, Iran

Z. Altamimi and X. Collilieux (eds.), *Reference Frames for Applications in Geosciences*, International Association of Geodesy Symposia 138, DOI 10.1007/978-3-642-32998-2_17,

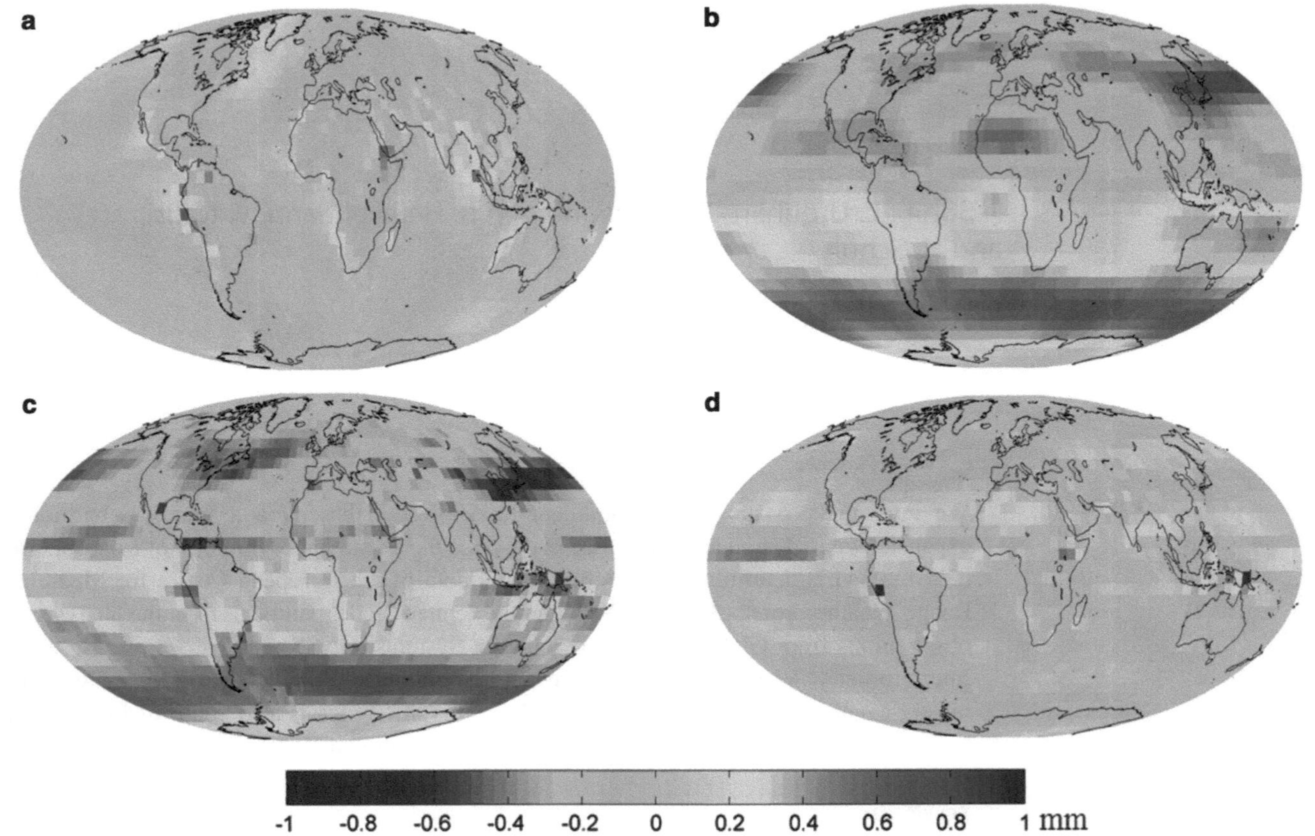

Fig. 17.1 Upper left (**a**): 5°× 5° east–west gradients. Lower left (**b**): 5°× 5° north–south gradients. Upper right (**c**): Spherical harmonic expansion up to degree and order 9 of north–south gradients (APG). Lower right (**d**): Residuals of north–south gradients, i.e. (**b**)−(**c**)

and they recommended to use $C = 0.0032$ if the total (hydrostatic plus wet) gradients are to be estimated. Based on scale heights of 13 km and 3 km for the hydrostatic and wet parts of the atmosphere, Chen and Herring (1997) determined the coefficients $C = 0.0031$ and $C = 0.0007$ if used for the hydrostatic and wet parts, respectively. Comparisons show that Eq. 17.1 by MacMillan (1995) agrees better with Eq. 17.2 if the 'wet' coefficient $C = 0.0007$ is applied. Since we are concerned here mainly with static gradients which we assume to be mostly of hydrostatic nature, and since $\cot(e)$ is singular for observations at the horizon, we decided to use Eq. 17.2 by Chen and Herring (1997) with C = 0.0032 to model gradients.

In GNSS and VLBI analysis, gradients are typically estimated without using a priori information for the azimuthal asymmetry of the delays. However, a priori gradients plus constraints (pseudo-observations loosely constraining the estimated gradients to the a priori gradients or their relative variation) are applied for VLBI observations of the early days of VLBI history (beginning of the 1980s) when only few observations to southern sources could be carried out. MacMillan and Ma (1997) describe a systematic influence of (constrained) a priori gradients on the terrestrial and celestial reference frames, in particular on the sources at southern declinations. Their list of constant a priori gradients derived from weather models of the Data Assimilation Office (DAO) at Goddard Space Flight Center (Schubert et al. 1993) is widely used by VLBI analysis centers.

At http://ggosatm.hg.tuwien.ac.at/ time series of north–south and east–west gradients with a time resolution of 6 h as derived from operational data of the European Centre for Medium-Range Weather Forecasts (ECMWF) are provided, and Böhm and Schuh (2007) showed that fixing those gradients improves baseline length repeatabilities compared to fixing zero gradients, i.e. compared to not estimating residual gradients. However, they also showed that the estimation of gradients always improves baseline length repeatabilities compared to fixing a priori gradients. It should be mentioned here that both types, constant gradients at VLBI stations by MacMillan and Ma (1997) and 6 h gradients at space geodetic sites by Böhm and Schuh (2007), are derived from integration over horizontal gradients of refractivity along the site vertical and not by rigorous 3D ray-tracing, as e.g. described by Hobiger et al. (2008).

In Sect. 2 we describe the determination of a static a priori gradient model based on a spherical harmonic expansion up to degree and order 9, and its application in the analysis of

VLBI and GNSS observations is discussed in Sects. 3 and 4. However, there is one important remark that we would like to make right at the beginning: there is no influence on station or source coordinates by a priori gradients if gradients are estimated without constraints and the same gradient mapping function is used for a priori and estimated gradients.

2 A Priori Gradient Model (APG)

We started with climatology data from the 40 years ECMWF Re-analysis (ERA40), i.e., we had 12 pressure level datasets (with 23 vertical levels for each month) with a horizontal resolution of 1°. These monthly datasets are averages over the respective months of all years contained in ERA40 (1957–2001). For all 12 months and at grid points with a horizontal resolution of 5° we determined the total troposphere delays at 5° outgoing elevation angle towards north, east, south, and west, as well as in zenith direction by ray-tracing. The average over the four azimuths per grid point together with the delay in zenith direction was then used to remove the symmetric part ΔL_0 in Eq. 17.2, and the north–south and east–west gradients were determined at every grid point and for every month applying the coefficient $C = 0.0032$. A time varying model was first considered but no evident time varying component was found. Therefore a static field was derived by averaging the north–south and east–west gradients over all 12 months.

Figure 17.1a shows the mean east–west gradients, and it is evident that the gradients are rather small. However, we can see pronounced east–west gradients along north–south oriented coastlines, e.g. at the west coast of South America. Figure 17.1b shows the mean north–south gradients and there is the prominent feature of negative north–south gradients in the northern hemisphere and positive north–south gradients in the southern hemisphere, which is due to the higher extension of the troposphere above the equator. Next, we expanded the gradients into a spherical harmonic expansion up to degree and order 9 by a least-squares adjustment, and exemplarily Fig. 17.1c shows this expansion for the north–south gradients. Figure 17.1d illustrates the residuals of the least-squares adjustment, i.e., the difference between Fig. 17.1b, c. Some unexpected features of the original grid are revealed in Fig. 17.1d which might be due to computational artifacts from the ray-tracing. However, the spherical harmonic expansion – due to its limited resolution – does not include those isolated features or outliers and thus can safely be used for the investigations described below. The spherical harmonic expansion called APG is available at http://ggosatm.hg.tuwien.ac.at/DELAY/SOURCE/.

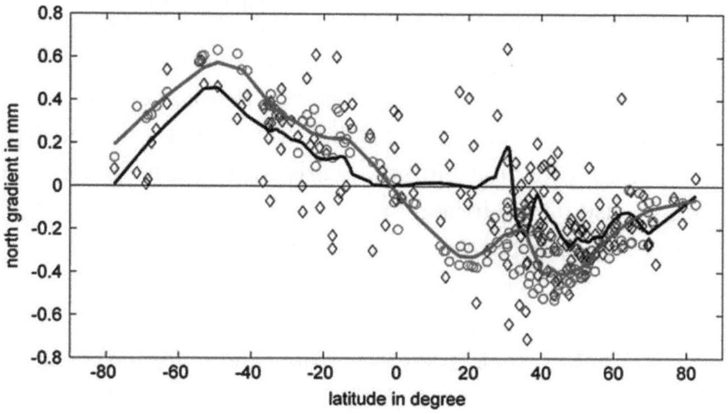

Fig. 17.2 GPS estimates of north–south gradients (*black diamonds*) plotted versus latitude in degrees, and APG (*red circles*) evaluated for the same GPS sites. The bold lines depict the respective smoothed values over 20° latitude

3 GPS Analysis

We analyzed a global GPS network of stations for 2007 and 2008 with the Bernese software package (Dach et al. 2007). Orbits, Earth orientation parameters, and station coordinates were estimated consistently. The cutoff elevation angle was set to 3°, and low elevation observations were down-weighted with $\cos^2 z$ where z is the zenith distance. In terms of troposphere delay modeling, the Vienna Mapping Functions 1 (Böhm et al. 2006) were used together with a priori hydrostatic zenith delays from data of the ECMWF to determine slant hydrostatic delays, and to estimate wet zenith delays every 2 h. If estimated, gradients were set up as 24 h piecewise linear functions and no constraints were applied.

In a first test we estimated the gradients applying the model by Chen and Herring (1997) with the coefficient $C = 0.0032$ without a priori gradients, and we compared the average of those estimates within the 2 years to the gradients that we determined with APG for the GPS stations (see Fig. 17.2). The scattering of gradients estimated with GPS is larger than the scattering of APG, and the GPS gradients are closer to zero at mid-northern latitudes. Figure 17.2 suggests that APG might be too large (in absolute sense) in those areas.

Next, we compared GPS station coordinate differences, again averaged for the two processed years, between two solutions estimated with different approaches. We concentrated here on the northern and radial difference in station coordinates, because the influence on east–west coordinates is significantly smaller (not shown). Figure 17.3a (upper plots) shows coordinate differences between solutions (1) without estimation of gradients and (2) with the estimation of gradients using the model by Chen and Herring (1997) with $C = 0.0032$. In both solutions, no a priori gradients were used. Figure 17.3a confirms that differences in north components are slightly negative in the southern hemisphere and slightly positive in the northern hemisphere as to be expected. The scatter in the radial component is larger than in northern direction but not clearly systematic. In the next step we used APG instead of zero gradients for solution (1), and we were expecting the systematic effects to decrease. However, as can be seen in Fig. 17.3b, the differences in the northern components were flipped with respect to zero, which implies that gradients from APG are larger than gradients estimated with GPS (see also Fig. 17.2). This kind of over-correction is even more obvious for the height differences where the radial components of stations close to the equator are increased by 5 mm if APG is fixed compared to estimating gradients.

We also investigated the impact of the gradient model on the estimation of gradients, i.e., we compared the model by MacMillan (1995) to the model by Chen and Herring (1997). In summary, the effect on station coordinates is systematic but generally smaller than 0.5 mm.

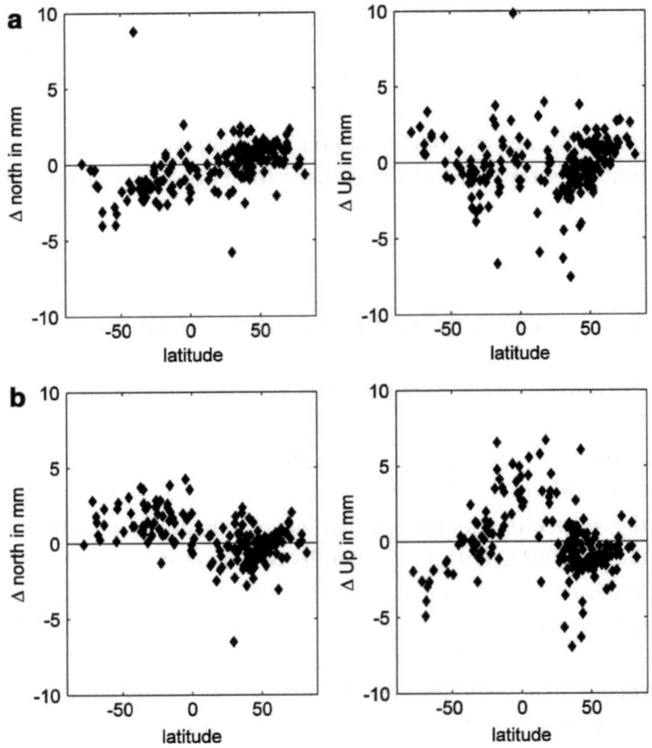

Fig. 17.3 Upper plots (**a**): Differences in north and radial station components between solutions (1) zero a priori gradients, no estimation of gradients, and (2) zero a priori gradients, estimation with model by Chen and Herring (1997) with C = 0.0032. Lower plots (**b**): Differences in north and radial station components between solutions (1) APG, no estimation of gradients, and (2) zero a priori gradients, estimation with model by Chen and Herring (1997) with C = 0.0032

4 VLBI Analysis

We also determined VLBI solutions of 24 h VLBI sessions with the software package Occam (Titov et al. 2004) for the years 1984–1990 with a focus on estimated declinations of source coordinates. We applied loose constraints (0.5 mm and 2 mm/sqrt(h)) on gradient estimates and took a closer look at declination changes when using different strategies for the gradients. In Fig. 17.4, red circles show the differences between solutions (1) zero a priori gradients, no estimation of gradients, and (2) zero a priori gradients, estimation with the model by MacMillan (1995). As expected there is a systematic effect on source declinations with increasing southern declinations. In the next step, we replaced the zero gradients with APG in solution (1) (black diamonds in Fig. 17.4), and – similar to the GPS results in Sect. 3 – there is an apparent over-correction with APG, i.e., estimated gradients with VLBI are generally smaller than gradients from APG. Finally, as illustrated with green crosses, we estimated gradients in solution (1) with the model by MacMillan (1995). This relieves the over-correction a bit, but the differences are not exactly zero for two reasons: First, APG uses the model by Chen and Herring (1997), and second, loose relative constraints were applied.

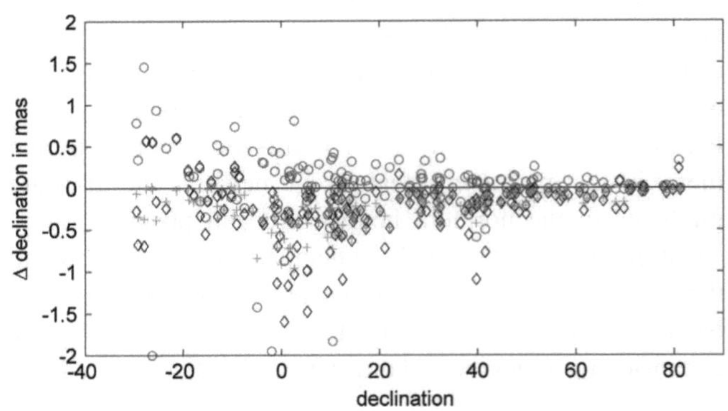

Fig. 17.4 Declination changes illustrate the difference between two solutions. Red circles: (1) Zero a priori gradients, no estimation of gradients minus (2) zero a priori gradients, estimation of gradients with the model by MacMillan (1995). Black diamonds: (1) APG, no estimation of gradients minus (2) zero a priori gradients, estimation of gradients with the model by MacMillan (1995). Green crosses: (1) APG, estimation of gradients with model by MacMillan (1995) minus (2) zero a priori gradients, estimation of gradients with the model by MacMillan (1995)

5 Discussion and Outlook

The comparison of gradients and the analysis of space geodetic observations have revealed that the gradients derived from ERA40 data of the ECMWF are larger than gradients estimated in GNSS and VLBI analysis. To resolve this discrepancy, more investigations need to be carried out. Although the model by Chen and Herring (1997) is conceptually correct for a synthetic distribution of refractivity, tests have to be made to investigate possible extensions to that model depending on climatological parameters like the spatial variation of hydrostatic and wet scale heights. In terms of GNSS and VLBI observations, the impact of GNSS antenna phase center models, GNSS multipath and the observation distribution on the sky need to be assessed carefully. Finally, an independent gradient model based on a different weather model, using a different ray-tracing algorithm and possibly ray-tracing at different elevation angles should be developed to confirm the model introduced here in Sect. 2.

Acknowledgements J. Böhm and V. Nafisi are grateful to the Austrian Science Fund (FWF) for supporting this research under project P20902-N10 (GGOS Atmosphere). V. Nafisi would also like to acknowledge Iran Ministry of Science, Research and Technology (MSRT) for funding part of his research at the Vienna University of Technology. L. Urquhart would like to thank the Natural Sciences and Engineering Research Council (NSERC) of Canada for providing funding. We like to acknowledge the International VLBI Service for Geodesy and Astrometry (IVS; Schlüter and Behrend 2007) and the International GNSS Service (IGS; Dow et al. 2009) for providing their data and products.

References

Böhm J, Schuh H (2007) Troposphere gradients from the ECMWF in VLBI analysis. J Geodesy 81(6–8):403–408. doi:10.1007/s00190-007-0144-2

Böhm J, Werl B, Schuh H (2006) Troposphere mapping functions for GPS and very long baseline interferometry from European Centre for Medium-Range Weather Forecasts operational analysis data. J Geophys Res 111:B02406

Chen G, Herring TA (1997) Effects of atmospheric azimuthal asymmetry on the analysis of space geodetic data. J Geophys Res 102 (B9):20489–20502. doi:10.1029/97JB01739

Dach R, Hugentobler U, Fridez P, Meindl M (eds) (2007) Bernese GPS software version 5.0. Astronomical Institute, University of Bern, Berne

Davis JL, Elgered G, Niell AE, Kuehn CE (1993) Ground-based measurement of gradients in the "wet" radio refractivity of air. Radio Sci 28(6):1003–1018. doi:10.1029/93RS01917

Dow JM, Neilan RE, Rizos C (2009) The international GNSS service in a changing landscape of global navigation satellite systems. J Geodesy 83(3–4):191–198. doi:10.1007/s00190-008-0300-3

Hobiger T, Ichikawa R, Kondo T, Koyama Y (2008) Fast and accurate ray-tracing algorithms for real-time space geodetic applications using numerical weather models. J Geophys Res 113:D20302. doi:10.1029/2008JD010503

MacMillan DS (1995) Atmospheric gradients from very long baseline interferometry observations. Geoph Res Lett 22(9):1041–1044. doi:10.1029/95GL00887

MacMillan DS, Ma C (1997) Atmospheric gradients and the VLBI terrestrial and celestial reference frames. Geoph Res Lett 24 (4):453–456. doi:10.1029/97GL00143

Petit G, Luzum B (eds) (2010) IERS Conventions (2010) IERS Technical Note 36, Frankfurt am Main: Verlag des Bundesamts für Kartographie und Geodäsie

Schlüter W, Behrend D (2007) The international VLBI service for Geodesy and Astrometry (IVS): current capabilities and future prospects. J Geodesy 81(6–8):379–387. doi:10.1007/s00190-006-0131-z

Schubert SD, Rood R, Pfaendtner J (1993) An assimilated data set for Earth science applications. Bull Am Meteorol Soc 74 (12):2331–2342

Titov O, Tesmer V, Böhm J (2004) OCCAM v. 6.0 software for VLBI data analysis. In: Vandenberg NR, Baver KD (eds) International VLBI service for Geodesy and Astrometry 2004 general meeting proceedings. NASA/CP-2004-212255, pp 267–271

Why Combining at the Observation Level?

18

D. Gambis, J.Y. Richard, R. Biancale, and C. Bizouard

Abstract

Space geodetic techniques have different strengths and weaknesses for recovering geodetic parameters which makes their combination useful. However they may have some systematic behaviour which can be detected and removed at the observation level. In order to review the interest in combining techniques at this level, a Working Group was set up in the course of 2009 in the frame of the International Earth Rotation and Reference Systems Service (IERS). A major task of the WG COL is to study methods and advantages of combining space geodetic techniques (DORIS, GNSS, SLR, VLBI), searching for an optimal strategy to solve for geodetic parameters. The first action of the Working Group was to organize an inter-comparison benchmark campaign to serve as a test. The period chosen is from August 10 to August 30, 2008. It includes the intensive CONT08 VLBI period.

The combination analyses are based on weekly (or daily for VLBI) combined SINEX files which contain normal equations of station coordinates, Earth Orientation Parameters from all space geodetic techniques, quasar coordinates for the VLBI technique and troposphere parameters for all techniques except SLR. The objectives of the present paper are twofold: first give an overview of the method, present the objectives and strategy of the newly born working group COL; second to present some preliminary results obtained by the Groupe de Recherche en Géodésie Spatiale (GRGS) for Earth orientation parameters.

Keywords

Geodetic space techniques • Combination at the observation level

1 IERS Working Group on Combination at the Observation Level (COL)

The method we use here is based on the combination of space geodetic techniques at the observation level that is being studied for several years (Yaya 2002; Coulot et al. 2007; Ray et al. 2005; Kudryashova et al. 2008; Gambis et al. 2009). Data from the different techniques are, in a first step, processed with different software packages GINS used by the GRGS (Chassaing and Roumiguier 1965; Bourda et al. 2007; Marty et al. 2011), DOGS (Gerstl et al. 2001) used by DGFI (Deutsches Geodätisches Forschungsinstitut), BERNESE (Dach et al. 2007) used by AIUB/BKG (Astronomical Institute of the University of Bern/Bundesamt für Kartographie und Geodäsie) and NAPEOS (Springer 2009) used by ESOC (European Space Operations Centre). The observation equations allow forming the so called normal equations (Angermann et al. 2004). In a second step normal equations derived from these processing are stacked on a weekly basis. This allows in particular the simultaneous estimation of both a Terrestrial reference frame realized

D. Gambis (✉) • J.Y. Richard • C. Bizouard
Observatoire de Paris, SYRTE/UMR 8630/GRGS, Paris, France
e-mail: daniel.gambis@obspm.fr

R. Biancale
CNES/GRGS, Toulouse, France

Z. Altamimi and X. Collilieux (eds.), *Reference Frames for Applications in Geosciences*,
International Association of Geodesy Symposia 138, DOI 10.1007/978-3-642-32998-2_18,

through station positions (TRF) and the set of Earth Orientation Parameters (EOP) as well as zenithal troposphere corrections. The troposphere gradients were derived by most of the analysis centers. The method seems quite straightforward, however in the individual data processing as well as in the combination process, a lot of difficulties arise, in particular those linked to systematic errors brought by the various techniques.

The main task of this IERS working group is to review the interest in combining techniques at the observation level for EOP, reference frames and troposphere parameters. It brings together groups able to do combinations at the observation level in order to improve homogeneity, precision and resolution of products. According to the WG charter (IERS working group on combination at the observation level), the detailed tasks are:

1. Promoting the approach within groups and the capability of processing jointly two or more techniques.
2. Establishing common processing standards for all techniques in order to guarantee homogeneity and consistency.
3. Studying appropriate weighting between techniques and the interest of using local ties or satellites tracked by several techniques.
4. Optimizing and unifying parameterization for instance for EOP or troposphere parameters in order to minimize globally the degree of freedom of the whole inverse system.
5. Elaborating benchmarks to intercompare results between groups from the same data set.
6. Ensuring the exchange format compatibility between techniques and with the international technique services and IERS.
7. Studying stabilization methods and looking for high temporal resolution of parameters.
8. Evaluating and comparing results to search for compatibility between groups.

In order to facilitate discussions and data exchanges between members an electronic forum (http://grgs.obspm.fr/forum/) and an FTP server for files exchanges ftp://hpiers.obspm.fr/iers/eop/grgs/ were set up at the Paris Observatory.

The main discussion topics concern the models used in the software packages, the exchange format Solution (Software/technique) Independent Exchange Format (SINEX) and the set of parameters estimated. The first action of the working group COL was to organize an intercomparison benchmark campaign to serve as a test. The period chosen is from August 10 to August 30, 2008. It includes the intensive CONT08 VLBI period, August 12–26, 2008. Combined SINEX have been delivered per week or per day for VLBI or GNSS sessions. The different participating groups (AIUB, DGFI, GFZ (Deutsches GeoForschungsZentrum), ESOC, and GRGS) processed the various observations and submitted the results in SINEX format version 2.02 (http://www.iers.org).

Table 18.1 Seven day Jason-2 orbit determination over a 7 day arc interval

Technique	Observations number	Residuals	Orbit WRMS
SLR + DORIS	2,247	4.2 cm	Reference
	109,614	.352 mm/s	
DORIS	109,884	.346 mm/s	10.6 cm
SLR	2,216	4.1 cm	12.1 cm

2 Equivalence of the Combination Between NEQ Level and Observation Level

Is there any difference between these two approaches? In the combination at the observation level the observations of the different techniques are simultaneously processed applying the least squares method while in the combination at the normal equation level the observations are first processed to derive normal equations which are then stacked in a second step. In order to compare these two approaches a series of simulations was performed using the GINS/DYNAMO software packages. SLR and DORIS observations on Jason-2 were processed over a 7-day orbit arc. Pole components were estimated in two ways.

1. By a consistent processing of the SLR and DORIS observation in the same run.
2. By two separate processing of SLR and DORIS observations whose normal equations were stacked afterwards.

In both cases, we used the same modeling, parameterizations and weighting. Discrepancies in the number of observations kept as well as the level of residuals are due to the editing procedure (three-sigma and under 12 deg. elevation) in the adjustment process.

Table 18.1 gives statistics for both cases: number of observations considered and the level of the observation residuals as well as the WRMS in the fitted orbit. The orbit fitted in the common SLR + DORIS processing is fixed in the separate processing of DORIS and SLR observations for the purpose of comparison. The table gives the statistics concerning the fit. The small differences arise from the different number of detected outliers.

We estimated the pole components according to both approaches: (1) Common SLR and DORIS processing in a single run. (2) Separate processing into two separate runs. It appears that the two approaches lead to similar results at the level of a few milliarcseconds. This justifies our assertion that the two approaches, i.e. combination at the observation level and combination at the normal equation level are equivalent.

3 Multi-Technique Processing by GRGS

3.1 Algorithm Description

The data processing for each technique is performed using the "GRGS" software package *GINS* which allows to process observations of all types of geodetic techniques and to generate normal equations for the combination process.

A priori dynamical and geometrical models include common standards as follows:

- EIGEN_GL04S gravity field model (Biancale et al. 2005) up to spherical harmonic degree and order 99 (limited to 12 for GPS satellites much less sensitive to the Earth gravity field) plus drifts and annual and semiannual periodic terms;
- Three body point mass attraction from the Sun, the Moon (with the J2 Earth's indirect effect) and from major planets according to the JPL DE405 ephemeris (Standish 1998);
- IERS 2003 solid Earth and pole tides;
- FES2004 ocean tide model (Lyard et al. 2006) up to degree and order 50 (up to 12 for GPS satellites) with 19 main waves and 67 sideband waves computed according to the admittance principle. The full model is applied for the loading effect;
- Atmospheric variation attraction from 3D-ECMWF pressure field at 6 h intervals up to degree 50, over land and considering the barotropic MOGD2D model over oceans (Carrere and Lyard 2003);
- Atmospheric tides derived from ECMWF model (Biancale and Bode 2002);
- Relativity corrections according to Schwarzschild, Lense-Thirring and geodetic precession models (IERS Conventions 2003);
- Direct solar radiation pressure, terrestrial albedo and the infra red pressure from ECMWF 6 h grids (Brankovic and Van Maanen 1985);
- Drag from the DTM94 atmospheric density model for low satellites (Berger et al. 1998);
- Station coordinates from ITRF 2005 (Altamimi et al. 2007) taking the ocean tide loading into account;
- Celestial frame (quasar coordinates) from ICRF2 (Fey et al. 2009);
- EOP from IERS C04 series (Gambis 2004; Bizouard and Gambis 2009) interpolated at 6 h intervals for pole coordinates and UT1 (to x, y and UT1 we add diurnal and semi-diurnal ocean tidal effects, given by IERS 2003 model) and at 12 h intervals for celestial pole offsets.

In a first step orbits and measurement residuals are computed iteratively. Outliers are rejected during this procedure. Then partial derivatives of EOP and station coordinates are computed together with all other ones (initial state vector, dynamical coefficients, tropospheric zenith delays) and normal equations are generated.

Individual normal equations are then handled by the DYNAMO software modules allowing permutation, reduction, stacking, solving with additional constraints on the selected parameters.

3.2 GRGS Analyses for Intensive CONT08 Campaign

The VLBI, GPS, DORIS and SLR weekly normal equations (NEQ) were generated over the 3 weeks of the intensive CONT08 VLBI campaign (http://ivs.nict.go.jp/mirror/program/cont08/) from August 10–30, 2008. The corresponding parameters to be fitted are:

- EOP pole coordinates, UT1-UTC both at 6 h intervals, and nutation offsets at 12 h intervals;
- Weekly stations coordinates;
- Troposphere zenith bias at 1 h intervals for VLBI, GPS and DORIS.

The GPS normal equations are based upon IGS data using 121 stations of the IGS tracking network (57 stations of them are collocated with DORIS, VLBI and/or SLR). The DORIS normal equations are derived from Doppler measurement done by SPOT-2, 4, 5 and ENVISAT satellites. The SLR normal equations result from the telemetry of the satellites LAGEOS 1 and 2. The VLBI normal equations are formed from VLBI sessions IVS-R1, IVS-R4 and CONT08.

The individual NEQ are first stacked over the 3 weeks for each technique after reduction of troposphere parameters and technique dependent parameters such as clocks and ambiguity parameters for GPS.

3.2.1 Calibration

Formal uncertainties associated with NEQ of one technique may be under or over-estimated. So we have calibrated mean weekly uncertainties of the techniques by applying Helmert's variance component analysis (Sahin et al. 1992). The corresponding relative weighting, and number of observations related to each technique, are given in Table 18.2. We estimated a scaling factor for the different techniques. The quantity dof is the degree of freedom: number of observations minus the number of unknown parameters. The Helmert's variance component analysis leads to the determination of a scaling factor to the quantity square (σ^2/dof) to obtain a variance factor close to the unit. Values listed in Table 18.2 show that these scaling factors are stable over the successive 3 weeks for the four techniques.

Table 18.2 Scaling factors computed by the Helmert's variance component analysis, number of observations and variance factor calculated with the rate of re-scaled weighted square sum of observation minus a-priori and dof (the degree of freedom) which is the number of observations minus the number of unknown parameters

	Week 1			Week 2			Week 3		
Technique	Scaling factor	Nb obs	σ^2/dof	Scaling factor	Nb obs	σ^2/dof	Scaling factor	Nb obs	σ^2/dof
GPS	0.14	806,609	48	0.15	818,851	47	0.15	785,384	45
VLBI	0.03	46,884	1224	0.03	66,610	1497	0.03	29,167	1584
SLR	1.44	3,110	0.5	1.38	2,549	0.56	1.26	2,990	0.7
DORIS	0.92	185,938	1.17	0.92	182,470	1.18	0.92	187,072	1.19

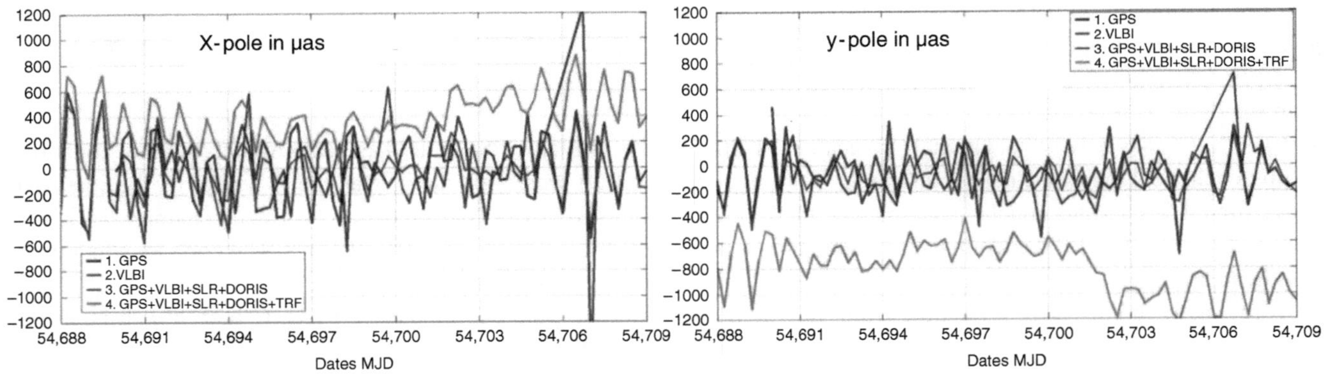

Fig. 18.1 x and y pole corrections at 6 h intervals. GPS (1), VLBI (2), Weighted Combination GPS + VLBI + SLR + DORIS (3). All these computations are performed with a fixed TRF. The last line (4) corresponds to the weighted solution GPS + VLBI + SLR + DORIS with TRF estimated. SLR and DORIS solutions are not displayed because their poor significance and large discrepancies with respect to C04. The significant bias with respect to the IERS C04 is due to some deficiency in the datum realization

Table 18.3 Weighted mean and standard deviation (SD) of 6 h pole corrections versus C04 series in µas

	x-Pole		y-Pole	
Technique	Mean (µas)	SD (µas)	Mean (µas)	SD (µas)
GPS	−10	270	−60	160
VLBI	−20	230	−90	210
DORIS	30	1,930	260	1,700
SLR	−30	1,030	−190	990
Comb	−10	150	−70	120
Comb + TRF	380	200	−790	190

The recalibrated NEQ are finally cumulated into a multi-technique NEQ which is then solved for using the Cholesky inverse method. More details on the GRGS combination multi geodetic techniques approach for EOP and TRF estimation are to be found in Richard et al. (2009).

3.3 Results for EOP

3.3.1 Pole Coordinates

They are obtained according to three different strategies:

- Stacking of the NEQs for each technique
- Weighted combination of the whole set of NEQ with station coordinates held fixed to those of the ITRF 2005
- The same as before with estimates of the station coordinates (with stability constraints). The minimum constraints are applied to the NEQ (the seven Helmert's transformation parameters are estimated). The common reference system is defined through stability constraints of 1 cm imposed on the three coordinates of all stations. Local ties are not applied to the collocation sites.

With a 6 h sampling, these corrections contain diurnal oscillations, as displayed on Fig. 18.1. Their standard deviation and bias are given in Table 18.3.

The combined solution (with or without estimating station coordinates) reproduces the GPS solution with correlation coefficients of 0.8 for x, 0.5 for y, and the VLBI solution with correlation coefficients of 0.5 for x and 0.6 for y. Since SLR and DORIS observations are in a too small amount (SLR) or present a degraded precision (DORIS), solved alone they cannot provide meaningful 6 h pole coordinates, as shown by their non-realistic standard deviation with respect to C04 pole coordinates. But added to VLBI and GPS, they can reinforce the quality of the EOP determination.

When station coordinates are estimated (once per week), appears a significant bias of a few hundred micro-arc-seconds, revealing some deficiency in the estimated terrestrial reference frame and the ITRF 2005 due to the fact that local ties were not applied in the process.

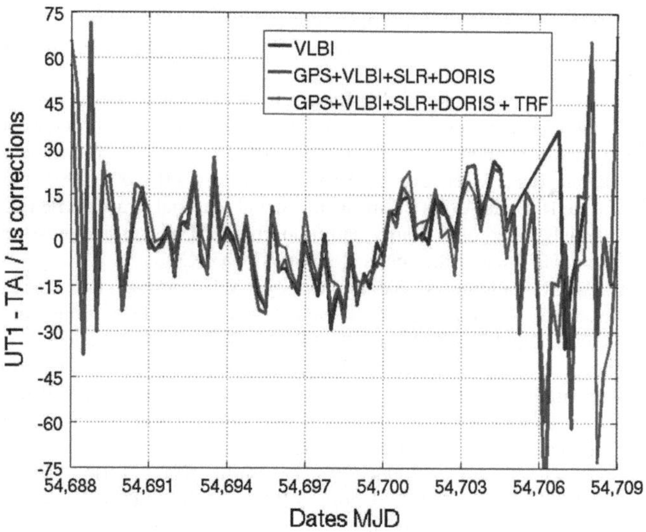

Fig. 18.2 UT1-TAI corrections at 6 h intervals: pure VLBI solution, multi-technique combinations with and without TRF estimation

Table 18.4 Weigthed mean and standard deviation of 6 h UT1-TAI corrections versus C04 series in μs

Technique	Mean (μs)	WRMS (μs)
VLBI	0.40	13
Comb	0.03	14
Comb + TRF	−0.01	22

3.3.2 UT1

Because of correlation with satellite orbital elements, the long term irregularities of UT1 (from a few days) cannot be derived from SLR, DORIS and GPS observations alone. The processing has to include VLBI data at least. This is done according to the following combinations:

- Stacking of the VLBI 3 weeks NEQs
- Stacking of the VLBI, GPS, SLR and DORIS NEQs
- Stacking of the VLBI, GPS, SLR and DORIS NEQs and estimates of the TRF

Offsets with respect to 6 h interpolated C04 series are displayed on Fig. 18.2, and corresponding statistics (standard deviation, bias) are given in Table 18.4. In the third week, we notice how satellite techniques fulfil the gap, which appears in the pure VLBI solution. That illustrates the interest of this combination.

The UT1 is not biased when estimating the terrestrial stations. Both VLBI and combined solutions match very well.

3.3.3 Nutation Offsets

Corrections for nutation offsets dX and dY are displayed on Fig. 18.3 at 12 h intervals and associated statistics are given in Table 18.5.

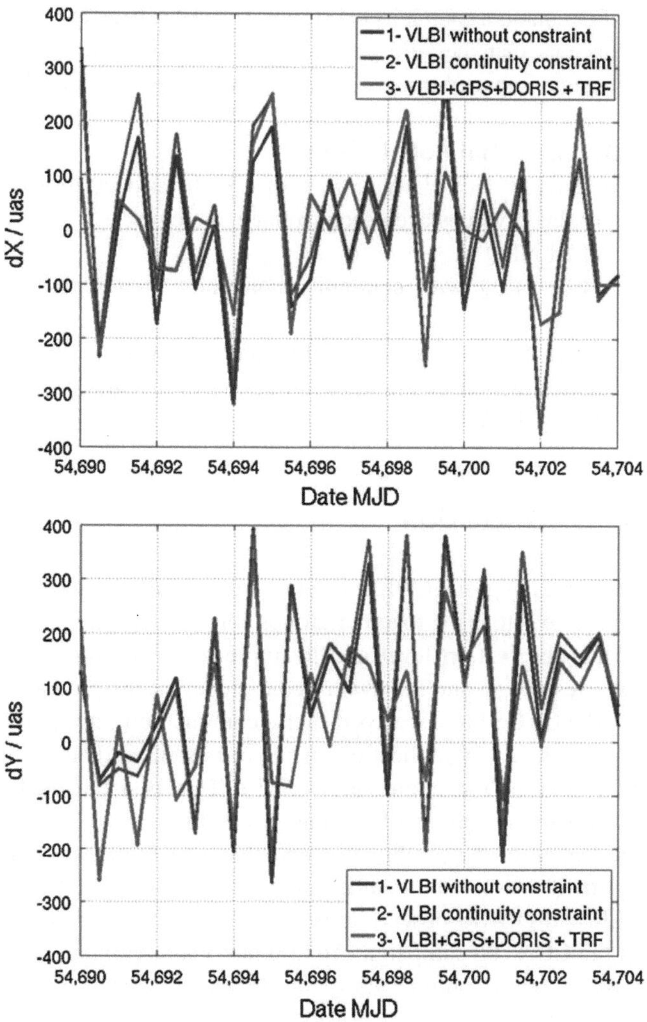

Fig. 18.3 Nutation corrections dX and dY at 12 h intervals: VLBI (1), VLBI with continuity constraint on nutation of 1 mas (2), VLBI + GPS + DORIS estimation of the TRF (3). The time span is restricted to CONT08 interval

Table 18.5 Corrections with respect to EOP 05 C04 series: weighted mean and standard deviation

Corrections	Technique	Mean	WRMS
dX	VLBI	−20	170
	VLBI with constraint	−10	170
	Combined + TRF	0	120
dY	VLBI	70	180
	VLBI with constraint	80	180
	Combined + TRF	49	150

The three VLBI solutions are very close to each other. Multi-technique combination VLBI + GPS + DORIS reproduce the VLBI solutions with a correlation coefficient of about 0.7.

For nutation, continuity constraints have been fixed to 1 mas. It appears that they are probably too loose since they have almost no impact. Stronger constraints should be considered.

3.3.4 Celestial Pole Offsets

We derived half daily corrections with respect to C04 interpolated values according to three strategies:

- VLBI NEQ
- VLBI NEQ with continuity constraint on nutation of 1 mas
- VLBI + GPS + DORIS NEQ stacking with continuity constraints on nutation, minimal constraints and estimation of the TRF

4 Discussion

The working group COL was set up in the course of 2009. Its objective is to bring together groups able to do combinations at the observation level and to improve the accuracy, the time resolution and the overall consistency of the product.

The GRGS has developed such a combination since 2005. One of the main applications is the consistent derivation of EOP and station coordinates (TRF). This is achieved by using various combination strategies, helping to understand the benefit of the combination and characterise the way it is optimal. We present results derived from the GPS, VLBI, SLR and DORIS observations gathered during the 3 weeks of the CONT08 campaign. At 12 h intervals we obtain short term nutation components (between 2 and 7 days), poorly caught by routine VLBI processing. Given at 6 h intervals, polar coordinates and UT1 unveil their diurnal oscillations of polar motion and UT1. This may help to refine the diurnal part of the theoretical ocean tide models, and determine atmospheric and non-tidal oceanic effects at these time scales. A critical point which needs to be investigated is the deficiency we find in the datum definition of the combined solution. This appears through the significant bias present in this combined pole combination.

Acknowledgement The authors were supported by the CNES under a specific grant. We thank S. Loyer, F. Deleflie, L. Soudarin, H. Capdeville and J.M. Lemoine for their contributions at various levels. ECMWF data used in this study have been obtained from the ECMWF Data Server. We thank the reviewers for their valuable suggestions and comments that helped to improve the paper.

References

Altamimi Z, Collilieux X, Legrand J, Garayt B, Boucher C (2007) ITRF2005: a new release of the international terrestrial reference frame based on time series of station positions and Earth orientation parameters. J Geophys Res 112:B09401. doi:10.1029/2007JB004949

Angermann D, Drewes H, Krügel M, Meisel B, Gerstl M, Kelm R, Müller H, Seemüller W, Tesmer V (2004) ITRS combination center at DGFI: a terrestrial reference frame realization 2003. In: Deutsche Geodätische Kommission, Reihe B, No. 313, München, ISBN 3769685938

Berger C, Biancale R, Ill M, Barlier F (1998) Improvement of the empirical thermospheric model DTM94, a comparative review of various temporal variations and prospects in space geodesy applications. J Geodesy 72(3):161–178. doi:10.1007/s001900050158

Biancale R, Bode A (2002) Mean annual and seasonal atmospheric tide models based on 3-hourly and 6-hourly ECMWF surface pressure data, GFZ scientific technical report STR05/XX

Biancale R, Lemoine JM, Balmino G, Loyer S, Bruinsma S, Perosanz F, Marty JC, Gégout P (2005) 3 years of decadal geoid variations from GRACE and LAGEOS data, CD-ROM, CNES/GRGS product

Bizouard C, Gambis D (2009) The combined solution C04 for Earth orientation parameters consistent with international terrestrial reference frame 2005. In: Drewes H (ed) International association of geodesy symposia series, vol 134. Geodetic reference frames, Munich, 9–14 Oct 2006. Springer, Berlin, 2009. ISBN: 978-3-642-00859-7, pp 265–270, doi: 10.1007/978-3-642-00860-3_41

Bourda G, Charlot P, Biancale R (2007) GINS: a new tool for VLBI Geodesy and Astrometry. In: Bouvier J, Chalabaev A, Charbonnel C (eds) SF2A-2007: Proceedings of the annual meeting of the French society of astronomy and astrophysics held in Grenoble, France, 2–6 July 2007, p 82

Brankovic C, Van Maanen J (1985) The ECMWF climate system, ECMWF Technical memorandum no. 109, 51pp

Carrere L, Lyard F (2003) Modeling the barotropic response of the global ocean to atmospheric wind and pressure forcing – comparisons with observations. Geophys Res Lett 30:1275. doi:10.1029/2002GL016473

IERS Conventions (2003) McCarthy DD, Petit G (eds). In: IERS Technical Note 32. Chapter 10.2, Frankfurt am Main, Verlag des Bundesamts für Kartographie und Geodäsie, 127pp. ISBN 3-89888-884-3

Coulot D, Berio P, Biancale R, Loyer S, Soudarin L, Gontier A-M (2007) Toward a direct combination of space-geodetic techniques at the measurement level: methodology and main issues. J Geophys Res 112:B5. doi:10.1029/2006JB004336

Dach R, Hugentobler U, Fridez P, Meindl M (2007) Bernese GPS software version 5.0. Astronomical Institute, University of Berne, Switzerland

Fey A, Gordon D, Jacobs CS (eds) (2009) The second realization of the international celestial reference frame by very long baseline interferometry, Verlag des Bundesamts für Kartographie und Geodäsie, Frankfurt am Main, IERS Technical Note 35

Gambis D (2004) Monitoring Earth orientation at the IERS using space-geodetic observations. J Geodesy 78(4–5):295–303. doi:10.1007/s00190-004-0394-1

Gambis D, Biancale R, Carlucci T, Lemoine JM, Marty JC, Bourda G, Charlot P, Loyer S, Lalanne L, Soudarin L (2009) Combination of Earth orientation parameters and terrestrial frame at the observation level. Springer Verlag series, Series international association of geodesy symposia, vol 134, Drewes (ed), pp 3–9, doi: 10.1007/978-3-642-00860-3_1

Gerstl M, Kelm R, Müller H, Ehrnsperger W (2001) DOGS CS Kombination und Lösung großer Gleichungssysteme. Manual VIII für DOGS Version 4.05. Internal report. DGFI, Munich

Kudryashova M, Snajdrova K, Weber R, Heikelmann R, Schuh H (2008) Combination of nutation time series derived from VLBI and GNSS, In: Finkelstein A, Behrend D (eds) IVS general meeting proceedings, pp 240–245. ISBN: 978-5-02-025332-2

Lyard F, Lefèvre F, Letellier T, Francis O (2006) Modelling the global ocean tides: a modern insight from FES2004. Ocean Dyn 56:394–415. doi:10.1007/s10236-006-0086-x

Marty JC, Loyer S, Perosanz F, Mercier F, Bracher G, Potier L, Capdeville H, Lemoine JM, Biancale R, Fund F (2011) GINS: the CNES/GRGS multi-GNSS scientific software. Proceedings ESA 3rd international colloquium scientific and fundament aspects of the Galileo programme, September 2011, CNES Internal Publication, Copenhagen

Ray J, Kouba J, Altamimi Z (2005) Is there utility in rigorous combinations of VLBI and GPS Earth orientation parameters? J Geodesy 79(9):505–511. doi:10.1007/s00190-005-0007-7

Richard J-Y, Bizouard C, Bourda G, Deleflie F, Gambis D, Loyer S, Soudarin L (2009) GRGS combination of the terrestrial frame and Earth orientation parameters at the observation level. In: M. Heydari-Malayeri (ed.) Contribution to ITRF2008 realization, SF2A proceedings, GRAAPH session, Paris, pp 115–119

Chassaing and Roumiguier (1965) Description d'un programme de correction différentielles utilisant l'intégration numérique, programme GIN (GINS), in French, AR/JPCH/0.339/CB/MT, CNES (ed), Brétigny sur Orge, 57pp

Sahin M, Cross PA, Sellers PC (1992) Variance component estimation applied to satellite laser ranging. Bull Géodesy 66:284–295

Springer T, Otten M, Romero N, Dow J (2008) "ESOC IGS, IDS, and ILRS (Re-) Processing", 16th International Workshop On Laser Ranging, Poznan Poland, Oct. 13–17, 2008

Springer TA (2009) NAPEOS mathematical models and algorithms. DOPS-SYS-TN-0100-OPS-GN, issue 1.0, Nov 5, 2009

Standish EM (1998) JPL planetary and lunar ephemerides. DE405/LE405, JPL IOM 312.F-98-048

Yaya Ph (2002) Apport des combinaisons de techniques astrométriques et géodésiques à l'estimation des paramètres d'orientation de la terre. Ph.D. thesis, Observatoire de Paris (ed), 230pp

Time-Correlated GPS Noise Dependency on Data Time Period

19

Alvaro Santamaría-Gómez, Marie-Noëlle Bouin, Xavier Collilieux, and Guy Wöppelmann

Abstract

GPS position time series contain time-correlated noise. The estimated parameters using correlated time series data, as station velocities, are then more uncertain than if the time series data were uncorrelated. If the level of the time-correlated noise is not taken into account, the estimated formal uncertainties will be smaller. By estimating the type and amplitude of the noise content in time series, more realistic formal uncertainties can be assessed.

However, time-correlated noise amplitude is not constant in long time series, but depends on the time period of the time series data. Older time series data contain larger time-correlated noise amplitudes than newer time series data. This way, shorter time series with older data time period exhibit time-correlated noise amplitudes similar to the whole time series. This paper focuses on the source of the time-correlated noise amplitude decrease from older to newer time series period data. The results of several tested sources are presented. Neither the increasing ambiguity fixation rate, nor the increasing number of tracking stations, nor the increasing number of observed satellites are likely the source of the noise reduction. The quality improvement of the equipment of both tracking network and constellation is likely the main source of the correlated noise evolution.

Keywords

GPS • Time series • Time-correlated noise

1 Introduction

Since more than a decade it is known that GPS station position time series contain time-correlated noise (Zhang et al. 1997; Mao et al. 1999). Noise parameters describing the type and the amplitude of the time-correlated noise content in position time series are usually estimated to reevaluate realistic uncertainties of the estimated parameters, such as station velocities (Williams 2003a; Wöppelmann et al. 2009), offset amplitudes (Williams 2003b), periodic signals (Bos et al. 2010), etc.

The analysis of the noise properties of the station position time series can also be used to get insights into the time-correlated noise sources of the GPS technique. This way, noise analyses have been used to estimate the amplitude of the station monument noise (King and Williams 2009), to

A. Santamaría-Gómez (✉)
Instituto Geográfico Nacional, Cerro de la Palera s/n, E-19141 Yebes, Spain

IGN/LAREG, Université Paris Diderot, 35 rue Hélène Brion, 75013 Paris, France
e-mail: asantamaria@fomento.es

M.-N. Bouin
Centre National de Recherches Météorologiques, Météo France, 13 rue du Chatellier, F-29228 Brest, France

X. Collilieux
IGN/LAREG, Université Paris Diderot, 35 rue Hélène Brion, 75013 Paris, France

G. Wöppelmann
Fédération de Recherche en Environnement pour le Développement Durable (FR 3097 CNRS), Université de La Rochelle/CNRS, 2 rue Olympe de Gouges, F-17000 La Rochelle, France

Z. Altamimi and X. Collilieux (eds.), *Reference Frames for Applications in Geosciences*, International Association of Geodesy Symposia 138, DOI 10.1007/978-3-642-32998-2_19,

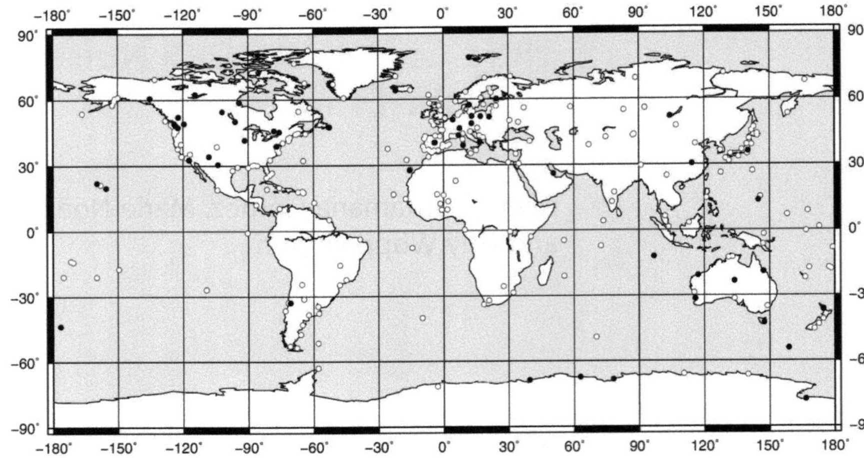

Fig. 19.1 ULR4 global tracking network. The solid dots represent the long-running 54 stations selected for the noise analysis

compare the performance of different monument designs (Beavan 2005) or to validate different GPS data processing strategies (Tregoning and Watson 2009).

In addition to these studies, investigating if the estimated properties of the time-correlated noise content are constant over time can also reveal some clues about its origin. Bos et al. (2010) already suggested that time-correlated noise nature is probably not time constant. Santamaría-Gómez et al. (2011) (SG11 hereinafter), using a power law plus white noise model demonstrated that time-correlated noise content is dependent on the data time period used in the analysis. They showed how older time series data periods contain larger time-correlated noise than newer. That is, the time-correlated noise content decreases within long position time series.

In SG11 three test were carried out in order to investigate the origin of the correlated-noise dependency on time series data period. By reprocessing 12 years of data of a global network composed of 54 long-running stations with a cut-of angle of 10º, they tested the role of the increment in the number of tracking stations and the role of the ambiguity fixation rate. A similar second reprocessing, but with a cut-off angle of 30º, was used to assess the impact of the low-elevation observations lost due to the L2 tracking problems of Turbo Rogue receivers under increasing ionospheric activity (IGS electronic mail message 2071, 1998).[1] The two first tested sources (increase of the tracking network and ambiguity fixation) were certainly rejected as the origin of the time-correlated noise dependency. The third test (lacking of low-elevation observations) did not provided significant results due probably to the fact that only half the observations were used in the GPS data processing. In addition, as shown by SG11, there was not substantial lacking of double difference observations before 2000 in the first reprocessed solution, where a cut-off angle of 10º was applied. SG11 concluded that the origin of the time-correlated noise dependency on data time period is likely related to the quality improvement of the tracking network equipment.

SG11 also confirmed that flicker noise is the dominant time-correlated noise type for long-term reprocessed global GPS solutions. Based on these results, we revisited in next Sects. 2 and 3 the noise analyses of SG11 using a combination of flicker plus white noise models to asses the dependency of the time-correlated noise amplitudes on data time period. Flicker noise amplitudes obtained in this study are easier to interpret, in terms of rate uncertainties for instance, because they do not depend on the spectral index change of the power law process used in SG11.

In addition to these three revisited tests, a third reprocessing was carried out in Sect. 4 and used to test a fourth possible source. We assessed in this study the role of the increasing number of observed satellites on the time-correlated noise evolution.

2 Noise Content Evolution Analysis

As in SG11, we used the GPS position time series from the University of La Rochelle (ULR) Analysis Center Consortium (ULR, IGN Spain and IGN France). Specifically, the height time series of the ULR4 solution were used to investigate the evolution of the flicker noise amplitude within the GPS position time series.

The ULR4 solution is based on a homogeneous reprocessing of a global network of 316 stations (see Fig. 19.1) from January 1996 to December 2008. The double-differenced phases approach implemented in the GAMIT/GLOBK package (Herring et al. 2008) was used. We applied absolute antenna phase calibration model (Schmid et al. 2007), a cut-off angle of 10º, VMF1 grids

[1] http://igscb.jpl.nasa.gov/pipermail/igsmail

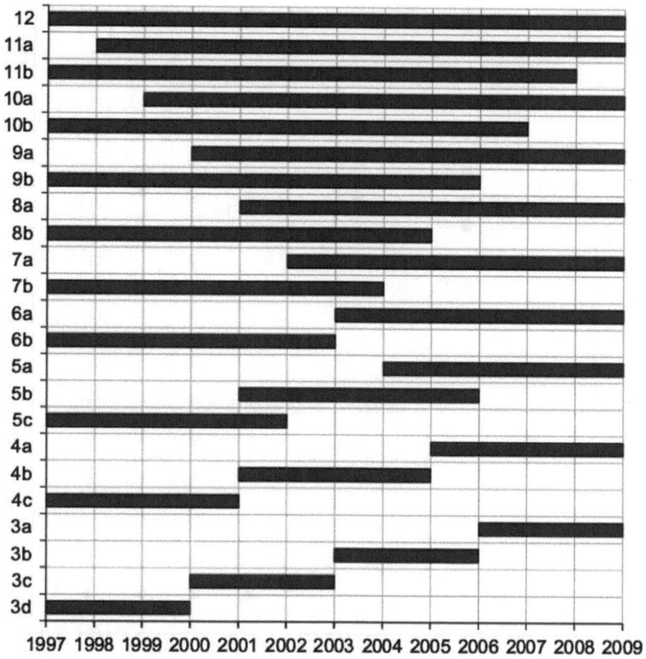

Fig. 19.2 The 23 windowed groups of time series used in this study. Each bar represents a group of time series with different length (from 12 to 3 years) and data time period (from 1997 to 2009)

(Boehm et al. 2006; Kouba 2007) for tropospheric delay, FES2004 model (Lyard et al. 2006) for ocean tide loading, but no higher ionospheric effects nor atmospheric loading were corrected for. More details on the GPS processing strategy can be found in Santamaría-Gómez et al. (2012).

A similar procedure of the noise content evolution analysis than in SG11 was performed here.

First, we selected 54 residual height time series (trends removed) from the ULR4 solution (see solid dots in Fig. 19.1) and from the analysis of the IGS reprocessed combined time series conducted for the International Terrestrial Reference Frame, ITRF2008 processing (Altamimi et al. 2011). These selected stations have time series longer than 12 years and at least 95 % of available data for the noise analysis. This way, the effect of data gaps on the flicker noise amplitude estimates is minimized. These 54 time series were then windowed into 23 groups. Each group (bars in Fig. 19.2) represents a different time series length (from 12 to 3 years) and a different data time period (from 1997 to 2009). Recent time series data ("a" groups in Fig. 19.2) were distinguished from older time series data ("b" to "d" groups in Fig. 19.2). Note that some of the time series groups of the same length have overlapping segments of data (e.g., groups 5a, 5b and 5c). Time series groups of 2 years length, used in SG11, were not used in this study due to the effect that missing data (up to 5 % maximum) could have on the noise estimates for such a short time series.

Second, as in SG11, we also estimated the median flicker noise amplitudes of the 23 groups using the height residual time series resulting from two reprocessed solutions (with cut-off angle of 10º and 30º, respectively) of a smaller global network composed of only the selected 54 stations. With these two reprocessed solutions, we revisit in this paper the three test of SG11: the role of the increasing tracking network (test A: cut-off of 10º, ambiguities fixed); the role of the ambiguity fixation (test B: cut-off of 10º, ambiguities free); and the role of the lacking of low-elevation observations before 2000 (test C: cut-off of 30º, ambiguities fixed).

Using the Maximum Likelihood Estimation technique implemented by Williams (2008) and a white plus flicker noise model, the median time-correlated noise amplitude of each time series group and data set was estimated. Note that contrary to SG11, the spectral index of the power law process used to estimate the time-correlated noise amplitude of the time series groups was held fixed here to −1. Therefore, only the correlated-noise amplitude dependence on the data time period was then considered in this paper.

The flicker noise amplitudes estimated here can be therefore useful to assess the station rate uncertainties for those solutions for which a dedicated noise analysis is not available. From numerical analysis the expression that relates flicker noise amplitudes with rate uncertainties is (Williams 2003a):

$$\sigma_r^2 \simeq \frac{9f^2}{16\Delta T^2(n^2-1)}$$

where f is the flicker noise amplitude, ΔT the time series length and n the number of data points.

3 Noise Results

The estimated flicker noise amplitudes of test A (54-station network) were 25 % larger with respect to the ULR4 solution (316-station network). This larger time-correlated noise level demonstrates that the number of tracking stations and their distribution has, to some extent, an impact on the time-correlated noise amplitude of global GPS solutions. This fact could be related to the poorer frame definition and orbital parameters estimation with a smaller global tracking network.

The estimated flicker noise amplitudes of test B were 22 % larger with respect to the test A, corroborating the effect that the fixation of ambiguities have on time-correlated noise content of GPS position time series (Tregoning and Watson 2009; King and Watson 2010).

The estimated flicker noise amplitudes of test C were 120 % larger with respect to the test A. This is probably due to the higher cut-off applied (30º) which resulted in using only half the observations to estimate the GPS parameters.

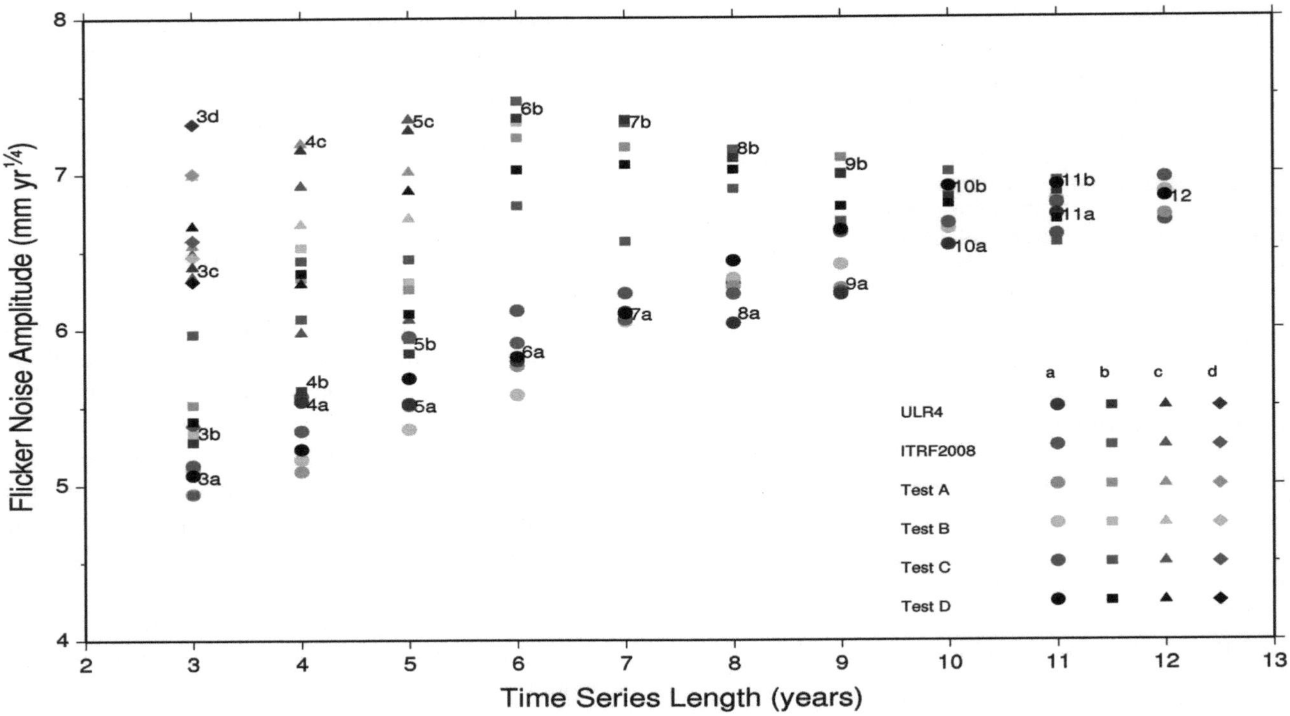

Fig. 19.3 Median flicker noise amplitudes computed over 54 stations for each of the 23 time series groups of Fig. 19.2 sorted by time series length. Blue: ULR4 solution; red: ITRF2008 solution; green: Test A; cyan: Test B; pink: Test C; black: Test D. Circles: "a" group; squares: "b" group; triangles: "c" group; diamonds: "d" group. Noise amplitudes correspond to ULR4 and ITRF2008 solutions. Test A, B and C noise amplitudes were scaled (see text). Labels correspond to the results of the ULR4 time series groups

Figure 19.3 shows the median flicker noise amplitudes of each of the 23 time series groups for all the five revisited data sets (ULR4, ITRF2008, test A, test B and test C). In this figure the flicker noise amplitudes of Test A, test B and test C were scaled to fit ULR4 flicker noise amplitudes with the above-mentioned factors. This figure confirms the time-correlated noise dependency on data time period found in SG11. The estimated median flicker noise amplitudes of Fig. 19.3 grow from "a" to "d" time series groups, that is, the flicker noise amplitude is systematically larger for older periods of time series data. This noise evolution pattern was found in all the data sets analyzed, although it was less significant for the test C, as for in SG11.

4 Testing Constellation Evolution

Understanding the source of the flicker noise content in GPS time series and why it becomes smaller in recent data is challenging. Ongoing with the investigation carried out in SG11, we suggest here to assess a fourth candidate source that may explain the flicker noise evolution over time.

SG11 showed how the observations (double differences) used to estimate the parameters in the reprocessing of the 54-station network (e.g., station coordinates) rose with time. Since the tracking network of such reprocessed solution remained almost constant, the origin of the increase of the number of available observations likely came from the improvement of both the tracking network and transmitting satellite equipment and/or from the increasing number of available satellites (from 24 in 1997 to 31 in 2008, see Fig. 19.4).

To investigate the impact of the increasing number of available satellites on the number of observations, and hence, on the time-correlated noise evolution, we performed in this study a new reprocessed solution (test D hereinafter). The global 54-station network was reprocessed with a cut-angle of 10º and fixing the ambiguities as for the test A. However, for this test, the number of observed satellites was held fixed nearly to 24. Knowing the daily observed satellites from test A, the exceeding number of satellites were automatically excluded from the processing. For each day, the most recently launched satellites were chosen for exclusion. This way we also tried to minimize the effect of the constellation quality evolution on the time-correlated noise content.

Figure 19.4 shows the number of double differences for the two reprocessed solutions of test A and test D. It also shows the number of tracking stations and the observed satellites in both tests. We can see in this figure that the number observed satellites in the test D were kept always close to 24. The difference in the number of observations between the test A and the test D are then due to the different number of observed satellites. Using a near-constant number of stations and

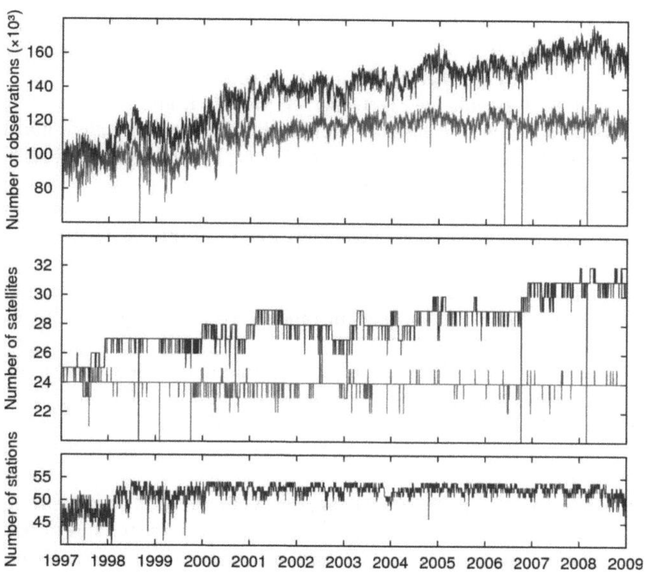

Fig. 19.4 Number of double differences (*top*), number of observed satellites (*middle*) and number of tracking stations (*bottom*) for test A (*black*) and test D (*grey*)

satellites, the number of available observations remains nearly constant for most of the analyzed period. A slight increment exists however between 1999 and 2002, with possibly an offset in the number of observations near mid-2000. Before and after the 1999–2002 period the number of observations in the test D remains constant. This slight increment might represent a quality improvement of the tracking network equipment as, for instance, the workaround found in mid-1999 for the Turbo Rogue L2 tracking problem (IGS electronic mail message 2336, 1999)[1].

Figure 19.3 also shows the median flicker noise amplitudes for the 23 windowed time series groups of test D. Flicker noise amplitudes for these 23 groups of test D resulted similar in the mean than for test A, so the same scale factor was applied in Fig. 19.3. As it is seen in this figure, a similar flicker noise differences were found between newer and older time series groups. This means that the increasing number of satellites in the constellation might be not responsible either for the reduction of the time-correlated noise amplitudes in long GPS time series. However, the flicker amplitude differences of test D seem to be reduced with respect to the other tests, specially for time series groups of 9 years and longer.

5 Conclusion

Long-term global GPS solutions contain time-correlated noise amplitude that is not time constant. It is dependent on the time series length but also on the data time period used to estimate the noise content. Using a flicker plus white noise model combination, this fact was corroborated here using the reprocessed time series from the ULR4 solution and from the ITRF2008.

Using a time-constant tracking network of 54 stations, three noise analysis were also revisited to investigate the source of the time-correlated noise amplitude evolution. We tested the influence of the tracking network evolution, the percentage of fixed ambiguities evolution and the loss of low-elevation observations from Turbo Rogue receivers under high ionospheric activity. Using a flicker plus white noise model, we confirmed however that none of these sources seems to be related to the flicker noise amplitude reduction in the time series.

The fact that, even when using a time-constant tracking network for those tests, the number of available daily observations (double differences) increases with time points to the quality improvement of the antennae and receivers of the tracking network and to the increasing number of satellites in the constellation. Both facts might be responsible of the time-correlated noise reduction over time.

To investigate if the increasing number of satellites are the main source of the noise content evolution we performed here a fourth test based on the reprocessing of a constant tracking network and a constant satellite constellation. Although flicker noise differences for this test are reduced with respect to the other tests, a similar noise pattern was found. Hence, the variable number of observed satellites could be rejected as the main source. It remains as a possible source the quality improvement of the tracking stations equipment (receivers and antennae) and of the transmitting satellites. This source has not been possible to verify due to the large amount of hardware and software changes usually performed on the stations.

Acknowledgments The authors acknowledge the invaluable technical support provided by Mikael Guichard, Marc-Henri Boisis-Delavaud and Frederic Bret from the IT Center of the University of La Rochelle. The University of La Rochelle computing infrastructure was partly funded by the European Union (Contract 31031–2008, European Regional Development Fund). This work was also feasible thanks to all institutions and individuals worldwide that contribute to make GPS data and products freely available (e.g., IGS, EPN, BIGF, GSI, RENAG, AMMA).

References

Altamimi Z, Collilieux X, Métivier L (2011) ITRF2008 : an improved 280 solution of the international terrestrial frame. J Geodesy 85:457–473, doi:10.1007/s00190-011-0444-4

Beavan J (2005) Noise properties of continuous GPS data from concrete pillar geodetic monuments in New Zealand and comparison with data from U.S. deep drilled braced monuments. J Geophys Res 110: B08410. doi:10.1029/2005JB003642

Boehm J, Werl B, Schuh H (2006) Troposphere mapping functions for GPS and very long baseline interferometry from European centre for medium-range weather forecasts operational analysis data. J Geophys Res 111:B02406. doi:10.1029/2005JB003629

Bos MS, Bastos L, Fernandes RMS (2010) The influence of seasonal signals on the estimation of the tectonic motion in short continuous GPS time-series. J Geodyn 49:205–209. doi:10.1016/j.jog.2009.10.005

Herring TA, King RW, McClusky SC (2008) Introduction to GAMIT/GLOBK, report, mass. Institute of Technology, Cambridge

King MA, Watson CS (2010) Long GPS coordinate time series: multipath and geometry effects. J Geophys Res 115:B04403. doi:10.1029/2009JB006543

King MA, Williams SDP (2009) Apparent stability of GPS monumentation from shortbaseline time series. J Geophys Res 114:B10403. doi:10.1029/2009JB006319

Kouba J (2007) Implementation and testing of the gridded Vienna mapping function 1 (VMF1). J Geodesy. doi:10.1007/s00190-007-0170-3

Lyard F, Lefevre F, Letellier T, Francis O (2006) Modelling the global ocean tides: modern insights from FES2004. Ocean Dynam 56:394–415. doi:10.1007/s10236-006-0086-x

Mao A, Harrison CGA, Dixon TH (1999) Noise in GPS coordinate time series. J Geophys Res 104(B2):2797–2816

Santamaría-Gómez A, Bouin M-N, Collilieux X, Wöppelmann G (2011) Correlated errors in GPS position time series: implications for velocity estimates. J Geophys Res 116:B01405. doi:10.1029/2010JB007701

Santamaría-Gómez A, Bouin M-N, Wöppelmann G (2012) Improved GPS data analysis strategy for tide gauge benchmark monitoring. In: S. Kenyon, M.C. Pacino, and U. Marti (eds.), Proceedings of the 2009 IAG Symposium, Buenos Aires, Argentina, 31 August–4 September 2009, doi:10.1007/978-3-642-20338-1_2

Schmid R, Steigenberger P, Gendt G, Ge M, Rothacher M (2007) Generation of a consistent absolute phase-center correction model for GPS receiver and satellite antennas. J Geodesy 81:781–798. doi:10.1007/s00190-007-0148-y

Tregoning P, Watson C (2009) Atmospheric effects and spurious signals in GPS analyses. J Geophys Res 114:B09403. doi:10.1029/2009JB006344

Williams SDP (2003a) The effect of colored noise on the uncertainties of rates estimated from geodetic time series. J Geodesy 76:483–494. doi:10.1007/s00190-002-0283-4

Williams SDP (2003b) Offsets in global positioning system time series. J Geophys Res 108(B6):2310. doi:10.1029/2002JB002156, 2003

Williams SDP (2008) CATS: GPS coordinate time series analysis software. GPS Solut 12(2):147–153. doi:10.1007/s10291-007-0086-4

Wöppelmann G, Letetrel C, Santamaría A, Bouin M-N, Collilieux X, Altamimi Z, Williams SDP, Martín Miguez B (2009) Rates of sea-level change over the past century in a geocentric reference frame. Geophys Res Lett 36:L12607. doi:10.1029/2009GL038720

Zhang J, Bock Y, Johnson H, Fang P, Williams S, Genrich J, Wdowinski S, Behr J (1997) Southern California permanent GPS geodetic array: error analysis of daily position estimates and site velocities. J Geophys Res 102:18,035–18,055

GPS-Specific Local Effects at the Geodetic Observatory Wettzell

20

P. Steigenberger, U. Hugentobler, R. Schmid, U. Hessels, T. Klügel, and M. Seitz

Abstract

Global Navigation Satellite Systems (GNSS) are important contributors to the realization of the International Terrestrial Reference System (ITRS). For the combination of different space geodetic techniques, terrestrial measurements between the corresponding reference points are necessary. Discrepancies between these so-called local ties on the one hand and the coordinate differences derived from space techniques on the other hand are a major limitation for the realization of the ITRS nowadays. In the past, these discrepancies have often been attributed to inaccurate terrestrial measurements. This paper shows that a major part of the differences can be explained by systematic GNSS-specific errors, if a global data analysis is simulated. One of the most important error sources for GNSS are interactions of the antenna with its immediate vicinity, primarily multipath.

At the Geodetic Observatory Wettzell (Germany), up to six GNSS permanent sites are operated in parallel at a distance of only a few meters. This antenna array is ideal to study the impact of local effects on the various GNSS observables and linear combinations. Comparisons of solutions obtained from different GNSS observables reveal cm-level discrepancies. Individual receiver antenna calibrations have an impact on the estimated station positions on the level of several millimeters. As other error sources dominate, their application does not lead to an improvement in all cases.

Keywords

Terrestrial reference frame • Local tie • Antenna phase center model • Global positioning system • Multipath

1 Introduction

At the Geodetic Observatory Wettzell, three of the space geodetic techniques used for the realization of the International Terrestrial Reference System (ITRS) are operated: Global Navigation Satellite System (GNSS) receivers, Satellite Laser Ranging (SLR), and Very Long Baseline Interferometry (VLBI). Doppler Orbitography and Radiopositioning Integrated by Satellite (DORIS) is the only technique not present at Wettzell due to signal interferences with VLBI. For the combination of the space geodetic techniques, terrestrial measurements (local ties) connecting the reference points of the different techniques are necessary. In the computation of the recent realization of

P. Steigenberger (✉) • U. Hugentobler • R. Schmid
Institut für Astronomische und Physikalische Geodäsie, Technische Universität München, Arcisstraße 21, Munich 80290, Germany
e-mail: steigenberger@bv.tum.de

U. Hessels • T. Klügel
Bundesamt für Kartographie und Geodäsie, Geodätisches Observatorium Wettzell, Sackenrieder Straße 25, Bad Kötzting 93444, Germany

M. Seitz
Deutsches Geodätisches Forschungsinstitut, Alfons-Goppel-Straße 11, Munich 80539, Germany

Z. Altamimi and X. Collilieux (eds.), *Reference Frames for Applications in Geosciences*, International Association of Geodesy Symposia 138, DOI 10.1007/978-3-642-32998-2_20,

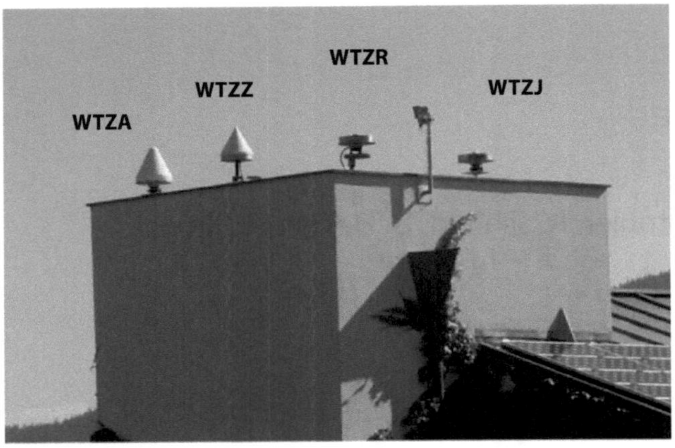

Fig. 20.1 GNSS permanent sites on top of the tower of the main building of the Geodetic Observatory Wettzell. A webcam to monitor snow coverage of the antennas is mounted between WTZR and WTZJ

the ITRS (ITRF2008), discrepancies between the local ties and the space geodetic techniques of up to 13.5 mm occurred (Seitz et al. 2013). This value is significantly larger than the precision that can be achieved with terrestrial as well as space geodetic observations. As several GNSS permanent sites are operated in parallel at Wettzell, a comparison of vectors obtained from different GNSS observables with the terrestrial measurements is possible. Differences between these solutions could indicate frequency-dependent systematic effects like near-field multipath (Dilßner et al. 2008). In this paper, all site-specific effects that could not clearly be identified are denoted as multipath, without knowing the exact mechanism how the signal is affected.

The first GPS permanent site at Wettzell was put into operation in 1991. The site WTZR installed in 1995 is still the one used by most of the analysis centers (ACs) of the International GNSS Service (IGS, Dow et al. 2009). Like most of the antennas at Wettzell, WTZR is mounted on the roof of a tower of the main building of the Geodetic Observatory, see Fig. 20.1. However, one antenna (WTZS) is mounted on a steel mast with a height of about 7.5 m above the ground, about 67 m north-northwest of the tower. All GNSS permanent sites operated at Wettzell are listed in Table 20.1. Most of the sites are part of the tracking network of the IGS. However, WTZL was part of the German GPS reference network DREF. This site was considered here, as individual antenna calibrations are available.

Rothacher et al. (2004) already analyzed the local GPS network at Wettzell. They could achieve sub-mm repeatabilities of the coordinate time series and detected significant discontinuities due to antenna changes. Section 2 discusses the GPS processing of the antenna array at the Geodetic Observatory Wettzell. Three different combined solutions are computed from L1 and L2 observations as well as from the ionosphere-free linear combination L3. These solutions are compared with terrestrial measurements in Sect. 3. The impact of individual receiver antenna calibrations is studied in Sect. 4.

2 GPS Processing

In order to study the discrepancies between local ties and GNSS-derived coordinates, an analysis strategy similar to that of the IGS ACs was applied. All available RINEX observation files between November 1997 and August 2010 were processed in daily batches with the current development version 5.1 of the Bernese GPS Software (Dach et al. 2007). As only a few sites are equipped with GLONASS-capable receivers, only GPS observations were considered for this analysis. For the time period until the end of 2008 the reprocessed orbits and Earth rotation parameters of the Center for Orbit Determination in Europe (CODE; Dach et al. 2009) were used, for 2009 and 2010 the operational products instead. The IGS05 coordinates of WTZA were chosen to be fixed, as this is the only site without discontinuities. For all other sites, baselines w.r.t. WTZA were formed and daily station coordinates were estimated.

Troposphere zenith delays were estimated as piecewise linear functions with a parameter spacing of 2 h using the Global Mapping Function (GMF; Boehm et al. 2006). The hydrostatic a priori delays were computed with pressure from the Global Pressure and Temperature model (GPT; Boehm et al. 2007) using the equation of Saastamoinen (1973). One pair of troposphere gradients per day was estimated in north-south and east–west direction. As it is not possible to determine absolute values for the tropospheric delays with such a small network, the estimated zenith delays of WTZR from the CODE solution were introduced and fixed. Observations were weighted with a weight of $w = \sin^2\varepsilon$ depending on the elevation ε, and an elevation cut-off angle of 3° was applied.

Ambiguities were fixed separately for L1 and L2 with the so-called Sigma method (Dach et al. 2007). The average resolution rate was 94 %. After an outlier detection based on residual screening, three different solutions were computed: L1 and L2 single-frequency solutions and a dual-frequency solution based on the ionosphere-free linear combination L3. Normal equations (NEQs) were saved to generate a combined solution. For these standard solutions, the official IGS antenna phase center model igs05.atx (Schmid et al. 2007) was used. A solution based on individual antenna calibrations is discussed in Sect. 4. The solution setup described above was chosen to be as similar as

Table 20.1 GNSS sites operated at the Geodetic Observatory Wettzell and their current usage by the IGS ACs for their final solutions. The sites are ordered by their start of operation. *Sol. no.* refers to solution numbers introduced due to discontinuities caused by antenna changes. For antenna/radome combinations marked with an asterisk, individual antenna calibrations are available

Site	Network	# ACs	Sol. no.	Antenna	Radome	Start	End
WTZR 14201M010	IGS	7	1	AOAD/M_T	NONE	02/1995	07/2002
			2	AOAD/M_T	NONE	07/2002	01/2009
			3	LEIAR25	LEIT	01/2009	06/2010
			4	LEIAR25.R3	LEIT⋆	06/2010	
WTZT 14201M011	IGS	–	1	TRM22020.00+GP	NONE	02/1997	12/1997
			2	TRM22020.00+GP	DOME	12/1997	11/1998
			3	TRM22020.00+GP	NONE	11/1998	01/2000
			4	TRM29659.00	NONE	01/2000	05/2005
WTZA 14201M013	IGS	3	Ref.	ASH700936C_M	SNOW	11/1997	
WTZZ 14201M014	IGS	4	1	ASH701073.1	SNOW	02/1999	06/2003
			2	TPSCR3_GGD	CONE	06/2003	
WTZJ 14201M012	IGS	2	1	TRM29659.00	NONE	07/2001	04/2002
			2	JPSREGANT_SD_E	NONE	04/2002	08/2005
			3	TRM29659.00	NONE⋆	08/2005	12/2009
			4	LEIAR25	LEIT⋆	12/2009	11/2010
WTZL 14201M022	DREF	–	1	LEIAX1202	NONE	03/2004	04/2007
			2	LEIAX1202GG	NONE	04/2007	12/2007
			3	LEIAT504GG	LEIS	12/2007	01/2008
			4	TPSCR.G3	TPSH⋆	01/2008	07/2008
			5	TPSCR3_GGD	CONE⋆	07/2008	09/2008
WTZS 14201M015	IGS	1	1	ASH701945G_M	SNOW	07/2005	02/2010
			2	LEIAR25.R3	NONE	02/2010	06/2010
			3	LEIAR25.R3	LEIT	06/2010	

possible to the global solution approach in order to study the systematic effects visible in the ITRF combination. Therefore, e.g., troposphere zenith delays were estimated for *all* sites but one although estimating one common troposphere zenith delay parameter would be sufficient for such a small network. Dilßner et al. (2008) showed that the estimation of troposphere parameters can cause a significant increase of systematic biases due to near-field multipath.

All antenna changes listed in Table 20.1 resulted in significant discontinuities that are clearly visible in the station coordinate time series. Therefore, independent sets of station coordinates (denoted by separate solution numbers) were estimated for the time intervals given in Table 20.1. After pre-eliminating the troposphere parameters, three separately combined solutions were computed by accumulating the L1, L2, and L3 NEQs for the full time interval from 1997 to 2010. As the formal errors of these combined solutions are by far too optimistic, realistic errors for the position estimates were computed from time series residuals according to Steigenberger et al. (2010). The errors for the height component range from 0.1 to 1.1, 1.4, and 1.9 mm for L1, L2, and L3, respectively, with a median value of 0.3 mm in all three cases. The comparisons with terrestrial measurements discussed in the next section are based on these combined solutions.

3 Comparisons with Terrestrial Measurements

The terrestrial measurements used here were performed in 2002 (Schlüter et al. 2005). They include all GNSS sites except for WTZS and are given in the file 14201_BKG_2002-266.SNX[1]. As WTZA was fixed for the GPS processing, the comparisons with the terrestrial measurements are also done w.r.t. WTZA. Height differences between GPS and terrestrial measurements in a local system centered at WTZA are given in Fig. 20.2. The largest height difference of more than 40 mm occurs for the L3 solution of WTZT when the site was equipped with an uncalibrated radome (WTZT, sol. no. 2, see Table 20.1). Ray et al. (2007) showed for the Fortaleza station that a major part of its GPS/VLBI local tie discrepancy could be explained by the effect of an uncalibrated radome. The height discrepancy could be reduced by 15.8 mm when the radome was removed. But also for antennas without radomes or with calibrated radomes, height discrepancies of up to 10 mm for L1 and L2 and of up to 15 mm for L3 are present. The L3 differences are particularly

[1] http://itrf.ensg.ign.fr/ties/ITRF2005/ITRF2005-SNX-localties.tar.gz

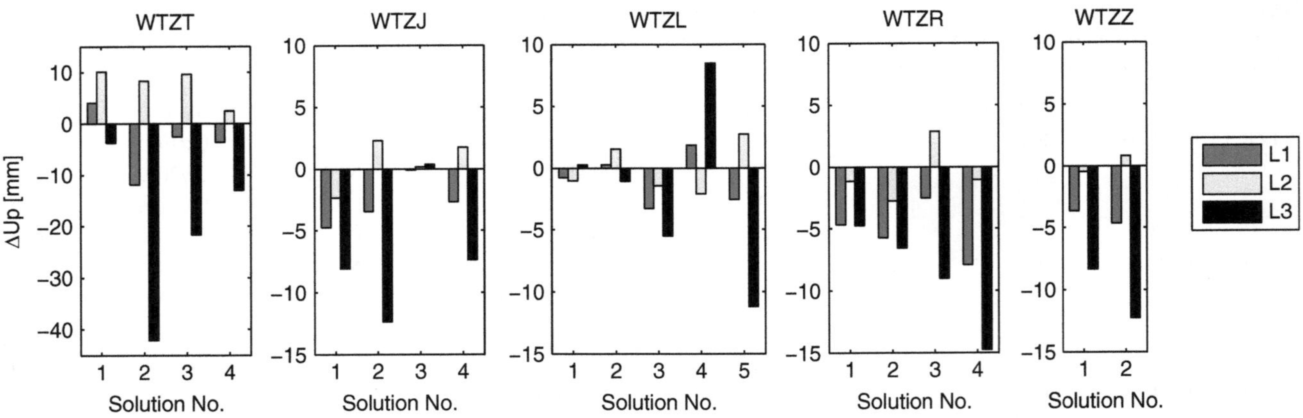

Fig. 20.2 Differences between L1, L2, or L3 GPS solutions and terrestrial measurements (Note the different scale for WTZT)

large for solutions with large differences between L1 and L2 as the latter are amplified by forming the ionosphere-free linear combination L3 (e.g., WTZR, sol. no. 4).

The horizontal L3 discrepancies (not shown here) are generally smaller than 6 mm except for WTZL (sol. no. 1 and 2) where they almost reach 12 mm. With about 4 mm, the horizontal L1 and L2 discrepancies are also the largest for the same two solutions, all other discrepancies are below 3 mm.

The differences shown in Fig. 20.2 exceed the sub-mm precision one can achieve with GPS on such short baselines by far. They also exceed the precision of the terrestrial measurements which is on the 1–2 mm level (based on the comparison of several terrestrial measurement campaigns from different years; Mähler et al. 2010).

The L3 differences are generally larger than the differences resulting from the single-frequency solutions. This is of particular importance as global solutions that are used as input for ITRF computations are always based on this linear combination. As the GPS-internal differences between the L1, L2, and L3 solutions are sometimes even larger than the discrepancies w.r.t. the terrestrial measurements, one can assume that they are caused by frequency-dependent systematic effects of the GPS technique. Deficiencies in the antenna calibration are one possible error source. Therefore, the impact of a more sophisticated antenna model will be discussed in the next section.

4 Individual Antenna Calibrations

For five antennas (marked with an asterisk in Table 20.1), individual receiver antenna calibrations are available besides the type-mean values from igs05.atx. All these calibrations are robot calibrations performed by Geo++ (Menge et al. 1998). As an example, Fig. 20.3 shows the L1 phase center differences between the type-mean and the individual

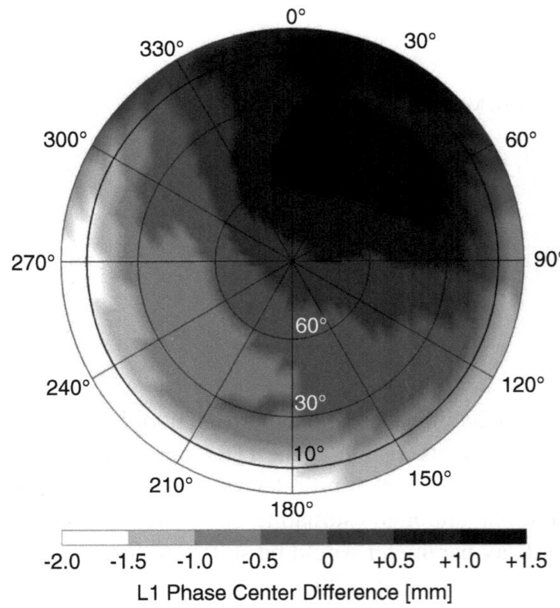

Fig. 20.3 L1 phase center differences between the type-mean values from igs05.atx and the individual calibration of the LEIAR25.R3 LEIT antenna at WTZR. The phase center variations are referred to one common phase center offset

calibration of the LEIAR25.R3 LEIT antenna of WTZR. The differences range from −2 to +1.5 mm. The L2 differences have a similar order of magnitude and shape. For the other antennas, the differences can reach ±4 mm with sometimes significant differences between L1 and L2.

The coordinate differences w.r.t. the terrestrial measurements already discussed in the previous section are now given for solutions using type-mean or individual antenna calibrations in Fig. 20.4. The solutions differ by up to 5 mm in the horizontal and the vertical component, except for WTZL (sol. no. 5) with a height difference of almost 13 mm. The phase center differences shown in Fig. 20.3 affect the horizontal coordinates of WTZR only on the sub-mm level whereas the height changes by about 5 mm. The

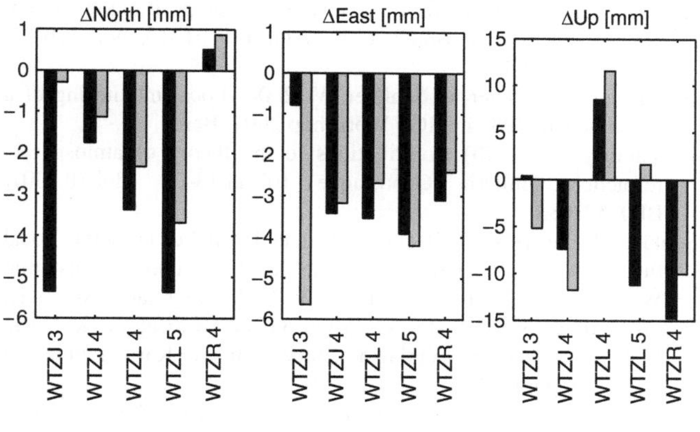

Fig. 20.4 Differences of L3 GPS solutions using type-mean (*black*) or individual (*gray*) receiver antenna calibrations w.r.t. the terrestrial measurements. Note the different scale of the height component. The solution numbers given after the site name refer to Table 20.1

discrepancies w.r.t. the terrestrial measurements decrease by one third for the height component when using the individual antenna calibration.

The largest differences for the horizontal components are visible for WTZJ (sol. no. 3). On the one hand, the discrepancy w.r.t. the terrestrial measurements gets smaller by more than 4 mm for the North component. On the other hand, it increases by about the same order of magnitude for the East component. The discrepancy for the WTZJ (sol. no. 3) height component also increases. Most other sites show a similar behavior: sometimes the agreement is better with individual calibrations, sometimes with the type-mean calibrations. In summary, no clear conclusions can be drawn, as obviously other error sources dominate.

5 Conclusions

The differences between the GPS-derived coordinates and the terrestrial measurements clearly exceed the precision of each technique. As even results derived from different frequencies and in particular from the ionosphere-free linear combination differ from each other by up to a few cm, it is reasonable to assume that a major part of the discrepancies w.r.t. the terrestrial measurements is caused by GPS-specific systematic errors. Deficiencies in the antenna phase center model could be a source for such systematic errors. The largest difference between terrestrial and GPS-derived coordinates is indeed caused by such a deficiency, namely by an uncalibrated radome. This is a worst case scenario for a geodetic co-location site that should be avoided in any circumstance. But also the application of individual antenna calibrations couldn't solve all the problems. The estimated station positions changed significantly by several mm compared to the solutions using type-mean calibrations. However, as obviously other error sources dominate, the agreement with the terrestrial measurements did not clearly improve.

Another possible error source are near-field multipath effects that most likely cause the frequency-dependent biases in the coordinate estimates. Dilßner et al. (2008) showed that reflecting objects in the vicinity of the antenna can bias the coordinate estimates by up to 15 mm and that the simultaneous estimation of troposphere parameters as necessary in global GNSS solutions can even enlarge those biases. According to Dilßner et al. (2008), the antenna near-field of Dorne Margolin type chokering antennas has a radius of about 1.4 m. For the Wettzell antenna array, possible multipath sources within this distance could be the railing of the tower that is covered with a metal sheet or a webcam mounted on a metal pole (see Fig. 20.1) to monitor the snow coverage of the antennas.

One possibility to account for the biases caused by near-field effects is the estimation of correction values from carrier phase residuals as it is, e.g., done for GNSS receiver antennas operated on LEO satellites (Montenbruck et al. 2009; Jäggi et al. 2009). The coordinates derived from terrestrial measurements could be fixed in order to determine residual maps from the analysis of the GPS observations. These corrections could be introduced for GPS solutions solving for empirically corrected coordinates. Another possibility would be a calibration system as proposed by Park et al. (2004) using a directional antenna to directly determine the multipath. Irrespective of the approach to be used, further studies are required to cope with the multipath problems at Wettzell. In particular a better understanding of the interaction of different antenna models, troposphere estimation, and multipath effects is desirable. A solution could also be to find a location at the Geodetic Observatory that is less affected by multipath.

Acknowledgements The authors would like to thank the IGS (Dow et al. 2009) for providing GPS observation data.

References

Boehm J, Niell A, Tregoning P, Schuh H (2006) Global Mapping Function (GMF): a new empirical mapping function based on numerical weather model data. Geophys Res Lett 33:L07304. doi:10.1029/2005GL025546

Boehm J, Heinkelmann R, Schuh H (2007) Short note: a global model of pressure and temperature for geodetic applications. J Geodesy 81 (10):679–683. doi:10.1007/s00190-007-0135-3

Dach R, Hugentobler U, Fridez P, Meindl M (eds) (2007) Bernese GPS Software version 5.0. Astronomical Institute, University of Bern, Switzerland

Dach R, Brockmann E, Schaer S, Beutler G, Meindl M, Prange L, Bock H, Jäggi A, Ostini L (2009) GNSS processing at CODE: status report. J Geodesy 83(3–4):353–365. doi:10.1007/s00190-008-0281-2

Dilßner F, Seeber G, Wübbena G, Schmitz M (2008) Impact of near-field effects on the GNSS position solution. In: Proceedings of ION GNSS 2008, Institute of Navigation, Savannah, pp 612–624

Dow JM, Neilan RE, Rizos C (2009) The International GNSS Service in a changing landscape of Global Navigation Satellite Systems. J Geodesy 83(3–4):191–198. doi:10.1007/s00190-008-0300-3

Jäggi A, Dach R, Montenbruck O, Hugentobler U, Bock H, Beutler G (2009) Phase center modeling for LEO GPS receiver antennas and its impact on precise orbit determination. J Geodesy 83 (12):1145–1162. doi:10.1007/s00190-009-0333-2

Mähler S, Schade C, Klügel T (2010) Local ties at the Geodetic Observatory Wettzell. Geophysical Research Abstracts 12, EGU2010-4815

Menge F, Seeber G, Völksen C,Wübbena G, Schmitz M (1998) Results of absolute field calibration of GPS antenna PCV. In: Proceedings of ION GPS-98, Institute of Navigation, Nashville, pp 31–38

Montenbruck O, Garcia-Fernandez M, Yoon Y, Schön S, Jäggi A (2009) Antenna phase center calibration for precise positioning of LEO satellites. GPS Solut 13(1):23–34. doi:10.1007/s10291-008-0094-z

Park KD, Elósegui P, Davis JL, Jarlemark POJ, Corey BE, Niell AE, Normandeau JE, Meertens CE, Andreatta VA (2004) Development of an antenna and multipath calibration system for Global Positioning System sites. Radio Sci 39:RS5002. doi:10.1029/2003RS002999

Ray J, Crump D, Chin M (2007) New global positioning system reference station in Brazil. GPS Solut 11(1):1–10. doi:10.1007/s10291-006-0032-x

Rothacher M, Lechner V, Schlüter W (2004) Local monitoring of a fundamental GPS site. IGS Workshop 2004, Bern

Saastamoinen J (1973) Contributions to the theory of atmospheric refraction. Bulletin Géodésique 107(1):13–34. doi:10.1007/BF02522083

Schlüter W, Zernecke R, Becker S, Klügel T, Thaller D (2005) Local ties between the reference points at the Fundamentalstation Wettzell. In: Richter B, Dick WR, Schwegmann W (eds) Proceedings of the IERS Workshop on site co-location, no. 33 in IERS Technical Note, Verlag des Bundesamts für Kartographie und Geodäsie, pp 64–70

Schmid R, Steigenberger P, Gendt G, Ge M, Rothacher M (2007) Generation of a consistent absolute phase center correction model for GPS receiver and satellite antennas. J Geodesy 81(12):781–798. doi:10.1007/s00190-007-0148-y

Seitz M, Angermann D, Drewes H (2013) Accuracy assessment of the ITRS 2008 realization of DGFI, DGFI2008 (this issue)

Steigenberger P, Seitz M, Böckmann S, Tesmer V, Hugentobler U (2010) Precision and accuracy of GPS-derived station displacements. Phys Chem Earth. doi: 10.1016/j.pce.2010.07.035 (in press)

Part III

Definition, Establishment, Maintenance and Integration of Regional Reference Frames

The First Insight into Station Velocities in Republic of Serbia

21

D. Blagojević and V. Vasilić

Abstract

The paper presents preliminary results of coordinate time series analysis in Serbian network of permanent GNSS stations. Analysis methodology is outlined and resulting station velocities are briefly commented. Despite the short time span used, the crustal deformation trend was clearly identified, confirming the potential of the network of permanent stations for local and regional geodynamical studies.

Keywords

Station velocities • Noise characteristics • Coordinate time series

1 Introduction

At the end of 2005, the network of 32 permanent GNSS stations was established in Republic of Serbia. The network covers the whole territory with station interdistance of about 60 km (see Fig. 21.1). Trimble equipment and network software were employed in order to support various positioning and surveying tasks.

The network of permanent stations is owned and managed by Republic Geodetic Authority (RGA), which is the reason that all GNSS antennas were mounted on roofs of official RGA administrative buildings across Republic of Serbia. But, despite the fact that station monumentation doesn't suit for precise geodynamical purposes, the research was conducted to estimate station velocities and hopefully discover trend and pattern in station movements.

For this purpose, station coordinates were determined in the form of weekly solutions using Bernese Software v5.0 and standard procedure for processing GPS observations, see Dach et al. (2007). Keeping station SUBO fixed, coordinates were obtained referring to ITRF2005 (Altamimi et al. (2007) and Altamimi (2006)) for the time span of about 1.4 year starting on January 1st 2006. Time series with 74 coordinate solutions along north and east direction were than reduced to first week solution for 27 out of 32 stations. Remaining stations were not used because of large gaps in data.

One of typical coordinate time series are station VRSA (see Fig. 21.2).

2 Noise Characteristics

Following Williams and Teferle (2004) and Williams et al. (2004), time series were subject of several standard analysis steps. Firstly, the data series were detrended and outliers were removed using three sigma criterion. Secondly, Lomb periodograms were calculated based on clean data sets (see Fig. 21.3). Annual and semiannual periods could clearly be identified in every periodogram.

In the sequel, the following power-law was assumed for the noise content:

$$P(f) = P_0 \left(\frac{f}{f_0} \right)^k \tag{21.1}$$

where k denotes spectral index, f is the temporal frequency, P_0 and f_0 being the normalizing constants (Mao et al. (1999) and Amiri-Simkooei et al. (2007)). Fitting the exponential

D. Blagojević (✉) • V. Vasilić
Department for Geodesy and Geoinformatics, Faculty of Civil Engineering, 73 Boulevard king's Aleksandar, The University of Belgrade, 11000 Belgrade, Serbia
e-mail: bdragan@grf.bg.ac.rs; tatic@grf.bg.ac.rs

Z. Altamimi and X. Collilieux (eds.), *Reference Frames for Applications in Geosciences*, International Association of Geodesy Symposia 138, DOI 10.1007/978-3-642-32998-2_21,

Fig. 21.1 Serbian network of permanent stations

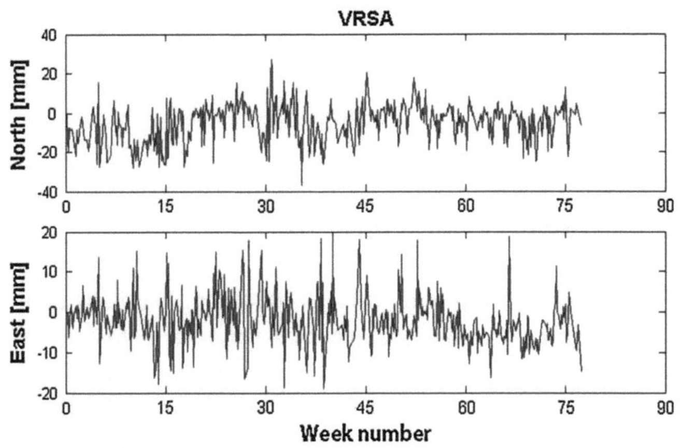

Fig. 21.2 Example of coordinate time series for station VRSA

curve to periodogram values in log–log space (see Fig. 21.4) gave the average spectral index of –0.20 and –0.22 for north and east direction respectively.

Since –1 < k < 1, it can be concluded that white noise dominates in time series. Corresponding white noise amplitudes along north and east direction were estimated to be ±7.8 mm and ±5.3 mm respectively.

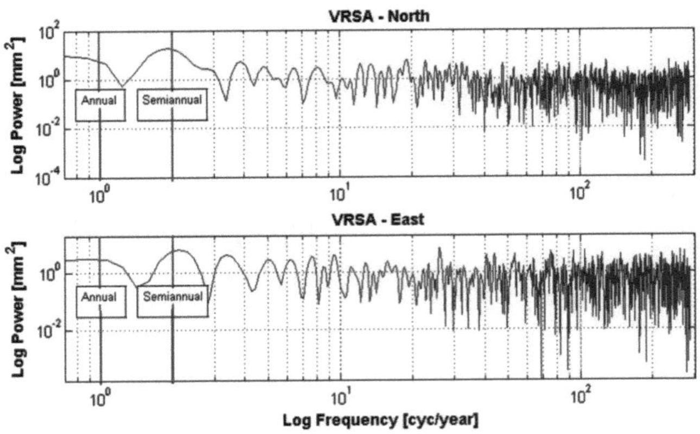

Fig. 21.3 Example of lomb periodogram with distinguished annual and semi-annual period

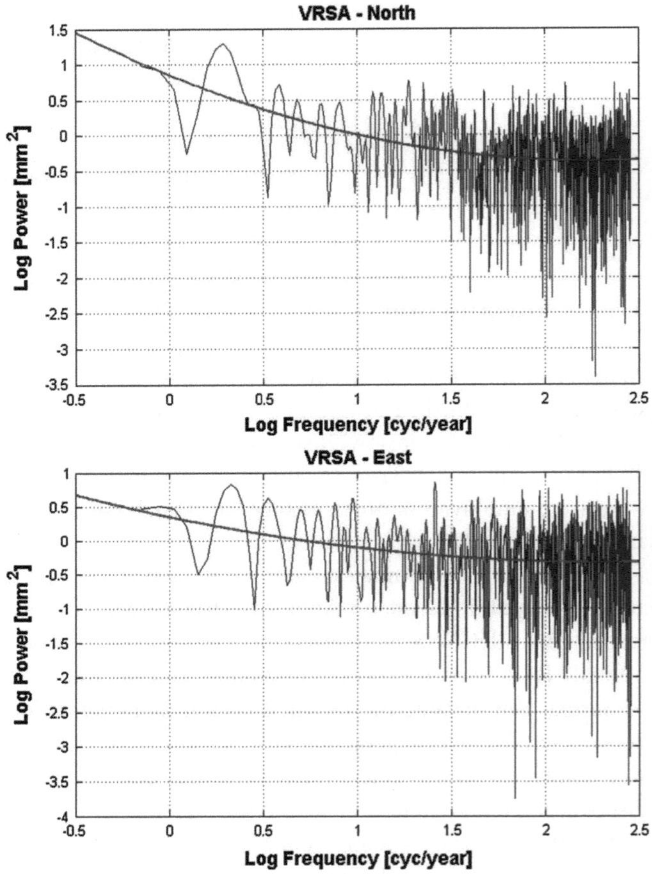

Fig. 21.4 Example of fitting exponential curve for spectral index determination

3 Station Velocities

Station movements along north and east direction, $x(t)$, were modeled relative to station SUBO according to:

Table 21.1 Relative station velocities with respect to station SUBO from 1.4 year measurements

STATION	Northing (mm/year)	Sigma (mm/year)	Easting (mm/year)	Sigma (mm/year)
BAJI	−1.643	0.782	2.488	0.542
BATO	−0.598	0.948	1.759	0.670
BEOG	−1.909	0.728	0.745	0.413
BOR	−0.863	1.074	1.505	0.994
BUJA	−2.927	1.146	1.698	0.967
DIMI	−1.192	1.181	3.125	1.015
INDJ	−6.198	0.856	5.762	0.566
IVAN	−0.093	1.139	2.645	0.697
KIKI	0.902	0.416	−0.258	0.276
KLAD	−0.478	0.981	0.558	0.821
KNJA	−1.461	1.116	4.648	0.931
KRUS	−2.570	1.297	0.846	0.975
KURS	−4.924	1.229	1.306	0.899
LESK	−4.299	1.319	2.354	0.960
LJIG	−4.067	1.028	−1.116	0.487
LOZN	−2.295	0.926	1.699	0.372
NEGO	−3.410	1.542	5.252	1.120
NPAZ	−3.151	1.104	0.937	0.637
NSAD	−4.208	0.845	0.084	0.283
SABA	−3.480	0.918	0.603	0.384
SID	−4.331	0.811	1.418	0.337
SOMB	−2.312	0.195	0.104	0.259
SRBO	−1.051	0.123	7.077	0.447
SUBO	0.000	0.000	0.000	0.000
VALJ	2.443	0.891	−1.461	0.962
VGRA	4.817	1.145	−1.932	0.873
VRSA	4.686	1.182	−2.802	0.751
ZITI	5.263	1.125	−2.679	0.663

$$x(t) = x(t_0) + v(t - t_0) + \sum a_i \cdot \cos \omega_i (t - t_0) + + \sum b_i \cdot \sin \omega_i (t - t_0) \tag{21.2}$$

where v denotes velocity, t_0 is initial time, and a_i, b_i, ω_i, $i = 1{,}2$, are amplitudes and frequencies referring to annual and semiannual periods in time series. Unknown parameters $x(t_0)$, v, a_i, b_i, relative velocities v being the most interesting, were estimated using least squares method because white noise dominates the noise content. Estimated relative velocity components for every network station are listed in Table 21.1, and relative velocity vectors are graphically depicted in Fig. 21.5. Values of 2.0 and 1.5 mm were obtained for average amplitudes of annual and semiannual terms respectively, with average estimated uncertainty of 0.6 mm.

It should be pointed out that all estimated quantities have relative nature because they refer to station SUBO situated in the northern part of Republic of Serbia.

Alternatively, relative station movements along north and east direction with respect to station SUBO, $x(t)$, were analyzed using simple model:

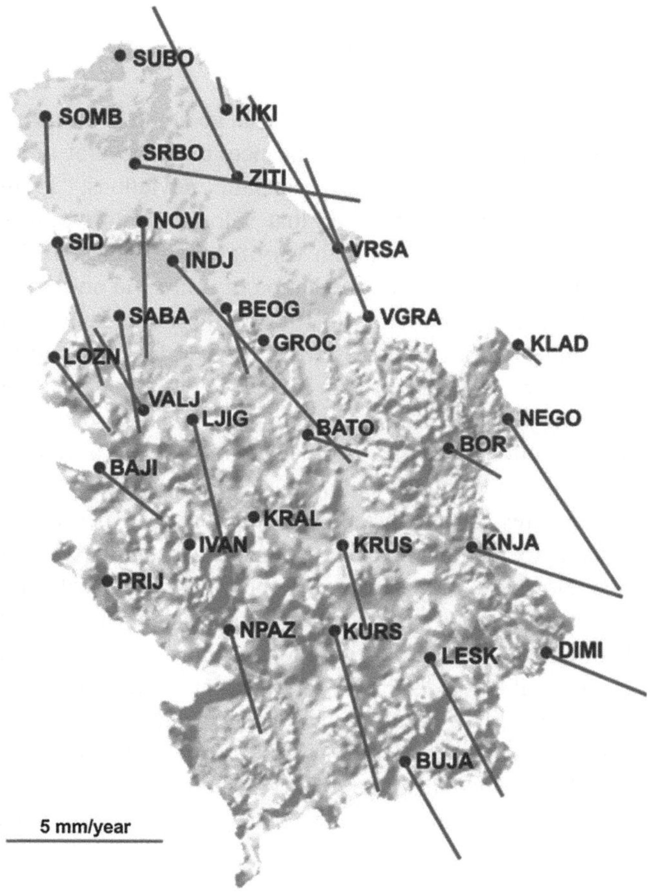

Fig. 21.5 Network station relative velocities with respect to station SUBO

$$x(t) = x(t_0) + v(t - t_0) \tag{21.3}$$

The purpose of this simplified procedure was to demonstrate the necessity of inclusion of periodic terms in time series analysis. The consequences on estimated station velocities were twofold. Both, the velocity values and their precision were affected.

The average velocity magnitude between simplified and complete solution has changed by 0.2 mm/year, but the variation of velocity differences has reached 0.6 mm/year. Likewise, average velocity direction has changed by 10^0 with variation of 13^0.

Furthermore, the precision of estimated velocities has increased by 7 % on average when annual and semiannual periodic terms were included in the model.

These results clearly confirm the sensitivity of velocities on model adequacy despite the fact that velocity changes were not statistically significant at 95 % or even 99 % confidence level. However, obtained differences in velocity vectors and their precision are primarily due to short time span considered during analysis, and should be substantially smaller when longer data sets are at disposal (Blewitt and Lavallee (2002)).

4 Discussion and Further Activities

It is quite obvious from Table 21.1 that about 80 % of station velocities have statistical significance, being several times larger than corresponding standard deviations. From Fig. 21.5 it can also be seen that velocity vectors mostly have north-east direction with very few exceptions. Explanation for quite different orientation of vectors in north-east region along the border with Romania was not found, but remaining vector in the western part of Serbia seems to fit into seismically interesting region. Velocity values of 3–6 mm/year are also quite common, although there are two exceptionally large. Their magnitudes were attributed to stability of buildings and not of the Earth crust.

These results clearly confirm the ability of permanent network to identify crustal movements accumulated over time. However, in the case of Serbian permanent network, the results have to be treated as a first insight into crustal deformations. In order to get more reliable information on tectonic situation, various activities are planned to be performed. First of all, much longer data sets will be analyzed. Station velocities will be additionally estimated by MLE method in order to get the utmost precision of results. Finally, the velocities pattern will be correlated with available seismic, tectonic, geologic and gravity data, and the impact on official Serbian reference frame stability will be assessed.

References

Altamimi Z (2006) Systèmes de référence terrestres: définition, réalisation, application à l'ITRF, état actuel et perspective. Dossier d'Habilitation à Diriger des Recherches. Université Pierre et Marie Curie, Paris.

Altamimi Z, Collilieux X, Legrand J, Garayt B, Boucher C (2007) ITRF2005: a new release of the international terrestrial reference frame based on time series of station positions and earth orientation parameters. J Geophys Res 112:B09401

Amiri-Simkooei AR, Tiberius C, Teunissen P (2007) Assessment of noise in GPS coordinate time series: methodology and results. J Geophys Res 112:B07413

Blewitt G, Lavallée D (2002) Effect of annual signals on geodetic velocity. J Geophys Res 107:B7

Dach R, Hugentobler U, Fridez P, Meindl M (eds.) (2007) Bernese GPS Software Version 5.0, Astronomical Institute. University of Bern, Bern, Switzerland

Mao A, Harrison CGA, Dixon TH (1999) Noise in GPS coordinate time series. J Geophys Res 104:2797

Williams S, Teferle N (2004) CGPS coordinate time series analysis strategy. EU-ESEAS-WP2-T2.2

Williams S, Bock Y, Fang P et al (2004) Error analysis of continuous GPS position time series. J Geophys Res 109:B03412

IAG WG SC1.3 on Regional Dense Velocity Fields: First Results and Steps Ahead

22

C. Bruyninx, J. Legrand, Z. Altamimi, M. Becker, M. Craymer, L. Combrinck, A. Combrink, J. Dawson, R. Dietrich, R. Fernandes, R. Govind, J. Griffiths, T. Herring, A. Kenyeres, R. King, C. Kreemer, D. Lavallée, L. Sánchez, G. Sella, Z. Shen, A. Santamaría-Gómez, and G. Wöppelmann

Abstract

The Working Group on "Regional Dense Velocity Fields" (see http://epncb.oma.be/IAG) of the International Association of Geodesy (IAG) aims at densifying the International Terrestrial Reference Frame and creating a dense velocity field based on regional and global GNSS networks. With the goal to generate a high-quality solution for a core network, several newly reprocessed global and regional cumulative position and velocity solutions were submitted to the Working Group. In order to find a consensus on discontinuity epochs for stations common to several networks (an issue which was problematic in previous submissions), the new submissions were restricted to contain only the core networks over which the analyst has full control so that ITRF2008 discontinuities could be applied. The 3D-RMS of the agreement of the new solutions with the ITRF2008 (after outlier rejection) varies between 0.6 and 1.1 mm/year; it is extremely good for some solutions, while others still require more iteration to reach the required level of agreement. A part of these disagreements has been identified and often originates in the use of different data time spans within the ITRF2008 and submitted solution.

In the upcoming year, the Working Group expects to generate and use a discontinuity database complementing the ITRF2008 set and identify/solve the sources of disagreements. In addition, several of the regional solutions will be reprocessed to embed the regional network in a global network and reduce the error induced by the network effect.

Keywords

GNSS • Reference frame • Velocity field

1 Introduction

The long-term goal of the IAG Working Group "Regional Dense Velocity Fields" is to provide a globally referenced dense velocity field based on GNSS observations to be used as a densification of the multi-technique global conventional reference frame, the ITRF (International Terrestrial Reference Frame).

The Working Group (WG) is embedded within IAG sub-commission 1.3 on "Regional Reference Frames" where it co-exists with the six regional reference frame sub-commissions for Europe, South and Central America, North America, Africa, South-East Asia and Pacific, and Antarctica (Drewes et al. 2008). The WG also closely links its activities with these regional sub-commissions, and regional coordinators have been appointed from the WG members. Their expertise, coordination role for their region, and their capability to generate a unique and unified cumulative solution for their region, including velocity solutions

C. Bruyninx (✉) • J. Legrand • Z. Altamimi • M. Becker • M. Craymer • L. Combrinck • A. Combrink • J. Dawson • R. Dietrich • R. Fernandes • R. Govind • J. Griffiths • T. Herring • A. Kenyeres • R. King • C. Kreemer • D. Lavallée • L. Sánchez • G. Sella • Z. Shen • A. Santamaría-Gómez • G. Wöppelmann
Royal Observatory of Belgium, Av. Circulaire 3, B-1180 Brussels, Belgium
e-mail: C.Bruyninx@oma.be

Z. Altamimi and X. Collilieux (eds.), *Reference Frames for Applications in Geosciences*, International Association of Geodesy Symposia 138, DOI 10.1007/978-3-642-32998-2_22,

from third parties (even campaigns), is a key element for the WG. More details on the WG can be found at its web site http://epncb.oma.be/IAG/.

In 2009, in reply to a first call for participation issued at the end of 2008, regional coordinators and analysts of global networks submitted cumulative velocity solutions to the WG. Several of the regional solutions were a combination of cumulative velocity solutions based on the permanent GNSS network operated by the sub-commission itself and third party velocity solutions. A first test combination of the individual solutions (Bruyninx et al. 2012) showed that the solutions could not be rigorously combined due to:

- Inconsistent discontinuity epochs and solution numbers for the frame-attachment sites (mostly ITRF2005 sites, Altamimi et al. 2007a) entailing large discrepancies at the common sites,
- Inconsistent station naming and DOMES numbering,
- Numerical instabilities caused by velocity constraints at sites with coordinate offsets.

In addition, using a European case study, Legrand et al. (2010a, 2012) showed that positions and velocities obtained from a regional GNSS network tied to the ITRF2005 using minimal constraints, can differ (up to 2 mm in the horizontal and 8 mm in the vertical for the positions and up to 0.5 mm/year in the horizontal and 2 mm/year in the vertical for the velocities) w.r.t. a global solution. In addition, when considering the residual velocity fields after removing the rigid block rotation, the velocity differences are considerably reduced but can still reach up to 0.8 mm/year in horizontal component (Legrand et al. 2010a, 2012).

The disagreement between regional and global positions and velocities is caused by the so-called “network effect”. Indeed, when using the minimal constraint approach, the alignment of the solutions to the reference solution is done using a similarity transformation. The advantage of these minimal constraints is that they preserve the original characteristics of the solution (Altamimi 2003) and do not deform its network geometry. However, in a regional network, the transformation parameters are correlated and sensitive to the reference stations used. The disagreement between regional and global solutions is amplified when the reference stations used in the regional solution cover a smaller geographical area or the different solutions to be combined exhibit large discrepancies at common sites. This means that sites showing different discontinuities, time spans or large non-linear signals should be treated with extreme care. The network effect, of course, challenges the provision of a consistent dense velocity field partly based on regional position/velocity solutions.

2 Methodology and Input

2.1 Approach

Upon the release of the ITRF2008 (Altamimi et al. 2011), the investigation done in Legrand et al. (2010b, 2012), verifying the agreement between regional and global GNSS solutions, was repeated using the ITRF2008 reference frame. Both regional and global solutions were tied to ITRF2008 and the associated ITRF2008 discontinuities were applied. The tests showed an improvement of the agreement (RMS of position/velocity differences) between these new regional and global solutions and the ITRF2008 by a factor of three compared to the ITRF2005 case. Thanks to this improvement, the disagreement between the global and regional position/velocity solutions is now reduced. It can nevertheless still reach 1 mm/year in the vertical and 0.5 mm/year in the horizontal.

These results demonstrate that in order to reduce network effects, it is essential:

- To have the best possible agreement between the solutions we want to combine (by e.g. using similar data span, outlier rejection and discontinuity epochs for the common stations as well as a similar analysis strategy),
- To increase as much as possible the coverage of each of the solutions we want to combine (best is global),
- To increase to a maximum extend the redundancy between regional and global solutions in order to mitigate individual problems at the common stations.

All contributors to the WG were therefore asked in the summer of 2010 to submit a new solution to the Working Group. This solution should preferably be a reprocessed solution (typically using the same absolute antenna models as used for the ITRF2008) and should apply ITRF2008 discontinuities during the stacking to increase the level of agreement of the solution with the ITRF2008.

In addition, three different network levels were introduced (see Fig. 22.1):

- Level 1: IGS/ITRF position/velocity solution.
- Level 2: Regional or global position/velocity solutions in SINEX format over which the contributor has “full control”, meaning that he/she has performed the stacking on the normal equation level, he/she can apply a pre-defined set of discontinuities and can provide access to the residual coordinate time series.
- Level 3: Typically sub-regional solutions, submitted by third parties (can be velocity-only solutions).

Submitters were requested to submit only level 2 solutions.

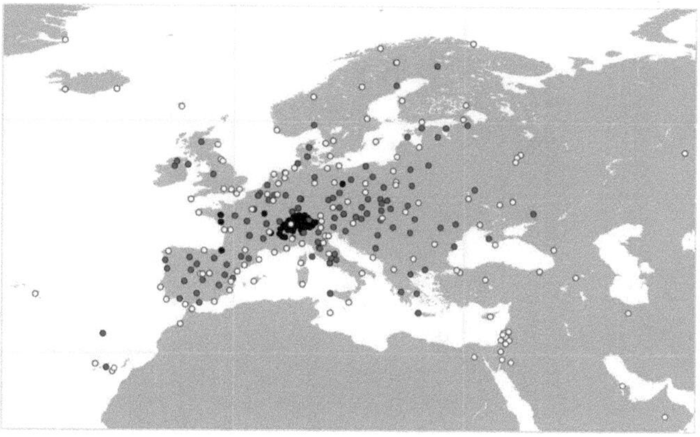

Fig. 22.1 Schematic view of Level 1 (sites in *white*), Level 2 (sites in *grey*) and Level 3 (sites in *black*) networks for Europe

2.2 New Position/Velocity Solutions

One new global solution was provided; it was computed by the ULR consortium (Université de La Rochelle, IGN France and IGN Spain, see Santamaría-Gómez et al. 2011). This position and velocity solution is a reprocessed solution covering the period 01/1996–12/2008 and stacked using the CATREF software (Altamimi et al. 2007b). The number of stations and solution numbers (based on the ITRF2008) are respectively 316 and 584.

The newly submitted regional solutions were:

- Asia and Pacific: This position/velocity solution is covering the period 01/1997–09/2010. The stacking was done using the CATREF software and the number of stations and solution numbers are respectively 155 and 223.
- Africa: This velocity-only solution covers the period 01/1996–08/2010 and includes 89 sites. The site velocities have been computed using a linear regression through daily-estimated site positions expressed in the ITRF2008.
- Europe: This position/velocity solution covers the period 06/1996–07/2009 and was generated using the CATREF software. It contains 234 sites and 427 solution numbers.
- Latin America and Caribbean: This position/velocity solution covers the period 01/2000–06/2010 and it was generated using the ADDNEQ2 program of the Bernese software (Beutler et al. 2007). It contains 183 sites and 215 solution numbers.

All of these solutions are based on a reprocessing, using absolute antenna models (igs05.atx), and applied, as much as possible, the ITRF2008 discontinuity list. The European solution is however not issued from an official reprocessing of the EUREF Permanent Network, but is based on a pilot reprocessing done by the Military University of Warsaw.

For the North America region, no new solution has been made available. Consequently, the level 2 solution used for this region is the NAREF + CBN solution covering the period 10/1994–11/2006. It was generated using the Sinex_combine V1.00 software and contains 728 sites and 871 solution numbers (based on the ITRF2005 discontinuities).

All of these solutions provided in total about 400 densification sites to the ITRF2008 (see Fig. 22.2).

3 Analysis of the Level 2 Solutions

As explained in Sect. 2.1, the combination of the different solutions can only be reliably done if they are compliant. Therefore, in preparation of a future combination, the compliance of each solution with the ITRF2008 discontinuities, positions and velocities was verified. As most solutions are free network/loosely constrained solutions, they were first transformed to be in the same frame as the ITRF2008 using a 14-parameter similarity transformation. The position/velocity differences between each of the transformed solutions and the ITRF2008 are shown in the maps in Figs. 22.3–22.8 and are classified as follows:

- Small black dots: the Level 2 densification stations included in the solutions and not included in ITRF2008,
- White dots: the ITRF2008 stations included in the solutions which satisfy the criteria and which consequently have good agreement with ITRF2008 (also called Selected)
  - Position differences: North and East < 5 mm, Up < 10 mm
  - Velocity differences: North and East < 1 mm/year, Up < 2 mm/year.
- Black dots: the ITRF2008 stations included in the solutions which do not show good agreement with ITRF2008, using the criteria mentioned above (also called Rejected),
- Grey dots: the ITRF2008 stations included in the solutions which do not show good agreement with ITRF2008 (again using the criteria mentioned above) and for which the data span in the solution differs from the data span in the ITRF2008 by more than 2.5 years (also called Rejected ($>$2.5 years)). The limit of 2.5 years has been chosen empirically.

Figure 22.3 illustrates the excellent agreement of the ULR solution with the ITRF2008. 232 among the 259 common sites with ITRF2008 fulfill the criteria. As requested from the submitters, in this first iteration of the ULR solution, the discontinuities applied for the ITRF2008 stations are fully compliant with the ITRF2008 discontinuities. However, as will be shown later in Sect. 4, using blindly the ITRF2008 discontinuities is also not optimal.

Also the European solution (Fig. 22.4) shows excellent agreement with ITRF2008. 100 among the 107 sites common with ITRF2008 fulfill the criteria. The European solution is almost completely compliant with the ITRF2008 discontinuities. The differences are mainly due to a different

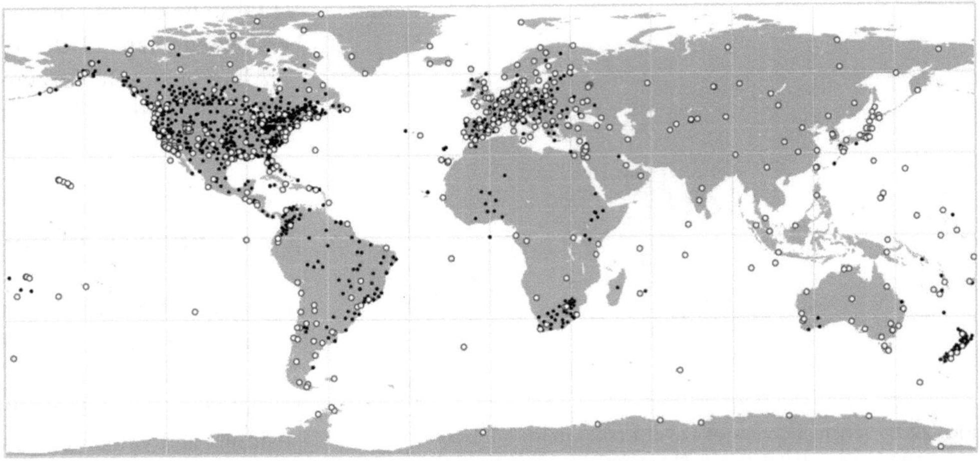

Fig. 22.2 In *white*: ITRF2008; in *black*: received Level 2 solutions

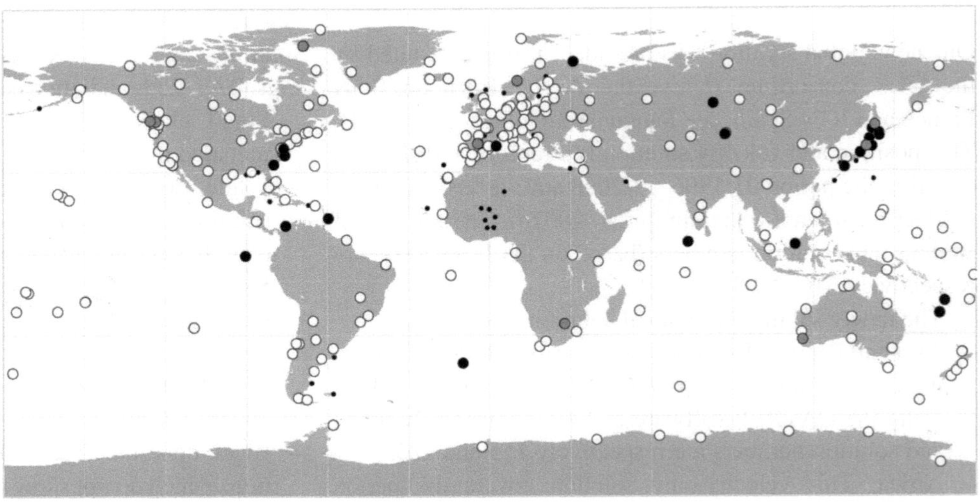

Fig. 22.3 Agreement of the ULR solution with the ITRF20008 (colour codes are explained in the text)

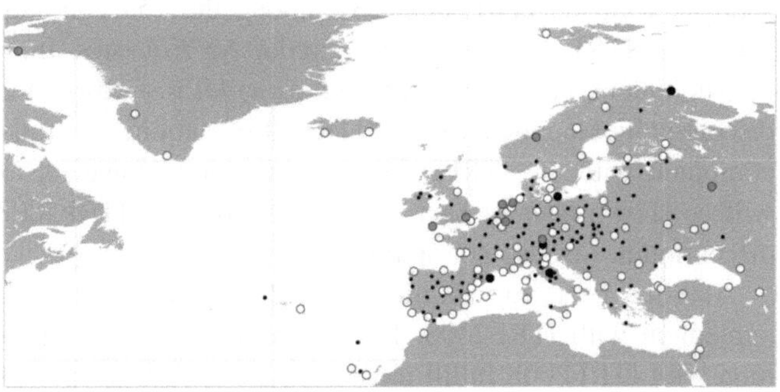

Fig. 22.4 Agreement of the European solution with the ITRF2008

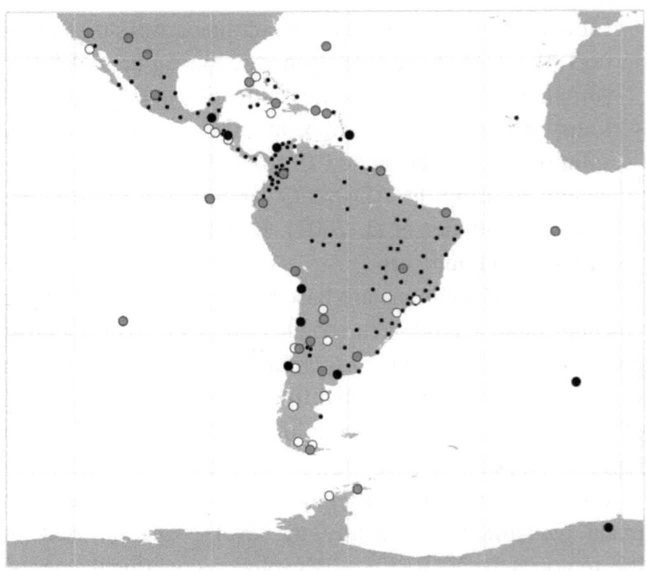

Fig. 22.5 Agreement of the South America and Caribbean solution with the ITRF2008

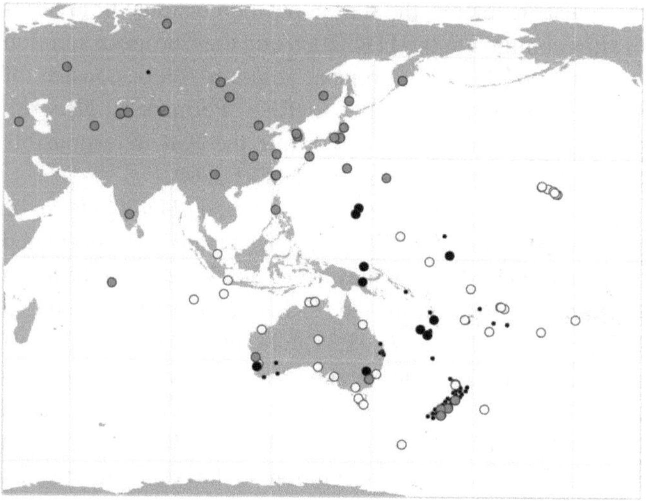

Fig. 22.6 Agreement of the Asia and Pacific solution with the ITRF2008

data span used in both solutions or may be related to the analysis itself (some examples are shown in Sect. 4).

Even if the Latin America and Caribbean solution applied exactly the same discontinuities as ITRF2008, only 65 % of the stations show good agreement with ITRF2008. 35 % of the stations have been rejected, and among them, 14 % show a difference in the data span larger than 2.5 years compared to ITRF2008.

The reason for these differences is under investigation and different leads will be checked in detail. First, there could be some inconsistencies between the ITRF2008 and Latin America and Caribbean: they have been computed using different software, resp. CATREF and ADDNEQ2. In addition, in the ITRF2008, nearby stations within the same site and multiple segments (in case of discontinuities) are constrained to have the same velocity, except for sites where geophysical events as earthquakes occur. However, the solution from Latin America and Caribbean estimated unconstrained velocity sets. Therefore, before performing the comparison, we constrained the velocities in the Latin America and Caribbean solution similarly to the ITRF2008 velocity constraints. Nevertheless, we did not yet check if the way such constraints are applied has an impact on the estimated velocities. Secondly, in presence of significant seasonal signals in this region (Tregoning et al. 2009), the velocity estimation is more sensitive to data span, size of the network, data cleaning.

Only 51.5 % of the stations from the Asia and Pacific solution passed the test. Figure 22.6 highlights that the majority of the stations located in the north of the network have been rejected. The first investigations revealed that these large differences can be explained by the difference in the data span considered in the Asia and Pacific solution with respect to ITRF2008. Indeed, almost 40 % of the

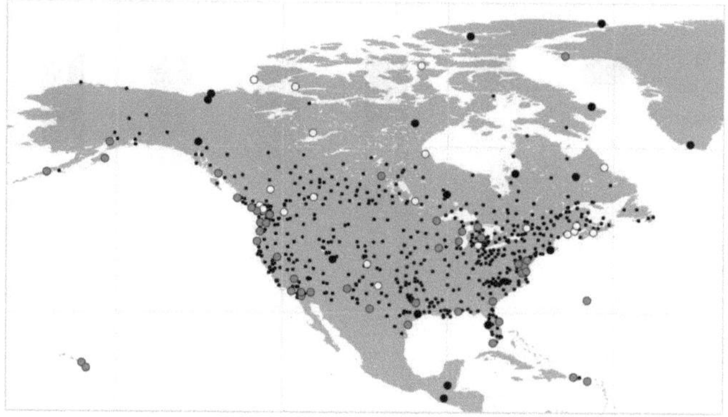

Fig. 22.7 Agreement of the North America solution with the ITRF2008

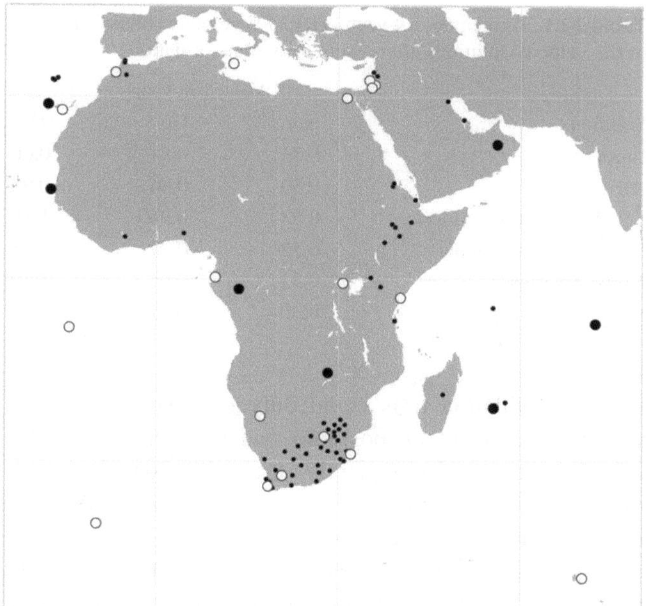

Fig. 22.8 Agreement of the Africa solution with the ITRF2008

rejected stations show a difference in the data span larger than 2.5 years. This issue will be solved in the next iteration of the Asia and Pacific solution which will include a larger time span for these stations.

Being the only non-reprocessed solution still applying ITRF2005 discontinuities, it is not surprising that only 49.4 % of the sites in the North America solution common with ITRF2008 have residuals which passed the chosen criteria. In addition, about 30 % of the rejected stations show a difference in the considered data span larger than 2.5 years of observations (Fig. 22.7). As the North America solution is presently being reprocessed, we expect a significant improvement when this new solution will be released.

Finally, the Africa solution cannot be considered as a full Level 2 solution as it is a velocity-only solution and it does not contain information on the solution epochs. This means

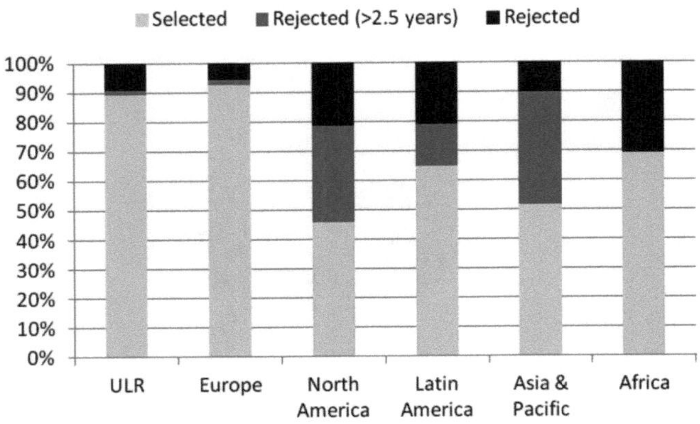

Fig. 22.9 Percentage of ITRF2008 stations included and rejected in each of the solutions

Table 22.1 Agreement between ITRF2008 velocity solution and each of the velocity solutions after a similarity transformation

	RMS Vel. (mm/year)		
Solution	Hor.	Up	3D
Europe	0.33	0.57	0.66
Africa	0.50	0.84	0.98
South and Central America	0.53	0.96	1.10
South-Asia and Pacific	0.52	0.77	0.93
North America	0.53	0.96	1.10
ULR	0.30	0.51	0.59

that the validation tests could only be performed on the velocities (and not on positions and velocities as for the other solutions). Consequently, the validation results are not comparable to those of the other solutions; they are only shown for the sake of completeness.

Figure 22.9 summarizes the number of ITRF2008 stations included or rejected in each of the solutions and Table 22.1 shows the RMS of the residuals that passed the criteria. It gives an indication of the agreement of each velocity solution with the ITRF2008 velocity field.

4 Discussion

After the first combination exercise done in 2009, five new solutions have been submitted to the working group in the summer of 2010. All of these solutions are now based on a consistent GNSS reprocessing using the igs05.atx absolute antenna models. In addition, following the lessons learned (Bruyninx et al. 2012), all of them applied the ITRF2008 discontinuities. This means that significant progress has been made to solve one of the main problems raised within the WG. However, as shown in the previous section, some solutions are still exhibiting significant differences with the ITRF2008 at some sites. In order to understand the origin of these disagreements, the station position residual time series of sites common to the ITRF2008 and the European solution were analyzed. This European solution solely consists of stations belonging to the EUREF Permanent Network (EPN, Bruyninx et al. 2009), while the IGS is responsible for the GNSS contribution to the ITRF2008.

We focus in this paper on the stations UNPG (Perugia, Italy) and VARS (Vardø, Norway) which are both considered as *rejected* in Fig. 22.4 as they have velocity residuals between the European solution and the ITRF2008 of respectively (0.34, 0.49, −2.20) and (−2.37, −1.31, 1.14) mm/year in North, East and Up. Figure 22.10 allows comparing the residual position time series of UNPG available from the ITRF2008 (from http://itrf.ign.fr/ITRF_solutions/2008/ITRF2008_ts.php) with the ones from the European solution. In the bottom of the plot, a horizontal color bar indicates the different solution numbers applied in each of the solutions. The usage of less data in the ITRF2008 compared to the European solution entails large differences in the estimated velocity. The UNPG velocity in the ITRF2008 is not representative for the true station behavior and will need to be removed from the ITRF2008 before combining it with the Level 2 networks. In addition, due to the usage of less data within the ITRF2008 solution, the station position discontinuity in April 2006, caused by an antenna change and seen by the European solution, will not be included in the ITRF2008 discontinuity list and will need to be added by the analyst himself.

A similar comparison is made for VARS in Fig. 22.11. It shows that within the ITRF2008, the station is performing significantly worse than within the European solution, which is problematic, does not give proper credit to the efforts of the station manager, and, again, causes discrepancies between the European solution and the ITRF2008. Both stations are not part of the IGS network. They should in principle show up as Level 2 stations, as information on station configuration and maintenance is formally not flowing to the global (IGS) level. However, these stations were picked up by IGS analysis centers at some point and they ended up in the IGS contribution to the ITRF2008 becoming at that point Level 1 stations. Due to the presence of these stations in the ITRF2008, they become possible frame-attachment sites which can be used to tie a regional solution to the ITRF2008. Especially, the station VARS, located in the North of Europe, could potentially be an important reference station when using minimal constraints to express the European regional solution in the ITRF as it is on the border of the European network.

From the examples shown, it is clear that a careful inspection and comparison of both ITRF2008 and regional time series is mandatory before using any site as a frame-attachment site. This adds an additional task to the WG which will need to verify and eventually discard some of the stations included in ITRF2008 before performing the combination. The reason for the lack of (good quality)

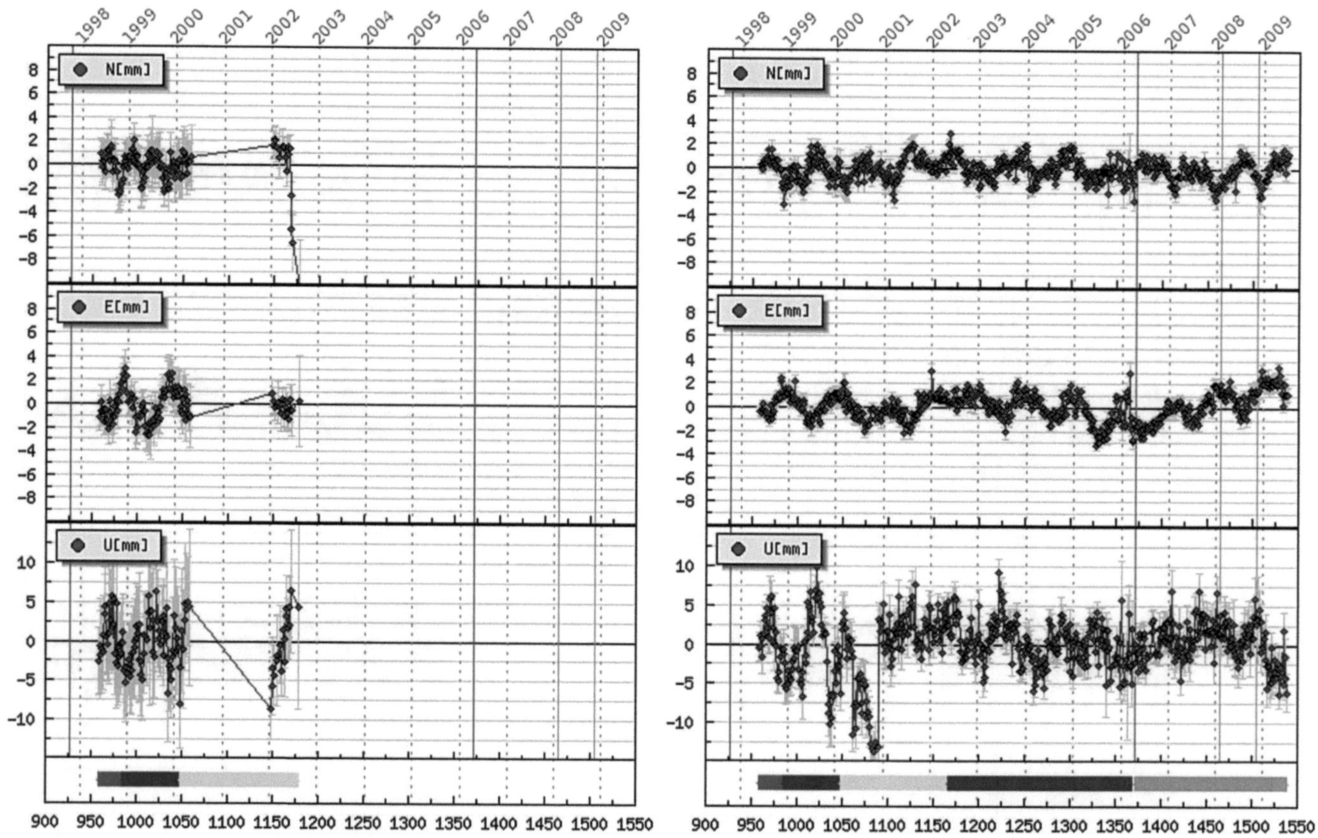

Fig. 22.10 Residual position time series (North, East, Up) of the station UNPG. *Left*: from ITRF2008, *right*: from European analysis. *Red* (*green*) vertical lines indicate a change in the station antenna/radome (receiver) configuration. The colored bars in the bottom of the plot indicate when a position discontinuity (identical colors indicate the same solution number) was introduced during the stacking

station results (for both UNPG and VARS) in the ITRF2008 solution is presently not yet understood and raises the need for more interaction between the regional reference frame sub-commissions and the IGS (International GNSS Service, Dow et al. 2009) and/or the ITRF product center in order to prevent from facing a similar situation in the next release of the ITRF.

5 Steps Ahead

A remaining problem is that the majority of the solutions submitted by the regional sub-commissions are based on regional network processing. In order to reduce the error induced by the network effect, these solutions should be embedded in a global analysis. For the regions of Asia and Pacific, Africa and North America such global solutions will become available in 2011. For South-America and Europe, however, the regional sub-commissions have no official plans to generate a global solution. Both regional groups offer as an alternative, as a first step, to combine their weekly regional solutions with the global weekly reprocessed solutions generated by the IGS or one of its Analysis Centers. In a second step, these weekly combined solutions will then be stacked and tied to the ITRF2008 taking advantage of the availability of a global set of reference stations.

On the other hand, Legrand et al. (2010b) have shown that the network effect on cumulative solutions is significantly reduced when global and regional solutions have a better agreement, which is the case when using the ITRF2008 and reprocessed GNSS solutions. This lets us hope for an improvement compared to the previously submitted ITRF2005 velocity fields.

Another possible way to go ahead could be to ask all contributors to submit weekly SINEX solutions instead of cumulative position/velocity solutions. These solutions could then be combined on the weekly level (using e.g. a strategy similar to described in Davis and Blewitt 2000) and the attribution of the discontinuity epochs will only have to be dealt with at the level of the final stacking, remedying both the network effect and the difficulty to find a common set of discontinuities. However, due to a lack of manpower, this idea is presently not considered as realistic. Moreover, the quality of the results that can be obtained from such a weekly combination must first be investigated.

The WG will, in any case, concentrate at the moment on the Level 2 solutions in order to try having a first

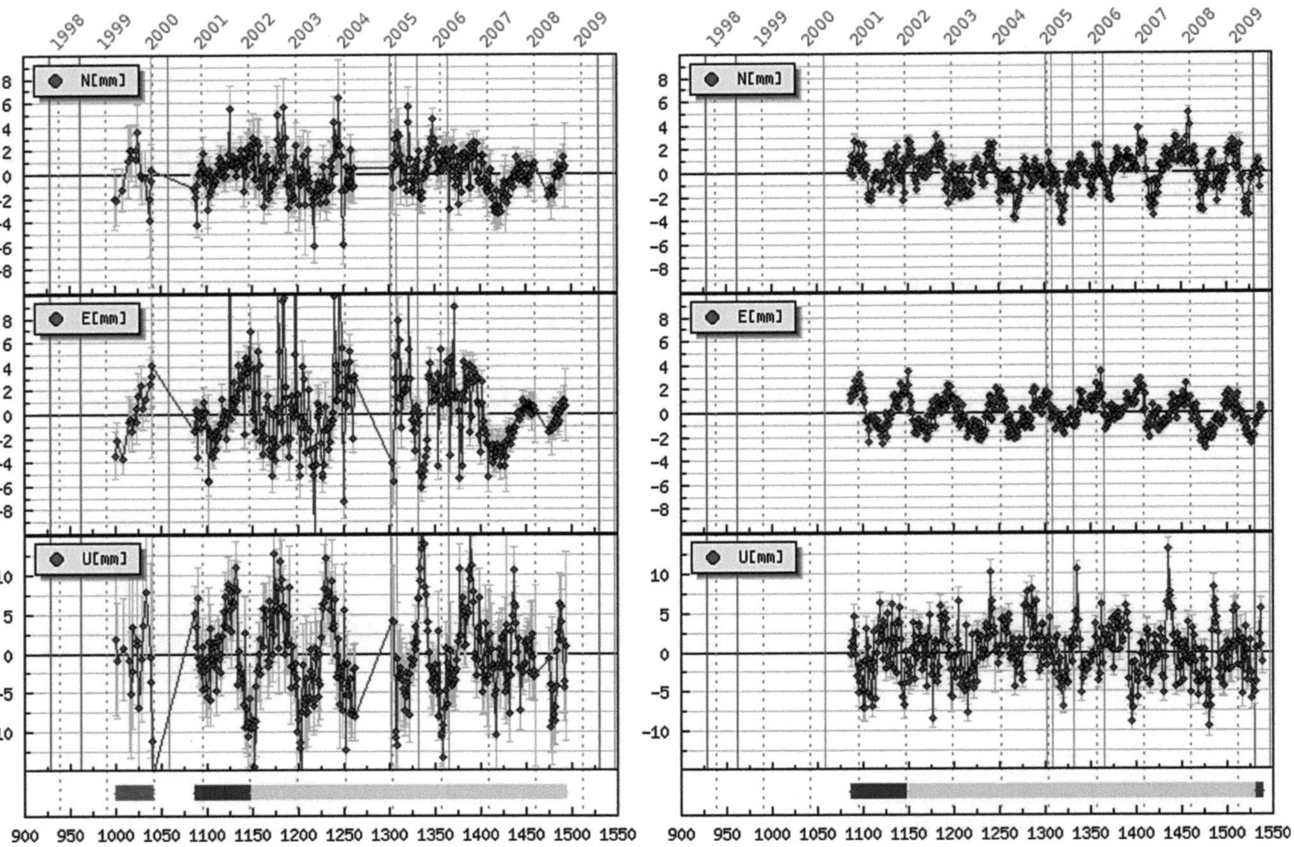

Fig. 22.11 Residual position time series of the station VARS (North, East, Up). Plot layout similar to Fig. 22.10. Although the station was only officially included in the EPN in GPS Week 1092, the IGS ACs already included earlier data in their solutions

densification of the ITRF2008 which is of good quality. As we have seen, blindly applying the ITRF2008 discontinuities should not be recommended. In order to provide an exhaustive list of discontinuities to be used within the WG, the contributors to this WG have agreed to document when/why they have chosen to deviate from the ITRF2008 discontinuity list and make available their associated residual coordinate time series which will then be centralized and intercompared within the WG. As soon as an agreement can be found on the discontinuities to be applied in the Level 2 solutions, and these Level 2 solutions can be successfully combined with the Level 1 solutions, then the WG will tackle the problem on how to integrate the third party (position and) velocity Level 3 solutions.

Acknowledgements The authors of this paper would like to thank the groups who submitted velocity solutions.

References

Altamimi Z (2003) Discussion on how to express a regional GPS solution in the ITRF, EUREF Publication No. 12, Verlag des Bundesamtes für Kartographie und Geodäsie, Frankfurt am Main, pp 162–167

Altamimi Z, Collilieux X, Legrand J, Garayt B, Boucher C (2007a) ITRF2005: a new release of the international terrestrial reference frame based on time series of station positions and Earth orientation parameters. J Geophys Res 112:B09401. doi:10.1029/2007JB004949

Altamimi Z, Sillard P, Boucher C (2007b) CATREF software: combination and analysis of terrestrial reference frames. LAREG, Technical, Institut Géographique National, Paris

Altamimi Z, Collilieux X, Métivier L (2011) ITRF2008: an improved solution of the international terrestrial reference frame. J Geodesy. doi:10.1007/s00190-011-0444-4

Beutler G, Bock H, Brockmann E, Dach R, Fridez P, Gurtner W, Habrich H, Hugentobler U, Ineichen D, Jaeggi A, Meindl M, Mervart L, Rothacher M, Schaer S, Schmid R, Springer T, Steigenberger P, Svehla D, Thaller D, Urschl C, Weber R (2007) In: Hugentobler U, Dach R, Fridez P, Meindl M (eds) Bernese GPS software version 5.0. Univerity of Bern, Bern, Switzerland

Bruyninx C, Altamimi Z, Becker M, Craymer M, Combrinck L, Combrink A, Dawson J, Dietrich R, Fernandes R, Govind R, Herring T, Kenyeres A, King R, Kreemer C, Lavallée D, Legrand J, Sánchez L, Sella G, Shen Z, Santamaría-Gómez A, Wöppelmann G (2012) A dense global velocity field based on GNSS observations: preliminary results. In: International Association of Geodesy Symposia 136, Geodesy for Planet Earth, pp. 19–26, doi:10.1007/978-3-642-20338-1_3, http://rd.springer.com/chapter/10.1007/978-3-642-20338-1_3, vol 136. Springer, Berlin

Bruyninx C, Altamimi Z, Boucher C, Brockmann E, Caporali A, Gurtner W, Habrich H, Hornik H, Ihde J, Kenyeres A, Mäkinen J, Stangl G, van der Marel H, Simek J, Söhne W, Torres JA, Weber G (2009) The European reference frame: maintenance and products. In: IAG symposia series, geodetic reference frames. Springer, Berlin, vol 134. pp 131–136, doi: 10.1007/978-3-642-00860-3_20, http://rd.springer.com/chapter/10.1007/978-3-642-00860-3_20

Davis P, Blewitt G (2000) Methodology for global geodetic time series estimation: a new tool for geodynamics. J Geophys Res 105 (B5):11083–11100

Dow JM, Neilan RE, Rizos C (2009) The international GNSS service in a changing landscape of global navigation satellite systems. J Geodesy 83:191–198. doi:10.1007/s00190-008-0300-3

Drewes H, Hornik H, Ádám J, Rózsa S (eds) (2008) The geodesist handbook 2008. J Geodesy, Springer, 82(11):661–846

Legrand J, Bergeot N, Bruyninx C, Wöppelmann G, Bouin M-N, Altamimi Z (2010a) Impact of regional reference frame definition on geodynamic interpretations. J Geodesy 49: 116–122. doi: 10.1016/j.jog.2009.10.002.

Legrand J, Bruyninx C, Bergeot N, Santamaría-Gómez A, Bouin MN, Wöppelmann G, Altamimi Z (2010b) Limitations of regional GNSS networks for geophysical and geodynamical studies. Presented at IAG commission 1 symposium 2010 reference frames for applications in geosciences (REFAG2010), http://iag.ign.fr/abstract/pdf/Legrand_REFAG2010.pdf

Legrand J, Bergeot N, Bruyninx C, Wöppelmann G, Santamaría-Gómez A, Bouin M-N, Altamimi Z (2012) Comparison of regional and global GNSS positions, velocities and residual time series. In: International Association of Geodesy Symposia 136, Geodesy for Planet Earth. Springer, Berlin, pp. 95–104, doi:10.1007/978-3-642-20338-1_12.

Santamaría-Gómez A, Bouin M, Collilieux X, Wöppelmann G (2011) Correlated errors in GPS position time series: implications for velocity estimates. J Geophys Res 116:B01405. doi:10.1029/2010JB007701

Tregoning P, Watson C, Ramillien G, McQueen H, Zhang J (2009) Detecting hydrologic deformation using GRACE and GPS. Geophys Res Lett 36:L15401

Local Ties at Fundamental Stations

23

Ulla Kallio and Markku Poutanen

Abstract

In this article we discuss and give an example of three different methods to measure the local tie between the IGS GPS antenna reference point (ARP) and the VLBI antenna reference point at the Metsähovi fundamental station. First we introduce traditional survey approach combined with a space intersection technique and then local tie based on kinematic GPS during geo-VLBI campaigns and finally local ties with static GPS measurements. We discuss the measurements and computations, problems and error sources encountered between the planning of the measurements and the end product, the local tie vector. Although millimetre precision can be achieved, our experience during tests at Metsähovi fundamental station shows that some techniques are very time consuming and impractical for a routine use. We aim for a more automated process in local tie measurement.

Based on our experiments and results, we propose that in the future, the VLBI antenna tie could be tracked permanently during geo-VLBI campaigns with attached GPS antennas.

Keywords

Geodetic ties • VLBI • GPS • Fundamental stations

1 Introduction

The global reference frame ITRF depends on the reliable products of IAG Services, and especially on the combination of different observing techniques. The existing infrastructure is provided by national agencies and therefore local infrastructure, instrumentation and its age differs from site to site.

The development will eventually lead to a number of "super stations", i.e. fundamental stations having most of space geodetic techniques. In coming years these stations will form the basic infrastructure for the global reference frame. One of the major challenges will be reliable ties between co-located techniques and ability to monitor temporal changes on the mm-level. The accuracy should be site-independent, consistent, reliably controlled, and traceable over long time periods, but also cost-effective and not too laborious to do it regularly.

What is the most suitable method to measure a local tie vector depends on the local circumstances. For example the radome over the radio telescope constrains the measurements remarkably. We cannot directly adopt methods used at some other fundamental stations; however many main principles are still common.

The reference point of a VLBI antenna has no physical marker but it can be determined as the intersection of the primary axis with the shortest vector between the primary and secondary axis (Dawson et al. 2007). The indirect measurements have been used in all cases in recent local ties reported in the ITRF web site (ITRF 2011).

Some points on the antenna structure were measured in different antenna positions, coordinates of them were computed and the centering parameters of a VLBI antenna were calculated from those coordinates. These include the coordinates of the reference point, axis offset, orientation and non-perpendicularity of the axes.

U. Kallio (✉) • M. Poutanen
Department of Geodesy and Geodynamics, Finnish Geodetic Institute, Geodeetinrinne 2, FI-02431 Masala, Finland
e-mail: Ulla.Kallio@fgi.fi

Z. Altamimi and X. Collilieux (eds.), *Reference Frames for Applications in Geosciences*, International Association of Geodesy Symposia 138, DOI 10.1007/978-3-642-32998-2_23,

Optimal number of points on the antenna structure and their best placement were studied by Kallio and Poutanen (Kallio and Poutanen 2010). Mathematical models to calculate the reference point were presented e.g. in Dawson et al (2007), Sarti et al. (2004), Lösler and Hennes (2008), Lösler (2009), Kallio and Poutanen (2009). In this paper we use the model introduced in (Kallio and Poutanen 2009, 2010).

The basic equation and the rotation matrix of our model are

$$X_0 + R_{\alpha,a}(E - X_0) + R_{\alpha,a}R_{\beta,e}p - X = 0 \qquad (23.1)$$

Rotation matrices $R_{i,j}$ are formed by applying the Rodrigues' rotation formula which uses rotation axis and angle. The rotated X is

$$X_r = \cos(\alpha)X + (1 - \cos(\alpha))aa^TX + \sin(\alpha)a \times X \qquad (23.2)$$

The steps needed to obtain the tie vector in ITRF depend on the measurement methods, calculation strategy and in which reference frame the calculation is performed; we refer here (ITRF 2011) and the "Proposed IERS Local Tie Report Layout" recommended by IERS Site Survey and Co-location Working Group 2 (IERS 2011).

2 Traditional Terrestrial Measurements

2.1 Main Principles

Tachymetry observations were taken using two face angle measurements and careful levelling of the instruments. We used only calibrated instruments. At least four sets of angles were observed using stable points. The height differences between the points were obtained using precise levelling. Besides the repeated measurements we tried to get the best possible geometric reliability which means that the triangle type network was preferred over a traverse type. The principle to use only stable points means that all points should be pillar points. In our measurements it was necessary to also use some temporary tripod points.

2.2 The Local Network

Measurements steps were: angle and distance measurements at the pillar network and at the small network inside the radome, angle and distance measurements to connect these networks, levelling of the height differences, space intersection measurements to the points in antenna structure in different antenna position, and GPS measurements at the pillar network.

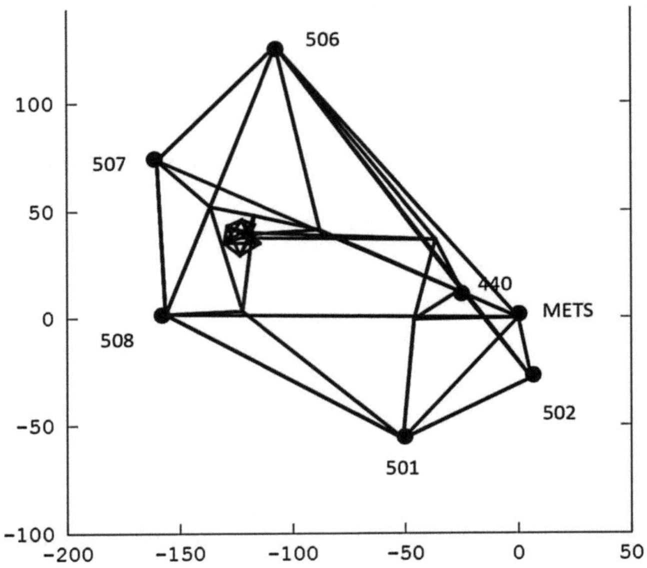

Fig. 23.1 The local network. 440, 501, 502, 506, 507and 508 are pillar points around the radio telescope, METS is the GPS antenna belonging to the IGS network. Scale is in metres

We measured a tachymeter network including the METS GPS ARP (Antenna Reference Point), six concrete pillar points and some older triangulation points monumented with steel bolts in the bedrock. The Metsähovi VLBI antenna is inside the radome and that is one of the main challenges in our local tie measurements. To connect the small network inside the radome and the pillar network outside we had to use some temporary points (Fig. 23.1).

This part of the local tie network was the weakest and most problematic in many ways. Firstl the temporary points were not stable. We assumed that they were stable enough during the short (5 days) measurement campaign we had for the connection. If we are to repeat the measurements after a few years we must use different points. Second it was problematic to find points where we have the line of sight to the inside of the radome. Anyhow, the network geometry and reliability improved from the earlier measurements when the connection was made from one point only (Jokela et al. 2009).

The horizontal and zenith angle together with distances between points were measured. Height differences between the points were levelled. For orientation of the network a 1 week GPS-campaign were carried out in autumn 2008. Afterwards the network was completed with additional vertical angle measurements and height difference determination between the METS ARP and two pillar points. Inside the radome we established six fixed tripods on the concrete wall around the antenna and together with three floor points (a pinhole and a steel bolt on the floor and an older screw point on the telescope concrete stand) they form the network. Angles and distances were measured between all possible points and the height difference between the floor points and wall points were levelled.

Table 23.1 The local tie vectors and their lengths between METS GPS and VLBI reference point and the axis offsets in the static GPS campaign, four kinematic GPS campaigns and tachymetric measurements. The standard deviations (std) of the vector components and axis offsets are included

Campaign	North (m)	std (mm)	East (m)	std (mm)	Up (m)	std (mm)	Length (m)	Axis offset (mm)	std (mm)
Static 2 h	37.6121	0.4	−122.3997	0.5	−14.6738	0.7	128.8863	−5.3	1.0
IVS-T2059	37.6086	0.2	−122.4009	0.2	−14.6781	0.4	128.8869	−4.5	0.6
EUROPE-97	37.6080	0.3	−122.4006	0.2	−14.6780	0.5	128.8864	−5.0	0.7
EUROPE-98	37.6071	0.3	−122.4019	0.2	−14.6776	0.6	128.8874	−5.2	0.9
IVS-T2061	37.6094	0.3	−122.3995	0.3	−14.6786	0.8	128.8859	2.0	1.1
Tachymetric	37.6105	0.1	−122.4015	0.1	−14.6722	0.2	128.8874	0.3	0.2
Combined	37.6095	0.1	−122.4006	0.1	−14.6751	0.2	128.8865	−2.1	0.3

We attached six reflective tape targets on the back of the VLBI dish. We planned to measure the points in 12 azimuth and 9 elevation position using a space intersection technique. Two TC2003 tachymeters were centered on the fixed tripods on the concrete wall in the radome around the VLBI antenna.

Measurements were possible only during bad weather conditions; otherwise the antenna was in normal use. As an example, when the radome was heated to melt the snow on the roof during our space intersection measurements, the temperature inside the radome varied 0–30 °C and the temperature gradients were too large to continue the measurements. At the end we got only 18 antenna positions and 68 points. Some targets were not visible in all antenna positions.

2.3 Calculation of the Local Network

We used ITRF2005 reference frame in our calculation. In order to get the orientation in the local terrestrial network we carried out a 1 week static GPS campaign. The network consisted of six pillars, some older benchmarks in the bedrock and the METS IGS GPS point as a fixed point. The GPS network was computed and adjusted with the Bernese GPS processing software v. 5.0 using the IGS precise orbits and ERP (Dach et al 2007).

The vector components from the fixed point METS to the other points with their covariance matrix together with horizontal and vertical angles and distances were further used as observations in the combined adjustment of the local terrestrial network.

The mathematical model used in the combined three dimensional network adjustment is presented e.g. in Leick (1995) and implemented in Octave by the first author. The model used in the VLBI antenna reference point adjustment is described in Kallio and Poutanen (2009). Angle observations and levelled height differences are referred to the plumb line as vertical. Angles were reduced to the local geodetic system using the deflections of vertical calculated from the FIN2005N00 geoid model (Bilker-Koivula and Ollikainen 2009).

Prior to the adjustment, levelled height differences between the pillar points and the points in the radome were converted to the ellipsoidal height differences using again the FIN2005N00 geoid model. The points on the concrete wall inside the radome were connected to the local network in same adjustment with the angle and distance observations made on the four temporary point. Height differences between METS GPS antenna and some pillar points were taken from the previous measurements (Jokela et al. 2009).

As the result we computed coordinates and the covariance matrix of the points in the network which we used in the determination of the coordinates of the points in VLBI antenna structure.

The space intersection measurements were adjusted to obtain coordinates and covariance matrix of the six targets in several antenna positions. Coordinates of the tachymeter points were constrained using the covariance matrix from the adjustment of the local network. After the adjustment of space intersection measurements the estimated standard deviations of the coordinates of the antenna points varied 0.5–1.5 mm. VLBI antenna azimuths were chosen so that we had a line of sight to the targets from the both tachymeters in nine different elevation position.

With 18 (of 108 planned) antenna positions we calculated the reference point, axis offset and the orientation of axes with our mathematical model (Kallio and Poutanen 2009). The local tie vector and axis offset with standard deviations are presented in Table 23.1.

3 Assessment of the Terrestrial Method

3.1 Orientation

One main source of uncertainty in local tie measurements using terrestrial observations is the orientation of the measurements to the reference frame. The shape and the scale of our local network are well determined and internally consistent when using angle and distance measurement and levelled height differences.

The vertical orientation of the instruments can be done by levelling them so that they refer to the local plumb line, and applying a geoid model for the deflections of vertical. There remains one orientation unknown per instrument setting in the adjustment and the orientation of the whole network needs extra information to remove the rank deficiency in the normal equation matrix. The information can be GPS vectors (coordinate differences from GPS network) or azimuths calculated from GPS coordinates or fixed points. We used the GPS vectors in our adjustment.

It is also possible to carry out a free network adjustment without GPS observations. Then the network is in an arbitrary frame and the three dimensional Helmert transformation parameters between the tachymeter network and the GPS network must be determined.

The second way is to choose an arbitrary Cartesian reference frame, for example defined by the axes of one instrument setting, and solve for the three dimensional orientation parameters (rotation matrix) as additional unknowns for each instrument setting referring to this local frame. Also in this case the three dimensional Helmert transformation parameters between the tachymeter network (local frame) and the GPS network must be determined.

In all cases the final solution dependents on the GPS measurements: How accurate is the orientation of the local campaign based on the GPS network? How accurate is the geoid model? Do we need in the future at least three permanent GPS stations for orientation of the local tie network? Do we need a more accurate local geoid model?

There is also a difference between the GPS antenna ARPs and the phase centres which are known from the antenna absolute calibration. It is uncertain how well the calibration values are suited for local circumstances.

3.2 Radome

The reliability and precision of observations of the antenna targets were restricted by the radome. The network geometry was moderate when tying the pillar network outside and the network inside the radome and in our case the connecting points were neither permanent nor stable. In the connection we used tripod points and the centring errors of the instrument were estimated larger than at pillar points. The variances of the centering were propagated in the adjustment to the variances of distance, angle and coordinate difference observations. We used 0.2–0.1 mm and 1–2 mm standard deviations for horizontal and vertical components of eccentricity, respectively.

3.3 Time Issues

There was no reasonable way to study the repeatability of the method. The time needed to carry out all the measurements in a proper way would take several weeks. We made measurements under the radome only when the normal use of the telescope was not possible (when raining or snowing). Better results can be achieved if all measurements are done in a short time period and under same environmental conditions. At the Metsähovi station (like in most other stations, too) even several weeks interruption is not reasonable or even possible.

3.4 Precision and Accuracy

In the computation of the local tie measurements the uncertainties in the distance, angle, coordinate difference, and height difference observations propagate in the network adjustment and from the adjustment of VLBI-antenna parameters to the end product, the local tie vector. Besides this we must include the errors in the centering and levelling of the instrument and errors in the geoid model. The precision of the local tie vector depends on the stochastic models of the adjustments, network geometry and geometry in the space intersection, number and placement of the targets and the number of antenna positions. The weighting of observations must be based on the real variances and covariances of the observations. If the calculations were made in a local reference, the errors in transformation to a global frame must be included. Additionally, data may have systematic errors which cannot be detected.

In our network adjustment the weight matrix is the inverse of the covariance matrix of the observations. Observations have individual variances. The covariances between the GPS-vector components and the covariances of height differences were included but otherwise observations were handled as stochastically independent.

4 Simultaneous Ties During Geo-VLBI Sessions

4.1 Instrumentation

Due to the time issues, we evaluated a test kinematic GPS approach during the normal use of the radio telescope. GPS data were collected during a regular geo-VLBI session and downloaded afterwards from receivers. We used two Ashtech Z-12 GPS receivers with calibrated Ashtech Dorne

Margolin type antennas. The GPS antennas were attached to the sides of the VLBI-antenna dish (Kallio and Poutanen 2009). The turnable antenna holders with counterweights forced the GPS antennas to point to the zenith regardless of the position of the radio telescope dish.

The VLBI telescope dish orientation readings (azimuth and elevation angles) were registered during the session. This information was needed in our computation.

4.2 Observations and Data Management

The results presented in this article have been computed from the data gathered during the following geo-VLBI-campaigns: IVS-T2059 (16.–17. 12.2008), EUROPE-97 (19.–20.1.2009), EUROPE-98 (25.–26.3.2009), and IVS-T2061 (7.–8.4.2009).

Using the information of the VLBI antenna orientation and the GPS antenna calibration tables we calculated the phase centre correction in the direction of each GPS satellite for every epoch and both frequencies. The observed phase values were corrected in the RINEX-files before the computation using our own software developed for the purpose. Computing the trajectory of the GPS antennas was made with Trimble Total Control (TTC) software using IGS precise ephemerides, On-The-Fly strategy and keeping IGS point METS fixed. Any additional antenna corrections were switched off. Only the fixed integer ambiguity solutions were accepted. The standard deviation of the accepted trajectory point coordinates varied 1–100 mm.

Some interpolations for azimuths and elevations were needed for synchronizing the GPS antenna trajectory and antenna position data. In interpolation the linear angular velocity of the antenna was assumed.

In our analysis we rejected about two thirds of the data as outliers due to different reasons: no fixed solution, pdop was larger than eight, there were less than six satellites, distance between the two GPS antennas attached on the radio telescope deviate too much from the median, or the standardized residuals of the solution were larger than three.

The radome and the shadowing by the telescope dish were the biggest problems during the measurements. The large dish blocked the signal and formed a source of multipath, as well as the support structure of the radome. We used only 30 s sampling interval of observations in this test, partly because it is also the interval for the Metsähovi IGS GPS and partly because the storage capacity of the receivers.

After these operations we had coordinates for GPS antennas, standard deviations of the coordinates and azimuth and elevation of the VLBI-antenna for every accepted epoch. The position of the reference point, axis offset and the orientation of the antenna were directly in ITRF2005 reference frame and no further transformations were needed.

The inverse of the covariance matrix of the observations – the coordinates and the telescope angle readings – were used as a weight matrix in the reference point computation. The TTC trajectory data includes standard deviations for trajectory coordinates for every epoch but not covariances between the components or trajectory points. That is why we used only the diagonal weight matrix when processing antenna parameters from the trajectory data. One solution to improve the stochastic model would be to calculate theoretical correlations between epochs for the trajectory points.

4.3 Advantages of the Kinematic Approach

One of the advantages of the kinematic method is that the normal use of the radio telescope is not interrupted and the GPS installation can be kept permanently there. The epoch of the local tie equals to the epoch of VLBI measurements and one can obtain a direct tie to a fixed GPS antenna, e.g. the collocated IGS network station. Moreover, one needs no specific arrangements but observations can be taken at any time with a minimal extra work.

The positions of the VLBI antenna during a geo-VLBI campaign depend on the distribution of the radio sources tracked. Variation in the azimuth and elevation is sufficient but there are still some holes in the distribution. An improvement would be to complete the GPS data set by rotating the VLBI-antenna before or after a geo-VLBI session in those positions which will be not observed during the VLBI session.

The angular velocity of the VLBI antenna seemed to be suitable for kinematic measurements when it tracked the radio source. When it changed the source the movement was too quick for fixed solutions.

The number of trajectory points is still high after outlier rejection which compensates slightly bigger uncertainty of one observation. Based on our simulations and results of test measurements we have shown that the reference point coordinates and the axis offset can be computed with mm precision using the kinematic approach (Kallio and Poutanen 2009, 2010). Repeatability of the method was also good. Final values are shown in Table 23.1.

5 Tie with a Static GPS

We carried out static GPS observations with 2 h sessions of 90 pre-planned VLBI antenna position in March and April 2009. Setup of the GPS antennas at the radio telescope was identical to that of the kinematic test. During the campaign some of the nearby pillar points were also occupied. The daily coordinates of them were calculated and then combined to one solution by stacking the normal equations.

The observations were processed with Bernese GPS software. In the processing we kept the coordinates of pillar points and METS IGS point fixed and estimated the coordinates for the two GPS antennas attached on the VLBI antenna. We used the calibration values for the GPS antenna offsets and phase variations and made the table of orientation of the antennas for each session for the Bernese software. A total of 110 two hour sessions were included in our analysis.

There should be 220 individual points for calculating the reference point and antenna elements but the data proved to be very noisy. During the Bernese processing we rejected several vectors and even more in outlier detection when calculating the antenna reference point. In the final calculation we had only 110 points – the loss of the measured antenna points was 50 %.

We could not use the same principle as we used with the kinematic trajectory data to reject points from both sides if one side was affected. There were too few data to reject more points than we already did. The standard deviations of accepted coordinates in 2 h static sessions varied between 0.77 and 23 mm most of them being less than 5 mm.

As a summary, we can conclude that the static method is not suitable in our case. The data seems to be no better than the kinematic trajectory data. We got considerably less points in over 9 day observation period than during one 24-h geo-VLBI campaign. Besides that, normal use of the telescope was interrupted.

6 Results

The local tie vectors from different techniques are shown in Table 23.1. We combined the results of the campaigns by stacking the normal equations of the adjustments and obtained the values of the local tie vector and axis offset designated "Combined". We have shown that the reference point coordinates and the axis offset can be computed on the mm precision using the kinematic approach. The difference in the reference point coordinates between the methods was maximum 5 mm.

7 Conclusion and Further Plans

The kinematic GPS proved to be the best strategy in our case. Based on this experiment we are currently repeating kinematic GPS measurements regularly during the geo-VLBI campaigns. In the future, a more automatic processing is needed for routinely undertaking the local tie measurements. Based on our results we propose that the method be adopted in other fundamental stations simultaneously with the geo-VLBI measurements.

In the future we plan to enlarge our local tie network at Metsähovi to include also SLR and Doris.

Acknowledgements The authors would like to thank the personnel of the Aalto University Metsähovi Radio Observatory for their invaluable help. This project was partly funded by the Academy of Finland, decision number 134952.

References

Bilker-Koivula M, Ollikainen M (2009) Suomen geoidimallit ja niiden käyttäminen korkeuden muunnoksissa, Geodeettinen laitos, tiedote 29, 2009 (in Finnish)

Dach R, Hugentobler U, Fridez P, Meindl M (eds) (2007) Bernese GPS software, version 5.0. Astronomical Institute, University of Bern, 612pp

Dawson J, Sarti P, Johnston GM, Vittuari L (2007) Indirect approach to invariant point determination for SLR and VLBI systems: an assessment. J Geodesy 81(6–8):433–441

IERS (2011) Proposed IERS local tie report layout, http://www.iers.org/nn_11306/IERS/EN/Organization/WorkingGroups/SiteSurvey/documents.html (24.1.2011)

ITRF (2011) Co-location survey: online reports, ITRF web site: http://itrf.ensg.ign.fr/local_surveys.php (24.1.2011)

Jokela J, Häkli P, Uusitalo J, Piironen J, Poutanen M (2009) Control measurements between the geodetic observation sites at Metsähovi. In: Drewes H (ed) Geodetic reference frames. IAG symposium, Munich, 9–14 Oct 2006. Springer, Berlin, pp 101–106

Kallio U, Poutanen M (2009) Can we really promise a mm-accuracy for the local ties on a geo-VLBI antenna. In: Proceedings of the IAG symposium geodesy for planet Earth, Buenos Aires, Aug 31–Sep 4, 2009, Manuscript no: OS1-Mo14

Kallio U, Poutanen M (2010) Simulation of local tie accuracy on VLBI antennas. In: Behrend D, Baver K (eds) VLBI2010: from vision to reality, proceedings of the sixth IVS general meeting, Feb 7–13, 2010, Hobart. NASA/CP-2010-215864. pp 360–364

Kallio U, Poutanen M (2012) Can we really promise a mm-accuracy for the local ties on a geo-VLBI antenna. In: Geodesy for Planet Earth. Proceedings of the 2009 IAG Symposium, Buenos Aires, Argentina, 31 August 31 - 4 September 2009. Eds. Kenyon, S; M. Pacino; U. Marti. International Association of Geodesy Symposia, Vol. 136. Springer Verlag. p. 35-42, DOI: 10.1007/978-3-642-20338-1_5

Leick A (1995) GPS satellite surveying. Wiley, New York. ISBN 0-417-30626-6

Lösler M (2009) New mathematical model for reference point determination of an azimuth-elevation type radio telescope. J Surv Eng 135 (4):131–135

Lösler M, Hennes M (2008) An innovative mathematical solution for a time-efficient vs reference point determination. In: Measuring the changes, 2008. 4th IAG symposium on geodesy for geotechnical and structural engineering, Lisbon, May 12–15, 2008

Sarti P, Sillard P, Vittuari L (2004) Surveying co-located space geodetic instruments for ITRF computation. J Geodesy 78(3):210–222

Long-Term Stability of the SIRGAS Reference Frame and Episodic Station Movements Caused by the Seismic Activity in the SIRGAS Region

24

L. Sánchez, W. Seemüller, H. Drewes, L. Mateo, G. González, A. da Silva, J. Pampillón, W. Martínez, V. Cioce, D. Cisneros, and S. Cimbaro

Abstract

The western part of the SIRGAS region is an extremely active seismic area because it is located in the plate boundary zone of six tectonic plates, namely the Pacific, Cocos, Nazca, North American, Caribbean, and South American plates. The frequent occurrence of earthquakes causes episodic station movements, which affect the long-term stability of the SIRGAS reference frame. Normally, these episodic events are taken into account in the frame realisation by introducing new position and, optionally, velocity parameters for the affected stations. However, this is not enough to guarantee the high precision required in a reference frame such as SIRGAS. Additional analyses about the post-seismic behaviour of the reference stations are necessary to allow the precise transformation between pre-seismic and post-seismic (deformed) frames. According to this, the paper presents an evaluation of the long-term stability of the SIRGAS reference frame including the comparison of the different SIRGAS realisations and the analysis of station displacements caused by earthquakes in the SIRGAS region. Special care is given to the events happened in

L. Sánchez (✉) • W. Seemüller • H. Drewes
Deutsches Geodätisches Forschungsinstitut (DGFI), München, Germany
e-mail: sanchez@dgfi.badw.de

L. Mateo
Instituto Argentino de Nivología, Glaciología y Ciencias Ambientales – CONICET, Mendoza, Argentina

G. González
Instituto Nacional de Estadística y Geografía (INEGI), Aguascalientes, Mexico

A. da Silva
Instituto Brasileiro de Geografia e Estatistica (IBGE), Rio de Janeiro, Brazil

J. Pampillón
Servicio Geográfico Militar (SGM), Montevideo, Uruguay

W. Martínez
Instituto Geográfico Agustín Codazzi (IGAC), Bogota, Colombia

V. Cioce
Laboratorio de Geodesia Física y Satelital, Universidad del Zulia, Maracaibo, Venezuela

D. Cisneros
Instituto Geográfico Militar (IGM), Quito, Ecuador

S. Cimbaro
Instituto Geográfico Nacional (IGN), Buenos Aires, Argentina

Z. Altamimi and X. Collilieux (eds.), *Reference Frames for Applications in Geosciences*, International Association of Geodesy Symposia 138, DOI 10.1007/978-3-642-32998-2_24,

Arequipa (on 2001-06-23, M = 8.4) and Chile (on 2010-02-27, M = 8.8). The analysis is based on the SIRGAS Continuously Operating Network (SIRGAS-CON). Beside analysing the station position time series and estimating the displacement vectors of the SIRGAS reference stations, some recommendations to mitigate the impact of this kind of events in the use of SIRGAS as a reference frame are formulated.

Keywords
SIRGAS reference frame • Reference frame deformation • Pre- and post-seismic frame realisation • Arequipa earthquake • Chile earthquake

1 Introduction

Terrestrial reference frames supporting precise positioning based on global navigation satellite systems (GNSS) must be consistent with the reference frame in which the orbits of the GNSS satellites are determined. At present, the conventional reference frame is the ITRF (International Terrestrial Reference Frame), which is computed and maintained by the International Earth Rotation and Reference Systems Service (IERS). According to the IERS conventions (Petit and Luzum 2010), the International GNSS Service (IGS) determines and provides the GNSS satellite ephemeris referring to the ITRF (Dow et al. 2009). Users applying IGS orbits for processing GNSS positioning have also to introduce terrestrial reference stations referring to the ITRF. The accessibility to this frame at regional and local levels is guaranteed through (1) regional (continental) densifications of the global frame, and (2) national densifications of the continental frames. Following this hierarchy, SIRGAS (Sistema de Referencia Geocéntrico para las Américas) is realised by a regional densification of the ITRF in Latin America and the Caribbean (Brunini et al. 2012), and it is further extended to each country by the national reference networks.

2 SIRGAS Realisations

Initially, SIRGAS was realised by means of two continental GPS campaigns:

1. SIRGAS95 including 58 stations distributed over South America observed for 10 days in May 1995 and resulting in station positions referred to the ITRF94, epoch 1995,4 (SIRGAS 1997).
2. SIRGAS2000 with 184 stations including the SIRGAS95 points and additional stations located in the Caribbean, Central, and North America. It was measured during 10 days in May 2000 and its station positions refer to ITRF2000, epoch 2000,4 (Drewes et al. 2005).

Today, SIRGAS is realised by a network of about 230 continuously operating GNSS stations. This so-called SIRGAS-CON network (Fig. 24.1) replaces the first two SIRGAS realisations and allows a permanent monitoring of the frame.

SIRGAS-CON is weekly processed by the SIRGAS Analysis Centres; main products of this processing are (Brunini et al. 2012): loosely constrained weekly solutions for station positions to be included in the IGS global polyhedron and in multi-year solutions of the network; and weekly station positions aligned to the ITRF for further applications in Latin America. The weekly free normal equations are solved using a set (see Fig. 24.2) of the same ITRF reference stations selected by the IGS to compute the GNSS orbits, i.e. the IGS reference frame (Kouba 2009), at present, the IGS05 (http://www.igs.org/network/refframe.html). The datum realisation is given by constraining the weekly positions of the IGS reference stations determined within the IGS weekly combination (solutions igsyyPwwww.snx, Dow et al. 2009).

To estimate the kinematics of the SIRGAS reference frame, a cumulative (multi-year) solution is computed (updated) every year, providing epoch positions and constant velocities for stations operating longer than 2 years. The coordinates of the multi-year solutions refer to the latest available ITRF and to a specified epoch, e.g. the most recent SIRGAS-CON multi-year solution SIR10P01 refers to ITRF2008, epoch 2005.0 (Seemüller et al. 2010). SIR10P01 includes all the weekly normal equations provided by the SIRGAS analysis centres from January 2, 2000 (GPS week 1,043) to June 5, 2010 (GPS week 1,586) and provides positions and velocities for 183 reference stations (Fig. 24.2). Its precision was estimated to be ~±0.5 mm (horizontal) and ~±0.9 mm (vertical) for the station positions at the reference epoch, and ~±0.2 mm/a (horizontal) and ~ ±0.4 mm/a (vertical) for the constant velocities, respectively. Precision is understand here as the solution repeatability within the SIRGAS-CON network processing. The comparison of the SIRGAS coordinates with the ITRF2008 coordinates for common points delivers the accuracy of the solution; i.e. the measure of the solution uncertainty with respect to the ITRF2008. This was assessed to be N = 0.8 ± 5.0 mm, E = 0.3 ± 3.6 mm and Up = −4.9 ± 8.6 mm

Fig. 24.1 SIRGAS reference frame (status October 2010)

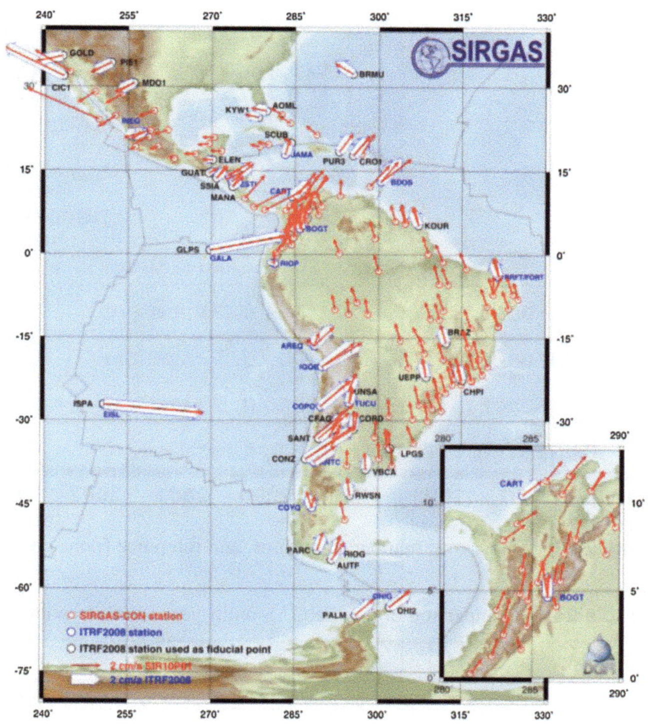

Fig. 24.2 Horizontal velocities of the SIR10P01 multi-year solution. Velocities of ITRF2008 stations are included for comparison

for station positions and $V_N = -0.1 \pm 1.1$ mm/a, $V_E = -0.1 \pm 1.1$ mm/a and $V_{Up} = 0.0 \pm 2.2$ mm/a for station velocities.

On this basis, most of the Latin-American countries adopted SIRGAS as the official national reference frame with station positions associated to a certain (conventional) epoch and to be extrapolated to any other epoch using the velocities, or – if not available for new stations – by applying the SIRGAS weekly solutions.

As reference frame, SIRGAS is the backbone for implementing applications associated to geo-referenced data such as land management, geo-spatial data infrastructures, property lines, country boundaries, etc. These applications need coordinates compatible in time and with long-term stability, i.e. all station positions referring to the measurement epoch and to the current ITRF shall be precisely transformed to the conventional epoch and to the ITRF adopted within the national reference frame. According to this, this study (1) evaluates the sustainability of the different SIRGAS realisations (SIRGAS95, SIRGAS2000 and multi-year solutions of SIRGAS-CON); (2) presents the episodic station movements caused by the seismic activity in the SIRGAS region since 2000; and (3) formulates a strategy to reduce the lack of precision due to the transformation between the pre-seismic and post-seismic (deformed) frame realisation.

3 Sustainability of the SIRGAS Reference Frame

The former SIRGAS realisations (SIRGAS95 and SIRGAS2000) as well as the multi-year solutions of the SIRGAS-CON network include those models, standards, and strategies widely applied at the time in which they were computed; e.g. different ITRF solutions, ocean tide loading, a priori ionosphere models for ambiguity resolution, relative corrections for the phase centre variations until 2006 and absolute corrections afterwards, etc.; see for SIRGAS95 (SIRGAS 1997), for SIRGAS2000 (Drewes et al. 2005), and for SIRGAS-CON (e.g. Seemüller et al. 2008, 2010a, Seemüller 2009).

Figure 24.3 summarizes the main characteristics of the different SIRGAS multi-year solutions. It should be noted that solutions computed since 2007 include the reprocessed normal equations for the weeks before November 2006 (GPS week 1,399). This reprocessing takes into account the IGS05 as reference frame and the antenna absolute phase centre corrections provided by the IGS (model igs05_1525.atx, see: http://igscb.jpl.nasa.gov/igscb/station/general/pcv_archive/).

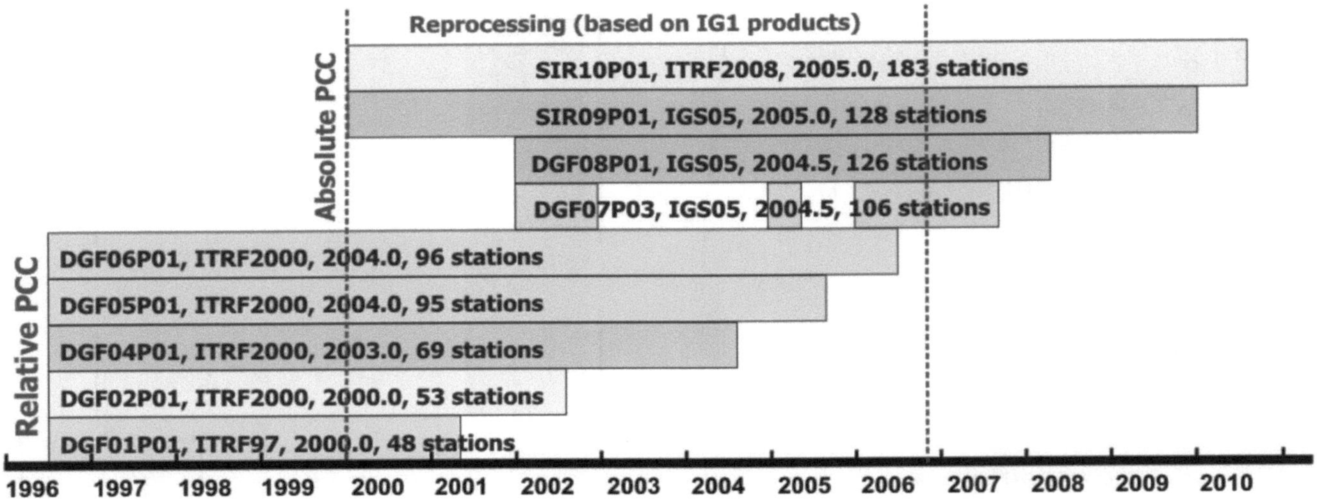

Fig. 24.3 Time spans, number of stations, and reference frame considered in the different SIRGAS multi-year solutions

Table 24.1 Comparison of the SIRGAS multi-year solutions with the ITRF2008

Solution	Comparison with the ITRF2008						
	Common stations with ITRF2008	Position deviations: Offsets ± RMS			Velocity deviations: Offsets ± RMS		
		N[mm]	E[mm]	h[mm]	VN[mm/a]	VE[mm/a]	Vh[mm/a]
DGF01P01	27	-16,3 ± 8,0	7,2 ± 19,5	27,9 ± 16,2	-0,4 ± 2,6	3,1 ± 4,7	1,3 ± 4,5
DGF02P01	24	-2,4 ± 3,7	-2,5 ± 5,8	4,0 ± 13,9	1,1 ± 1,6	1,4 ± 2,1	-3,7 ± 6,7
DGF04P01	35	-0,4 ± 4,3	-3,4 ± 5,0	1,3 ± 14,9	1,9 ± 2,3	1,3 ± 2,1	0,1 ± 3,6
DGF05P01	34	0,2 ± 3,8	-2,0 ± 5,0	0,1 ± 13,1	1,8 ± 2,1	1,1 ± 2,1	1,2 ± 3,6
DGF06P01	32	0,0 ± 3,9	-1,7 ± 4,9	1,1 ± 12,3	2,0 ± 2,2	1,0 ± 1,9	0,8 ± 3,0
DGF07P03	22	-1,3 ± 5,1	0,9 ± 6,2	-4,4 ± 19,5	0,5 ± 1,3	-0,4 ± 1,3	0,5 ± 2,7
DGF08P01	28	-3,2 ± 5,1	1,1 ± 8,9	-8,0 ± 10,0	0,5 ± 1,3	-0,5 ± 1,6	1,0 ± 2,3
SIR09P01	34	0,3 ± 4,0	-0,6 ± 6,7	-5,1 ± 12,0	0,3 ± 1,0	0,0 ± 1,1	-0,2 ± 1,9
SIR10P01	74	0,8 ± 5,0	0,3 ± 3,6	-4,9 ± 8,6	-0,1 ± 1,1	-0,1 ± 1,1	0,0 ± 2,2

To evaluate the sustainability of the SIRGAS realisations, following steps were considered:

1. The different SIRGAS multi-year solutions are compared with the ITRF2008. For this purpose, the SIRGAS-CON multi-year solutions are transformed to ITRF2008 and the coordinate comparison is done for epoch 2000,0.
2. The SIRGAS95 and SIRGAS2000 realisations are compared with the latest SIRGAS-CON multi-year solution (SIR10P01). This comparison is done in the ITRF2008 reference frame and at the conventional epochs of the former realisations, i.e. 1995,4 and 2000,4.

In both cases, the transformation parameters presented in the IERS Conventions (Table 4.1, Petit and Luzum 2010) are applied and stations affected by earthquakes are excluded (see Sect. 4).

Table 24.2 Comparison of SIRGAS95 and SIRGAS2000 with the multi-year solution SIR10P01

Realization	Comparison with SIR10P01			
	Common stations	Position deviations: Offsets ± RMS		
		N[mm]	E[mm]	h[mm]
SIRGAS95	19	-21,3 ± 4,9	-18,7 ± 4,2	5,8 ± 18,3
SIRGAS2000	53	-0,3 ± 8,6	0,1 ± 7,5	-6,2 ± 10,3

Results (Tables 24.1 and 24.2) show a very good consistency between the different SIRGAS realisations. The largest discrepancies (~2 cm) were detected for the SIRGAS realisations referring to ITRF94 and ITRF97. Realisations referring to ITRF2000 and IGS05 have an agreement better

Table 24.3 Seismic events with high impact in the SIRGAS frame since 2000

Location	Date	M	Coordinate change	Affected stations
Mexicali, Mexico	2010-04-04	7,2	23 cm	MEXI
Chile	2010-02-27	8,8	1 to 305 cm	See table 5
Costa Rica	2008-01-08	6,1	2 cm	ETCG
Martinique	2007-11-29	7,4	1 cm	BDOS, GTK0
Copiapo, Chile	2006-04-30	5,3	2 cm	COPO
Tarapaca, Chile	2005-06-13	7,9	6 cm	IQQE
Managua, Nicaragua	2004-10-09	6,9	1 cm	MANA
Arequipa, Peru	2001-06-23	8,4	52 cm	AREQ
El Salvador	2001-02-13	7,8	4 cm	SSIA

than ±5 mm. This reflects the expected improvement of the frame as consequence of longer time series of station positions and the better new models, standards, and analysis strategies applied today.

4 Impact of Seismic Events on the SIRGAS Reference Frame

The western part of the SIRGAS region is located in the plate boundary zone between the Pacific, Cocos, and Nazca plates in the west and the North American, Caribbean, and South American plates in the east. The interaction of these drifting plates causes an extremely high seismic activity in this area, generating episodic station movements (Table 24.3) and deformations in the SIRGAS reference frame. The precise determination and modelling of co-seismic and post-seismic displacements is necessary to guarantee:

1. The reliability of the SIRGAS weekly positions estimated for the week when a seismic event occurs;
2. The appropriate transformation of station positions between the pre-seismic and the post-seismic (deformed) reference frame;
3. The long-term stability of the SIRGAS reference frame.

According to this, always when a strong earthquake shakes the SIRGAS region, the SIRGAS Analysis Centres attempt to process as soon as possible the available GNSS measurements to estimate the impact on the reference frame. The usual procedure is based on the computation of free daily normal equations, which include IGS reference stations located outside the SIRGAS region, i.e. in Europe, North America, Africa, and Antarctica. These external IGS stations are used for the datum definition in the solution of the normal equations and as fiducial points for the calculation of a similarity transformation between the pre-seismic and post-seismic networks. By comparing daily station positions and the geometry of the network before and after the earthquake, it is possible to determine displacements of the SIRGAS reference stations associated to the seism. In the same way, the analysis of station position time series allows to estimate further post-seismic movements and/or significant changes in the constant velocity of the affected stations. As examples, the displacements generated by the earthquakes of Arequipa, Peru (on 2001-06-23), and Chile (on 2010-02-27) are described in the following.

The basic procedure for the determination of daily station positions before and after any earthquake is:

- Elevation mask and data sampling rate are set to 3 ° and 30 s, respectively.
- Absolute calibration values for the antenna phase centre corrections published by the IGS are applied.
- Satellite orbits, satellite clock offsets, and Earth orientation parameters are fixed to the combined IGS weekly solutions (Dow et al. 2009).
- Phase ambiguities for L1 and L2 are solved by applying the quasi ionosphere free (QIF) strategy of the Bernese software (Dach et al. 2007).
- Periodic site movements due to ocean tide loading are modelled according to the FES2004 ocean tide model (Letellier 2004). The corresponding values are provided by M.S. Bos and H.-G. Scherneck at http://129.16.208.24/loading/.
- Zenith delays due to tropospheric refraction (∼wet part) are estimated at a 2 h interval within the network adjustment and mapped to zenith using the Niell (1996) wet mapping function. The a priori zenith delays (∼dry part) are modelled using the Saastamoinen (1973) model and mapped to zenith using the Niell (1996) dry mapping function.
- Daily free normal equations are computed by applying the double difference strategy (Bernese Software 5.0, Dach et al. 2007). The baselines are created taking into account the maximum number of common observations for the associated stations. Afterwards, loosely constrained daily solutions for station positions are generated by constraining all station coordinates to ±1 m. These loosely constrained solutions are used to compare the network geometry before and after the earthquake.
- The daily normal equations are separately solved with respect to selected IGS reference stations located outside the SIRGAS region. The geodetic datum is defined by constraining the coordinates of the reference stations to the positions calculated within the IGS weekly combinations (igsyyPwwww.snx or iglyyPwwww.snx for reprocessed products). The applied constraints guarantee that the coordinates of the reference stations do not

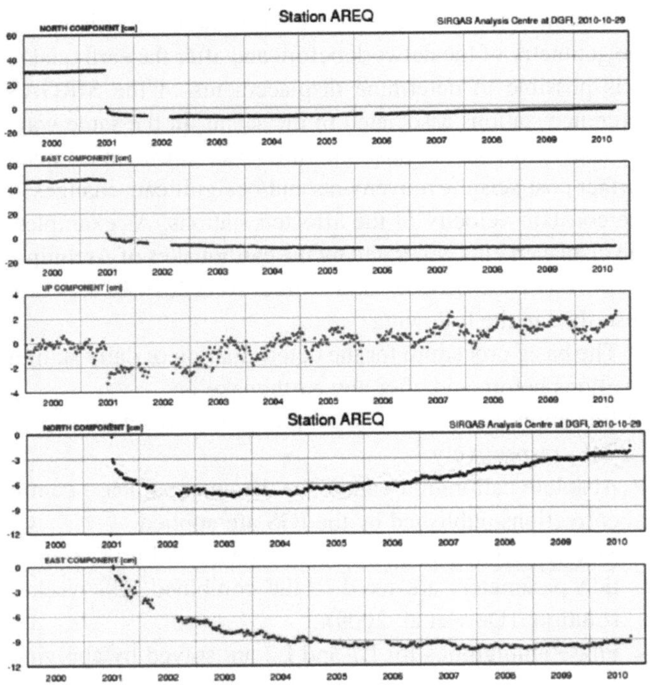

Fig. 24.4 Time series of station positions for AREQ. Lower pictures show a zoom of post-seismic displacements in the north and east components

Table 24.4 Station velocities estimated for AREQ before and after the earthquakes occurred in June and July 2001

Time span		Velocities [m/a]	
from	to	North	East
1996-06-03	2001-06-24	0,0118	0,0126
2001-06-25	2001-08-07	-0,4919	-0,4611
2001-08-08	2001-09-11	-0,0390	-0,1174
2001-09-12	2002-08-25	-0,0202	-0,0400
2002-08-26	2007-12-03	0,0042	-0,0056
2007-12-04	2010-06-06	0,0089	0,0033

change more than 1.5 mm within the SIRGAS adjustment (Sánchez et al. 2012). In this study, the IGS05 reference frame is applied.

- Finally, residual position time series are generated. For this, the daily solutions are transformed to a pre-seismic reference frame (e.g. the SIRGAS weekly solution for the previous week to the earthquake or the latest available SIRGAS-CON multi-year solution) by a seven-parameter similarity transformation. The residual time series are analysed to detect discontinuities or changes in the movement trend of the SIRGAS sites.

Applying this strategy we analysed the GPS measurements registered by the station AREQ (Arequipa, Peru) before and after the two earthquakes (with magnitudes 8.4 and 7.6) on 2001-06-23 and 2001-07-07. It was possible to determine co-seismic displacements of 52.3 cm and 3.5 cm in SW-direction. The station position time series (Fig. 24.4) shows that the station moved horizontally for more than 2 years in the opposite direction as was usual before the earthquake. During the weeks following the earthquakes, velocities of 1.8 mm/day and 1.0 mm/day, respectively, were estimated (Kaniuth et al. 2002). Afterwards, the station presents a creeping movement with a slowing down post-seismic velocity until December 2007, when its trend has again the same direction and a similar velocity as in the pre-seismic period (Table 24.4).

The biggest impact produced by an earthquake on the SIRGAS reference frame corresponds to the seism (M = 8.8) that occurred on 2010-02-27 in Chile. 23 SIRGAS reference stations moved more than 1.5 cm (Table 24.5). The largest displacements were detected between latitudes 30°S and 40°S from the Pacific to the Atlantic coast (Fig. 24.5). Results show that the station CONZ (Concepción, Chile) initially moved 305.4 cm in the SW-direction. In the 2 weeks following the first earthquake, additional post-seismic movements of about 10 cm were identified. Until now (October 2010), this station moved 9 cm more in the E-direction (Fig. 24.6)

Table 24.5 Residuals after a similarity transformation between the station positions computed for the GPS weeks 1572 (2010-02-24) and 1573 (2010-03-03). For station ANTC positions of week 1539 (2009-07-08) were used

	Residuals		
Station	N (mm)	E (mm)	Up (mm)
ANTC	195	−808	−12
AZUL	1	−23	3
BCAR	3	−16	3
CFAG	−18	−33	1
CONZ	−682	−2977	−47
CSLO	−21	−36	6
IGM1	−5	−18	8
LHCL	16	−66	4
LPGS	−2	−17	−1
MA01	37	−83	9
MZAC	−48	−109	15
MZAE	−39	−105	17
MZAS	−27	−210	18
RWSN	8	−10	3
SANT	−145	−257	−21
SL01	−21	−64	1
SRLP	3	−58	8
UCOR	−11	−22	1
UNRO	−7	−19	−3
UNSJ	−18	−29	−1
UYMO	−1	−11	10
VALP	−13	−76	−137
VBCA	8	−30	4

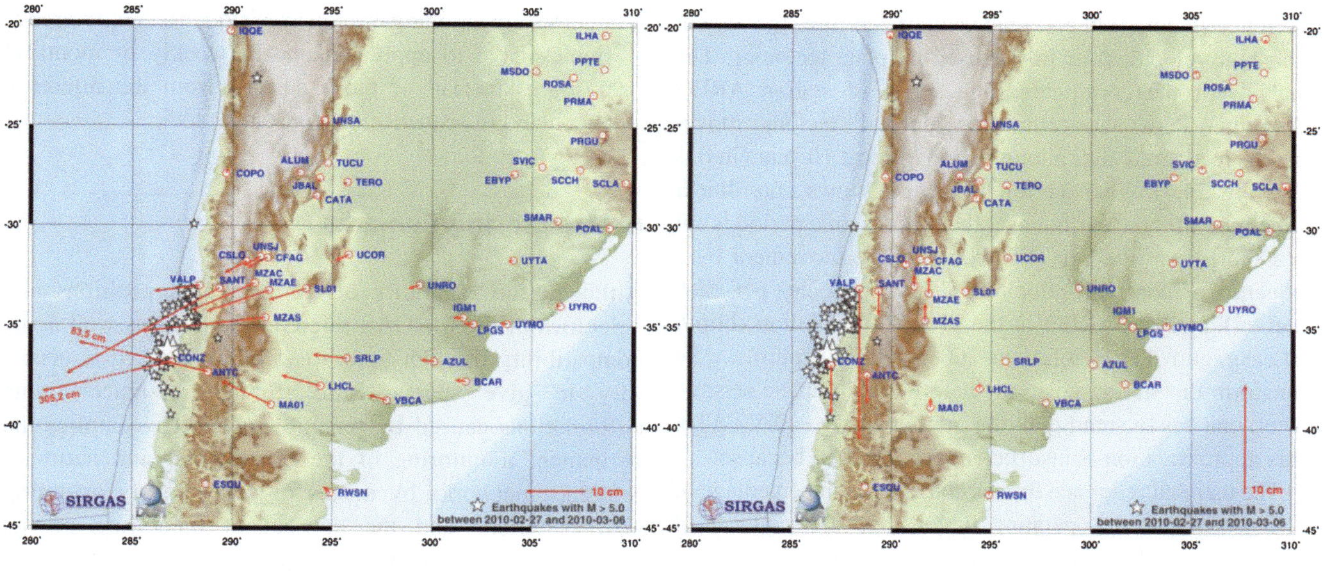

Fig. 24.5 Station displacements caused by the earthquake occurred on 2010-02-27 in Chile

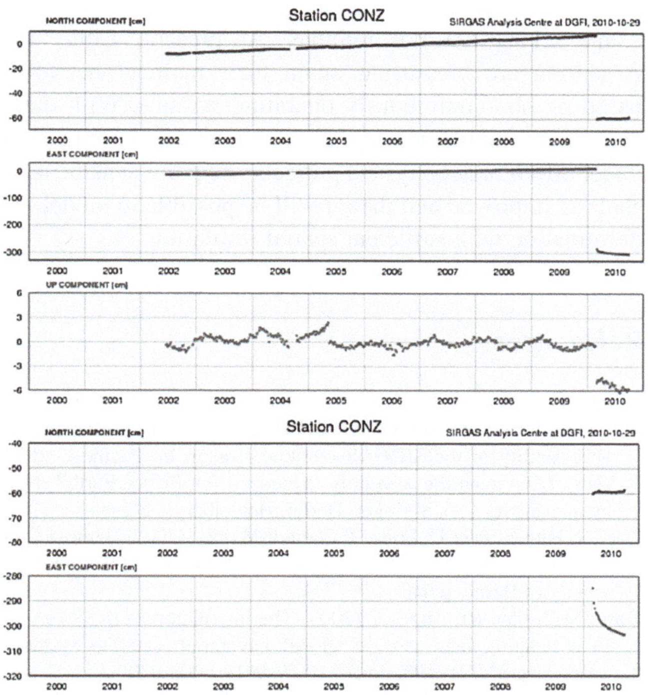

Fig. 24.6 Time series of station positions for CONZ. Lower pictures show a zoom of post-seismic displacements in the north and east components

Table 24.6 Station velocities estimated for CONZ after the earthquake occurred on 2010-02-27

Time span		Velocities [m/a]	
from	to	North	East
2010-03-21	2010-05-08	-0,0111	-0,4043
2010-05-09	2010-07-17	-0,0071	-0,1488
2010-07-18	2010-10-02	0,0315	-0,0856

Strong vertical co-seismic displacements were also identified in Concepción (CONZ), Santiago (SANT), Valparaíso (VALP) and the Province of Mendoza (MZAC, MZAE, MZAS) in Argentina. Stations located in the west of the Andes moved downwards, stations located in the east moved upwards (Fig. 24.5). Time series of station positions for CONZ (Fig. 24.6) and other close stations (ANTC, SANT, VALP, MZAS) shows that the post-seismic displacement in the east component presents the same behaviour observed in station AREQ. Table 24.6 summarizes the linear trends of the post-seismic time series for the north and east components. At present, we are not yet able to determine the effect of this earthquake on the reference frame because it is necessary to analyse cumulative solutions of more than 2 years after its occurrence.

5 Modelling Seismic Effects Within Reference Frames

To guarantee the long-term stability of the reference frame, it is necessary to enable the transformation between the pre-seismic and the post-seismic frame realisations. This transformation cannot be done by usual approaches (like affine or seven-parameter similarity transformations) because the deformed network does not fulfil the similarity condition; i.e. changes occurred in the geometry of the network due to the earthquake cannot be sufficiently represented by means of rotations, translations or scale factor.

Effects of seismic events are normally taken into account by introducing new position and/or velocity parameters for the affected stations. Nevertheless, earthquakes of big magnitudes generate not only jumps in the position of the reference stations, but also change their "normal" movement (constant velocities) mainly caused by plate tectonics. The analysis of the coordinate time series of station AREQ (Arequipa, Peru) shows that this "normal" tectonic movement is recovered after a long period (almost 10 years in this case, Fig. 24.4). When a reference station shows a non-linear behaviour after the earthquake, the post-seismic period is cut into short time intervals ΔT_i to model the movement by a sequence of constant velocities V_i (one velocity per each interval), e.g. see AREQ in the ITRF2008 solution (http://itrf.ensg.ign.fr/ITRF_solutions/2008/ITRF2008.php). To transform the station positions before and after the seismic event, one has to sum up all the intervals ($\Delta X = \Sigma V_i * \Delta T_i$). This approximation is insufficient for SIRGAS because:

- The national reference frames contain a high percentage of non-continuously operating stations and the sequence of velocities after an earthquake cannot be reliably determined.
- The selection of the time intervals to compute constant velocities approximating the non-linear movement of a station is open to more than one interpretation; i.e. it cannot always be unambiguous and estimated velocities can significantly vary from one computation to other one.
- The determination of velocities for too short time intervals (less than 2 years) within the multi-year solutions of the SIRGAS reference frame is not reliable. This is because the seasonal (e.g. loading) signals contained in the station position time series are not considered up to now and they can mislead the velocity estimation (Brunini et al. 2012).
- The SIRGAS reference frame is composed of about 230 continuously operating stations; however, their geographical distribution does not provide the required density (coverage) to interpolate (model) the effects of the seismic events with high accuracy.

According to this, to mitigate the impact of this kind of events in the use of SIRGAS as reference frame in the different countries of the region, it is necessary:

- To continue improving the national reference frames by installing more continuously operating GNSS stations in order to precisely monitor possible deformations.
- After a strong seism like in Chile 2010, the reference networks composed by non-continuously operating stations must be re-measured as soon as possible.
- The transformation between the pre-seismic and the post-seismic frame realisations must be based on a deformation model derived from discrete (weekly) station positions. Usual network transformations (similarity or affine) cannot be applied.
- In stations not observed continuously, the post-seismic coordinate changes can be interpolated from the deformation model.
- In GNSS positioning of high precision, users of reference frames have to apply the epoch (weekly or monthly) positions instead of those derived from the reference epoch and (a sequence of) constant velocities.

6 Closing Remarks

Although the reliability of the estimated positions and velocities of the SIRGAS reference stations as well as its compatibility through time are demonstrated, it is necessary to give special care to the reference frame deformations caused by seismic events. This implies the permanent monitoring of the (continental and national) reference networks by means of continuously operating GNSS stations and the consequent modelling of the deformations caused by this type of events. Today, it is not possible to estimate a continuous deformation model representing the impact of the seism in Chile, because the SIRGAS reference stations are too sparse and the computations are not reliable. At present, Chile and Argentina are measuring again their national networks based on non-continuously operating stations. With these data and complementary GPS observations registered under different geophysical projects, the point densification shall be improved and then it will be possible to model the deformation with sufficient spatial resolution.

References

Brunini C, Sánchez L, Drewes H, Costa S, Mackern V, Martínez W, Seemüller W, and da Silva A (2012) Improved analysis strategy and accessibility of the SIRGAS reference frame. In: Pacino C et al. (eds) IAG scientific assembly "Geodesy for Planet Earth". IAG symposia, vol 136, Springer, Berlin, Heidelberg, pp 3–8

Dach R, Hugentobler U, Fridez P, Meindl M (eds) (2007) Bernese GPS software version 5.0 – Documentation. Astronomical Institute, University of Berne, p 640

Dow JM, Neilan RE, Rizos C (2009) The international GNSS service in a hanging landscape of global navigation satellite systems. J Geodesy 83:191–198. doi:10.1007/s00190-008-0300-3

Drewes H, Kaniuth K, Voelksen C, Costa S, Fortes LP (2005) Results of the SIRGAS campaign 2000 and coordinates variations with respect to the 1995 South American geocentric reference frame. In: Sanso F (ed) A window on the future of geodesy. IAG symposia, vol 128, Springer, Berlin, Heidelberg, pp 32–37

Kaniuth K, Müller H, Seemüller W (2002) Displacement of the space geodetic observatory Arequipa due to recent earthquakes. Z Vermessungswesen 127:238–243

Kouba J (2009) A guide to using international GNSS service products. Available at http://igscb.jpl.nasa.gov/igscb/resource/pubs/UsingIGSProductsVer21.pdf

Letellier T (2004) Etude des ondes de marée sur les plateaux continentaux. Thèse doctorale, Université de Toulouse III, Ecole Doctorale des Sciences de l'Univers, de l'Environnement et de l'Espace, p 237

Niell AE (1996) Global mapping functions for the atmosphere delay at radio wavelength. J Geophys Res 101:3227–3246

Petit G, Luzum B (eds) (2010) IERS conventions 2010. IERS technical note 36, Verlag des Bundesamtes für Kartographie und Geodäsie, Frankfurt, AM

Saastamoinen J (1973) Contribution to the theory of atmospheric refraction. Part II: refraction corrections in satellite geodesy. Bull Géod 107:13–34

Sánchez L, Seemüller W, Seitz M (2012) Combination of the weekly solutions delivered by the SIRGAS processing centres for the SIRGAS-CON reference frame. In: Pacino C et al. (eds) IAG scientific assembly "Geodesy for planet earth". IAG symposia, vol 136, Springer, Berlin, Heidelberg, pp 651–656

Seemüller W, Krügel M, Sánchez L, Drewes H (2008) The position and velocity solution DGF08P01 of the IGS regional network associate analysis centre for SIRGAS (IGS RNAAC SIR). DGFI report no. 79. DGFI, Munich. Available at www.sirgas.org

Seemüller W (2009) The position and velocity solution DGF06P01 for SIRGAS. In: Drewes H (ed) Geodetic reference frames. IAG symposia, vol 134, Springer, Berlin, Heidelberg, pp 167–172

Seemüller W, Sánchez L, Seitz M, Drewes H (2010) The position and velocity solution SIR10P01 of the IGS regional network associate analysis centre for SIRGAS (IGS RNAAC SIR). DGFI report no. 86, Munich, p 120. Available at www.sirgas.org

SIRGAS (1997) SIRGAS final report, working sroups I and II IBGE, Rio de Janeiro, p 96. Available at www.sirgas.org

Part IV

Interaction Between the Celestial and the Terrestrial Reference Frames

The Impact of the New IAU Resolutions on ICRF Definition and Realization

25

N. Capitaine

Abstract

Following the adoption of the International Celestial Reference System and Frame (ICRS and ICRF) by the IAU in 1997, several resolutions on reference systems have been passed by the IAU in 2000 and 2006 and endorsed by the IUGG in 2003 and 2007, respectively. These resolutions concern especially the transformation between the International Terrestrial Reference System (ITRS) and the Geocentric Celestial Reference System (GCRS) that is essential for realizing the ICRS from directions of extragalactic radio sources observed from the Earth by VLBI.

First, the IAU 2000 resolutions have refined the concepts and definition of the astronomical reference systems and parameters for Earth's rotation, and adopted the IAU 2000 precession-nutation. Then, the IAU 2006 resolutions have adopted a new precession model that is consistent with dynamical theories and have addressed definition, terminology or orientation issues relative to reference systems and time scales that needed to be specified after the adoption of the IAU 2000 resolutions. These in particular provide a refined definition of the pole and the origin on the equator as well as a rigorous definition of sidereal rotation of the Earth. These also allow an accurate realization of the celestial intermediate system that replaces the classical celestial system based on the true equator and equinox of date. There was an additional IUGG 2007 resolution for the terrestrial reference system. Finally, the IAU 2009. Transactions of the IAU XXVIIB. Rio de Janeiro, Corbett I (ed) vol 6. Cambridge University Press, pp 55–70) resolutions have adopted a new system of astronomical constants – including conventional values of the IAU 2000/2006 resolutions – and adopted the Second Realization, ICRF2, of the International Celestial Reference Frame.

This paper recalls the main aspects of these recent IAU resolutions as well as their consequences on the concepts, definitions, nomenclature and models that are suitable for modern realizations of reference systems. The impact of these resolutions on the definition and realization of the International Celestial Reference Frame (ICRF) is described.

Keywords

Earth rotation • Reference systems • Time

N. Capitaine (✉)
SYRTE, Observatoire de Paris, CNRS, UPMC, 61, avenue de l'Observatoire, 75014 Paris, France
e-mail: nicole.capitaine@obspm.fr

Z. Altamimi and X. Collilieux (eds.), *Reference Frames for Applications in Geosciences*,
International Association of Geodesy Symposia 138, DOI 10.1007/978-3-642-32998-2_25,

1 Introduction

IAU resolutions have been passed at successive General Assemblies (GA) from 1988 for preparing the adoption of a celestial reference system based on directions of extragalactic radio sources in order to replace the FK5, which was based on directions of stars. These resolutions were as follows:

- The IAU GA 1988 called for the use of extragalactic objects to define the celestial reference frame,
- The IAU GA 1991 adopted General Relativity as the fundamental theory, confirmed the 1988 Resolution and specified the continuity with existing stellar and dynamic realisations,
- The IAU GA 1994 adopted a list of some 600 extragalactic radio sources and formed a working group to define the positions.

These preliminary steps were successful and then, following the recommendation of the IAU Working Group on Reference Frames, the IAU GA 1997 resolved:

(a) that, as from 1 January 1998, the IAU celestial reference system shall be the International Celestial Reference System (ICRS) as specified in the 1991 IAU Resolution on reference frames and as defined by the International Earth Rotation Service (IERS),

(b) that the corresponding fundamental reference frame shall be the International Celestial Reference Frame (ICRF) constructed by the IAU Working Group on Reference Frames;

(c) that the Hipparcos Catalogue shall be the primary realisation of the International Celestial Reference System (ICRS) at optical wavelengths.

The adoption of the ICRS/ICRF was followed by a series of important resolutions on reference systems adopted by the IAU in 2000 and 2006 and endorsed by the IUGG in 2003 and 2007, respectively; in addition there was a specific IUGG (2007) resolution giving a definition of the International Terrestrial Reference System (ITRS), which strictly complies with the IAU resolutions.

In 2009, the IAU completed this set of international agreements by two other resolutions on the astronomical references, one of them being the adoption of the Second Realization, ICRF2, of the ICRF, which benefited from the improved concepts and models adopted by the previous resolutions.

It is important to note that the international coordination in this field, especially due to the existence of international scientific services, such as the IERS, or IVS, and international working groups (within the IAU, IERS and IVS), played a key role in the preparation of the IAU/IUGG resolutions, both for the modernization of the concepts and for their implementation.

The following sections report on the main aspects of these 2000–2009 resolutions and review their impact on the concepts, definitions, nomenclature and models that are suitable for modern realizations of reference systems, and especially the ICRF.

2 The Recent IAU Resolutions

2.1 The Significant Change from FK5 to ICRF

The adoption of the ICRS/ICRF and its optical counterpart constitutes a significant change in astronomy due to the sub-milliarcsecond precision of the positions of the reference points and the sub-milliarcsecond accuracy in the orientation of the axes. The ICRF was aligned with the FK5 at J2000, but no attempt was made to refer the positions of the sources to the mean pole and mean equinox at J2000. Therefore, since the adoption of the ICRF, the celestial reference frame is no longer dependent on the Earth's motion; further improvements of the ICRF are to be accomplished without introducing any global rotation. This corresponds to an historical abandonment of the link of the celestial reference system with the motion of the Earth.

2.2 The Category List of the IAU/IUGG 2000–2009 Resolutions on Reference Systems

After the adoption of the ICRS/ICRF, the purpose of the next series of resolutions that were passed between 2000 and 2009 was to improve definitions and procedures in order they are in agreement with the accuracy and properties of the ICRS and the precision of modern astro-geodetic observations. These recommended refined definitions regarding the celestial reference system and provided a new paradigm and high accuracy models to be used in the transformation from terrestrial to celestial systems.

These resolutions can be put into three different categories:

Resolutions on space reference systems

IAU 2000 Resolution B1.2: *The Hipparcos Celestial Reference Frame (HCRF).*

IAU 2000 Resolution B1.3: *Definition of BCRS and GCRS*

IAU 2000 Resolution B1.5: *Extended Relativistic framework for time transformation*

IAU 2006 Resolution B2: *Supplement to the* IAU 2000 *Resolutions on reference systems;* Rec 2: *Default orientation of the BCRS/GCRS*

IUGG 2007 Resolution 2: *Definition of the GTRS and ITRS*

IAU 2009 Resolution B3: *Adoption of ICRF2*

Resolutions on models for Earth rotation

IAU 2000 Resolution B1.6: IAU 2000 *Precession-Nutation Model*

IAU 2006 Resolution B1: *Adoption of the P03 Precession and definition of the ecliptic*

IAU 2009 Resolution B2: *Adoption of the* IAU 2009 *system of astronomical constants*

Resolutions on concepts and nomenclature

IAU 2000 Resolution B1.7: *Definition of Celestial Intermediate Pole*

IAU 2000 Resolution B1.8: *Definition and use of CEO and TEO*

IAU 2006 Resolution B2: *Supplement to the* IAU 2000 *Resolutions on reference systems;* Rec 1: *Harmonizing "intermediate" to the pole and the origin (CIP, CIO)*

Resolutions on time scales

IAU 2000 Resolution B1.9: *Re-definition of TT*

IAU 2006 Resolution B3: *Re-definition of TDB*

2.3 Main Recommendations of the IAU 2000 and IUGG 2003 Resolutions

The IAU 2000 resolutions adopted by the XXIV IAU General Assembly (August 2000) and endorsed by the XXIII IUGG General Assembly (July 2003), have made important recommendations on the space and time reference systems, the concepts, the parameters, and the models for the Earth's rotation. These resolutions resulted from the recommendations of the IAU ICRS Working Group and the IAU/IUGG Working Group on "Non-rigid Earth nutation theory."

The most important recommendations that concern the ICRF and the associated time coordinate are (1) adopting the HCRF as the realization of the ICRF at optical wavelengths, (2) specifying the systems of space-time coordinates for the solar system and the Earth within the framework of General Relativity and providing clear procedures for the transformation between the barycentric and geocentric coordinates and (3) re-defining Terrestrial Time (TT) as a linear function of Geocentric Coordinate Time (TCG).

The most important recommendations that concern the concepts, the parameters, and the models for the Earth's rotation are (1) the adoption of the IAU 2000 precession-nutation with submilliarcsecond accuracy and (2) the definition of the pole of the nominal rotation axis, and of new origins on the equator, the Earth Rotation Angle (ERA) and UT1.

2.4 Main Recommendations of the IAU 2006 and IUGG 2007 Resolutions

The IAU 2006 resolutions adopted by the XXVI IAU General Assembly (August 2006) and endorsed by the XXIV IUGG General Assembly (July 2007), supplement the IAU 2000 resolutions on reference systems. The most important recommendations that are relevant to the realization of the ICRF and the associated time coordinate are (1) the adoption of a new precession model as a replacement for the IAU 2000 precession in order to be consistent with both dynamical theory and the IAU 2000A nutation, (2) addressing definition, terminology or orientation issues relative to reference systems that needed to be specified after the adoption of the IAU 2000 resolutions, and (3) the redefinition of the Barycentric Dynamical Time (TDB) as a linear function of Barycentric Coordinate Time (TCB).

The adoption of a new precession model was recommended by the IAU Working Group on "Precession and the ecliptic" (Hilton et al. 2006), while the new terminology associated with the IAU 2000/2006 resolutions, along with some additional definitions related to them, were recommended by the IAU Working Group on "Nomenclature for Fundamental Astronomy" (Capitaine et al. 2007).

2.5 Main Recommendations of the IAU 2009 Resolutions

The IAU 2009 resolutions adopted by the XXVII IAU General Assembly (August 2009) have completed the 2000–2006 set of international agreements by two important recommendations for the ICRF: (1) adopting the IAU 2009 System of astronomical constants, consistent with the current measurement accuracy, as recommended by the IAU Working Group "Numerical Standards for Fundamental astronomy", and (2) adopting a new version, called ICRF2, of the International Celestial Reference Frame, which was compiled by the IERS/IVS Working Group and recommended by the IAU Working Group "ICRF2."

3 Consequences of the Resolutions on the Concepts, Definitions and Models

3.1 Improved Definitions of Space Reference Systems for Astronomy and Geodesy

As specified by IAU 2000 Resolution B1.3, the Barycentric Celestial Reference System (BCRS) should be used, with Barycentric Coordinate Time (TCB), as a global coordinate system for the solar system. In contrast, the Geocentric Celestial Reference System (GCRS) should be used, with Geocentric Coordinate Time (TCG), as a local coordinate system for the Earth, e.g. for the Earth's rotation and precession-nutation of the equator. The spatial orientation of the GCRS is derived from that of the BCRS. Consequently, the GCRS is "kinematically non-rotating" so that Coriolis terms (that come mainly from geodesic precession) have to be considered when dealing with equations of motion in that system. The BCRS-to-GCRS transformation was specified as an extension of the Lorentz transformation for the space and time coordinates (see Soffel et al. 2003).

Thus the BCRS/GCRS are theoretically defined, but without any constraint on the orientation of the BCRS axes. Their default orientation was specified by IAU 2006 Resolution B2 that recommends that the BCRS orientation be such that for all practical applications, unless otherwise stated, the BCRS is assumed to be oriented according to the ICRS axes.

The IAU 2000/2006 resolutions on reference systems have been endorsed by IUGG (IUGG 2003 Resolution 4 and 2007 Resolution 2). In addition, IUGG recommended a terrestrial counterpart (IUGG 2007 Resolution 2), which stipulates that the Geocentric Terrestrial Reference System (GTRS) is defined in agreement with IAU 2000 Resolution B1.3, and the International Terrestrial Reference System (ITRS) as "the specific GTRS for which the orientation is operationally maintained in continuity with past international agreements."

Thanks to these IAU/IUGG resolutions, the theoretical definition, as well as the orientation of the BCRS, GCRS, GTRS and ITRS, are specified in a clear and consistent way, which is essential for realizing both the ICRF and the ITRF, as well as for determining the Earth rotation parameters.

3.2 Improved Definitions of Time Coordinates for Astronomy and Geodesy

The IAU 2000/2006 resolutions have clarified the definitions of both Terrestrial Time (TT) and Barycentric Dynamical Time (TDB).

The re-definition of TT by IAU 2000 Resolution B1.9 is such that TT is a time scale differing from TCG by a constant rate, which is a defining constant:

$TCG - TT = L_G \times (JD - 2443144.5\ 0) \times 86400$, with $L_G = 6.969290134 \times 10^{-10}$.

In a very similar way, the re-definition of TDB by IAU 2006 Resolution B3 is a linear transformation of TCB, the coefficients of which are defining constants:

$TCB - TDB = L_B \times (JD - 2443144.50) \times 86400 + TDB_0$

$(L_B = 1.550519768 \times 10^{-8}; TDB_0 = -6.55 \times 10^{-5}$ s).

The consequence is that TT and TDB, which may be for some practical applications of more convenient use than TCG and TCB, respectively, can be used with the same rigorous approach. This applies in particular to the solutions of the Earth's rotational equations that are usually expressed in TT and the solar system ephemerides (necessary for computing the luni-solar and planetary torque acting on Earth's rotation) that are usually expressed in TDB. This also applies to the realization of the ICRF, using TT-compatible VLBI space-coordinates.

It is important to note that, although TT and TDB have rigorous IAU definitions, the IAU/IUGG recommended time coordinates for the solar system and the Earth are TCB and TCG, respectively. Note also that there is currently a significant effort in astronomy for using TCB instead of TDB (e.g. for solar system ephemerides and pulsar timing).

3.3 Improved Definitions for Earth Rotation

1. *The Celestial Intermediate Pole*
   IAU 2000 Resolution B1.7 specifies that the pole of the nominal Earth's rotation axis is the *Celestial Intermediate Pole* (CIP). It is defined as the intermediate pole in the ITRS to GCRS transformation, separating nutation from polar motion by a specific convention in the frequency domain. The CIP definition in fact is an extension of the 1980 definition of the Celestial Ephemeris Pole (CEP) (see Fig. 25.1) in order to best realize the pole in the high frequency domain.
   The convention defining the CIP is such that (see Fig. 25.2): (1) the GCRS CIP motion includes all the terms with periods greater than 2 days in the GCRS (i.e. frequencies between −0.5 cycles per sidereal day (cpsd) and + 0.5 cpsd); (2) the ITRS CIP motion, includes all the terms outside the retrograde diurnal band in the ITRS (i.e. frequencies less than −1.5 cpsd or greater than −0.5 cpsd). This allows us to clarify the models to be used for high frequency polar motion versus those to be used for the GCRS motion of the pole. With that convention, the celestial motion includes precession, nutations with periods greater than 2 days, the free core nutation (FCN), plus the offsets; in contrast, nutations with periods lower than 2 days should be included in model for the pole motion in the ITRS.
2. *The Earth Rotation Angle*
   IAU 2000 Resolution B1.8 recommends using "non-rotating origins" (Guinot 1979) as origins on the CIP equator in the GCRS and ITRS; they were re-named *Celestial and Terrestrial Intermediate Origins* (CIO and TIO), respectively by IAU 2006 Resolution B2. Their kinematical property (see Fig. 25.3) provides a very straightforward

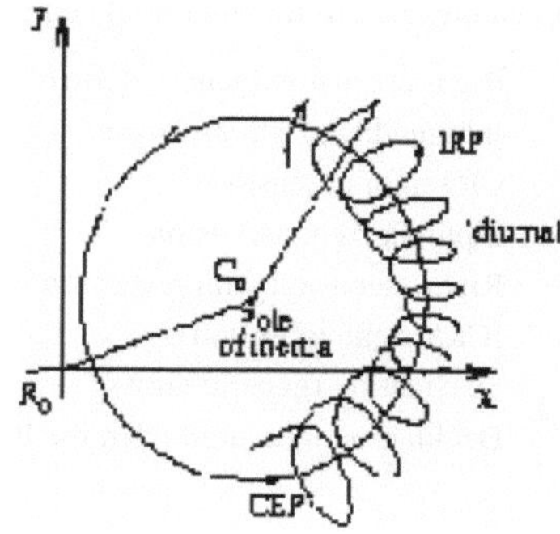

Fig. 25.1 Displacements of the celestial ephemeris pole (CEP) and instantaneous pole of rotation (IRP) with respect to the Earth's pole of inertia

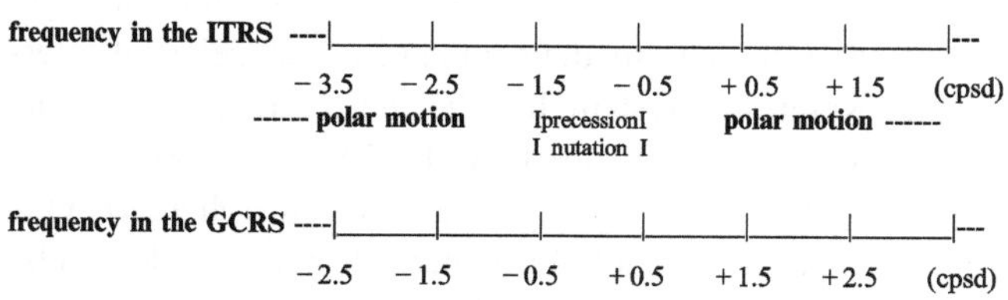

Fig. 25.2 Frequency convention (in cycles per sidereal day) for the definition of the celestial intermediate pole (CIP) (note that, due to Earth rotation: $\sigma_{ITRS} = \sigma_{GCRS} - 1$).

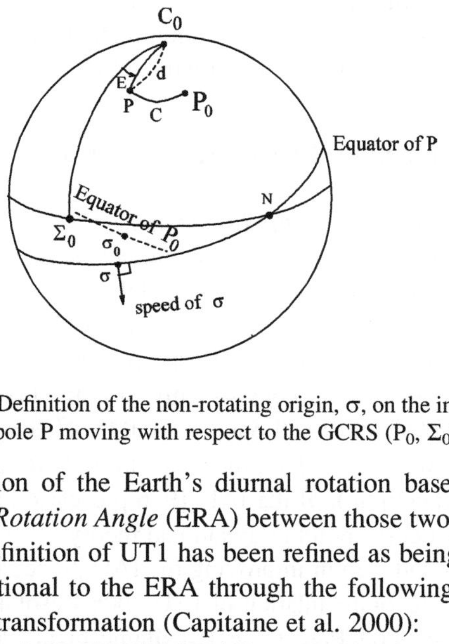

Fig. 25.3 Definition of the non-rotating origin, σ, on the intermediate equator of pole P moving with respect to the GCRS (P_0, Σ_0)

definition of the Earth's diurnal rotation based on the *Earth Rotation Angle* (ERA) between those two origins. The definition of UT1 has been refined as being linearly proportional to the ERA through the following conventional transformation (Capitaine et al. 2000):

ERA = 2π[0.7790572732640 + 1.00273781191135448 (Julian UT1 date – 2451545.0)].

The CIO (σ) is at present very close to the GCRS x-origin, Σ_0, and almost stationary in longitude, while the equinox (γ) to which Greenwich sidereal time, GST, refers is moving at about 50″/year in longitude. The CIO based procedure allows a clear separation between precession-nutation and the ERA, which is not model-dependent. In contrast, precession and nutation are mixed with Earth's rotation into the equinox based expression

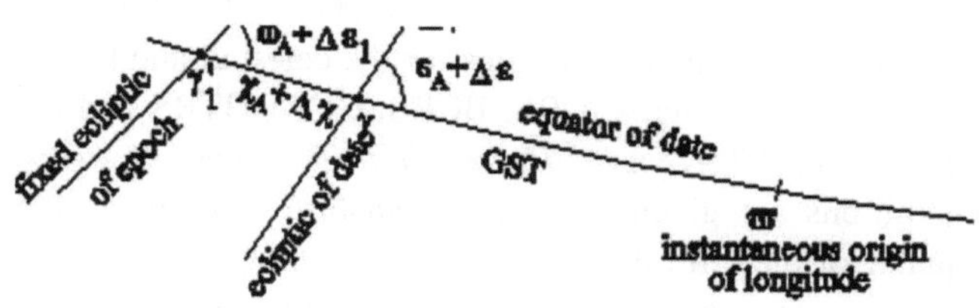

Fig. 25.4 The equinox based parameters for Earth rotation

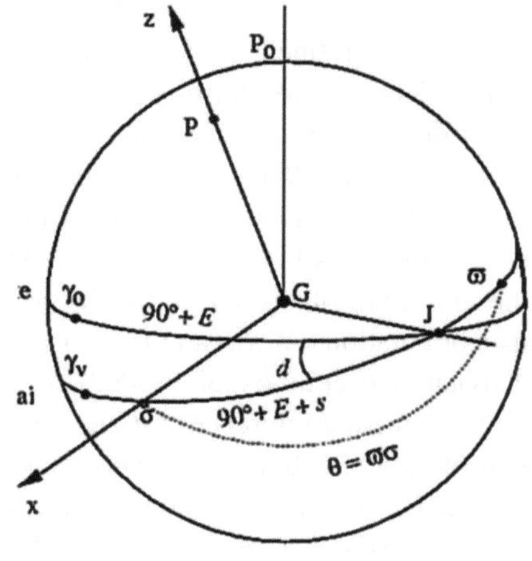

Fig. 25.5 The CIO based parameters for Earth rotation; E and d are the polar coordinates of the CIP (P) in the GCRS and θ is the ERA

for GST (Fig. 25.4), which includes the accumulated precession and nutation in right ascension.

3. *The motion of the CIP in the GCRS*

According to IAU 2000 Resolution B1.8, the ITRS to GCRS transformation should be specified by the position of the CIP in the GCRS, the position of the CIP in the ITRS, and the ERA. The GCRS direction of the CIP unit vector (Fig. 25.5), which includes precession, nutation and the frame bias, thus replaces the classical precession and nutation quantities (Fig. 25.4).

3.4 Improved Nomenclature for Fundamental Astronomy

The IAU Working Group on "Nomenclature for Fundamental Astronomy" (IAU NFA WG) made a number of recommendations on terminology (see Capitaine et al. 2007). It also produced the "IAU 2006 NFA Glossary" including a set of detailed definitions (compliant with GR) that best explain all the terms required for implementing the IAU 2000 resolutions; it also contains new definitions proposed by the WG, including those formally endorsed by the IAU in 2006 and the IUGG in 2007. The IAU 2000/2006 resolutions have provided the appropriate terminology for the pole, the Earth's angle of rotation, the longitude origins and the related reference systems. The NFA Glossary includes in particular definitions for the celestial and terrestrial reference systems ICRS, BCRS, GCRS, ITRS and the *Celestial and Terrestrial Intermediate Reference Systems.* Definitions are given for the *intermediate equator* as the equator of the CIP, the CIO and TIO origins, the *CIO and TIO locators*, s and s', for positioning those origins in the GCRS, the *equation of the origins* (EO), as the distance between the CIO and the equinox along the intermediate equator, and the time scales TCB, TDB, TCG and TT.

A few examples of refined definitions provided in the IAU 2006 NFA Glossary for terms related to the ICRF definition and realization are given below.

International Celestial Reference System (ICRS): the idealized barycentric coordinate system to which celestial positions are referred. It is kinematically non-rotating with respect to the ensemble of distant extragalactic objects. It has no intrinsic orientation but was aligned close to the mean equator and dynamical equinox of J2000.0 for continuity with previous fundamental reference systems. Its orientation is independent of epoch, ecliptic or equator and is realized by a list of adopted coordinates of extragalactic sources.

International Celestial Reference Frame (ICRF): a set of extragalactic objects whose adopted positions and uncertainties realize the ICRS axes and give the uncertainties of the axes. It is also the name of the radio catalog whose 212 defining sources is currently the most accurate realization of the ICRS. Note that the orientation of the ICRF catalog was carried over from earlier IERS radio catalogs and was within the errors of the standard stellar and dynamic frames at the time of adoption. Successive revisions of the ICRF are intended to minimize rotation from its original orientation. Other realizations of the ICRS have specific names (e.g. Hipparcos Celestial Reference Frame).

Barycentric Celestial Reference System (BCRS): a system of barycentric space-time coordinates for the solar system within the framework of General Relativity with metric tensor specified by the IAU 2000 Resolution B1.3. Formally, the metric tensor of the BCRS does not fix the coordinates completely, leaving the final orientation of the spatial axes undefined. However, according to IAU 2006 Resolution B2, for all practical applications, unless otherwise stated, the BCRS is assumed to be oriented according to the ICRS axes.

Geocentric Celestial Reference System (GCRS): a system of geocentric space-time coordinates within the framework of General Relativity with metric tensor specified by the IAU 2000 Resolution B1.3. The GCRS is defined such that the transformation between BCRS and GCRS spatial coordinates contains no rotation component, so that GCRS is kinematically non-rotating with respect to BCRS. The equations of motion of, for example, an Earth satellite, with respect to the GCRS will contain relativistic Coriolis forces that come mainly from geodesic precession. The spatial orientation of the GCRS is derived from that of the BCRS, that is (cf. IAU 2006 Resolution B2), unless otherwise stated, by the orientation of the ICRS.

Table 25.1 NFA WG Nomenclature for equatorial coordinates

α	Right ascension (generic term)
α_i	Intermediate right ascension
	CIO right ascension
α_e	Equinox right ascension
	Right ascension with respect to the equinox
α_{ICRS}	ICRS right ascension
δ	Declination (generic term)
δ_{ICRS}	Declination measured from the ICRS equator

Celestial Intermediate Origin (CIO): origin for right ascension on the intermediate equator in the Celestial Intermediate Reference System. It is the non-rotating origin in the GCRS that is recommended by the IAU 2000 Resolution B1.8, where it was designated the Celestial Ephemeris Origin. The name Celestial Intermediate Origin was adopted by IAU 2006 Resolution B2. The CIO was originally set close to the GCRS meridian and throughout 1900–2100 stays within 0.1 arc seconds of this alignment.

A change with respect to the usual equinox based nomenclature concerns the equatorial coordinates the nomenclature of which has been associated, in addition to the equinox, with the use of the new origins, or with the ICRS (see Table 25.1).

3.5 Improved Precession-Nutation

1. *The improvement of the IAU model*
   IAU 2000 Resolution B1.6 recommends the adoption of the new precession-nutation model that is designated IAU 2000A corresponding to the model of Mathews et al. (2002), denoted MHB2000. The precession part of the IAU 2000A model consists only in corrections, $\delta\psi_A = -0.2996''$/century and $\delta\omega_A = -0.02524''$/century to the precession rates (in longitude and obliquity referred to the J2000.0 ecliptic), of the IAU 1976 precession and hence does not correspond to dynamical theory.
   The second step in improving the IAU precession model was the recommendation of IAU 2006 Resolution B2 to adopt the P03 Precession (Capitaine et al. 2003) as a replacement for the precession part of the IAU 2000A precession-nutation, beginning on 1 January 2009. Details for implementing the IAU 2006/2000 precession-nutation have been given by Capitaine and Wallace (2006), and Wallace and Capitaine (2006).
   All the procedures, data and software for implementing the IAU 2000/2006 space-time coordinates, parameters and paradigm, nomenclature and models for Earth's rotation have been made available in Chap. 5 of the IERS Conventions 2003, and, in its final form in the

2010 version (see at: http://www.iers.org/IERS/EN/Publications/TechnicalNotes/tn36.html) and the Standards Of Fundamental Astronomy (SOFA) (Wallace 1998).

2. *Main features of the IAU 2000A Nutation*

The IAU 2000A nutation is based on the REN2000 rigid Earth nutation of Souchay et al. (1999) for the axis of figure. The latter is expressed as a series of luni-solar and planetary nutations in longitude $\Delta\psi$ and obliquity $\Delta\varepsilon$ referred to the ecliptic of date, composed of "in-phase" and "out-of-phase" components with their time variations, with arguments that are functions of the fundamental arguments of the nutation theory.

The rigid Earth nutation was transformed to the non-rigid Earth nutation by applying the MHB2000 "transfer function" to the REN2000 series of the corresponding prograde and retrograde nutations.

The sub-diurnal terms due to the imperfect axial symmetry of the Earth are not part of the solution, so that the axis of reference of the nutation model is compliant with the definition of the CIP. The MHB transfer function is based on the solution of the linearized dynamical equation of the wobble-nutation problem. Seven *Basic Earth Parameters* (BEP) were treated as adjustable for fitting the theoretical outputs to the VLBI.

This improves the IAU 1980 theory of nutation by taking into account the effect of mantle anelasticity, ocean tides, electromagnetic couplings produced between the fluid outer core and the mantle, as well as between the solid inner core and fluid outer core, and the consideration of nonlinear terms. The axis of reference is the axis of maximum moment of inertia of the Earth ignoring time-dependent deformations. The geodesic nutation contributions to the annual, semiannual and 18.6-year terms from Fukushima (1991) are part of the model.

The IAU 2000A nutation includes, as the REN2000 series, 678 lunisolar terms and 687 planetary terms. The resulting nutation is expected to have an accuracy of about 10 μas for most of its terms. On the other hand, the FCN (see Sect. 3.3) being a free motion, which cannot be predicted rigorously, is not considered a part of the IAU 2000A model; this limits the accuracy in the computed direction of the celestial pole in the GCRS to about 0.3 mas.

The MHB2000 Basic Earth parameters fitted to VLBI data are:

- The Earth's dynamical flattening e/(1 + e), which is a scale factor for the precession rate and nutation amplitudes,
- The deformability parameters of the whole Earth and the core under tidal forcing,
- Three real and imaginary parts of the complex coupling constants of the electromagnetic, core/mantle and fluid core/inner core couplings,
- The dynamical flattening of the core, which is a resonance factor for the nutation amplitudes.

3. *Main features of the IAU 2006 precession*

The IAU 2006 precession (Capitaine et al. 2003) provides improved polynomial expressions up to the fifth degree in time, both for the precession of the ecliptic and the precession of the equator.

The precession of the equator was derived from the dynamical equations expressing the motion of the mean pole about the ecliptic pole. Consequently, the IAU 2006 precession is consistent with a dynamical theory. The convention for separating precession from nutation, as well as the integration constants used in solving the equations, has been chosen in order to be consistent with the IAU 2000A nutation. This includes corrections for the perturbing effects in the observed quantities.

In particular, the IAU 2006 value for the precession rate in longitude is such that the corresponding Earth's dynamical flattening is consistent with the MHB value for that parameter (taking into account the change by 42 mas of the J2000 mean obliquity with respect to the IAU 2000 value). Moreover, the IAU 2006 precession includes the Earth's J_2 rate effect ($dJ_2/dt = -3 \times 10^{-9}$/century), mostly due to post-glacial rebound, which was not taken into account in the IAU precession models previously. The contributions to the IAU 2006 precession rates for the 2nd order effects, the J_3 and J_4 effects of the luni-solar torque, the J_2 and planetary tilt effects, as well as the tidal effects are from Williams (1994), and the non-linear terms are from MHB2000. The geodesic precession (i.e. $p_g = 1.9198830$/century) is from Brumberg et al. (1992).

The maximum discrepancy of VLBI celestial pole offsets with respect to IAU 2006/2000, expressed as GCRS direction of the CIP and corrected for the FCN, are currently of the order of 0.1 mas/century in the coefficients of the secular terms and a few tens of microarcseconds in the amplitude of the 18.6-year nutation.

4 Impact on the ICRF Definition and Realization

The second version, called ICRF2, of the International Celestial Reference Frame, was compiled by the IERS/IVS Working Group and recommended by the IAU Working Group "ICRF2"; it was adopted by IAU 2009 Resolution B3 as the fundamental realization of the ICRS. The aim was to improve the realization with densification of the frame and a more precise definition of the axes.

Such an improvement in the realization of the ICRS, especially for the densification of the frame, was made possible thanks to the very large number of VLBI and VLBA observations available. Improvement in the accuracy results from the improved accuracy of VLBI observations along with advances in modeling and estimation. These have benefited, in many aspects, from the improvements in the concepts, models and procedures that have been introduced by the recent IAU resolutions, and especially the 2000/2006 resolutions.

One important aspect for the accuracy of the ICRF is the use of the IAU 2000/2006 definitions and procedures for the barycentric and geocentric space-time coordinates and the transformation between them.

Another important aspect is the orientation issue. ICRF2 was aligned to ICRF1 by using a set of stable sources common to both ICRF2 and ICRF1-ext2. The purpose was to minimize rotation from the original ICRF orientation. This complies with the IAU 1997 resolution recommending that further improvements of the ICRF be accomplished without introducing any global rotation. This also complies with IAU 2006 resolution B2 on the default orientation of the BCRS/GCRS. Moreover, it is important to note that, according to the IAU 2000/2006 resolutions, (1) the geodesic precession/nutation have been taken into account and (2) the frame biases between the model and the GCRS have been introduced in a rigorous way. This ensures that the GCRS is without any time-dependent rotation with respect to the BCRS and that the orientation of the ICRF is not dependent on the Earth's orientation at epoch.

The way the varying celestial Earth's orientation (i.e. of the ITRS in the GCRS) is taken into account is also essential. The refined definition of the pole as well as the use of the new paradigm recommended by the resolutions for the terrestrial-to-celestial coordinate transformation allow an accurate estimation of the Earth's rotation and precession-nutation separately. This is important for the accuracy of the ICRF realization. Thanks to special efforts, the IAU 2000/2006 expressions have been developed in order that the equinox based paradigm can benefit from the clear separation between the Earth's angle of rotation and precession-nutation offered by the CIO based representation and consequently provide the same accuracy.

5 Summary

The IAU 2000, IAU 2006, IUGG 2003 and IUGG 2007 resolutions on reference systems have clarified the definition of the reference systems in the framework of general relativity and modified the way the Earth orientation (i.e. the transformation between the ITRS and the GCRS) is expressed.

The IAU 2000, IAU 2006 and IAU 2009 resolutions have adopted high accuracy models and conventions for expressing the relevant quantities for the transformation from terrestrial to celestial systems.

The concepts, nomenclature, models and conventions in fundamental astronomy have been improved by these resolutions in order to be the best suitable for the best accurate realizations of the reference systems.

All these improvements offered by the recent IAU Resolutions have contributed to the improvement of the definition and realization of the ICRF, the second version of which has been adopted by the IAU in 2009 as the fundamental realization of the ICRS.

References

Brumberg V, Bretagnon P, Francou G (1992) In: Capitaine N (ed) The proceedings of the Journées Systèmes de reference spatio-temporels, Observatoire de Paris, pp 141–148

Capitaine, N and the IAU NFA WG (2007) In: van derHucht KA (ed) Transactions of the IAU XXVIB, vol 14, pp 474–475

Capitaine N, Guinot B, McCarthy DD (2000) Definition of the Celestial Ephemeris origin and of UT1 in the International Celestial Reference Frame. A&A 355(1), pp. 398–405

Capitaine N, Wallace PT (2006) Precession-nutation procedures consistent with IAU 2006 resolutions. A&A 450: pp. 855–872, doi:10.1051/0004-6361:20054550

Capitaine N, Wallace PT, Chapront J (2003) Expressions for IAU 2000 precession quantities, A&A 412(2), pp. 567–586, doi:10.1051/0004-6361:20031539

IERS Conventions (2003) IERS technical note 32. In: McCarthy DD, Petit G (eds) Frankfurt am Main: Verlag des desamts für Kartographie und Geodäsie, 2004

Fukushima T (1991) Geodesic Nutation A&A 244(1), pp. L11–L12

Guinot B (1979) Basic Problems in the Kinematics of the Rotation of the Earth in Time and the Earth's Rotation. In: McCarthy DD, Pilkington JD (eds) Time and the Earth's rotation. Dordrecht D. Reidel Publishing Company, pp. 7–18

Hilton J, Capitaine N, Chapront J et al (2006) Report of the International Astronomical Union Division I Working Group on Precession and the Ecliptic. Celest Mech Dyn Astr 94(3), pp. 351–367, doi:10.1007/s10569-006-0001-2

IAU (1997) In: Andersen J. (ed) Transactions of the IAU Vol. XXIII B; Kluwer Academic Publishers, ISBN 0-7923-5588-1

IAU (2000) In: Rickman H. (ed) Transactions of the IAU Vol. XXIV B, ASP, ISBN 1-58381-087-0

IAU (2006) In: van der Hucht KA (ed) Transactions of the IAU Vol. XXVIB, Cambridge University Press, ISBN 978-0-521-85606-5

IAU (2006) NFA glossary of the IAU working group on nomenclature for fundamental astronomy. http://syrte.obspm.fr/iauWGnfa

IAU (2009) In: Corbett I (ed) Transactions of the IAU XXVIIB, Cambridge University Press, ISBN 9780-521-76831-3. Cambridge University Press, pp 55–70

IUGG (2003) IUGG Resolutions. http://www.iugg.org/resolutions/sapporo03.pdf

IUGG (2007) IUGG Resolutions. http://www.iugg.org/resolutions/perugia07.pdf

Mathews PM, Herring TA, Buffett BA (2002) Modeling of nutation and precession: New nutation series for nonrigid Earth, and insights

into the Earth's Interior. J Geophys Res 107(B4). doi:10.1029/2001JB000390
Soffel M, Klioner SA, Petit et al. (2003) The IAU 2000 Resolutions for Astrometry, Celestial Mechanics, and Metrology in the Relativistic Framework: Explanatory Supplement, AJ 126, 6, pp. 2687–2706, doi:10.1086/378162
Souchay J, Loysel B, Kinoshita H, Folgueira M (1999) Corrections and new developments in rigid Earth nutation theory: III. Final tables REN-2000 including crossed-nutation and spin-orbit coupling effects. A&AS 135(1), pp. 111–131, doi:10.1051/aas:1999446
Wallace PT (1998) SOFA: Standards of Fundamental Astronomy. In: Andersen J (ed) Highlights of astronomy, vol 11A. Kluwer Academic Publishers, p. 191
Wallace PT, Capitaine N (2006) Precession-nutation procedures consistent with IAU 2006 resolutions. A&A 459(3): pp. 981–985, doi:10.1051/0004-6361:20065897
Williams JG (1994) Contributions to the Earth's obliquity rate, precession, and nutation. AJ 108(2): pp. 711–724, doi:10.1086/117108

Effects of ICRF2 on the TRF, CRF, and EOP

26

David Gordon, Chopo Ma, Dan MacMillan, Sergei Bolotin, Karine Le Bail, and John Gipson

Abstract

The ICRF2 became official on Jan 1, 2010. It includes positions of 3414 compact radio astronomical sources observed with VLBI, a fivefold increase over the first ICRF. ICRF2 was aligned with the ICRS using 138 stable sources common to both ICRF2 and ICRF-Ext2. Maintenance of ICRF2 is to be made using 295 defining sources chosen for their historical positional stability, minimal source structure, and sky distribution. The switchover to ICRF2 has had some small effects on the terrestrial reference frame (TRF), celestial reference frame (CRF) and Earth orientation parameter (EOP) solutions from VLBI. A CRF based on ICRF2 shows a relative rotation of ~40 μas with respect to ICRF, mostly about the Y-axis. Small shifts are also seen in the EOP, the largest being ~11 μas in X_{pole}. Some small but insignificant differences are also seen in the TRF.

Keywords

ICRF • ICRF2 • Terrestrial reference frames • Celestial reference frames • Earth orientation parameters

1 Introduction

ICRF was the first realization of the International Celestial Reference Frame by VLBI (Ma et al. 1997, 1998). It used VLBI data from August 1979 through July 1995. It was adopted by the IAU in 1997 and became official on 1 Jan 1998. Its stability and precision represented an ~10 fold improvement over the previous stellar reference frame, the FK5 (Fricke et al. 1988). It initially contained positions of 608 sources, and used 212 'defining' sources to define the axes orientation. Figure 26.1 shows the ICRF sources. Two extensions were later made, adding 109 additional sources (IERS 1999; Fey et al. 2004).

D. Gordon (✉) • D. MacMillan • S. Bolotin • K. Le Bail • J. Gipson
NVI Inc. and NASA/GSFC, Code 698.2, Greenbelt, MD 20771, USA

C. Ma
NASA/GSFC, Code 698.2, Greenbelt, MD 20771, USA
e-mail: David.Gordon-1@nasa.gov

ICRF2 (IERS 2009) was the next step. It used VLBI data through March 2009. It was adopted by the IAU in 2009 and became official on 1 Jan 2010. It yields an approximately five to sixfold improvement in precision and an approximately twofold improvement in stability over ICRF. It contains positions of 3,414 sources, of which 1,448 were observed in multiple VLBI sessions and 1966 were observed in single VLBI sessions. Figure 26.2 shows the ICRF2 sources observed in multiple sessions. Table 26.1 gives a short comparison of the two reference frames.

The major weaknesses of ICRF were the uneven sky distribution of the defining sources, positional instability of the VLBI phase centers of many of the defining sources, source structure effects, and possibly systematic modeling errors (Sokolova and Malkin 2007). Improvements in VLBI observing and modeling, and the quadrupling in the amount of data from 1995 to 2009 allowed greater scrutiny of source stabilities and source structure (Fey and Charlot 1997). This allowed picking the most stable sources in all parts of the sky as defining sources, resulting in a much more even sky distribution of the ICRF2 defining sources. The ICRF2

Z. Altamimi and X. Collilieux (eds.), *Reference Frames for Applications in Geosciences*,
International Association of Geodesy Symposia 138, DOI 10.1007/978-3-642-32998-2_26,

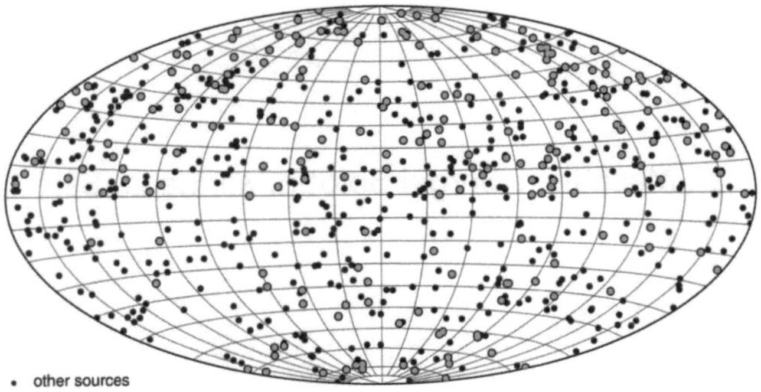

Fig. 26.1 The 608 ICRF sources

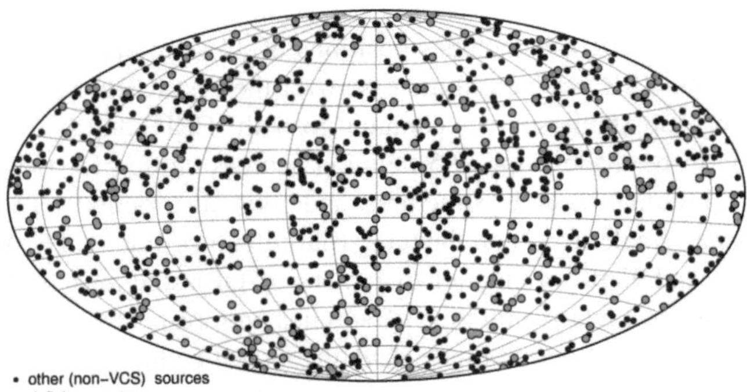

Fig. 26.2 The 1448 ICRF2 multiple session sources

Table 26.1 ICRF versus ICRF2

	ICRF	ICRF2
# VLBI observations	~1.6 million	~6.5 million
# Defining sources	212	295
# Total sources	608	3,414
Noise floor	~250 μas	~40 μas
Axis stability	~20 μas	~10 μas

work showed that only 97 of the original 212 ICRF defining sources were stable enough and without significant structure to qualify as ICRF2 defining sources and only 24 of those were at southern declinations. Thus, the ICRF may not have been as stable as originally estimated.

2 ICRF2 Versus ICRF Based Solutions

The goal of this study is to determine the effect the switchover from ICRF to ICRF2 has on the terrestrial reference frame (TRF), the celestial reference frame (CRF), and Earth orientation parameter (EOP) results from VLBI solutions. The latest VLBI solutions are based on ICRF2 positions, whereas previous solutions were based on ICRF positions. In these solutions, the positions of the defining sources are initially set to their ICRF or ICRF2 catalog positions. Then most or all source positions are solved for globally with the constraint that there be no net rotation of the defining sources as a group. In ICRF2-based solutions, there is also a set of 39 'special handling' sources (the most unstable sources) whose positions are solved for as arc parameters (for each session) in order to avoid distortions of the reference frame. These solutions also usually solve for global site positions, site velocities, and daily EOP.

Table 26.2 TRF differences, ICRF2-based versus ICRF-based

	X	Y	Z
Translation (mm)	−0.08 ± 0.17	−0.25 ± 0.18	+0.26 ± 0.16
Rotation (μas)	+17.4 ± 7.1	+2.9 ± 6.8	−0.7 ± 4.9

3 Comparison of ICRF2 and ICRF Based Solutions

We generated and compared ICRF and ICRF2 based solutions to show how the switch to ICRF2 has affected the TRF, the CRF, and the EOP values from VLBI solutions. The ICRF2-based solution was our current (gsf2010a) IVS solution. The ICRF-based solution used the same solution setup and data, except that it used the 212 ICRF defining sources and their ICRF positions, and the 39 unstable special handling sources were not given special treatment. These solutions did not use the 26 sessions from the six VLBA Calibrator Surveys (Petrov et al. 2008), which contain most of the single session sources. The ICRF-based solution would have been our latest quarterly solution if there were no ICRF2.

3.1 TRF Comparisons

The two sets of site positions and velocities were compared, and a seven-parameter fit was made to their differences (ICRF2 − ICRF). The translation (in mm) and rotation (in μas, where 1 mm = 32.2 μas at the Earth's equator) differences of the site positions are shown below in Table 26.2.

These differences are quite small and have an insignificant scale factor (.007 ± .022 ppb). For comparison, we looked at the variations in the TRF from several quarterly GSFC VLBI (ICRF-based) solutions over the past 10 years (Sect. 5). The differences seen among those solutions are typically an order of magnitude greater than the differences found here.

3.2 EOP Comparisons

We also compared daily Earth orientation parameters between the two solutions. The differences (ICRF2 − ICRF)

Table 26.3 EOP differences, ICRF2-based versus ICRF-based

	Shift	Drift ($year^{-1}$)	WRMS	R1/R4 uncertainties
X_p (μas)	11.1 ± .8	−1.8 ± .2	47.5	~40–150
Y_p (μas)	−4.0 ± .7	3.3 ± .1	40.5	~40–150
UT1 (μs)	−0.5 ± .1	.07 ± .01	2.8	~1.5–4.0
dX (μas)	37.6 ± .8	−0.4 ± .1	47.3	~30–100
dY (μas)	20.8 ± .8	0.1 ± .1	45.5	~30–100
X_p rate(μas/day)	2.3 ± 2.2	0.2 ± .4	125.	~120–300
Y_p rate (μas/day)	−2.2 ± 2.1	0.0 ± .4	122.	~120–300
UT1 rate(μs/day)	.05 ± .09	−.01 ± .02	5.2	~4–10

Table 26.4 CRF differences, ICRF2-based versus ICRF-based

X (μas)	Y (μas)	Z (μas)
+17.8 ± 0.5	−38.8 ± 0.5	+3.6 ± 0.4

are shown in Table 26.3. The EOP shifts shown here are no greater than the typical uncertainties seen in our weekly R1 and R4 sessions (last column of Table 26.3), and are also similar to the shifts seen between our quarterly VLBI solutions.

We also made an Allan variance study of the EOP differences between the ICRF and ICRF2 solutions compared to IGS EOP. The Allan variances of the X_{pole} and Y_{pole} differences show no significant differences between the ICRF and ICRF2 values.

3.3 CRF Comparisons

The source catalogs from the two solutions show a small relative rotation. Using 1167 common sources, we get the following rotation angles, shown in Table 26.4.

This rotation represents a difference of around twice the estimated ICRF axis stability. Though not significant, it merits further investigation and explanation. [The CRF axes are defined such that the X axis is towards 0 h RA, 0 ° Declination; Y is towards 6 h RA, 0 °Declination; and Z is towards +90 ° Declination.]

4 ICRF2 Alignment with ICRF

ICRF2 came from the gsf008a solution, which was an ICRF-based solution. ICRF2 defining sources were selected based on positional stability, low structure index (Fey and Charlot 1997), and sky distribution. It was desired to align the ICRF2 defining sources with the ICRF defining sources. However, there were only 97 common defining sources and 73 of those were in the northern half of the sky. To improve the distribution of sources used for alignment, an additional 41 ICRF2 defining sources were selected. All had ICRF-Ext2 positions and 35 of them were at southern declinations. Thus, ICRF2 is considered to be aligned with ICRF-Ext2. The rotation angles applied to gsf008a to obtain ICRF2 (+23.3, −33.5, +7.8 μas; IERS 2009) are very similar to those found in this study.

Table 26.5 CRF differences: ICRF2 quarterly 2010a versus ICRF quarterly 2009a

	X (μas)	Y (μas)	Z (μas)
2010a/2009a	+18.1 ± .8	−38.8 ± .8	+6.2 ± .6

5 Stability of ICRF Solutions

We compared several ICRF-based source catalogs from quarterly GSFC TRF/CRF/EOP solutions over the past 10 years. These catalogs were from the following solutions:

- 2000a – ICRF-based (oldest quarterly).
- 2002c – ICRF-based.
- 2005b – ICRF-based.
- 2007c – ICRF-based.
- 2009a – ICRF-based.
- 2010a – ICRF2-based.

Comparing 2010a (first GSFC ICRF2-based quarterly) to 2009a (last GSFC ICRF-based quarterly), we get the relative rotation given in Table 26.5 between the CRF solutions, which is similar to Table 26.4, as expected. Table 26.6 gives the corresponding translation and rotation differences from a seven-parameter fit of the TRF site positions, where the scale difference is 0.03 ± 0.04 ppb.

Also, four comparisons were made between various ICRF-based quarterly solutions. The CRF comparisons are given in Table 26.7. The translation and rotation differences from seven-parameter fits are given in Tables 26.8 and 26.9.

The corresponding TRF scale differences are: −0.14 ± 0.22, −0.46 ± 0.17, −0.47 ± 0.12, and −0.44 ± 0.10 ppb. Differences between the various solutions are a result of different data sets, modeling changes, and

Table 26.6 TRF differences: ICRF2 quarterly 2010a versus ICRF quarterly 2009a

2010a/2009a	X	Y	Z
Translation (mm)	+1.5 ± 0.3	+1.9 ± 0.3	−0.7 ± 0.3
Rotation (μas)	+34.3 ± 11.6	−0.7 ± 11.0	+23.6 ± 8.1

Table 26.7 CRF differences among quarterly ICRF-based solutions

	X (μas)	Y (μas)	Z (μas)
2009a/2000a	−37.6 ± 4.9	+49.5 ± 4.9	+1.9 ± 4.0
2009a/2002c	−34.7 ± 3.5	+18.8 ± 3.5	−6.1 ± 2.8
2009a/2005b	+3.6 ± 2.2	+17.5 ± 2.2	+15.1 ± 1.8
2009a/2007b	−18.6 ± 1.6	+21.0 ± 1.6	+7.0 ± 1.3

Table 26.8 TRF translation differences among quarterly ICRF-based solutions

.	X (mm)	Y (mm)	Z (mm)
2009a/2000a	−10.0 ± 1.6	−1.6 ± 1.7	+23.6 ± 1.6
2009a/2002c	−1.7 ± 1.3	+3.3 ± 1.3	+1.5 ± 1.2
2009a/2005b	−1.2 ± 0.9	+1.9 ± 0.9	+2.9 ± 0.9
2009a/2007b	−0.1 ± 0.8	+4.7 ± 0.8	+1.9 ± 0.7

Table 26.9 TRF rotation differences among quarterly ICRF-based solutions

	X (μas)	Y (μas)	Z (μas)
2009a/2000a	+242.2 ± 66.5	+48.8 ± 64.1	+51.8 ± 44.5
2009a/2002c	+90.0 ± 51.8	+93.5 ± 49.4	+18.2 ± 34.9
2009a/2005b	+229.3 ± 37.4	+16.8 ± 35.6	−76.7 ± 25.2
2009a/2007b	+263.1 ± 30.6	+44.1 ± 29.2	−92.8 ± 20.6

different à priori's. The CRF and TRF differences are mostly larger than that seen between the ICRF2-based and ICRF-based solutions. In contrast, the changes seen in switching to ICRF2 do not look unusual in comparison.

6 ICRF2 Versus ICRF-Ext2 Comparison

Because ICRF2 was aligned with ICRF-Ext2 (and not strictly with ICRF), two additional solutions were made. The first solution held all ICRF2 sources fixed (not solved for) to their ICRF2 positions (except the special handling and single session sources). The second held all 717 ICRF-Ext2 sources fixed to their ICRF-Ext2 positions. Table 26.10 shows a comparison of their EOP. The overall shifts are very small and are all less than the typical uncertainties in the EOP values.

Table 26.11 gives the corresponding effect on the TRF. Differences are below the 2 mm level, and do not appear large in comparison to the quarterly differences.

Table 26.10 EOP differences, ICRF2-fixed versus ICRF-Ext2-fixed

	Shift	Drift (year^{-1})	WRMS
X_p (μas)	−4.1 ± 0.7	−9.1 ± 0.1	43.4
Y_p (μas)	1.7 ± 0.5	1.4 ± 0.1	28.1
UT1 (μs)	−1.7 ± .03	−0.5 ± .01	1.8
dX (μas)	0.1 ± 0.5	−1.4 ± 0.1	32.1
dY (μas)	8.3 ± 0.5	−1.3 ± 0.1	27.6
X_p rate (μas/day)	1.2 ± 1.5	−0.2 ± 0.3	87.5
Y_p rate (μas/day)	−3.1 ± 1.3	0.6 ± 0.2	79.8
UT1 rate (μs/day)	0.24 ± .05	−.03 ± .01	2.8

Table 26.11 TRF differences, ICRF2-fixed verus ICRF-Ext2-fixed

	X	Y	Z
Translation (mm)	+0.38 ± 0.25	−1.39 ± 0.27	+0.50 ± 0.24
Rotation (μas)	+4.2 ± 10.3	−34.3 ± 10.0	+60.1 ± 7.1

7 Conclusions

In the switchover to ICRF2, differences in the terrestrial reference frame are very small, and less than has been seen between various VLBI quarterly solutions over the past 10 years. Some small systematic EOP differences are seen, but again, these are no larger than the differences typically seen between various quarterly solutions. There are also some small rotations seen in the celestial reference frame solutions. This is primarily a result of the lack of stability of many of the ICRF defining sources and their uneven sky distribution, which prevented a strict alignment of the two respective sets of defining sources. ICRF2-based CRF solutions can be expected to show greater stability in future solutions than was seen for ICRF-based solutions because of the greater stability of the ICRF2 defining sources and their more even sky distribution.

Acknowledgements We would like to thank the International VLBI Service for Geodesy and Astrometry (IVS) and all its components, which have made the ICRF2 and this study possible. We would also like to thank the National Radio Astronomy Observatory and the VLBA for observing time and support which greatly improved the quality and the quantity of the data used in the ICRF2 and in this study. NRAO is an NSF facility operated under cooperative agreement by Associated Universities, Inc.

References

Fey AL, Charlot P (1997) VLBA observations of radio reference frame sources. II. Astrometric suitability based on observed structure. Astrophys J Suppl 111:95–142

Fey AL, Ma C, Arias EF, Charlot P, Feissel-Vernier M, Gontier A-M, Jacobs CS, Li J, MacMillan DS (2004) The second extension of the international celestial reference frame: ICRF-EXT 1. Astron J 127:3587–3608

Fricke W, Schwan H, Lederle T, Bastian U, Bien R, Burkhardt G, Mont BD, Hering R, Jährling R, Jahreiß H, Röser S, Schwerdtfeger H-M, Walter HG (1988) Fifth fundamental catalogue (FK5). Part I: the basic fundamental stars, vol 32. Veröff. Astron. Rechen-Institut, Heidelberg, pp 1–106

IERS (1999) 1998 international Earth rotation service annual report. Observatoire de Paris, Paris

IERS (2009) IERS technical note 35, The second realization of the international celestial reference frame by very long baseline interferometry. In: Fey AL, Gordon D, Jacobs CS (eds) Verlag des Bundesamts für Kartographie und Geodäsie, Frankfurt am Main. (http://www.iers.org/TN35)

Ma C, Arias EF, Eubanks TM, Fey AL, GontierA-M, Jacobs CS, Sovers OJ, Archinal BA, Charlot P (1997) The international celestial reference frame realized by VLBI. In: Ma C, Feissel M (eds) IERS technical note 23, Definition and realization of the international celestial reference system by VLBI astrometry of extragalactic objects, Observatoire de Paris, Paris (http://www.iers.org/TN23)

Ma C, Arias EF, Eubanks TM, Fey AL, Gontier A-M, Jacobs CS, Sovers OJ, Archinal BA, Charlot P (1998) The international celestial reference frame as realized by very long baseline interferometry. Astron J 116:516–546

Petrov L, Kovalev YY, Fomalont EB, Gordon D (2008) The sixth VLBA calibrator survey: VCS6. Astron J 136:580–585

Sokolova J, Malkin Z (2007) On comparison and combination of radio source positions. AA 474:665–670

Systematic Inconsistencies Between VLBI CRF and TRF Solutions Caused by Different Analysis Options

27

R. Heinkelmann and V. Tesmer

Abstract

We assess the systematics between Very Long Baseline Interferometry (VLBI) terrestrial and celestial reference frames (TRF and CRF) solutions caused by different analysis options. Comparisons are achieved by sequential variation of options relative to a reference solution, which fulfills the requirements of the International VLBI Service for Geodesy and Astrometry (IVS) analysis coordination. Neglecting the total NASA/GSFC Data Assimilation Office (DAO) a priori gradients causes the largest effects: Mean source declinations differ up to 0.2 mas, station positions are shifted southwards, and heights are systematically larger by up to 3 mm, if no a priori gradients are applied. The effect is explained with the application of gradient constraints. Antenna thermal deformations, atmospheric pressure loading, and the atmosphere pressure used for hydrostatic delay modeling still exhibit significant effects on the TRF, but corresponding CRF differences (about 10 μas) are insignificant. The application of NMF atmosphere mapping functions can systematically affect source declinations up to 30 μas, which is between the estimated axes stability (10 μas) and the mean positional accuracy (40 μas) specified for the ICRF2. Further significant systematic effects are seasonal variations of the terrestrial network scale (±1 mm) neglecting antenna thermal deformations, and seasonal variations of station positions, primarily of the vertical component up to 5 mm, neglecting atmospheric loading. The application of NMF instead of VMF1 can result in differences of station heights up to 6 mm, but no overall global systematic can be found. Using constant atmosphere pressure values for the determination of hydrostatic zenith delays systematically deforms the TRF: station height differences mostly show the same sign with absolute values exceeding 1 mm.

Keywords

Very long baseline interferometry (VLBI) • Terrestrial reference frame (TRF) • Celestial reference frame (CRF) • Interactions of the reference frames

1 Introduction

State of the art terrestrial reference frames determined by space-geodetic techniques involve various intra- and inter-technique combination steps. The quality and interpretability of the combined frame depend to a large extend on the consistency of the solutions used for the combination. Identifying and quantifying the systematic effects of analysis options on the reference frames is a major issue for the assessment of the consistency of the frames in a combination

R. Heinkelmann (✉)
Deutsches Geodätisches Forschungsinstitut (DGFI), Alfons-Goppel-Street 11, Munich D-80539, Germany
e-mail: heinkelmann@dgfi.badw.de

V. Tesmer
OHB-System AG, Universitätsallee 27-29, Bremen D-28359, Germany

Z. Altamimi and X. Collilieux (eds.), *Reference Frames for Applications in Geosciences*,
International Association of Geodesy Symposia 138, DOI 10.1007/978-3-642-32998-2_27,

solution, but it is also necessary for understanding the quality of frames determined by a single solution, such as the celestial reference frame. In this paper, we highlight the effects of some of the most recent analysis options:

- VLBI antenna thermal deformation,
- Atmospheric pressure loading,
- Meteorological data,
- Atmosphere mapping functions, and
- A priori atmosphere gradients,

and review remaining deficiencies. We want to avoid a theoretical discussion and prefer the empirical sequential application of the various options in VLBI analyses. Consequently, our methodology is to display differences between each individual 'test' solution and a common 'reference' solution, which is described in the next section. The absolute results of the analyses are not relevant in this context and thus are not displayed in the paper. For more details about absolute results of DGFI's reference frames we refer the interested reader to Tesmer (2002) or to Heinkelmann et al. (2006). The applied analysis options are discussed in Sect. 3.

2 Reference Solution

The DGFI VLBI reference solution in this context is a "core" solution, i.e. stations, radio sources, and observing sessions are selected to form a stable representative subset without rarely observed radio sources—in particular, without the sources solely observed by the VLBA calibrator surveys (VCS)—and without mobile and rarely observing sites, all of which would otherwise add higher-than-average noise to the solution that could mask small systematic effects. The solution includes 2,990 daily sessions between 1984.0 and 2008.0 with 53 radio telescopes observing 779 sources. Single baseline sessions were excluded. A station is not considered for the analysis of a session, if it had less than 100 valid group delay observations. All of the 779 ICRF2 (IERS 2009) sources were observed in at least three sessions. The reference solution analysis options meet the requirements of operational VLBI analysis for input to the International VLBI Service for Geodesy and Astrometry (IVS) Earth orientation parameters (EOPs) combination specified by the IVS Analysis Coordinator.

2.1 A Priori Data

The a priori source coordinates are taken from ICRF2. The a priori station coordinates are taken from the DGFI global terrestrial reference frame (TRF) solution "dgf2009a". For the a priori transformation from the celestial reference frame (CRF) to the TRF the IERS C04 05 series (Bizouard and Gambis 2009) are used and appropriately interpolated considering tidal variations of the Earth's rotation by Eanes (IERS 2004). A priori nutation angles are calculated with the precession-nutation equations of IAU 2000A based on the MHB 2000 model (Mathews et al. 2002) using the classical equinox-based transformation scheme and corresponding corrections. Meteorological data are derived from the European Centre for Medium-Range Weather Forecasts (ECMWF)[1] 40-years reanalysis, ERA-40 (Uppala et al. 2005), or the ECMWF operational analysis, after 2002.

2.2 Station Position Modeling

Solid Earth tides and pole tides are considered following the IERS Conventions 2003 (IERS 2004). Deformations due to ocean loading are taken from the FES2004 model (Lyard et al. 2006). Atmospheric pressure loading is modeled by global convolution (Petrov and Boy 2004).

2.3 Troposphere Model

Troposphere delay and bending are modeled as (Davis et al. 1985)

$$\begin{aligned} L(\alpha,\varepsilon) = {} & mf_h(\varepsilon)\cdot ZHD + \\ & + mf_g(\varepsilon)\cdot(G_{N,apr}\cdot\cos\alpha + G_{E,apr}\cdot\sin\alpha) + \\ & + mf_w(\varepsilon)\cdot ZD_{est} + \\ & + mf_g(\varepsilon)\cdot(G_{N,est}\cdot\cos\alpha + G_{E,est}\cdot\sin\alpha) \end{aligned} \quad (27.1)$$

where the hydrostatic and wet mapping functions (mf_h, mf_w) are the Vienna mapping functions, VMF1 (Böhm et al. 2006b), determined with data from the ECMWF weather model and are considered known; the gradient mapping function (mf_g) is taken from MacMillan (1995). The a priori delay equals the zenith hydrostatic delay given by the Saastamoinen model. The non-hydrostatic zenith delay (ZD_{est}) is estimated along with geodetic and other parameters. The a priori gradients ($G_{N,apr}$, $G_{E,apr}$) can be set to zero, then the total atmosphere gradients can be estimated. If non-zero a priori gradients are applied, e.g. according to MacMillan and Ma (1998), residual gradients ($G_{N,est}$, $G_{E,est}$) can be estimated.

2.4 Relativistic Scale and Gravitational Bending

The theoretical delay is computed according to the "consensus model" (IERS 2004) without scaling the observed time delay from terrestrial time (TT) to geocentric coordinate time (TCG). Post-Newtonian gravitational signal deflection is considered for the Sun and the Earth.

[1] http://www.ecmwf.int/

2.5 VLBI Antenna Model

For the delays caused by antenna axis offsets[2] and station eccentricities[3] the official values specified by the IVS are applied. Thermal expansion is compensated for by the model provided by Nothnagel (2008) with reference temperatures derived from the Global Pressure and Temperature model, GPT (Böhm et al. 2007a).

2.6 Parameterization

While station coordinates, source coordinates, EOP, and atmosphere gradients and gradient rates are estimated once per session, clock parameters and atmosphere zenith delays are modeled as piece-wise linear polynomials with 1 h resolution. The reference solution applies a local no-net-rotation (NNR) and no-net-translation (NNT) condition including the coordinates of all sites as well as a NNR condition for the source coordinates.

3 Effects Caused by Different Analysis Options

The study of effects caused by analysis options is of major concern to the technique services for the determination of consistent products and the assessment of the quality of the products. Independent of whether an individual or a combined solution is finally provided to the user community, the results will depend on the applied models, parameterizations, and other analysis options. In this context, a number of analysis options have already been tested: Tesmer et al. (2006, 2007), Schuh et al. (2007), Tesmer (2007), and Heinkelmann et al. (2009). Some of the scenarios were repeated or extended for the computations, which led to the ICRF2. In this section some systematic effects caused by analysis options are presented. Each specific "test" solution is compared against the "reference" solution, which was described in Sect. 2.

3.1 VLBI Antenna Thermal Deformations

A conventional model on VLBI antenna thermal deformation was released by Nothnagel (2008) and was integrated into and then tested with our VLBI analysis software OCCAM. The model considers different antenna mounts, materials, dimensions of the construction parts, and time lags of the reaction of the construction on variations of the ambient temperature. Thermal expansions of all antennas can be corrected for with the proposed model. However, the model does not account for the difference between temperatures inside and outside radomes covering antennas. Stations WESTFORD, HAYSTACK (USA), ONSALA60 (Sweden), METSAHOV (Finland), or YEBES (Spain), need to be considered individually because no temperatures inside the radomes are available. To derive inside temperatures from the outside temperatures reported in IVS log-files or NGS-files some individual models are available, while others are missing. For WESTFORD, for example, it is possible to approximate inside temperatures with a simple approach provided by Niell (pers. com.):

$$t_{\mathrm{inside}} = 20°\mathrm{C} + 0.6 \cdot (t_{\mathrm{outside}} - 20°\mathrm{C}) \qquad (27.2)$$

For ONSALA60 an advanced thermal deformation model is available and was presented e.g. by Wresnik et al. (2007). To allow for a consistent application of thermal expansion correction, methods to derive inside temperatures should be available for all antennas covered by radomes.

The thermal expansion of radio telescopes is a small effect: local horizontal station coordinates differ by less than 1 mm. The largest variations happen in the vertical component at those sites with large telescopes and large annual temperature variations, such as ALGOPARK (Canada) or GILCREEK (Alaska, USA), where the annual variations can reach 8 mm from peak to peak. Since primarily vertical site coordinates are affected, modeling the thermal deformation causes large annual network scale differences (Fig. 27.1). The annual amplitude varies among the years because different stations have participated in the observing networks over the last 25 years. ALGOPARK and GILCREEK have ceased observations at the end of 2006 and 2005, respectively. Thereafter the annual network scale variations are significantly smaller (Fig. 27.1).

Another topic in this context is the definition of the reference temperature for the expansion. Theoretically, the reference temperature equals the ambient temperature at which the dimensions of the antenna structure are given, i.e. the thermal deformation is zero. Since those theoretical temperatures are unavailable, it became common practice to use site-dependent mean ambient temperatures as reference temperatures. The new conventional model (Nothnagel 2008) recommends site-dependent mean ambient temperatures derived from GPT as reference. If this reference temperature does not equal the site-dependent mean temperature, a systematic shift can be expected, primarily in the vertical component.

When considering thermal expansion corrections another significant effect becomes obvious: the station coordinates depend to a certain extent on the station network. The network dependency arises from the single-difference principle

[2] http://vlbi.geod.uni-bonn.de/IVS-AC/Conventions/antenna-info.txt
[3] http://gemini.gsfc.nasa.gov/solve_save/ECCDAT.ecc

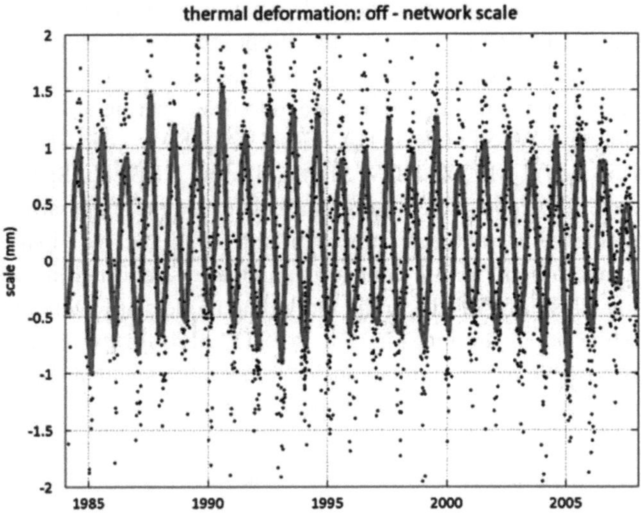

Fig. 27.1 Network scale variations (mm) applying thermal deformation modeling of VLBI antennas

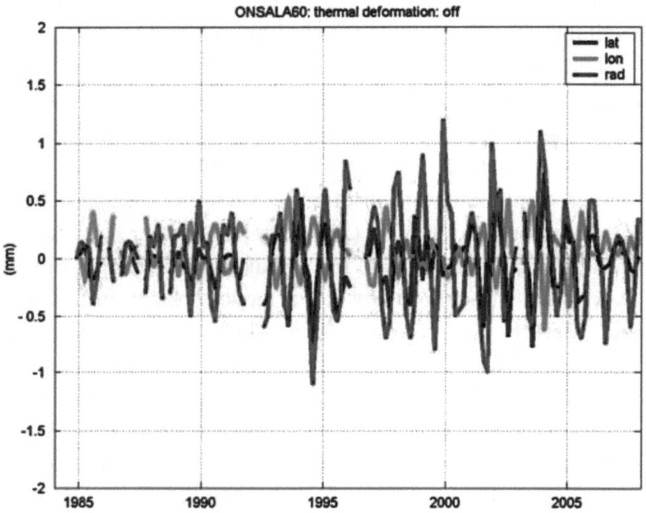

Fig. 27.2 Median coordinate differences at Onsala, Sweden, when applying the thermal deformation model at the other sites

of VLBI observations and the necessary NNT and NNR conditions applied to networks with a small number of stations. Figure 27.2 shows the median coordinate differences at ONSALA60, where annual variations with amplitudes larger than 1 mm can be seen. Since at this site no thermal expansion was modeled, while at the other sites it was applied, the effects must have propagated from the other sites of the observing network. Hence, even if the local effect might be small, we recommend considering all available corrections at the observation level to prevent the distribution of unconsidered local effects among the sites. Without considering corrections at the observation level, the interpretation of adjusted station coordinates as solely local signals is not fully admissible.

Despite lacking models for deriving inside-radome temperatures the application of this antenna thermal expansion correction model is mandatory for operational IVS analyses. Other significant antenna-dependent effects, in particular gravitational antenna deformations as introduced and discussed in detail by e.g. Sarti et al. (2009, 2010), Abbondanza and Sarti (2010), have not entered IVS recommendations up to now. We recommend the release or the recommendation of an appropriate model to correct for antenna gravitational deformation through the IVS.

3.2 Atmosphere Pressure Loading

Atmosphere pressure loading has been discussed by various groups before. Early models (Rabbel and Schuh 1986) considered a region of 2,000 km around a site for which atmosphere pressure loading deformations were determined. At this stage no sufficient correlation could be found between modeled and observed atmosphere loading deformations. Eight years later van Dam and Herring (1994) and MacMillan and Gipson (1994) found signatures of atmosphere surface loading signals in VLBI observations for the first time. Baseline length repeatabilities as well as estimates of vertical station components were improved through the application of atmosphere loading corrections. Finally, a significant improvement for the determination of atmosphere loading deformations came through the utilization of global reanalysis numerical weather models for geodesy, such as the NCEP/NCAR reanalysis (Kalnay et al. 1996). This model was used for the derivation of the atmosphere pressure loading model of Petrov and Boy (2004). We enabled this for the reference solution and disabled it for the test solution.

Atmosphere loading causes signals in all three local station components. While primarily station heights reflect signals with up to 5 mm amplitudes, the horizontal components show signals with amplitudes of about 2–3 mm. At some sites, for example at FORTLEZA, Brazil, (not shown here) annual signals can be found not only in the vertical component but also in the horizontal components of about half the size. Since the signals have mainly annual periodicity but vary in phase among the sites, a systematic annual variation is present in the session-wise network scale time-series (Böhm et al. 2009) but it is not as distinct as for thermal deformation. The differences in estimated source positions are noise-like and rarely exceed 10 μas.

3.3 Atmosphere Pressure Data Used for Zenith Hydrostatic Delay Modeling

The meteorological data used for routine VLBI analyses are surface air pressure and temperature records at the geodetic

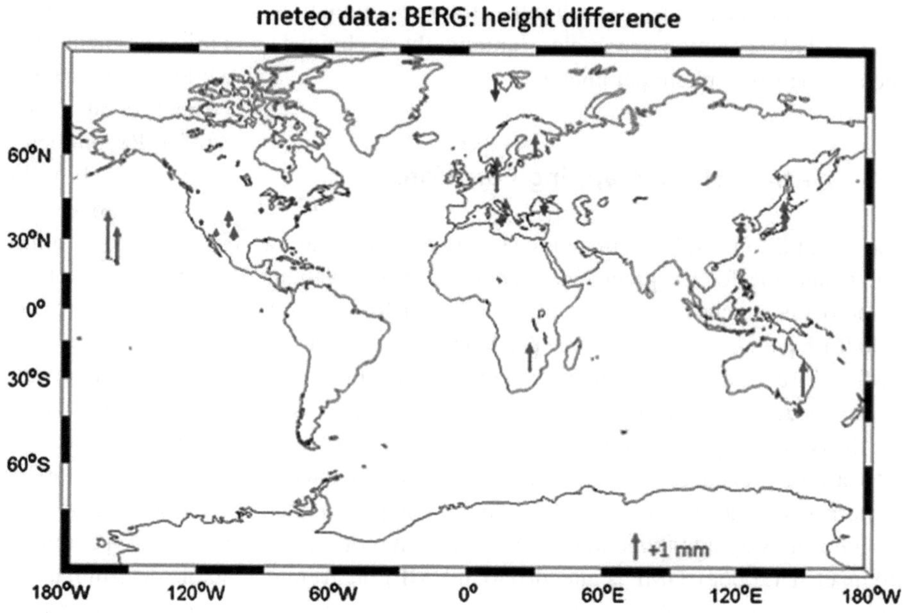

Fig. 27.3 Height differences when using Berg (1948) pressure instead of ECMWF pressure for zenith hydrostatic delay modeling. Negative differences in *blue* and positive differences in *red*

observing sites. While the temperature records enter antenna thermal deformation models (Sect. 3.1), the pressure can be used to derive loading corrections (Sect. 3.2) and precise a priori zenith hydrostatic delays. In this section the effect of atmosphere pressure is considered for hydrostatic delay modeling.

Corrections have to be applied for referring the raw readings from the location of the met sensor to the phase center of the geodetic instrument. Thus, the position of the met sensor needs to be known relative to the phase center in a local horizontal system. While horizontal gradients are rather small and corrections are usually neglected for the existing horizontal distances of up to about 10 km, vertical pressure gradients are on the order of 1 hPa per 8 m height and need to be considered.

The effects of applying various sets of pressure data were quantified before: Tesmer et al. (2006) compared the in-situ pressure observations provided by the IVS in addition to the observation data with constant pressure values given through a general meteorological model (Berg 1948), which is still widely used for GNSS analysis, e.g. by the Bernese GPS Software; and Heinkelmann et al. (2009) additionally discussed the use of GPT and pressure values directly obtained from the ECMWF weather model for application in VLBI analysis. Extensive comparisons and combinations of zenith delays (Heinkelmann et al. 2007) led Heinkelmann (2009) to present state of the art analysis recommendations including the pressure data. For precise analyses the pressure provided by IVS can be used but has to be homogenized (Heinkelmann et al. 2005) for long-term analyses. As an alternative, pressure values can be taken from numerical weather models, such as the ECMWF.

Other models, such as the standard meteorological model of Berg (1948) or GPT are not recommended for the analysis of space-geodetic techniques at radio-wavelengths, if high precision is required.

Switching from ECMWF pressure data to locally observed values provided by IVS causes slightly different station positions (not shown here): most of the height differences are negative with a maximum absolute value of about 0.5 mm. Horizontal differences are insignificant. Differences in the celestial coordinates show no systematics but mean values reach up to 10 μas in both components, declination and right ascension (times cosine of declination), at about −20 ° declination. Larger differences can be found, if standard pressure from the Berg (1948) model is used instead of ECMWF pressure (Fig. 27.3). The majority of height differences are positive and several height differences exceed 1 mm. The celestial coordinate differences, however, stay at about the same level with maximal mean values of 10 μas.

The use of accurate surface pressure values plays a significant role for TRF computations. Since the IERS Terrestrial Reference Frame (ITRF) is computed by combining several space-geodetic techniques at radio-wavelengths, care should be taken that the contributing solutions use precise and homogeneous pressure data. The same met data should be used among the techniques to ensure consistency. While current TRF combinations rely on the so-called local tie measurements for connecting the reference markers of the various techniques to each other, future combinations may include common parameters among the techniques, such as the troposphere parameters.

3.4 Atmosphere Mapping Functions

The neutral atmosphere mapping functions can play a significant role for the precise data analysis of observations at radio frequencies. Tesmer et al. (2007) and Böhm et al. (2007b) have shown that the Vienna Mapping Functions 1, VMF1, are the state-of-the-art mapping functions for global and regional analyses. Nevertheless, some analysts still use the Niell Mapping Functions (NMF; Niell 1996) and some software or procedures cannot work with large amounts of numerical weather model data, thus a more compact model, the Global Mapping Functions, GMF, was introduced by Böhm et al. (2006a). GMF is intended to fit VMF1 on average and to work as a possible substitute in case where no proper numerical weather model data are available.

Mapping function errors mostly affect station heights because of the correlations between site vertical and zenith wet delay estimates. Station height positions differ up to about 3 mm using GMF instead of VMF1; about twice the effect can be found for NMF versus VMF1. Largest absolute positional differences are at stations with short time-series where differences in the station height velocity estimates show the opposite sign and thus, partly recover the positional height differences. Horizontal differences are below the 1 mm level for GMF versus VMF1 and for NMF versus VMF1 with up to 2 mm, they are again about twice as large. In spite of the rather large individual effects, there is no systematic common to all sites of the TRF when using different mapping functions. In terms of the CRF, GMF with respect to VMF1 shows no systematic difference. The mean declinations as well as mean scaled right ascensions are below 10 μas. Applying NMF instead of VMF1 causes mean differences in the order of maximal 20 μas (declinations) and about 30 μas (scaled right ascensions) both increasing towards the celestial south pole (Figs. 27.4 and 27.5).

About 15 years after the release of NMF, the IERS Conventions 2010 now recommend VMF1 for all desired applications. For TRF as well as CRF computations it is essential that any analysis center providing results or normal equations for the determination of IERS or other relevant products uses these mapping functions.

3.5 A Priori Atmosphere Gradients

Among all the effects investigated in this paper, a priori atmosphere gradients can exhibit the largest systematics on TRF as well as CRF determinations. Despite their relevance, a priori gradients have been accordingly treated rather lately, while comparisons have already been reported earlier: Tesmer et al. (2006), Schuh et al. (2007), and Tesmer (2007). Recommendations in the IERS Conventions, however, have only followed very recently (IERS 2010).

When modeling neutrosphere gradients for VLBI data analysis, constraints are introduced for the quantity itself (gradient offset) and its temporal variability (gradient rates) to ensure non-singularity of the equation system. Such events (singularity or close-to-singularity in a numerical sense) can occur when observations are absent during a time interval or if the geometrical space-time distribution of observations is insufficient for the determination of gradients. A priori gradients are important for VLBI analysis because of both, the constraints, and the possible lack of information coming from the observations themselves. For a stable determination of north–south and east–west gradients a large number of observations in various directions and under varying elevation angles within a rather short duration is necessary. This necessity is not always fulfilled in case of VLBI observing networks/schedules. In particular in the early years before about 1990, very often only one European site observed together with two to three North-American sites. For such a geometrical constellation the distribution of observations for the European site towards north, east, and south is insufficient. Gradient constraints designed for this type of network are not representative for the geodetic-optimized observation schedules applied nowadays, for IVS-R1/-R4 types of sessions for instance, but were not always changed accordingly. Therefore, gradient offsets tend to be over-constrained for the analysis of modern VLBI types of sessions.

There are physical reasons for neutrosphere gradients: the atmosphere bulk shows a general decrease from the equator to the poles comparable to the flattening of solid Earth. Consequently, for stations in mid northern or southern latitudes, relatively constant north–south gradients are expected (Böhm et al. this issue). Average wet atmosphere gradients may be found at straight coastlines, with wet air over the ocean and dry air over the continent, such as the situation for TIGO at Concepción, Chile, or at Shanghai, China. MacMillan and Ma (1998) have developed a model for a priori gradients already in the 1990s.

Ten years later, Böhm and Schuh (2007) have determined another a priori gradient series, and recently Böhm et al. (this issue) have presented a new a priori gradient model: APG. However, it is recommended by the IERS (2010) to additionally estimate gradients.

Applying DAO total a priori gradients (MacMillan and Ma 1998) and estimating or not estimating gradients affects the station vertical positions up to 4 mm and the horizontal components up to about the same extent. There are no

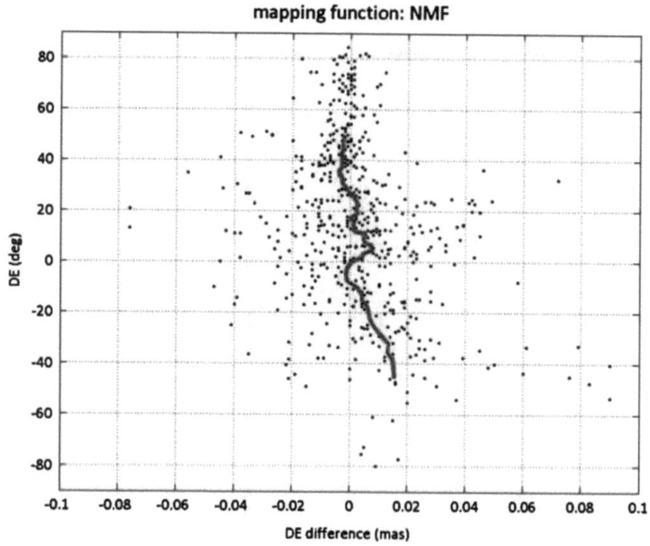

Fig. 27.4 Differences in declination (mas) when using NMF instead of VMF1. A moving average of 30 radio sources is displayed in *red* and shows an increase in declination towards the celestial south pole

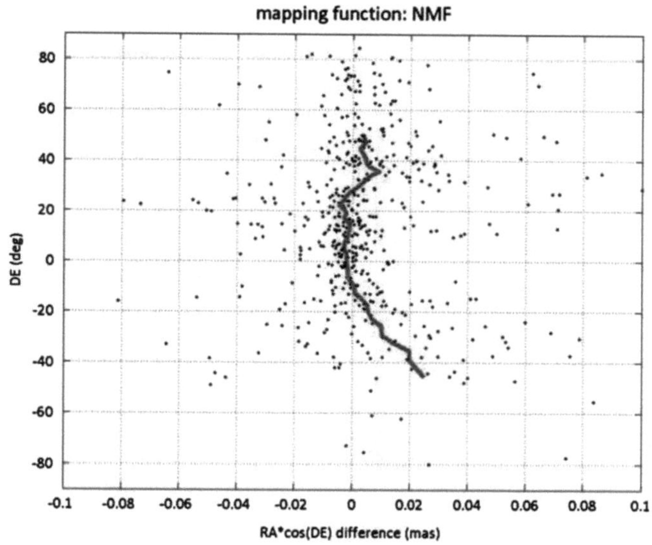

Fig. 27.5 Differences in right ascension times cosine of declination (mas) when using NMF instead of VMF1. The *red line* shows a moving average of 30 radio sources. The differences tend to increase towards the celestial south pole

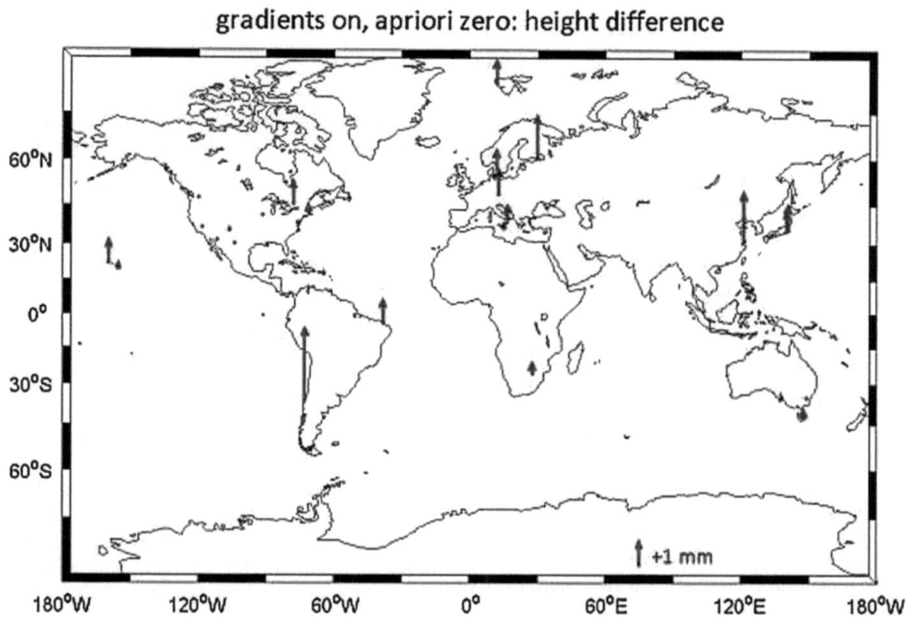

Fig. 27.6 Height differences between zero and DAO a priori gradients when gradients are estimated. *Red arrows* show an increase and *blue arrows* a decrease in height. There is an overall increase in height

systematics in the CRF components, but the mean differences show clear maxima in the order of 10 μas at declinations around −20 ° to −30 °. Without application of a priori gradients compared to using DAO a priori gradients the TRF is systematically deformed yielding larger station heights of up to about 3 mm (Fig. 27.6) and the horizontal positions are shifted southwards with a slight increase towards the terrestrial south pole (Fig. 27.7). The larger station heights also cause an overall increase in the network scale (not shown here). Celestial coordinates also show significant effects: declinations systematically increase towards the equator (Fig. 27.8), with maximal mean differences of about 200 μas. Without application of a priori gradients and without estimating gradients yields the largest systematics compared to the reference (DAO gradients with estimation of gradient parameters), however, all IVS

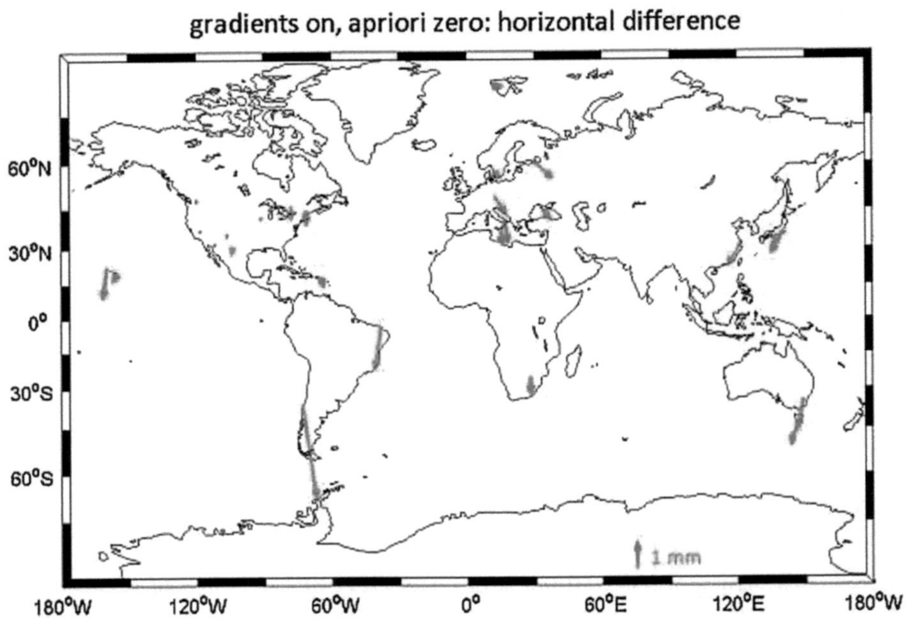

Fig. 27.7 Horizontal position differences between zero and DAO a priori gradients when gradients are estimated. The majority of sites are shifted southwards

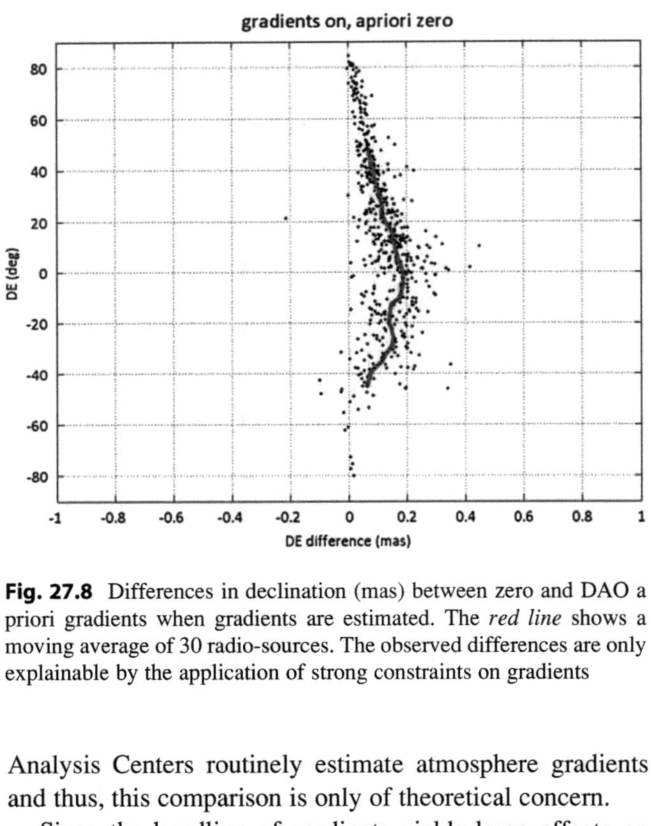

Fig. 27.8 Differences in declination (mas) between zero and DAO a priori gradients when gradients are estimated. The *red line* shows a moving average of 30 radio-sources. The observed differences are only explainable by the application of strong constraints on gradients

Analysis Centers routinely estimate atmosphere gradients and thus, this comparison is only of theoretical concern.

Since the handling of gradients yields large effects on both TRF and CRF, we strongly recommend applying non-zero a priori gradients and estimating residual gradients in VLBI analyses. An optimal parameterization and the optimal size of the weights of the gradient constraints need to be assessed. The optimizations, however, are likely to apply to a certain type of session only. It is questionable whether recommendations concerning the gradient parameterization can be given in general.

Acknowledgements We thank Dan MacMillan and the other reviewers and acknowledge IVS for providing excellent VLBI data.

References

Abbondanza C, Sarti P (2010) Effects of illumination functions on the computation of gravity-dependent signal path variation models in primary focus and Cassegrainian VLBI telescopes. J Geodesy 84:515–525

Berg H (1948) Allgemeine meteorologie. Dümmler, Bonn

Bizouard C, Gambis D (2009) The combined solution C04 for Earth orientation parameters consistent with international terrestrial reference frame 2005. In: Drewes H (ed) Proceedings of the IAG symposia 134. Springer, Heidelberg, pp 265–270

Böhm J, Schuh H (2007) Troposphere gradients from the ECMWF in VLBI analysis. J Geodesy 81:403–408

Böhm J, Niell AE, Tregoning P, Schuh H (2006a) Global mapping function (GMF): a new empirical mapping function based on numerical weather model data. Geophys Res Lett 33:L07304

Böhm J, Werl B, Schuh H (2006b) Troposphere mapping functions for GPS and very long baseline interferometry from European Centre for Medium-Range Weather Forecasts operational analysis data. J Geophys Res 111:B02406

Böhm J, Heinkelmann R, Schuh H (2007a) Global pressure and temperature (GPT): a spherical harmonic expansion of annual pressure and temperature variations for geodetic applications. J Geodesy 81:679–683

Böhm J, Mendes-Cerveira PJ, Schuh H, Tregoning P (2007b) The impact of mapping functions for the neutral atmosphere based on numerical weather models in GPS data analysis. In: Rizos C, Tregoning P (eds) Proceedings of the IAG symposia 130. Springer, Berlin, pp 837–843

Böhm J, Heinkelmann R, Mendes-Cerveira PJ, Pany A, Schuh H (2009) Atmospheric loading corrections at the observation level in VLBI analysis. J Geodesy 83:1107–1113

Böhm J, Urquhart L, Steigenberger P, Heinkelmann R, Nafisi V, Schuh (this issue) A priori gradients in the analysis of space geodetic observations. In: Proceedings of the IAG symposia "Reference frames for applications in geosciences". Springer, Heidelberg

Davis JL, Herring TA, Shapiro II, Rogers AEE, Elgered G (1985) Geodesy by radio interferometry: effects of atmospheric modeling errors on estimates of baseline length. Rad Sci 20(6):1593–1607

Heinkelmann R (2009) IVStrop: status and recommendations of the IVS rapid troposphere combination. In: Bourda G, Charlot P, Collioud A (eds) Proceedings of the 19th EVGA working meeting, pp 180–182

Heinkelmann R, Böhm J, Schuh H (2005) Homogenization of surface pressure recordings and its impact on long-term series of VLBI tropospheric parameters. In: Vennebusch M, Nothnagel A (eds) Proceedings of the 17th EVGA working meeting, pp 74–78, http://www.evga.org/files/2005EVGA-proc_Noto.pdf

Heinkelmann R, Böhm J, Schuh H, Tesmer V (2009) The effect of meteorological input data on the VLBI reference frames. In: Drewes H (ed) Proceedings of the IAG symposia 134, Springer, Berlin, pp 245–251

Heinkelmann R, Böhm J, Schuh J, Tesmer V (2006) Global VLBI solution IGG05R01. In: Behrend D, Baver K (eds) Proceedings of the 4th IVS general meeting. NASA/CP-2006-214140, pp 42–46, ftp://ivscc.gsfc.nasa.gov/pub/general-meeting/2006/pdf/heinkelmann2.pdf

Heinkelmann R, Böhm J, Schuh H, Bolotin S, Engelhardt G, MacMillan DS, Negusini M, Skurikhina E, Tesmer V, Titov O (2007) Combination of long time-series of troposphere zenith delays observed by VLBI. J Geodesy 81:483–502

IERS (2004) IERS conventions (2003). In: McCarthy DD, Petit G (eds) IERS technical no. 32, BKG, Frankfurt

IERS (2009) The second realization of the international celestial reference frame by very long baseline interferometry (2009). In: Fey A, Gordon D, Jacobs CS (eds) IERS technical no. 35. BKG, Frankfurt

IERS (2010) IERS conventions (2010). In: Petit G Luzum B (eds) IERS technical no. 36. BKG, Frankfurt

Kalnay E, Kanamitsu M, Kistler R et al (1996) The NCEP/NCAR 40-year reanalysis project. Bull Am Meteorol Soc 77:437–471

Lyard F, Lefevre F, Letellier T, Francis O (2006) Modelling the global ocean tides: modern insights from FES2004. Ocean Dynam 56:394–415

MacMillan DS (1995) Atmospheric gradients from very long baseline interferometry observations. Geophys Res Lett 22:1041–1044

MacMillan DS, Gipson JM (1994) Atmospheric pressure loading parameters from very long baseline interferometry observations. J Geophys Res 99:18081–18087

MacMillan DS, Ma C (1998) Using meteorological data assimilation models in computing tropospheric delays at microwave frequencies. Phys Chem Earth 23:97–102

Mathews PM, Herring TA, Buffett BA (2002) Modeling of nutation and precision: new nutation series for nonrigid Earth and insights into the Earth's interior. J Geophys Res 107(B4). doi:10.1029/2001JB000390

Niell AE (1996) Global mapping functions for the atmosphere delay at radio wavelength. J Geophys Res 101:3227–3246

Nothnagel A (2008) Conventions on thermal expansion modeling of radio telescopes for geodetic and astrometric VLBI. J Geodesy 83:787–792

Petrov L, Boy J-P (2004) Study of the atmospheric pressure loading signal in very long baseline interferometry observations. J Geophys Res 109:B03405. doi:10.1029/2003JB002500

Rabbel W, Schuh H (1986) The influence of atmospheric loading on VLBI-experiments. J Geophys 59:164–170

Sarti P, Abbondanza C, Vittuari L (2009) Gravity-dependent signal path variation in a large VLBI telescope modeled with a combination of surveying methods. J Geodesy 83:1115–1126

Sarti P, Abbondanza C, Petrov L, Negusini M (2010) Height bias and scale effect induced by antenna gravitational deformations in geodetic VLBI data analysis. J Geodesy. doi:10.1007/s00190-010-0410-6

Schuh H, Heinkelmann R, Sokolova R, Tesmer R (2007) Interaction of celestial and terrestrial reference frames. IAG sub-commission 1.4 final report, May 2007, pp 46–51

Tesmer V (2002) VLBI solution DGFI01R01 based on least-squares estimation using OCCAM 5.0 and DOGS-CS. In: Vandenberg N, Baver K (eds) Proceedings of the 2nd IVS general meeting. NASA/CP-2002-210002, pp 295–299

Tesmer V (2007) Effect of various analysis options on VLBI-determined CRF. In: Böhm J, Pany A, Schuh H (eds) Proceedings of the 18th EVGA working meeting, pp. 103–110, http://www.evga.org/files/2007EVGA-proc_Vienna.pdf

Tesmer V, Böhm J, Heinkelmann R, Schuh H (2006) Impact of analysis options on the TRF, CRF and position time series estimated from VLBI. In: Behrend D, Baver K (eds) Proceedings of the 4th IVS general meeting. NASA/CP-2006-214140, pp 243–251

Tesmer V, Böhm J, Heinkelmann R, Schuh H (2007) Effect of different tropospheric mapping functions on the TRF, CRF, and position time-series estimated from VLBI. J Geodesy 81:409–422

Uppala SM, Kallberg PW, Simmons AJ (2005) The ERA-40 Reanalysis. Q J Roy Meteorol Soc 131:2961–3012. doi:10.1256/qj.04.176

vanDam T, Herring TA (1994) Detection of atmospheric pressure loading using very long baseline interferometry measurements. J Geophys Res 99(B3):4505–4517

Wresnik J, Haas R, Böhm J, Schuh H (2007) Modeling thermal deformation of VLBI antennas with a new temperature model. J Geodesy 81:423–432

The Celestial Reference Frame at X/Ka-band (8.4/32 GHz)

28

C.S. Jacobs, J.E. Clark, C. García-Miró, M.B. Heflin, S. Horiuchi, V.E. Moll, L.J. Skjerve, and O.J. Sovers

Abstract

A celestial reference frame at X/Ka-band (8.4/32 GHz) has been constructed using fifty-one 24-h sessions with the Deep Space Network. We report on observations which have detected 436 sources covering the full 24 h of right ascension and declinations down to −45°. Comparison of this X/Ka-band frame to the S/X-band (2.3/8.4 GHz) ICRF2 shows wRMS agreement of 200 micro-arcsec (μas) in $\alpha cos\delta$ and 290 μas in δ. There is evidence for zonal errors at the 100 μas level. Known errors include limited SNR, lack of phase calibration, troposphere mismodelling, and limited southern geometry. The motivations for extending the ICRF to frequencies above 8 GHz are to access more compact source morphology for improved frame stability, to provide calibrators for phase referencing, and to support spacecraft navigation at Ka-band.

Keywords

Reference systems • Catalogs • Astrometry • Celestial reference frame • ICRF • Interferometry • VLBI • Radio continuum • Ka-band • Galaxies: active galactic nuclei • Quasars • Blazars

1 Introduction

For over three decades now, radio frequency work in global astrometry, geodesy, and deep space navigation has been done at S-band (2.3 GHz) and X-band (8.4 GHz). While this work has been tremendously successful in producing 100 μ as level global astrometry (e.g. Ma et al. 2009) and sub-cm geodesy, developments made over the last decade have made it possible to consider the merits of moving to a new set of frequencies. In this paper we present global astrometric results from X/Ka (8.4/32 GHz) observations.

Advantages: Moving the observing frequencies up by approximately a factor of four has several advantages. For our work in the Deep Space Network, the driver is the potential for higher data rates for telemetry signals to probes in deep space. Other advantages include (1) the spatial distribution of flux becomes significantly more compact (Charlot et al. 2010) lending hope that the positions will be more stable over time, (2) Radio Frequency Interference (RFI) at S-band would be avoided, (3) Ionosphere and solar plasma effects on group delay and signal coherence are reduced by a factor of 15!

Disadvantages: While these are very significant advantages, they do not come without a price. The change from 2.3 / 8.4 GHz to 8.4 / 32 GHz moves one closer to the water vapor line at 22 GHz and thus increases the system temperature from a few Kelvins per

C.S. Jacobs (✉) • J.E. Clark • M.B. Heflin • L.J. Skjerve • O.J. Sovers
Jet Propulsion Laboratory, California Institute of Technology, 4800 Oak Grove Dr, Pasadena, CA 91109, USA
e-mail: Christopher.S.Jacobs@jpl.nasa.gov

C. García-Miró • V.E. Moll
Ingenieria y Servicios Aeroespaciales, Instituto Nacional de Técnica Aeroespacial/NASA, Madrid Deep Space Communication Complex, Paseo del Pintor Rosales, 34 bajo, Madrid E-28008, Spain

S. Horiuchi
C.S.I.R.O. Astronomy and Space Science/NASA, PO Box 1035, Tuggeranong ACT, AU 2901, Australia

Z. Altamimi and X. Collilieux (eds.), *Reference Frames for Applications in Geosciences*, International Association of Geodesy Symposia 138, DOI 10.1007/978-3-642-32998-2_28,

atmospheric thickness up to 10–15 K per atmosphere or more. Thus one becomes much more sensitive to weather. Furthermore, the sources themselves are in general weaker and many sources are resolved. Also, with the observing wavelengths shortened by a factor of 4, the coherence times are shortened so that practical integration times are a few minutes or less—even in relatively dry climates. The shorter wavelengths also imply that the antenna pointing accuracy requirements must be tightened by the same factor of 4. The combined effect of these disadvantages is to lower the system sensitivity. Fortunately, advances in recent years in recording technology make it feasible and affordable to offset these losses in sensitivity by recording more bits. Thus while most of the X/Ka data presented in this paper used the same overall 112 Mbps bit rate as previous S/X work, recent data were taken at a 4 times higher rate with an increase to 8 times higher rate hoped for within the next year.

This paper is organized as follows: We will describe the observations, modelling, and present the results. Next, we will estimate the accuracy by comparing to the S/X-based ICRF2 (Ma et al. 2009) including a look at zonal errors. This will be complemented by a discussion of the error budget and the potential for improving the geometry of our network by adding a southern station.

2 The VLBI Observations

The results presented here are from 51 Very Long Baseline Interferometry (VLBI) observing sessions of ~24 h duration done from July 2005 until April 2010 using NASA's Deep Space Stations (DSS) 25 or 26 in Goldstone, California to either DSS34 in Tidbinbilla, Australia or DSS 55 outside Madrid, Spain to form interferometric baselines of 10,500 and 8,400 km length, respectively. We recorded VLBI data simultaneously at X-band (8.4 GHz) and Ka-band (32 GHz). Initially, sampling of each band was at 56 Mbps while more recent passes used 160/288 Mbps at X/Ka. Each band used a spanned bandwidth of ~360 MHz. The data were filtered, sampled, and recorded to the Mark4 or Mark5A VLBI systems. The data were then correlated with the JPL BlockII correlator (O'Connor 1987) or the JPL SOFTC software correlator (Lowe 2005). Fringe fitting was done with the FIT fringe fitting software (Lowe 1992). This procedure resulted in 12,860 pairs of group delay and phase rate measurements covering the full 24 h of right ascension and declinations down to $-45°$. Individual observations were about 1–2 min in duration.

3 Modelling

The above described set of observations were then modelled using the MODEST software (Sovers et al. 1998). A priori Earth orientation was fixed to the MHB nutation model (Mathews et al. 2002) and the empirically determined UT1-UTC and Polar Motion of the Space 2008 series (Ratcliff and Gross 2010). The celestial frame was aligned to the ICRF2 defining sources (Ma et al. 2009) using a No-Net-Rotation constraint (Jacobs et al. 2010). Station velocities were estimated; station locations were estimated with a 1 cm constraint per component to a decades-long S/X-band VLBI solution.

4 Results

In all, we detected 436 extragalactic radio sources which covered the full 24 h of RA and Declinations down to $-45°$. In Fig. 28.1 these sources are plotted using Hammer's (1892) equal-area projection to show their locations on the sky. RA $= 0$ is at the center. The ecliptic plane is shown by the dashed blue-gray line and the Galactic plane is indicated by the yellow-red dashed line. The sources are color coded according to their 1-σ formal declination uncertainties with the value ranges indicated in the figure's legend. Note that the declination precision drops as one moves toward the south. This is a result of having significantly less data on the California to Australia baseline combined with the need to observe sources closer to the horizon as declination moves south thus incurring greater error from higher system temperatures and tropospheric mis-modelling.

5 Accuracy: X/Ka Versus S/X Comparisons

Experience shows that formal uncertainties tend to underestimate true errors. An independent estimate of position errors was obtained by comparing our X/Ka-band positions to the S/X-based ICRF2. For 372 common sources, the differences of X/Ka minus S/X are shown for $\Delta\alpha\cos\delta$ in Fig. 28.2 and for $\Delta\delta$ in Fig. 28.3. Weighted RMS (wRMS) differences are ~ 200 μas in $\alpha\cos\delta$ and ~ 290 μas in δ.

6 Zonal Errors

Section 5 gave a measure of overall coordinate agreement. We now turn to differences which are systematically correlated as a function of position on the sky. Figure 28.4

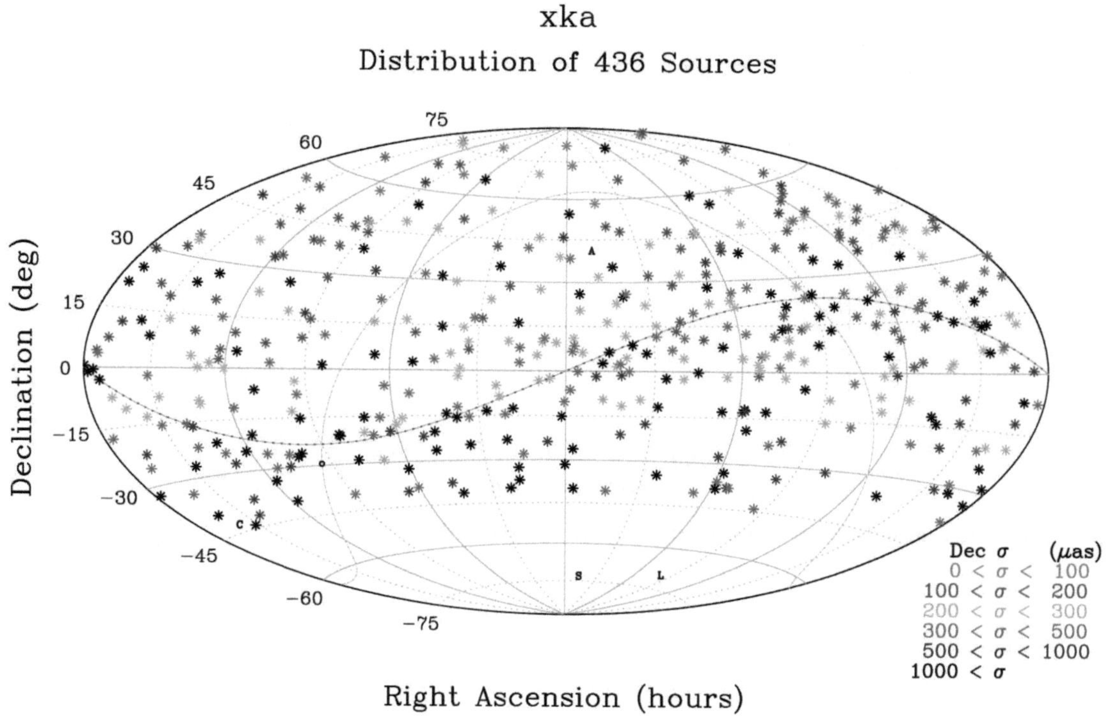

Fig. 28.1 Distribution of 436 X/Ka-band sources detected to date. Symbols indicate 1-σ formal declination uncertainties as defined in the legend at *lower right*. $(\alpha, \delta) = (0, 0)$ is at the center. The ecliptic plane is indicated by a *dashed sinusoidal curve*. The galactic plane is indicated by the Ω-*shaped curve*. Note the trend for decreasing declination precision moving southward. Local galactic neighborhood indicated by *A, C, S, L*: Andromeda, Centaurus-A, Small & Large Magellanic clouds (none observed at X/Ka)

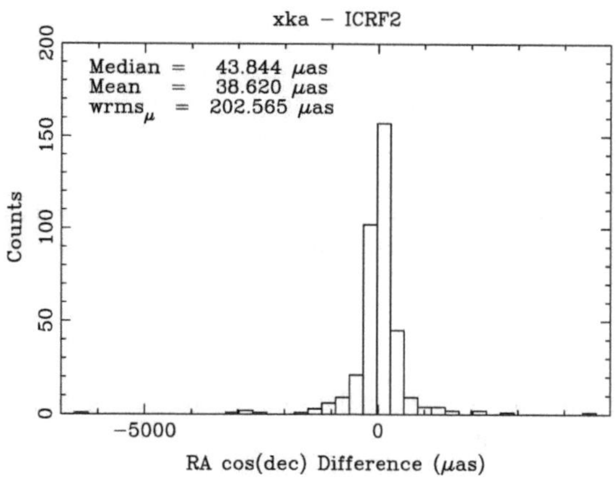

Fig. 28.2 X/Ka $-$ S/X: $\Delta\alpha\cos\delta$

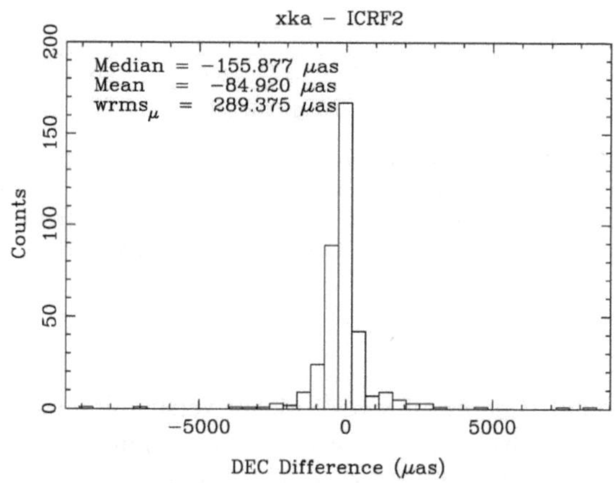

Fig. 28.3 X/Ka $-$ S/X: $\Delta\delta$

shows the mean arclength differences vs. arclength in the sense (X/Ka $-$ S/X) peaking a bit over 100 μas. For another measure of zonal error we look at α and δ shifts versus α and δ:

$\alpha\cos\delta$ versus $\alpha = 4.3 \pm 1.7$ μas/h

$\Delta\delta$ versus $\alpha = 2.0 \pm 1.2$ μas/h

$\Delta\alpha\cos\delta$ versus $\delta = 0.3 \pm 0.5$ μas/degree

$\Delta\delta$ versus $\delta = 1.5 \pm 1.0$ μas/degree

The most significant slope is $\Delta\alpha\cos\delta$ versus α at 2.5 σ. Note that the use of full correlations had a significant effect on the determination of these slopes.

7 Discussion of Error Budget

Having assessed the size of errors in our positions using the much larger ICRF2 S/X data set as a standard of accuracy, we now discuss the major contributions to the errors in the X/Ka measurements: SNR, instrumentation, and troposphere. Figure 28.5 shows the weighted RMS group delay vs. the Ka-band SNR. We conclude that for SNR $<$ 15 dB, the thermal error dominates the error budget. For higher SNRs, troposphere and instrumentation errors become more

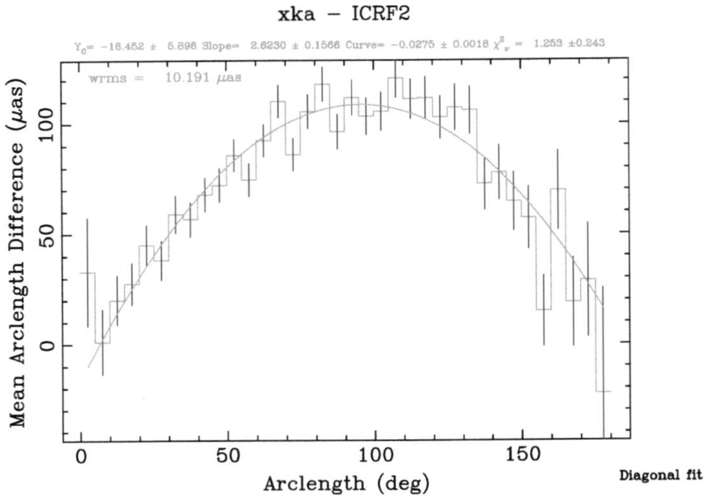

Fig. 28.4 Zonal errors: mean arc differences versus arclength for X/Ka–S/X(ICRF2)

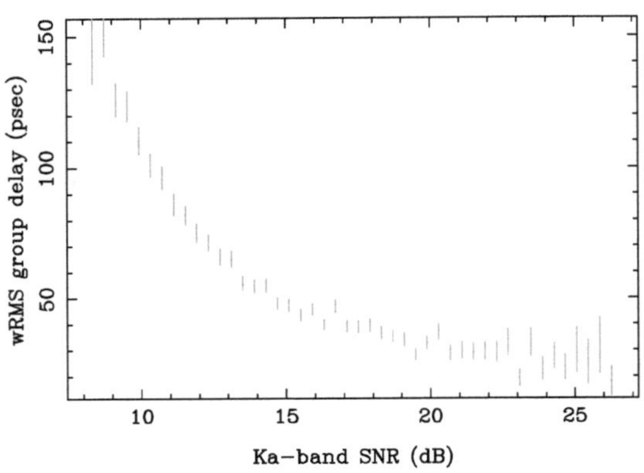

Fig. 28.5 The wRMS residual group delay versus Ka-band SNR. Thermal error dominates the VLBI residuals for SNR < 15 dB. As SNR increases past that point, a noise floor of ≈ 30 psec from tropospheric and instrumental errors is asymptotically approached

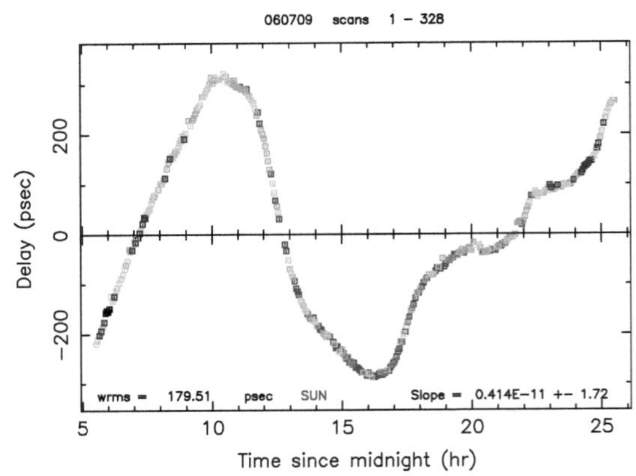

Fig. 28.6 Ka-band proto-type phase calibrator group delays versus time from 09 Jul 2006. Diurnal variation is driven by thermal changes in cables and other instrumentation. Color code indicates the sun angle (in order closest to farthest: *orange, red, green, cyan, purple, black*)

important. Binning of wRMS delay versus airmass thickness shows that troposphere is not the dominant error due to the generally low SNRs just mentioned. However, the phase rates (which carry much less weight in the fit) are dominated by errors from tropospheric mismodelling, thus hinting that troposphere will become more important as our SNR improves with increased data rates. Lastly, we have errors from un-calibrated instrumentation. A proto-type phase calibrator was developed in order to calibrate the signal path from the feed to the sampler (Hamell et al. 2003). Test data shown in Fig. 28.6 indicate an approximately diurnal instrumental effect with ~ 180 psec (5.4 cm) RMS. Although the data themselves can be used to estimate instrumental parameters which partially characterize this effect, operational phase calibrators are being built in order to make direct reliable calibrations of the instrumentation.

8 Southern Geometry

Besides the three classes of measurement errors described above, our reference frame suffers from a very limited geometry—we have only one station in the southern hemisphere. In order to better understand this limitation, we simulated the effect of adding a second southern station (Bourda et al. 2010). Data from 50 real X/Ka sessions (Fig. 28.7) were augmented by simulated data (Fig. 28.8) for 1000 group delays each with SNR = 50 on a ~ 9,000 km baseline: Australia to S. America or S. Africa. The resulting

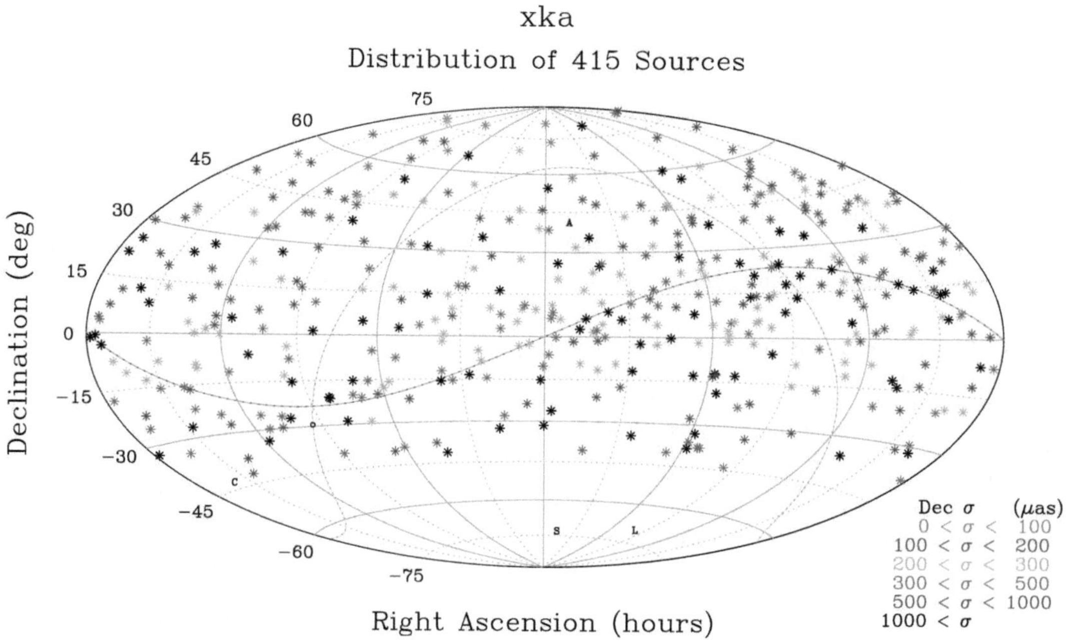

Fig. 28.7 Real X/Ka data from 50 sessions using two baselines: CA-Spain and CA-Australia

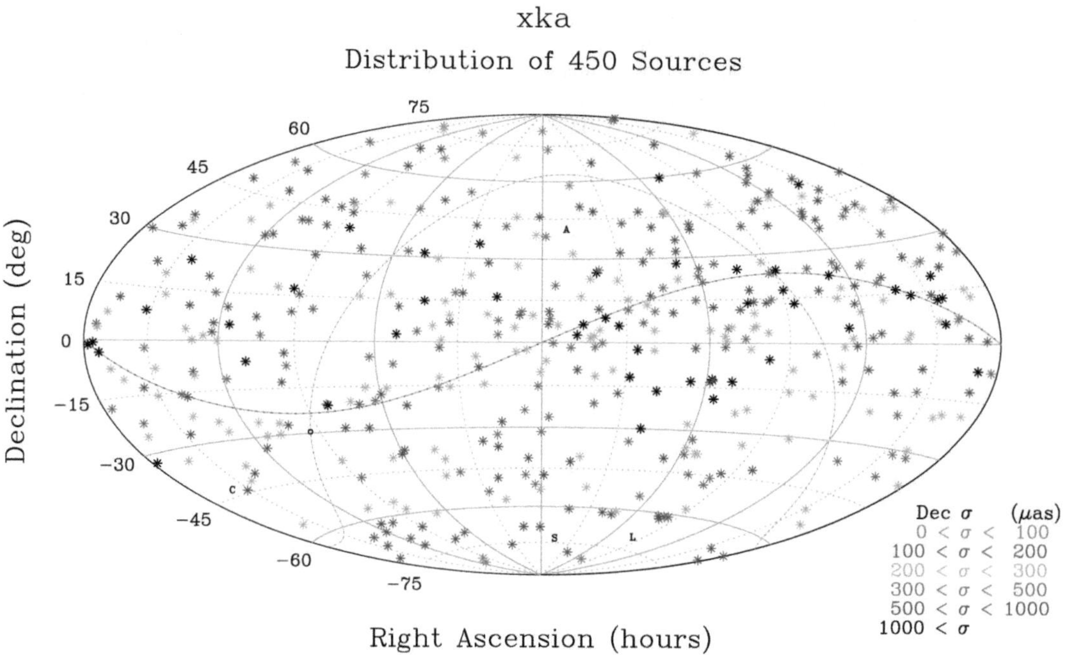

Fig. 28.8 After adding 1,000 delays from a simulated third baseline in the south, the southern cap would be covered with sources of ~ 200 μas precision (color code same as Fig. 28.1)

solution extended Declination coverage to the south polar cap region: −45 to −90°. Precision in the south cap region was ~ 200 μas (1 nrad) and in the mid south precision was 200–1,000 μas, all with just a few days observing. We conclude that adding a second southern station would greatly aid our X/Ka frame's accuracy. In fact, the resulting four station network should compete well in astrometric accuracy with the historical S/X network and its ICRF2.

9 Conclusion

The S/X-based ICRF has now been extended to four times higher frequency to X/Ka-band (8.4/32 GHz). A total of 436 sources have been successfully detected at Ka-band. For the 372 sources common to X/Ka and the S/X-based ICRF2, we find positional agreement of 200 μas (1 nrad) in $\alpha\cos\delta$ and

290 μas (1.4 nrad) in δ with zonal errors of 100–150 μas (0.5–0.75 nrad). Improvements in data rates and instrumental calibration are projected to allow better than 200 μas (1 nrad) accuracy within the next few years. Simulations of adding another southern station predict better than 200 μas accuracy for the southern polar cap within a very short time of adding data from an all southern baseline. This gives hope that better than 100 μas accuracy over the full sky might be achieved within a few years of adding a southern baseline.

Acknowledgements We are thankful for thorough reviews of this paper by Bill Petrachenko and an anonymous referee who corrected several errors and improved the clarity of the presentation. Any remaining deficiencies are solely the responsibility of the authors. The research described herein was performed at the Jet Propulsion Laboratory of the California Institute of Technology, under a contract with the National Aeronautics and Space Administration. Government sponsorship acknowledged.

References

Bourda G, Charlot P, Jacobs CS (2010) Future radio reference frames and implications for the Gaia link. In: Proceedings of ELSA conference: Gaia at the frontiers of astrometry, Sevres, France. http://wwwhip.obspm.fr/gaia2010/IMG/pdf/Poster_Bourda.pdf

Charlot P, Boboltz DA, Fey AL, Fomalont EB, Geldzahler BJ, Gordon D, Jacobs CS, Lanyi GE, Ma C, Naudet CJ, Romney JD, Sovers OJ, Zhang LD (2010) The celestial reference frame at 24 and 43 GHz II. Imaging Astron J 139(5):1713. doi:10.1088/0004-6256/139/5/1713

Hamell R, Tucker B, Calhoun M (2003) Phase calibration generator. NASA JPL IPN Progess report, 42–154, pp 1–14. http://tmo.jpl.nasa.gov/progress_report/42-154/154H.pdf

Hammer E (1892) Über die Planisphäre von Aitow und verwandte Entwürfe, insbesondere neue flächentreue ähnlicher Art. Petermanns Mitt 38:85–87

Jacobs CS, Heflin MB, Lanyi GE, Sovers OJ, Steppe JA (2010) Rotational alignment altered by source position correlations. In: Behrend D, Baver KD (eds) IVS 2010 general meeting proceedings, NASA/CP-2010-215864, IVS Analysis Session, Hobart, Tasmania, Australia. http://ivscc.gsfc.nasa.gov/publications/gm2010/jacobs2.pdf

Lowe ST (1992) Theory of post-blockII VLBI observable extraction. JPL Publication, Pasadena, CA, pp 92–97, http://ntrs.nasa.gov/archive/nasa/casi.ntrs.nasa.gov/19940009399_1994009399.pdf

Lowe ST (2005) SOFTC: a software VLBI correlator, JPL section 335 internal document. Jet Propulsion Laboratory, Pasadena, CA

Ma C et al. (2009). In: Fey AL, Gordon D, Jacobs CS IERS Technical note 35: the 2nd realization of the ICRF by VLBI, IERS, Frankfurt, Oct 2009. http://www.iers.org/nn_11216/IERS/EN/Publications/TechnicalNotes/tn35.html

Mathews PM, Herring TA, Buffet BA (2002) Modeling of nutation and precession: New nutation series for nonrigid Earth and insights into the Earth's interior. J Geophys Res 107:B4. doi:10.1029/2001JB000390

O'Connor T (1987) Introduction to the blockII correlator hardware. JPL Internal Publication, Pasadena CA

Ratcliff JT, Gross RS (2010) Combinations of Earth orientation measurements: space 2008, COMB2008, and POLE 2008. JPL Publication, Pasadena, CA, pp 10–14, http://hdl.handle.net/2014/41512

Sovers OJ, Fanselow JL, Jacobs CS (1998) Astrometry, geodesy with radio interferometry. Expts Models Results Rev Mod Phys 70 (4):1393–1454. doi:10.1103/RevModPhys.70.1393

Systematic Errors of a VLBI Determined TRF Investigated by Simulations

29

L. Plank, H. Spicakova, J. Böhm, T. Nilsson, A. Pany, and H. Schuh

Abstract

In this study, we investigate the influence of different analysis setup options for the processing of VLBI measurement data from 2002 until 2010 to derive the terrestrial reference frame (TRF). For estimating the consequent changes of the TRF, the simulation tool of the Vienna VLBI Software (VieVS) is used by applying different a priori models. We show that neglecting atmosphere loading causes a systematic annual scale variation of ± 0.3 mm, and that the application of the cubic model recommended in the most recent IERS Conventions for the mean pole introduces a scale change of -0.6 mm over 8.5 years. The effects of antenna thermal deformation on the TRF are ± 0.5 to 1 mm/year in translation and ± 2 mm/year in scale. No systematic effects are found for the different troposphere mapping functions tested. Besides systematic, annual, and episodic impacts on the coordinates, we focus on possible changes in the scale of the reference frames.

Keywords

Terrestrial reference frame • VLBI analysis • Vienna VLBI software VieVS

1 Introduction

In standard Very Long Baseline Interferometry (VLBI) analysis, certain models are applied in order to account for tidal and non-tidal station motions, antenna thermal deformation, and propagation effects in the atmosphere. These models are listed in the IERS Conventions (McCarthy and Petit 2004; Petit and Luzum 2010) and need to be consistently applied by all analysis centres contributing to the ITRF. From time to time these models get revised, i.e., new and more accurate models are developed in parallel to the improved accuracy of the observations.

For selected VLBI observations from the years 2002 through 2010 we investigate the effect of recently adopted correction models on a VLBI determined terrestrial reference frame (TRF). In Sect. 2 we describe the data used, the corresponding VLBI networks, the Vienna VLBI Software (VieVS), and we give details about the processing steps performed to derive our results. Basically, the simulation module of the software enables us to create zero-input observation files, which are entered into the global solution of VieVS. This allows the determination of a new TRF by reducing and stacking the normal equations of all processed single sessions. By changing the analysis set-up and subsequent comparison with a reference solution, we demonstrate the separate effect of each single model on the derived reference frames. The results presented in Sect. 3 give the effects of the new cubic model for the mean pole (Sect. 3.1), the thermal antenna deformation model (Sect. 3.2), and the difference between various troposphere mapping functions (Sect. 3.3). In Sect. 3.4 we show variations in the TRF scale induced by a model for atmosphere loading, which has not been adopted for standard IERS analysis so far.

L. Plank (✉) • H. Spicakova • J. Böhm • T. Nilsson • A. Pany • H. Schuh
Institute of Geodesy and Geophysics, Vienna University of Technology, Gusshausstrasse 27-29, Vienna 1040, Austria
e-mail: lucia.plank@tuwien.ac.at

Z. Altamimi and X. Collilieux (eds.), *Reference Frames for Applications in Geosciences*, International Association of Geodesy Symposia 138, DOI 10.1007/978-3-642-32998-2_29,

2 Processing

2.1 Vienna VLBI Software VieVS

The Vienna VLBI Software (VieVS) (Böhm et al. 2011) has been developed at the Institute of Geodesy and Geophysics of the Vienna University of Technology. The programming language Matlab is used, providing many built-in functions which ease the programming effort and shorten the coding significantly compared to, e.g. the Occam software package (Titov et al. 2004), that was formerly used at the institute. With the first version released in 2010, VieVS is up to date with the latest IERS Conventions (Petit and Luzum 2010). Designed for routine analysis of geodetic VLBI observations, the modular structure of VieVS allows the straightforward implementation of additional tools, e.g. the Vie_glob tool for the global solution and the Vie_sim module that can be used to simulate observations. Further information about the software VieVS and details of additional modules can be found at http://vievs.hg.tuwien.ac.at and the references given there.

2.2 Investigated Time Span and Data

The observational configuration consists of all rapid turnaround sessions (R1 and R4) which the International VLBI Service for Geodesy and Astrometry (IVS, Schlüter and Behrend 2007) has been conducting twice a week since 2002. This gives a total time series of about 850 24-h-experiments over the last 8.5 years, corresponding to about 20 % of the VLBI sessions that contribute to ITRF2008 (Böckmann et al. 2009). R1 and R4 sessions are regularly observed every Monday and Thursday by a network of up to eight globally distributed stations. Normally, the R1 and R4 networks include six core stations each and additionally two other stations from the IVS network, with the R1 networks being even more variable than R4. As described in the following, no real observation data is used in the present study, but the configuration of actually scheduled observations provides a realistic geometry for our simulations.

2.3 Computation Procedure

The module Vie_sim of VieVS can be used to generate artificial delay observables for a predefined observation schedule. In our case we use it to create VLBI observation files with observables identical to the calculated time delays from VieVS, getting zero-input files for the R1 and R4 sessions. For the purpose of controlling the processing, we first create a reference solution by running the simulated data files with the default settings and solving for the reference frame. As expected, the resulting corrections for the TRF are zero. Now, by changing a single option for processing the zero-input data files, the influence of this particular effect on the TRF can be investigated.

By setting the observed minus computed delay equal zero in the first step, we do not have to consider outliers, clock breaks, and further error sources existing in real measurements which could falsify our investigations. In a last step we perform a standard least-squares adjustment (Vie_lsm) for each session and calculate the seven Helmert-parameters with relation to the reference solution. Alternatively, we apply the global solution tool Vie_glob to determine global station positions and velocities. In the latter, parameters for the clocks, zenith wet delays, troposphere gradients, and Earth orientation parameters are reduced, the reduced normal equation systems are stacked, and the station positions and velocities are estimated applying no-net-translation (NNT) and no-net-rotation (NNR) conditions.

3 Results

3.1 Mean Pole Model

Rotational deformation, due to the deviation of the Earth rotation axis from the Earth figure axis, causes variations of station coordinates, which is called pole tide effect, and hence has to be included in VLBI analysis. Conventionally, the instantaneous pole coordinates should be corrected for the secular wander of the mean pole. Until recently (IERS Conventions 2003), the mean pole representing the secular variation was modelled by a linear formula. In the 2010 Conventions, a new model for the conventional mean pole is described as a cubic polynomial valid for the time period 1976.0–2010.0 and a linear model for extrapolation after 2010.0. In Fig. 29.1 we show the corresponding change in the scale component of the TRF when the new cubic model instead of the formerly used linear one was applied to all R1 and R4 sessions. A total scale change of −0.6 mm for the investigated time span, corresponding to a scale rate of −0.07 mm/year, is the consequence, which is a non-negligible effect that is close to the GGOS goal of 0.1 mm/year.

Additionally, the impact on the scale rate depends on the VLBI network. In Fig. 29.2 the resultant scale variations of the IERS model 2010 with relation to the IERS model 2003 are shown separately for R1 (top) and R4 (below) sessions, with two rather stable but different observing networks. As an indicator for the stability of the network, sessions with same network volume are plotted in the same colour.

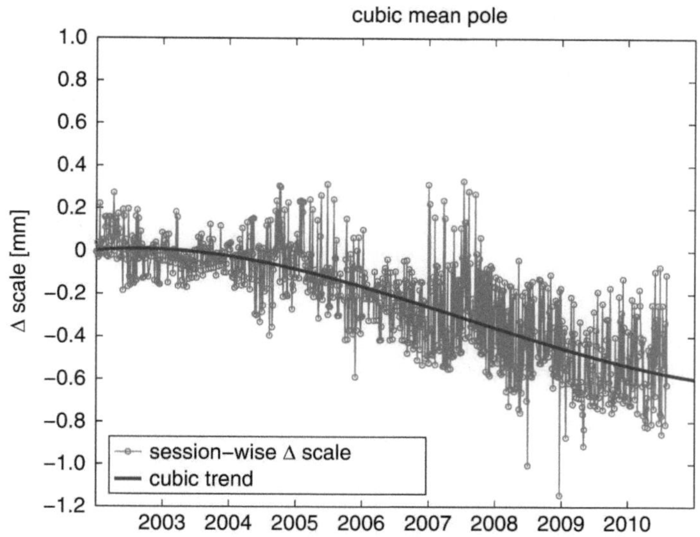

Fig. 29.1 Change in the scale component of the TRF series induced by the new cubic model (IERS Conventions 2010) for the mean pole with relation to the former linear one (IERS Conventions 2003)

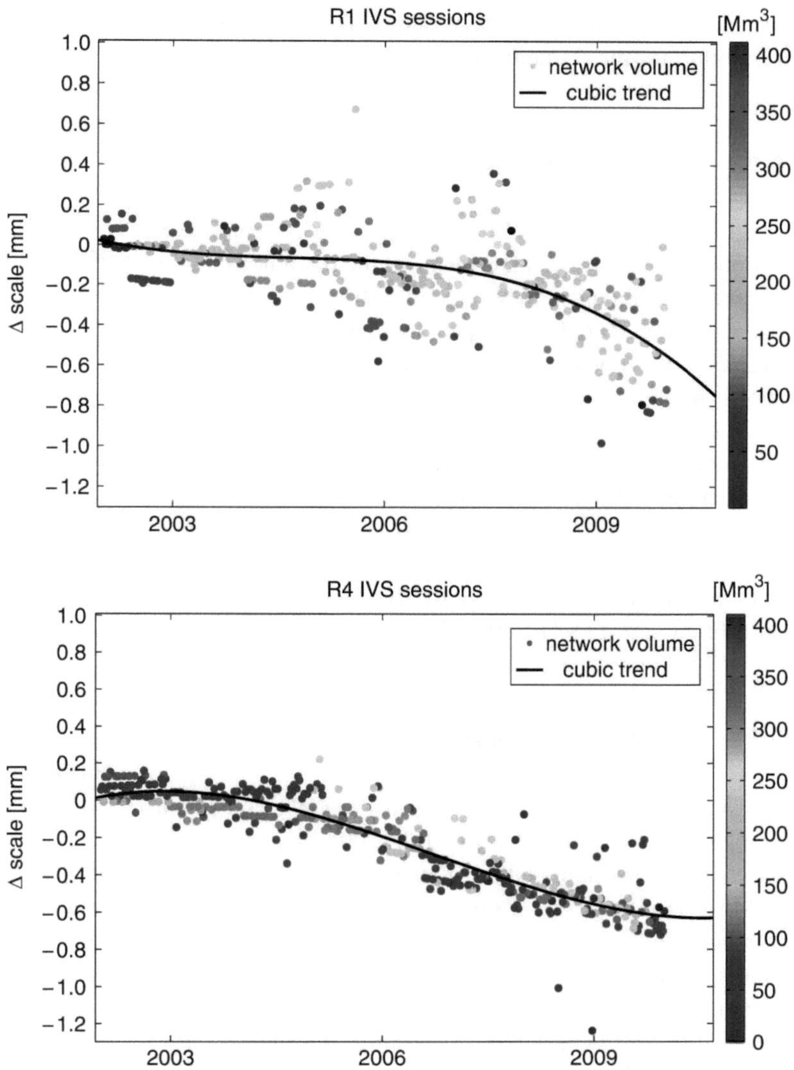

Fig. 29.2 Change in the scale component of the TRF series induced by the new cubic model (IERS Conventions 2010) for the mean pole with relation to the former linear one (IERS Conventions 2003), for the IVS R1 (*top*), respectively R4 (*below*) sessions. Same colours represent same network volume

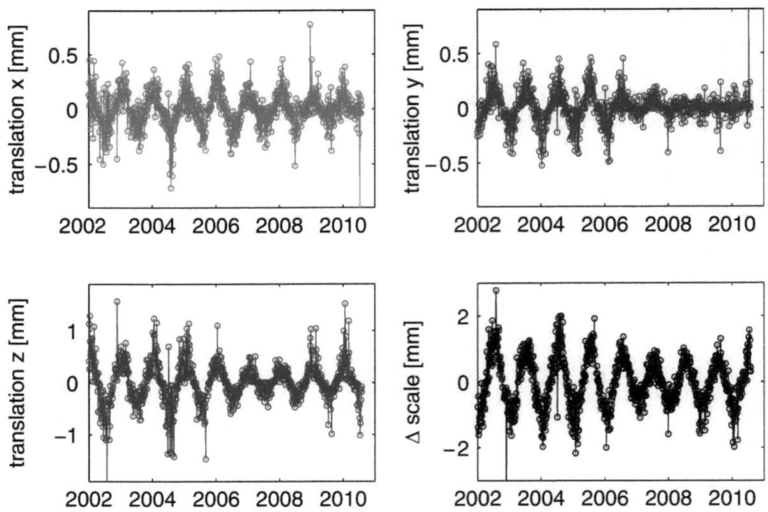

Fig. 29.3 Impact of antenna thermal deformation on the session-wise derived TRF

Network volume denotes the volume of the convex hull of the participating stations. The considerably larger scatter for the R1 sessions is most likely due to the higher network variability for the R1 compared to the R4 sessions.

3.2 Antenna Thermal Deformation

Thermal expansion of VLBI antennas due to temperature variations can cause displacements of the reference point up to several millimetres. In VieVS, this effect is accounted for by using the model developed by Nothnagel (2008). In Fig. 29.3 the impact on the session-wise derived TRF is shown when neglecting this effect during VLBI.

A clear annual signal is observed with variations of ±0.5 mm for the translation in x (blue, upper left) and y (green, upper right), ±1 mm for the translation in z (red, lower left), and ±2 mm for the scale component (black, lower right) of a seven-parameter Helmert transformation with relation to the reference, the VLBI Terrestrial Reference Frame 2008 (VTRF2008, Böckmann et al. 2009).

3.3 Troposphere Mapping Functions

In Chap. 9 of the IERS Conventions, models to calculate propagation delays are described. In 2007 this section was completely rewritten, and the Vienna Mapping Functions 1 (VMF1) (Böhm et al. 2006a) were recommended as the standard model to map the zenith delay to the actual observation angle.

For VMF1, time series for each station with 6 h time resolution, based on numerical weather models, are used. By default VMF1 is applied in VieVS. An alternative are the empirical Global Mapping Functions (GMF) (Böhm et al. 2006b) which require only the station coordinates and the time of observation as input parameters. In Fig. 29.4 the local station displacements for Ny-Ålesund are plotted when using GMF instead of VMF1 for the investigated VLBI sessions. Though the effect can reach a few millimetres for single sessions, the average value is zero and no secular or harmonic systematic variations can be seen. This conclusion holds for all investigated VLBI stations and is valid in general, as was shown by Steigenberger et al. (2009) for GPS. Hence the choice between VMF1 and GMF does not cause systematic effects in the TRF.

3.4 Atmosphere Loading

At the moment, ITRF solutions are calculated without correcting for atmosphere loading at the observing sites. For our reference solution we took atmosphere loading effects into account, namely S1 and S2 tidal atmospheric pressure loading and non-tidal atmosphere loading with a time resolution of 6 h, as provided by Petrov and Boy (2004). In a second run atmosphere loading was switched off and the resulting effect in the scale component of the derived TRF is shown in Fig. 29.5. The session-wise scale variations shown in light blue reach maximum differences of several millimetres. Plotted in red is the effect of atmosphere loading smoothed over 50 days, the black line represents a fitted signal with a period of 365.25 days and an amplitude of 0.3 mm. The observed scale variations are clearly systematic and too big to be neglected for future realizations of the terrestrial reference frame.

Damping of atmosphere loading corrections can be seen in Fig. 29.6. In red we show the station height variations for Wettzell if atmosphere loading is neglected during the analysis. Compared to the radial component of the atmosphere loading model plotted in black, the damping is clearly

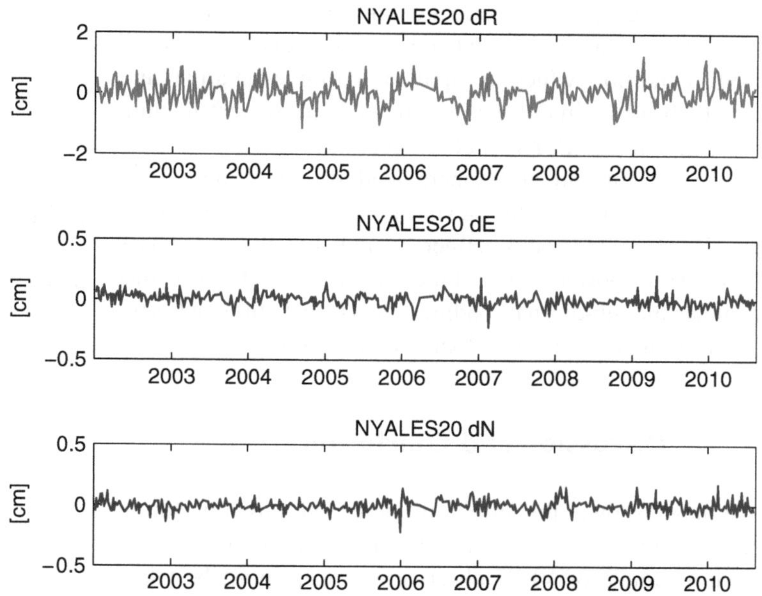

Fig. 29.4 Station displacements (radial, east, north) at Ny-Ålesund, induced by the use of GMF alternatively to VMF1

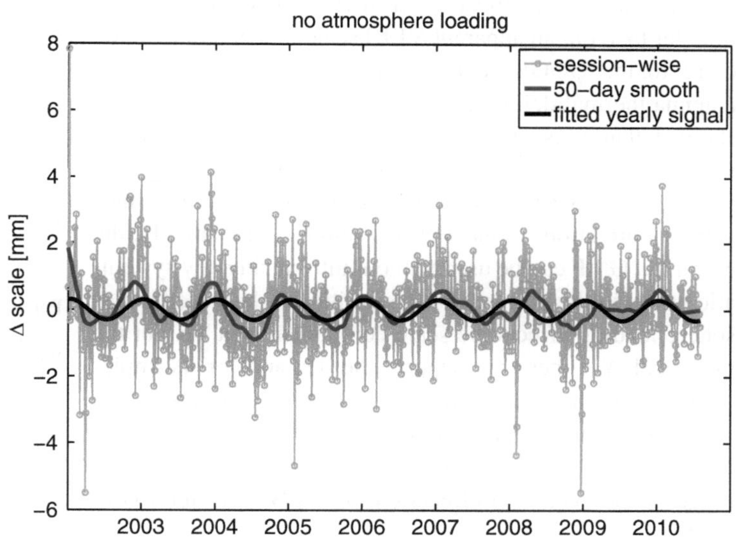

Fig. 29.5 Scale variations induced by neglecting atmosphere loading

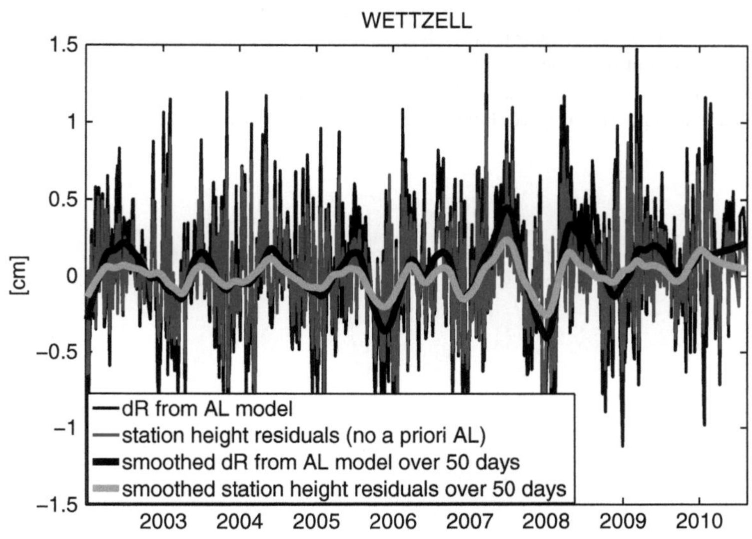

Fig. 29.6 The variation of station heights (*red*) and smoothed over 50 days (*green*) compared to the corresponding radial component of the atmosphere pressure loading model (*thin black*) and their smoothed values over 50 days (*bold black*)

visible, meaning that the effect of atmosphere loading is not fully reflected in the station heights. This is due to the NNT/NNR condition which is applied during estimation and which absorbs parts of the investigated atmosphere loading deformation (see Böhm et al. 2009).

We recommend atmosphere loading corrections to be applied at the observation level in VLBI analysis. The geophysical signals should neither be added to the coordinate time series a posteriori nor at the stacking level, because this neglects all sub-diurnal loading effects, that can reach several millimetres; see also Böhm et al. (2009).

4 Discussion and Conclusion

We have shown the effect of new and possibly future standard models on the terrestrial reference frame derived from a sample of VLBI observations. Care must be taken with the adoption of the new cubic model for the mean pole recommended in the most recent IERS Conventions (2010), in order to avoid an apparent VLBI scale rate of a magnitude close to the GGOS goal of 0.1 mm/year. The effect of antenna thermal deformation is well known by now and we showed that it is absolutely necessary to correct for the thermal expansion of VLBI antennas when processing long time series. VMF1 is the standard model to convert atmospheric corrections to the elevation angle. Alternatively, the empirical GMF can be used without introducing any systematic errors in the derived station positions. Neglected atmosphere loading affects the scale component of a TRF at the mm-level. We therefore support the demand for its implementation in the analysis for future TRF realizations.

Though our investigated data sample covers barely a decade, the presented results are representative to show the range of the evaluated effects on the VLBI contribution to the ITRF.

Acknowledgements The authors would like to thank the editor and the two reviewers for their useful advice and help to improve our paper. This contribution to the REFAG 2010 was enabled by the receipt of the IAG Travel Award for Lucia Plank. Hana Spicakova is recipient of financial support during her PhD study at TU Vienna by Mondi Austria Privatstiftung and Andrea Pany receives a DOC-fForte fellowship of the Austrian Academy of Sciences at the Institute of Geodesy and Geophysics, Vienna University of Technology. Tobias Nilsson is grateful to the Deutsche Forschungsgemeinschaft (DFG) (project SCHUH 1103/3-2).

References

Böckmann S, Artz T, Nothnagel A (2009) VLBI terrestrial reference frame contributions to ITRF2008. J Geodesy 84(3):201–219. doi:10.1007/s00190-009-0357-7

Böhm J, Werl B, Schuh H (2006a) Troposphere mapping functions for GPS and very long baseline interferometry from European Centre for Medium-Range Weather Forecasts operational analysis data. J Geophys Res 111:B02406. doi:10.1029/2005JB003629

Böhm J, Niell A, Tregoning P, Schuh H (2006b) Global mapping function (GMF): a new empirical mapping function based on data from numerical weather model data. Geophys Res Lett 33:L07304. doi:10.1029/2005GL025546

Böhm J, Heinkelmann R, Mendes Cerveira PJ, Pany A, Schuh H (2009) Atmospheric loading corrections at the observation level in VLBI analysis. J Geodesy 83(11):1107–1113. doi:10.1007/s00190-009-0329-y

Böhm J, Böhm S, Nilsson T, Pany A, Plank L, Spicakova H, Teke K, Schuh H (2011) The new Vienna VLBI software VieVS. In: Kenyon S, Pacino MC, Marti U (eds) Proceedings of the 2009 IAG symposium, Buenos Aires, International Association of Geodesy Symposia 136. ISBN 978-3-642-20337-4

McCarthy DD, Petit G (2004) IERS conventions 2003. IERS technical note, no. 32, Verlag des BKG, Frankfurt am Main

Nothnagel A (2008) Short note: conventions on thermal expansion modelling of radio telescopes for geodetic and astrometric VLBI. J Geodesy 83(8):787–792. doi:10.1007/s00190-008-0284-z

Petit G, Luzum B (eds) (2010) IERS conventions 2010. IERS technical note, no. 36, Verlag des BKG, Frankfurt am Main (in print)

Petrov L, Boy JP (2004) Study of the atmospheric pressure loading signal in very long baseline interferometry observations. J Geophys Res 109:B03405

Schlüter W, Behrend D (2007) The international VLBI service for geodesy and astrometry (IVS): current capabilities and future prospects. J Geodesy 81(6–8):379–387. doi:10.1007/s00190-006-0131-z

Steigenberger P, Böhm J, Tesmer V (2009) Comparison of GMF/GPT with VMF1/ECMWF and implications for atmospheric loading. J Geodesy 83(10):943–951. doi:10.1007/s00190-009-0311-8

Titov O, Tesmer V, Böhm J (2004) OCCAM v.6.0 software for VLBI data analysis. In: Vandenberg NR, Baver K (eds) International VLBI service for geodesy and astrometry 2004 general meeting proceedings, NASA/CP-2004-212255, pp 267–271

Part V

Definition and Establishment of Vertical Reference Systems

Influence of Vertical Datum Inconsistencies on Gravity Field Modelling

30

Z. Fašková, R. Čunderlík, K. Mikula, and R. Tenzer

Abstract

Precise gravity field modelling is essential for a unification of local vertical datums (LVDs) and realization of the World Height System. The quality of terrestrial gravimetric measurements has substantial impact on the accuracy of detailed geoid/quasigeoid models. The precision of their positions, especially their vertical components, is of the same importance as precision of gravity itself. Therefore inconsistencies due to shifts and tilts of LVDs can distort precise solutions.

In this paper we present how inconsistencies of vertical positions of input terrestrial gravity data can influence numerical solutions obtained by the finite element method and finite volume method. Considering information from satellite missions, we solve the geodetic BVP with mixed boundary conditions (BCs) in the 3D domain above the Earth's surface. This space domain is bounded by the Earth's surface at the bottom, one spherical artificial boundary outside the Earth at altitude of a satellite mission and four side artificial boundaries. All numerical solutions are fixed to the satellite only geopotential model on all artificial boundaries, where the Dirichlet BCs are imposed. On the Earth surface the oblique derivative BC in the form of surface, gravity disturbances is prescribed. In our numerical experiments we compare numerical solutions with and without considering the corrections from the shifts and tilts of LVDs in the input surface gravity disturbances. We study how the corrected solutions backward-influence estimations of the shifts and tilts of LVDs. Our experiments are performed in areas of Australia, New Zealand and Great Britain.

Keywords

Boundary value problem with mixed boundary conditions • Finite element method • Finite volume method • Inconsistencies of local vertical datums

1 Introduction

The unification of local vertical datums (LVDs) and the realization of the World Height System (WHS) are very important tasks of recent geodesy. They are usually performed on a basis of global geopotential models (GGMs); cf. Burša et al. (1999, 2004, 2007), Sánchez (2005). The low frequency part of GGMs is very precisely obtained from satellite missions. It is fully independent from LVDs but it has insufficient resolution for a unification of LVDs, especially in small regions. The high frequency part

Z. Fašková (✉) • R. Čunderlík • K. Mikula
Department of Mathematics and Descriptive Geometry, The Slovak University of Technology, Faculty of Civil Engineering, Radlinského 11, Bratislava 813 68, Slovakia
e-mail: faskova@math.sk; cunderli@svf.stuba.sk; mikula@math.sk

R. Tenzer
National School of Surveying, University of Otago, 310 Castle Street, Dunedin, New Zealand
e-mail: robert.tenzer@otago.ac.nz

Z. Altamimi and X. Collilieux (eds.), *Reference Frames for Applications in Geosciences*, International Association of Geodesy Symposia 138, DOI 10.1007/978-3-642-32998-2_30,

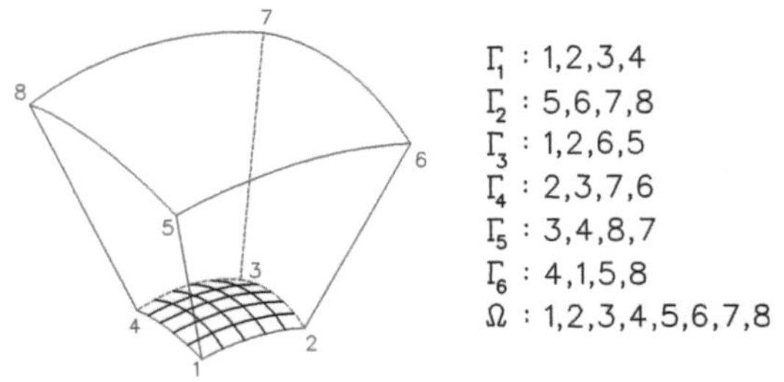

Fig. 30.1 Geometry of computational domain

requires reliable terrestrial and altimetry-derived gravity data. Since vertical information of terrestrial data is mostly related to LVDs, such input data are globally inconsistent. Therefore, a combination with satellite GGMs is usually used to overcome this drawback. It results in combined GGMs with significantly improved accuracy, e.g., EGM2008 up to 2160, Pavlis et al. (2008). However, in case of SH-based models as well as models, achieved by the remove-compute-restore technique (Tscherning 1978), a residual part is fully influenced by inconsistencies in vertical information due to shifts and tilts of LVDs.

This paper presents an approach based on a numerical solution to the BVP with mixed BCs using the finite element method (FEM), cf. Meissl (1981), Shaofeng and Dingbo (1991), Fašková et al. (2009), and the finite volume method (FVM), cf. Eymard et al. (2003), Mikula and Sgallari (2003). Numerical solutions are fixed to the satellite GGMs at altitude of a satellite mission. On the Earth's surface we generate surface gravity disturbances from available terrestrial gravimetric data. We try to reconstruct their ellipsoidal heights using national geoid/quasigeoid models and corrections from shifts and tilts of LVDs estimated from the GPS-levelling test. In this way, we reduce inconsistencies of input data due to LVDs. Consequently, the obtained numerical solutions, which are less dependent on LVDs, can provide new estimates of their shifts and tilts with respect to a chosen W_0 defining the World Height System.

2 Formulation of the BVP with Mixed BC

Let us consider the linearized fixed gravimetric BVP (cf. Koch and Pope (1972), Holota (1997), (2005), Čunderlík et al. (2008)):

$$\Delta T(\boldsymbol{x}) = 0, \qquad \boldsymbol{x} \in R^3 - \Omega^E, \tag{30.1}$$

$$\langle \nabla T(\boldsymbol{x}), s(\boldsymbol{x}) \rangle = -\partial g(\boldsymbol{x}), \qquad \boldsymbol{x} \in \partial\Omega^E, \tag{30.2}$$

$$T = O\left(|\boldsymbol{x}|^{-1}\right) \qquad as \qquad \boldsymbol{x} \to \infty. \tag{30.3}$$

where $T(\boldsymbol{x})$ is the disturbing potential, $\delta g(\boldsymbol{x})$ is the gravity disturbance, $\boldsymbol{s}(\boldsymbol{x}) = -\nabla U(\boldsymbol{x}) / |\nabla U(\boldsymbol{x})|$ and Ω^E is the Earth volume.

We construct an artificial boundary $\Gamma_2 \in \partial\Omega^E$ away from the Earth and restrict our computations to a partial domain Ω depicted in Fig. 30.1.

The surface gravity disturbance in Eq. 30.2 represents the oblique derivative BC. In order to get the Neumann BC we project the oblique derivative BC onto the normal to Γ_1; see Čunderlík et al. (2008). On the upper spherical boundary $\Gamma_2 = \{x;\ |x| = R\}$ and side boundaries $\Gamma_{3,\ldots,6} \in \partial\Omega$, the Dirichlet BC is prescribed. Then our BVP is defined as

$$\Delta T(x) = 0, \qquad x \in \Omega, \tag{30.4}$$

$$\frac{\partial T}{\partial n_{\Gamma_1}} = -\delta g^*(\boldsymbol{x}) = -\delta g \cos\mu, \qquad on \quad \Gamma_1, \tag{30.5}$$

$$T(x) = T_{SAT}, \qquad on \qquad \Gamma_i, \qquad i = 2, \ldots, 6 \tag{30.6}$$

where T_{SAT} represents the disturbing potential generated from a satellite GGM and μ is the angle $\angle\,(\boldsymbol{n}_\Gamma(\boldsymbol{x}), \boldsymbol{s}(\boldsymbol{x}))$.

3 Solution of the BVP by FEM

In accord with Reddy (1993), we divide the domain Ω into finite elements. Then one constructs the weak formulation of the differential equation on every element

$$\int_{\Omega^e} \nabla w \cdot \nabla T \, dx\, dy\, dz - \int_{\Gamma^e} w q_n d\sigma = 0, \tag{30.7}$$

where w is a weight function, Ω^e is a finite element, Γ^e is an element boundary and q_n denotes an inter-element fluxes. Over each part, one seeks an approximation to the solution as a linear combination of nodal values t_j^e and approximation functions Ψ_j^e; i.e.,

$$T(x, y, z) \approx T^e(x, y, z) = \sum_{j=1}^{N} t_j^e \Psi_j^e(x, y, z) \tag{30.8}$$

where N is a number of element nodes. Then we consider Ψ_i^e, $i = 1, \ldots, N$ for w and apply approximation (30.8) in Eq. 30.7 to get the linear system of the equations for each element Ω^e. After rearrangement, we obtain

$$\sum_{j=1}^{N} t_j^e \underbrace{\int_{\Omega^e} \nabla\Psi_i^e \nabla\Psi_j^e \, dx\, dy\, dz}_{K_{ij}} - \underbrace{\int_{\Gamma_1^e} \Psi_i^e q_n \, d\sigma}_{Q_i} = 0, \quad (30.9)$$

where $\boldsymbol{K} = [K^e{}_{ij}]_{N\times N}$ is an element stiffness matrix and $\boldsymbol{Q} = [Q^e{}_i]_N$ represents a vector of fluxes through element faces. Finally, we assemble the parts to obtain the solution to the whole by using the balance of the inter-element fluxes, continuity of numerical solution on inter-element boundaries and taking the BCs into account. More details about the solution of the GBVP by FEM can be found for instance in Fašková (2008) and Fašková et al. (2009).

4 Solution of the BVP by FVM

Let us have $(n_1 \times n_2 \times n_3)$ discrete points and embed them into the finite volume mesh τ where the central point of each volume is one of these points. The elements of τ will be called control volumes and for any $(p,q) \in \tau^2$ with $p \neq q$, we denote by $e_{pq} = p \cap q$ their common interface. The measure of e_{pq} is labeled as $m(e_{pq})$ and $\boldsymbol{n}_{pq}$ denotes the unit vector normal to e_{pq}, oriented from p to q. We apply Green's theorem to Eq. 30.4 to get a local mass balance

$$-\int_{p} \Delta T \, dx\, dy\, dz = \int_{\partial p} \frac{\partial T}{\partial n} d\sigma = \sum_{q\in N_P} \int_{e_{pq}} \frac{\partial T}{\partial n_{pq}} d\sigma, \quad (30.10)$$

where p represents volume and N_p are its neighbors. In order to derive discrete finite volume numerical scheme, we approximate the normal derivative along the boundary of p in Eq. 30.10 by $\partial T/\partial n_{pq} \approx (T_q - T_p)/d_{pq}$, where d_{pq} is a distance between centers of p and q. In such way, we get discrete scheme for Eq. 30.4 in the form

$$\sum_{q\in N_p} \frac{m(e_{pq})}{d_{pq}} (T_p - T_q) = 0, \quad (30.11)$$

where $m(e_{pq})/d_{pq}$ is the transmissivity coefficient. More information about FVM can be found for instance in Eymard et al. (2003), Mikula and Sgallari (2003), or Fašková (2008).

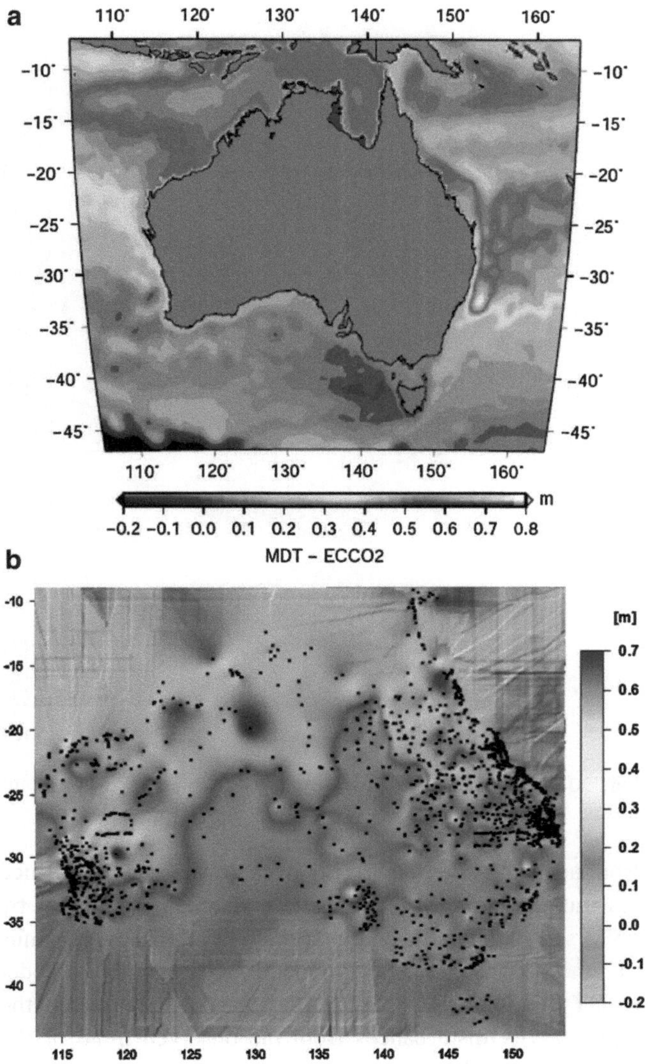

Fig. 30.2 (**a**) ECCO2 mean dynamic topography around Australia (average in 2007); (**b**) shift and tilt of AHD evaluated at GPS-levelling points using the FVM solution with original data and $W_o = 63{,}626{,}856.0\ m^2\ s^{-2}$

5 Numerical Experiments

We have performed three numerical experiments in areas of Australia, Great Britain and New Zealand. In all experiments the gravity disturbances at oceans have been generated from the DNSC08-GRAV altimetry-derived gravity anomalies (Andersen et al. 2009) and the disturbing potential has been computed from the ITG-GRACE03S satellite geopotential model up to degree 180 (Mayer-Gürr 2007) that is fully independent from LVDs.

In the first numerical experiment in Australia, the gravity disturbances on lands have been derived from simple

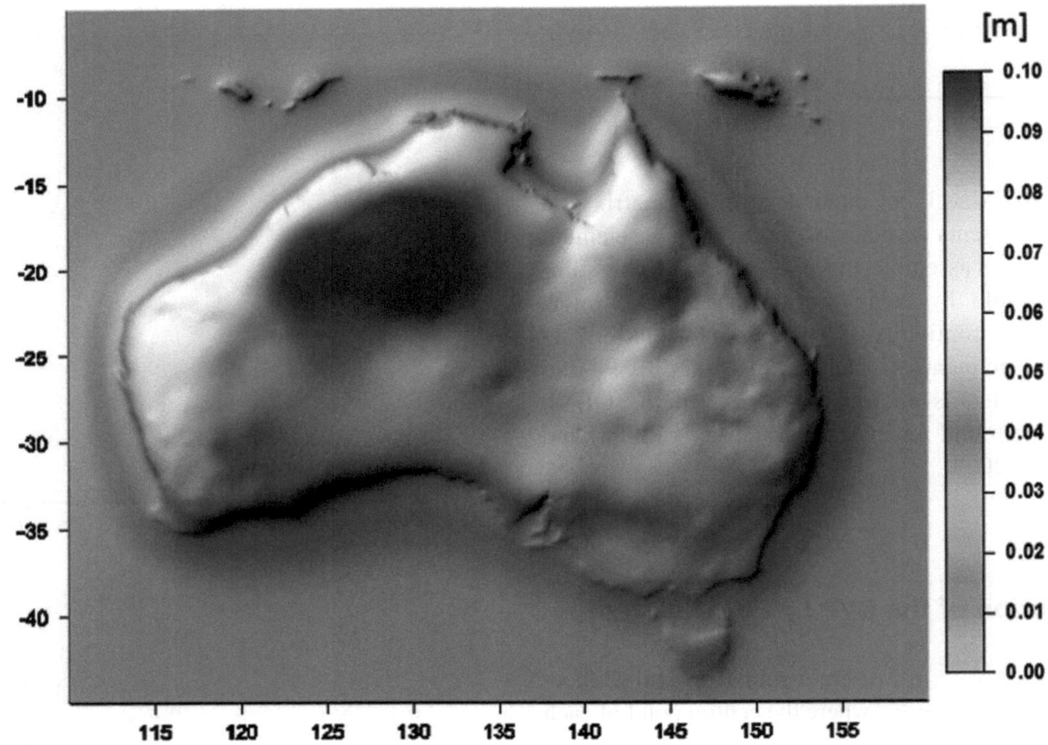

Fig. 30.3 A comparison between numerical solutions by FVM in Australia with and without considering corrections from the shifts and tilts of AHD

Bouguer anomalies on a grid over Australia (Geoscience Australia). The GPS-levelling test of our first solution by FVM shows the north–south oriented trend that corresponds to the ECCO2 mean dynamic topography model (MDT-ECCO2) (Fig. 30.2). Then we have corrected the input gravity disturbances from the observed trend of the Australian Height Datum (AHD) assuming that the vertical information of the original gravity data was influenced by these systematic tendencies. Figure 30.3 depicts that changes in ellipsoidal heights from 0.2 to 1.2 m (Fig. 30.2b) cause changes in quasigeoid up to 0.1 m. It is partly due to the fact that both solutions are fixed to ITG-GRACE03S on the artificial boundaries (Chap. 2). Figure 30.4 shows the GPS-levelling test of the corrected FVM solution.

Statistical characteristics of the GPS-levelling test (Table 30.1) indicate that corrections of input data from the shift and tilt of AHD slightly improve the precision of the corrected solution (the standard deviation is improved by 1 cm). However, it has more significant influence on the estimation of the AHD offset to the considered W_0 value (the mean value decreases by 5 cm).

We have done a similar experiment in the area of Great Britain. Here, the input surface gravity disturbances on land have been generated from the DNSC08-GRAV dataset. Again, we have compared two solutions with and without considering corrections from the shift and tilt of the LVD. In this case we focus on the impact of the W_0 value used in the GPS-levelling test. The numerical experiments were performed using FEM implemented in the ANSYS software (ANSYS).

Figure 30.5 and Table 30.2 depict the GPS-levelling test of the obtained FEM solutions. In case of $W_0 = 63{,}626{,}856.0\ \mathrm{m}^2\ \mathrm{s}^{-2}$, the mean value −0.432 m and the standard deviation 0.056 m changed after applying the corrections to −0.402 m and 0.048 m, respectively. In case of $W_0 = 63{,}626{,}853.0\ \mathrm{m}^2\ \mathrm{s}^{-2}$, the mean value −0.739 m changed to −0.687 m and the standard deviation 0.057 to 0.048 m. This confirms our experience that the corrections of input data from inconsistencies of LVDs have more significant influence on estimations of the LVD offsets than on the overall accuracy. Different values of W_0 considered in the GPS-levelling test give different changes of the mean values, namely3.0 cm in case of $W_0 = 63{,}626{,}856.0\ \mathrm{m}^2\mathrm{s}^{-2}$ and 5.2 cm in case of $W_0 = 63{,}626{,}853.0\ \mathrm{m}^2\mathrm{s}^{-2}$ (Table 30.2).

The last numerical experiment is in area of New Zealand. Here the input gravity disturbances on land have been generated from discrete gravimetric measurements (GNS Science). Our goal is to find a numerical solution by FEM, which

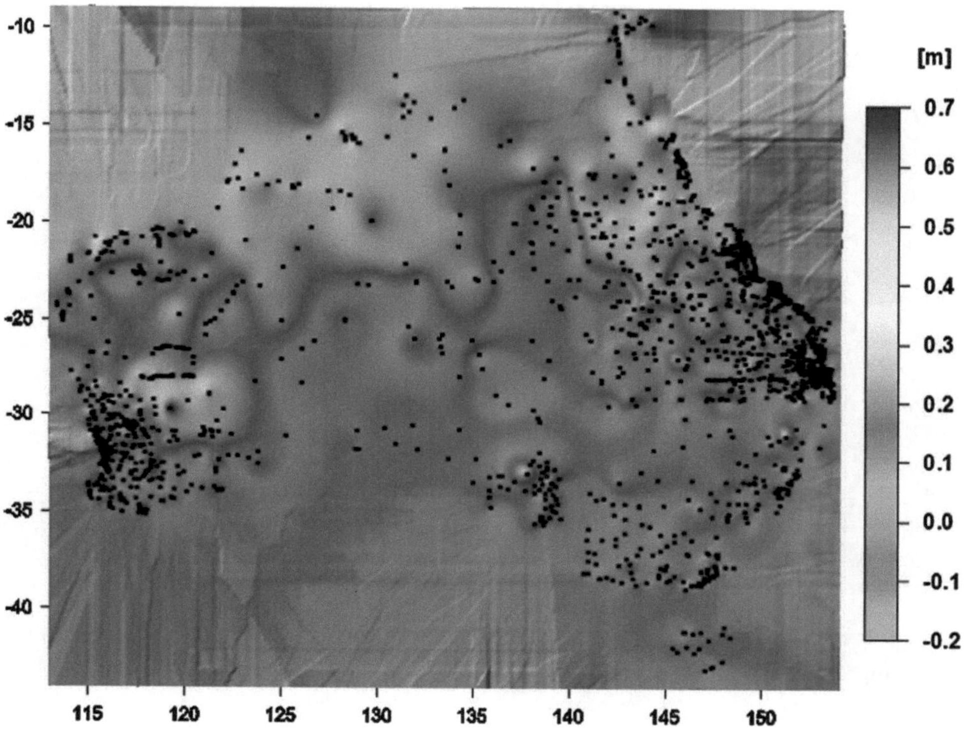

Fig. 30.4 Shift and tilt of AHD evaluated at GPS/Levelling points using the FVM solution with corrected data and $W_o = 63{,}626{,}856.0\ m^2\ s^{-2}$

Table 30.1 GPS-levelling test of the FVM solutions in Australia before and after removing the trend from MDT-ECCO2 with respect to $W_o = 63{,}626{,}856.0\ m^2\ s^{-2}$

Input data	Original		Corrected	
Trend	— (m)	Removed (m)	— (m)	Removed (m)
Max	0.840	0.512	0.748	0.420
Min	−0.321	−0.526	−0.345	−0.598
Mean	0.160	−0.157	0.110	−0.207
Std	0.171	0.116	0.161	0.109

best corresponds to MDT-ECCO2. In the first case, we have reconstructed ellipsoidal heights of the terrestrial gravimetric measurements using the NZGeoid09 national quasigeoid model (Amos and Featherstone 2009). The obtained FEM solution, similarly to NZGeoid09, does not correspond well to MDT-ECCO2, showing an evident shift and tilt on the Southern Island (Fig. 30.6). Therefore, we have tried to correct the ellipsoidal heights using information from MDT-ECCO2 considering correcting of the normal-orthometric heights from the gravity field. The recomputed FEM solution shows better agreement with MDT-ECCO2 (compare Fig. 30.7a with Fig. 30.6c). Naturally, the GPS-Leveling test is worse for the recomputed solution (Fig. 30.8, Table 30.3) due to the fact that normal-orthometric heights in the mountainous Southern Island show systematic tendencies up to 0.7 m. Consequently, estimates of the shifts and tilts of the LVDs in New Zealand from our recomputed FEM solution differ considerably from the ones obtained from our previous FEM solution (Table 30.3) or from NZGeoid09 (Amos and Featherstone 2009).

6 Conclusions

The presented numerical solutions to the geodetic BVP with mixed boundary conditions using FEM or FVM are efficient for a combination of satellite and terrestrial gravity data. All solutions are fixed to the satellite only geopotential models at altitude of the satellite mission, which partly reduces the impact of inconsistencies in LVDs. Numerical experiments in Australia, Great Britain and New Zealand confirm that the accuracy of vertical positions of terrestrial gravity data is of the same importance as the accuracy of gravity itself. Corrections of input data from inconsistencies of LVDs are necessary to get gravity field models that are not related to

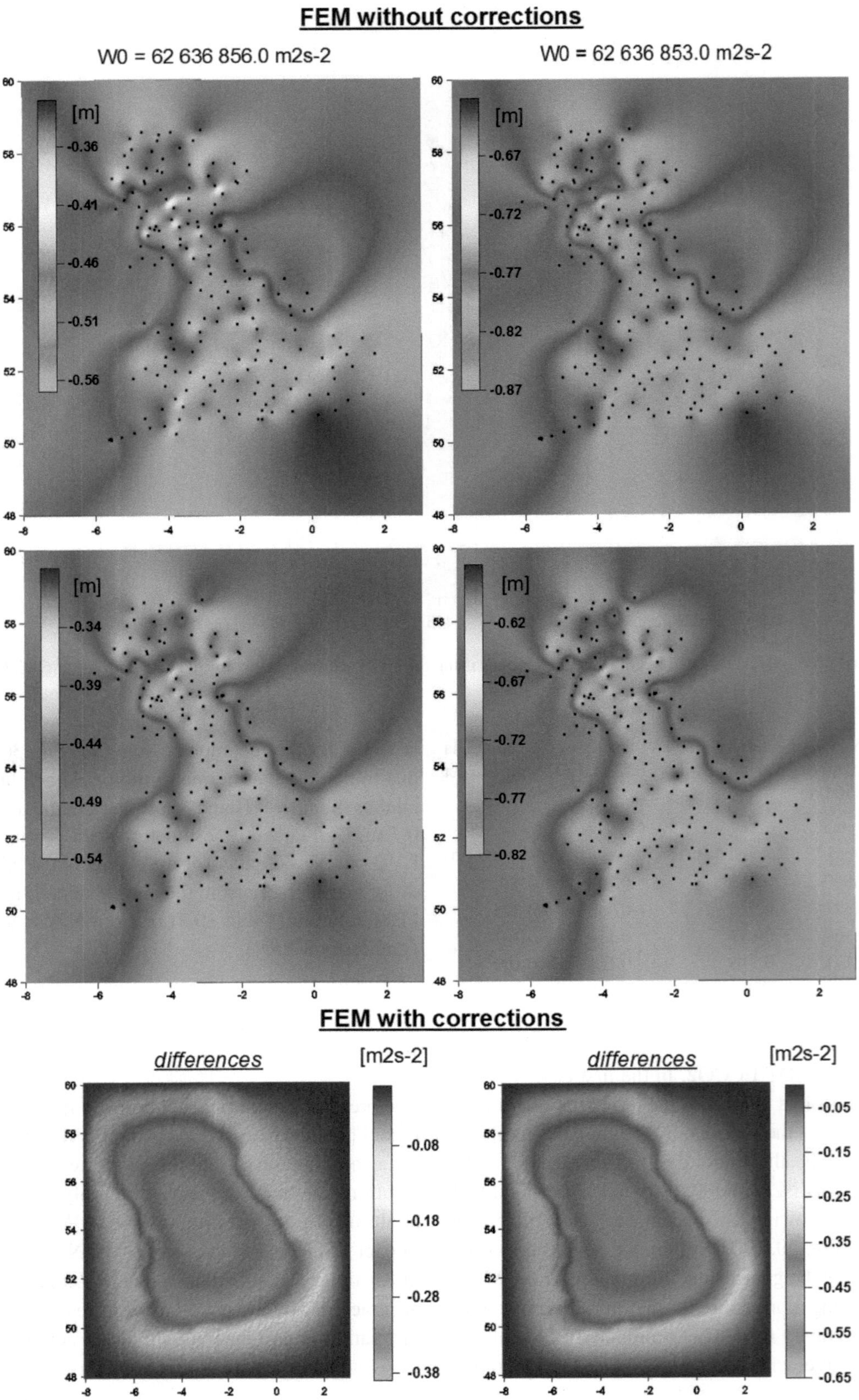

Fig. 30.5 GPS-levelling test of the FEM solutions in Great Britain obtained without and with corrections to ellipsoidal heights with respect to $W_0 = 63\ 626\ 856.0\ m^2\ s^{-2}$ and $W_0 = 63\ 626\ 853.\ 0\ m^2\ s^{-2}$, and differences between them

Table 30.2 GPS-levelling test of the FEM solutions in Great Britain

W_0	$63{,}626{,}856.0\ m^2\ s^{-2}$		$63{,}626{,}853.0\ m^2\ s^{-2}$	
Input data	Original (m)	Corrected (m)	Original (m)	Corrected (m)
Max	−0.229	−0.304	−0.534	−0.591
Min	−0.778	−0.519	−1.084	−0.803
Mean	−0.432	−0.402	−0.739	−0.687
Std	0.056	0.048	0.057	0.048

Table 30.3 GPS-levelling test of the FEM solutions in New Zealand

Input data	Original (m)	Corrected (m)
Max	0.680	1.062
Min	−0.439	−0.073
Mean	0.170	0.318
Std	0.123	0.192

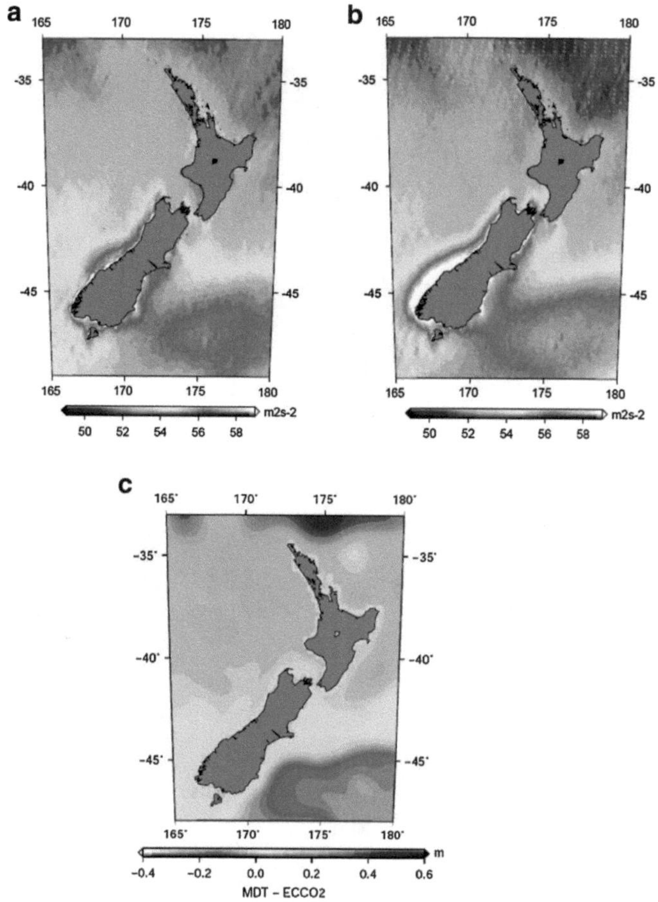

Fig. 30.6 Geopotential on the DNSC08 mean sea surface around New Zealand computed (**a**) from the FEM solutions using original data, (**b**) from NZGeoid09 (the constant $63{,}626{,}800.0\ m^2\ s^{-2}$ is removed), and (**c**) the ECCO2 mean dynamical topography model (average in 2007)

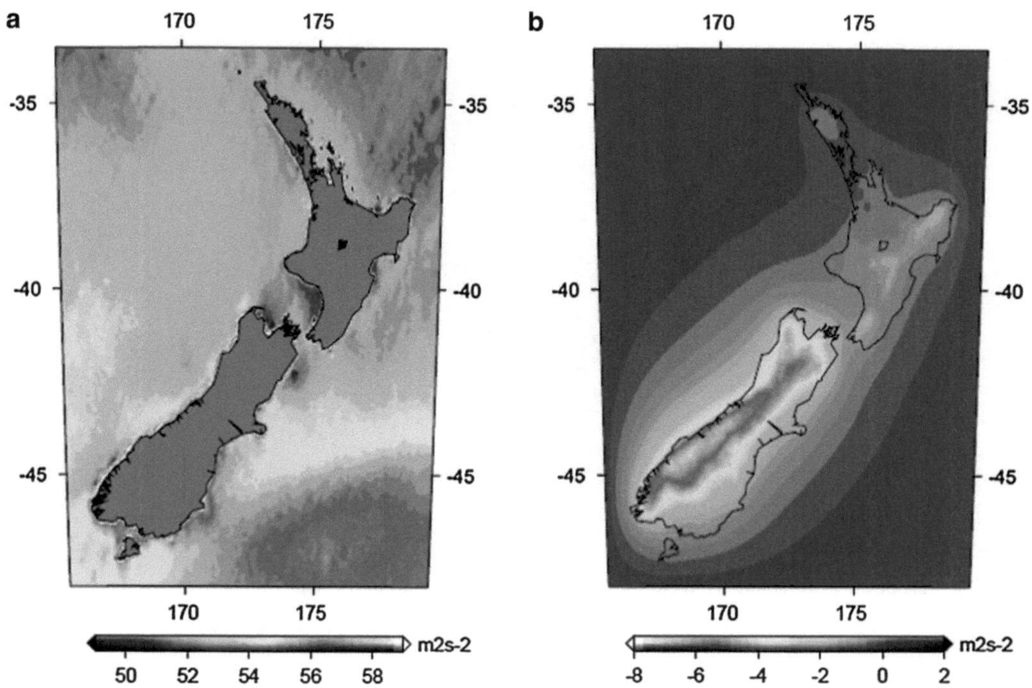

Fig. 30.7 (**a**) Geopotential on the DNSC08 mean sea surface around New Zealand computed from the FEM solutions using corrected data (the constant $63{,}626{,}800.0\ m^2\ s^{-2}$ is removed), (**b**) differences between both FEM solutions

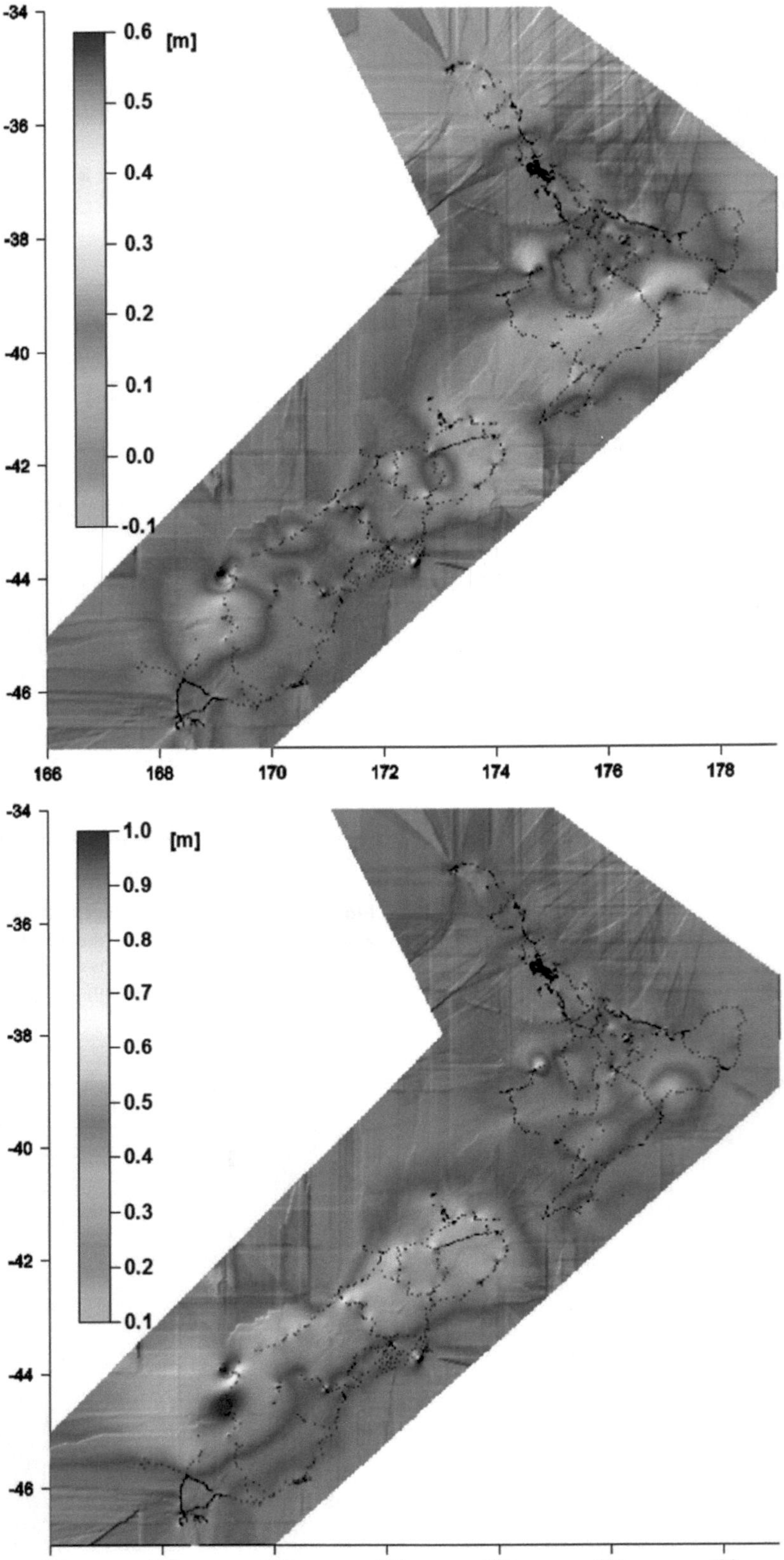

Fig. 30.8 GPS-levelling test of the FEM solutions in New Zealand obtained without (*above*) and with corrections (*below*) to ellipsoidal heights

LVDs. Such models provide new estimations of shifts and tilts of LVDs that can improve their relation to the WHS. The GPS-levelling test of the obtained results shows that the corrections of input data have more significant influence on the estimates of the LVD offsets than on the overall accuracy.

Acknowledgement The authors gratefully thank the providers of all data used, especially for the opportunity to access the new GPS-levelling dataset in Australia supplied by the Intergovernmental Committee on Surveying and Mapping (ICSM) members and compiled by Geoscience Australia. The work has been supported by the grants VEGA 1/0269/09, grant VEGA 1/1063/11 and the project APVV-0184-10.

References

Amos MJ, Featherstone WE (2009) Unification of New Zealand's local vertical datums: iterative gravimetric quasigeoid computations. J Geodesy 83(1):57–68

Andersen OB, Knudsen P, Berry P (2009) The DNSC08GRA global marine gravity field from double retracked satellite altimetry. J Geodesy 84(3):191–199

ANSYS, online tutorial. www.ansys.com

Burša M, Kouba J, Kumar M, Muller A, Radej K, True SA, Vatrt V, Vojtišková M (1999) Geoidal geopotential and world height system. Stud Geophys Geodaet 43:327–337

Burša M, Kenyon S, Kouba J, Šima J, Vatrt V, Vítek V, Vojtišková M (2004) A global vertical reference frame based on four regional vertical datums. Stud Geophys Geodaet 48(3):493–502

Burša M, Kenyon S, Kouba J, Šima J, Vatrt V, Vítek V, Vojtišková M (2007) The geopotential value W_0 for specifying the relativistic atomic time scale and a global vertical reference system. J Geodesy 81(2):103–110

Čunderlík R, Mikula K, Mojzeš M (2008) Numerical solution of the linearized fixed gravimetric boundary value problem. J Geodesy 82(1):15–29

Eymard R, Gallouet T, Herbin R (2003) Finite volume methods. In: Ciarlet PG, Lions JL (eds) Handbook of numerical analysis. Elsevier, North-Holland, pp 713–1018 ISBN: 978-0-444-50350-3

Fašková Z (2008) Numerical methods for solving geodetic boundary value problems. Ph.D. thesis, SvF STU Bratislava, Slovakia

Fašková Z, Čunderlík R, Mikula K (2009) Finite element method for solving geodetic boundary value problems. J Geodesy 84(2):135–144

Holota P (1997) Coerciveness of the linear gravimetric boundary-value problem and a geometrical interpretation. J Geodesy 71(10): 640–651

Holota p (2005) Neumann's boundary-value problems in studies on earth gravity field: weak solution. 50 Years of research institute of GEODESY, Topography and Cartography. Prague, vol 50: pp 34, 49–69

Koch KR, Pope AJ (1972) Uniqueness and existence for the godetic boundary value problem using the known surface of the earth. Bull Geod 46:467–476

Mayer-Gürr T (2007) ITG-Grace03s: the latest GRACE gravity field solution computed in Bonn, Presentation at GSTM + SPP, Potsdam

Meissl P (1981) The use of finite elements in physical geodesy. Report 313, Geodetic Science and Surveying, The Ohio State University

Mikula K, Sgallari F (2003) Semi-implicit finite volume scheme for image processing in 3D cylindrical geometry. J Comput Appl Math 161(1):119–132

Pavlis NK, Holmes SA, Kenyon SC, Factor JK (2008) An Earth gravitational model to degree 2160: EGM2008. General Assembly of EGU, Vienna, Austria

Reddy JN (1993) An introduction to the finite element method, 2nd edn. McGraw-Hill, Singapore

Sánchez L (2005) Definition and realisation of the SIRGAS vertical reference system within a globally unified height system. Dynamic Planet, Cairns, 2005, oral presentation

Shaofeng B, Dingbo C (1991) The finite element method for the geodetic boundary value problem. Manuscr Geodet 16:353–359

Tscherning CC (1978) Collocation and least squares methods as a tool for handling gravity field dependent data obtained through space research techniques. In: Hieber S, Guyenne TD (eds) On space oceanography, navigation and geodynamics European workshop. European Space Agency, Paris, pp 141–149

The Role of a Conventional Transformation Scheme for Vertical Reference Frames

31

C. Kotsakis

Abstract

A conventional transformation between different realizations of a vertical reference system is an important tool for geodetic studies on precise vertical positioning and physical height determination. Its main role is the evaluation of the consistency for co-located vertical reference frames (VRFs) on the basis of some fundamental 'datum perturbation' parameters. Our scope herein is to discuss a number of key issues related to the formulation of such a VRF transformation model and to present a few examples from its practical implementation in the comparison of various existing vertical frames over Europe.

Keywords

Vertical reference frames • Physical height transformation

1 Introduction

The comparison of terrestrial reference frames (TRFs) that are established by different observation techniques and/or optimal estimation strategies is a common task in geodetic studies, constituting either a research goal in itself or an auxiliary step for other applications that depend on precise geometric positioning. Such a comparison is typically based on the *linearized similarity transformation* (e.g., Leick and van Gelder 1975) which is a standard tool for the evaluation of Earth-fixed TRFs on the basis of datum-perturbation parameters that are inherently associated with the definition of geodetic terrestrial reference systems (Altamimi et al. 2007). After a least-squares fitting of this transformation model over a network of common points, a set of estimated parameters is obtained that quantify the origin, orientation and scale consistency of the underlying TRFs in terms of their relative translation/rotation/scale variation. The aforementioned scheme provides a geodetically meaningful framework for comparing and combining spatial reference frames, and also for assisting their quality assessment through a suitable de-trending of their systematic discrepancies in order to identify any localized distortions in their respective coordinate sets.

To a large extent, a similar situation occurs also in geodetic applications related to the establishment of *vertical reference frames* for physical height determination. Several realizations of a vertical reference system (VRS) may be available over a regional or even continental network, originating from separate leveling campaigns, alternative data combination schemes and different adjustment strategies. As an example, let us consider a set of national leveling benchmarks, that is part of the United European Leveling Network (UELN). At least three different vertical reference frames (VRFs) co-exist in this regional leveling network, whose physical heights are respectively obtained from the EVRF00 and EVRF07 continental solutions (Ihde and Augath 2001; Sacher et al. 2008) and by the (usually older) national adjustment of the primary height network in the underlying country. If, in addition, GPS data are available at the particular UELN stations, then more VRFs could emerge through the synergetic use of gravimetric geoid models that enable the conversion of observed geometric heights to physical heights.

C. Kotsakis (✉)
Department of Geodesy and Surveying, Aristotle University of Thessaloniki, Thessaloniki 541 24, Greece
e-mail: kotsaki@topo.auth.gr

Z. Altamimi and X. Collilieux (eds.), *Reference Frames for Applications in Geosciences*, International Association of Geodesy Symposia 138, DOI 10.1007/978-3-642-32998-2_31,

An objective comparison among different VRFs needs to be based on a conventional transformation model that quantifies the inconsistencies in the realization of a vertical reference system from co-located physical height datasets. The adopted model must resemble the role of the linearized similarity transformation that is used in TRF studies while its associated parameters should reflect the vertical datum disturbance implied by the corresponding VRFs. Eventually, the utmost role of the underlying height transformation is to be used for generating a combined optimal VRF solution from individual realizations that are jointly merged into a unified vertical frame by postulating appropriate minimum constraints to the datum-related parameters of the height transformation model.

The aim of this paper is to discuss some general aspects about the formulation of a conventional height transformation model for vertical frame studies, and to present a few examples from its practical use in the comparison of various VRFs over Europe.

2 Height Transformation Schemes in Practice

Various transformation schemes for physical heights exist in geodetic practice, mostly in support of gravity field modeling and vertical positioning with heterogeneous data. Typical examples include the reduction of physical heights to a conventional permanent tide system and/or to a reference time epoch due to temporal variations caused by various geodynamical effects (Mäkinen and Ihde 2009; Jekeli 2000), the conversion from normal to orthometric heights (and vice versa), and the determination of apparent height variations due to a known geopotential offset in the zero-height level of the underlying vertical datum.

Moreover, a number of modeling schemes have appeared in the geodetic literature for the optimal fitting of co-located height datasets and the inference of hidden systematic differences between them. The treatment of these problems relies on the implementation of a height transformation model that is selected on the basis of (mostly) empirical criteria. A well-known example is the combined adjustment of ellipsoidal, geoid and leveled heights over a terrestrial control network, which represents a common procedure that has been applied under different objectives in numerous geodetic studies. In the context of our discussion herein, such an adjustment task can be perceived in terms of a generalized transformation scheme for physical heights:

$$H_i' - H_i = \mathbf{a}_i^{\mathrm{T}}\mathbf{x} + s_i + v_i \qquad (31.1)$$

where H_i and H_i' denote the orthometric (or normal) heights obtained from levelling measurements and GPS/geoid (or quasi-geoid) data, respectively. Their systematic discrepancies are modelled by a low-order parametric model and (optionally) a spatially-correlated zero-mean signal, whereas v_i contains the remaining random errors in the height data. The estimated values of the unknown parameters $\mathbf{x}$ and the predicted values of the stochastic signals s_i are obtained from the least-squares (LS) inversion of Eq. 31.1 over a number of control stations using some a-priori information for the data noise level and the signal covariance function (Kotsakis and Sideris 1999).

Several choices have been used in practice for the parametric component $\mathbf{a}_i^{\mathrm{T}}\mathbf{x}$ in Eq. 31.1, none of which has ever assumed the role of a geodetically meaningful model of the systematic differences between the underlying VRFs; that is, between the levelling frame $\{H_i\}$ and the GPS/geoid frame $\{H_i'\}$. In most cases, the suitability of the adopted model is judged by the reduction of the sample variance of the adjusted height errors $\{v_i\}$ within the test network, and not by the physical or geometrical meaning (if any) of its parameters. In fact, the estimated values of $\mathbf{x}$ have never been of any actual importance in geodetic studies, other than offering a more or less arbitrary parametric description for the spatial trend of the height differences H_i'–H_i.

It is worth noting that the use of the classic *four-parameter model*:

$$\mathbf{a}_i^{\mathrm{T}}\mathbf{x} = x_o + x_1 \cos\varphi_i \cos\lambda_i + x_2 \cos\varphi_i \sin\lambda_i + \\ + x_3 \sin\varphi_i \qquad (31.2)$$

may be viewed, to some extent, as an attempt to infer 'datum perturbations' between the physical height frames $\{H_i\}$ and $\{H_i'\}$. Such a viewpoint relies on the equivalent form of Eq. 31.1

$$N_i' - N_i = \mathbf{a}_i^{\mathrm{T}}\mathbf{x} + s_i + v_i \qquad (31.3)$$

where N_i and N_i' denote the geoid undulations from a gravimetric model and GPS/levelling data, respectively. If the four-parameter model is used into Eq. 31.1, then the systematic part of the differences H_i'–H_i is essentially described, in view of Eq. 31.3, through *a 3-D spatial shift* (x_1, x_2, x_3) *and a scale change* (x_o) *between the associated zero-height reference surfaces of the physical heights*. This perspective may also be evoked for the comparison of vertical frames that are obtained exclusively from terrestrial levelling, without the external aid of GPS heights and gravimetric geoids.

The aforementioned four-parameter model was often used in older studies as a basic tool for estimating geodetic datum differences from heterogeneous height data; especially for assessing the geocentricity of TRFs based on Doppler-derived and gravimetrically-derived geoid undulations and also for determining the Earth's optimal equatorial radius from geometric and physical heights (e.g., Schaab and Groten 1979; Grappo 1980; Soler and van Gelder 1987). These tasks require a global data distribution otherwise the 3-D translation parameters (x_1, x_2, x_3) become highly correlated with the zero order term (x_o), and their adjusted values may be entirely unrealistic from a physical point of view. For that reason, the LS inversion of Eq. 31.1 will not always produce a geodetically meaningful solution for the individual components of the four-parameter model—not even for the estimated height bias x_o; for some examples, see Kotsakis and Katsambalos (2010). Moreover, the drawback of this model for VRF evaluation studies is that it compares the zero-height surfaces between two vertical frames with respect to a (fictitious) geocentric reference system, *without* considering the most important element in vertical datum realization: a geopotential reference value W_o and its possible variation between different VRFs.

The estimation of the (usually unknown) zero-height level W_o that is inherently linked to any vertical frame can be carried out through various strategies based on 'external' geopotential information and space geodetic measurements at a number of leveling benchmarks (e.g., Burša et al. 2001; Ardalan et al. 2002). In this way, any VRF is comparable to another, not necessarily co-located, VRF' in terms of the estimated geopotential difference (δW_o) of their zero-height levels. However, such a value is affected by the errors in the adopted geopotential model and thus it may give a misleading assessment of the zero-height consistency between the tested VRFs.

Furthermore, a detailed comparison of vertical frames should take into account their scale variation due to systematic differences in their associated measurement techniques and modeling assumptions. In fact, one should not forget that the fundamental height constraint $h - H - N = 0$ requires not only the 'origin consistency' among the heterogeneous height types, but also their reciprocal vertical scale uniformity.

3 Basic Formulation of a Conventional VRF Transformation

An objective assessment of the consistency between VRFs requires a conventional model describing their systematic discrepancies over a common group of control stations. The parameters $\mathbf{x}$ of such a model:

$$H_i' - H_i = f(\mathbf{x}) + v_i \tag{31.4}$$

should quantify the (actual and/or apparent) vertical datum perturbations induced by the physical height datasets $\{H_i\}$ and $\{H_i'\}$ while the remaining residuals, after a LS adjustment of Eq. 31.4, indicate the relative accuracy level of the corresponding vertical frames.

Note that a VRF is a realization of a 1-D terrestrial coordinate system with respect to an equipotential surface of Earth's gravity field. The latter defines a conventional zero-height level relative to which vertical positions (geopotential numbers and their equivalent physical heights) can be obtained by various geodetic techniques and terrain modeling hypotheses. Hence, the key role of Eq. 31.4 is to appraise the variation of the reference equipotential surface and the vertical metric scale, which both signify the fundamental datum constituents for vertical positioning within every physical height frame.

Two essential parameters should be incorporated in $f(\mathbf{x})$, namely a VRF *translation parameter* in the form of a geopotential disturbance δW_o, and a VRF *scale-change parameter* in the form of a unitless factor δs reflecting the scale difference between the corresponding height frames. In case of dynamic VRFs, the time derivatives of the above parameters need also to be considered when transforming physical heights between different epochs or vertical velocities from a VRF to another VRF$'$. In contrast to the Helmert transformation scheme that is used in geometric TRFs, there are no rotational terms within the VRF transformation model $f(\mathbf{x})$ since the frame orientation aspect is not a geodetically meaningful characteristic of vertical reference systems.

3.1 General Remarks

The notion of the 'scale' in a vertical reference system is often linked to the geopotential value W_o that is adopted for defining absolute vertical coordinates (geopotential numbers) and their equivalent physical heights on the Earth's surface. Specifically, the VRS scale is explicitly related to an equipotential surface realized by the combination of a mean sea surface topography model and a global gravity field model, in accordance with the classic Gauss-Listing definition of the geoid (Ihde 2007). This is a simplified approach to quantify the *average size* of the reference surface for vertical positioning, since the ratio of the geocentric gravitational constant to the adopted reference geopotential level:

$$R = GM/W_o \tag{31.5}$$

yields the mean radius of the geoid, which itself defines a physical metric for the geocentric spatial position of terrestrial points with zero heights! Obviously, any change of W_o induces an apparent offset to the terrestrial physical heights, which can be perceived as an indirect 'scaling effect' due to the changed spatial dimension of their zero-height reference surface.

The previous viewpoint aims at the standardization of the Earth's global scale in terms of the physical parameters GM

and W_o, and it is not related to the notion of a scale variation between different realizations of a VRS. In fact, a change of W_o is related to a transformation from a conventional height 'origin' to another one, whereas the scope of a VRF scale change is to account for the systematic discrepancy of the vertical metric scale realized by alternative heighting techniques and datasets when determining physical height differences. Both types of VRF perturbation (origin and scale) are feasible and they may co-exist in the joint analysis of vertical frames.

3.2 The Effect of δW_o

Changing the zero-height level of a VRF means that a new vertical frame with a different equipotential surface will be used as a reference for physical heights. Such a transformation is described through a single parameter (δW_o) reflecting the geopotential disturbance of the zero-height equipotential surface with respect to a conventional representation of the Earth's gravity field. The effect on the VRF geopotential numbers is a simple offset equal to δW_o while for the VRF orthometric or normal heights it takes the form of a nonlinear and spatially inhomogeneous variation according to the following power series expansion:

$$H_i' - H_i = \frac{\delta W_o}{g_i} - \frac{\partial g}{\partial H}\frac{\delta W_o^2}{2g_i^3} + \dots \tag{31.6}$$

for the case of *orthometric* heights, or

$$H_i' - H_i = \frac{\delta W_o}{\gamma_i} - \frac{\partial \gamma}{\partial H}\frac{\delta W_o^2}{2\gamma_i^3} + \dots \tag{31.7}$$

for the case of *normal* heights. The terms g_i and $\partial g/\partial H$ denote the actual gravity and its vertical gradient on the geoid, or more precisely on the equipotential surface associated with the initial orthometric height. Also, γ_i and $\partial\gamma/\partial H$ denote the normal gravity and its vertical gradient on the reference ellipsoid which is associated with the initial normal height.

For practical purposes, both Eqs. 31.6 and 31.7 can be replaced by the simplified linearized formula:

$$H_i' - H_i = \frac{\delta W_o}{\gamma_i} \tag{31.8}$$

since their second, and higher, order terms have a negligible contribution (<1 mm) for reasonably low values of δW_o (up to 1–2 gpu, or equivalently up to 10–20 $m^2\ s^{-2}$). Moreover, the geoidal gravity g_i in Eq. 31.6 may be safely replaced by the normal gravity γ_i on the reference ellipsoid, causing an approximation error into the transformed orthometric height that is below the mm level, even for gravity anomaly values up to 500 mGal.

3.3 The Effect of δs

In contrast to geometric Cartesian TRFs, the assessment of a systematic scale difference between VRFs is not a straightforward issue. The effect of a scale change on the physical heights depends on the way we (choose to) handle the Earth's gravity field and its equipotential surfaces under a uniform spatial re-scaling. The underlying problem is similar to the TRF similarity transformation of GPS heights with respect to a reference ellipsoid, where the latter may or may not 'follow' the spatial scale variation that is imposed by the TRF transformation (Soler and van Gelder 1987; Kotsakis 2008).

Starting from the fundamental differential formula (Heiskanen and Moritz 1967, p. 50)

$$dW = -g\ dH \tag{31.9}$$

where g denotes the magnitude of the gravity vector, the following relationship can be obtained:

$$dH = \frac{g_x}{g}dx + \frac{g_y}{g}dy + \frac{g_z}{g}dz = -\frac{gradW \cdot \mathbf{dr}}{g} \tag{31.10}$$

which gives the vertical (physical height) metric in terms of a weighted combination of the Euclidean metric components with respect to an Earth-fixed spatial coordinate system. The associated weights are the normalized geopotential gradients and they represent the influence of the Earth's gravity field on the physical height scale.

Considering that the geopotential signal and its gradient vector remain invariant under a uniform scale change $\mathbf{dr}' = (1 + \delta s)\ \mathbf{dr}$, then the resulting effect on the physical heights over the Earth's surface is expressed through a simple linear re-scaling:

$$H_i' = (1 + \delta s)H_i \tag{31.11}$$

The above formula provides the basis for assessing the scale difference between VRFs relative to a fixed reference surface; note that zero-height points are preserved by the scaling transformation of Eq. 31.11. In essence, the differential factor δs absorbs the (linear part of) topographically-correlated discrepancies between $\{H_i\}$ and $\{H_i'\}$, which cause an apparent scale variation between their corresponding VRFs.

4 Least-Squares Adjustment of the VRF Transformation Model

Based on the discussion given in Sect. 3, a conventional VRF transformation model can be formulated in terms of the linearized expression:

$$H_i' - H_i = \frac{\delta W_o}{\gamma_i} + \delta s H_i \tag{31.12}$$

where the meaning of each term has already been explained in previous paragraphs. Essentially, the above model represents the 1D-equivalent of the similarity transformation for vertical positions (physical heights) from a VRF to another VRF′.

The LS adjustment of Eq. 31.12 over a network of m control points leads to the following system of normal equations (NEQs):

$$\begin{bmatrix} \mathbf{q}^T\mathbf{P}\mathbf{q} & \mathbf{q}^T\mathbf{P}\mathbf{d} \\ \mathbf{d}^T\mathbf{P}\mathbf{q} & \mathbf{d}^T\mathbf{P}\mathbf{d} \end{bmatrix} \begin{bmatrix} \delta\hat{W}_o \\ \delta\hat{s} \end{bmatrix} = \begin{bmatrix} \mathbf{q}^T\mathbf{P}(\mathbf{d}' - \mathbf{d}) \\ \mathbf{d}^T\mathbf{P}(\mathbf{d}' - \mathbf{d}) \end{bmatrix} \tag{31.13}$$

where the vectors $\mathbf{d}$ and $\mathbf{d}'$ contain the known physical heights (orthometric or normal) with respect to different vertical frames, i.e.,

$$\mathbf{d} = [H_1 \quad \cdots \quad H_m]^T \quad \mathbf{d}' = [H'_1 \quad \cdots \quad H'_m]^T$$

while $\mathbf{P}$ is a weight matrix for their differences, and the auxiliary vector $\mathbf{q}$ is defined as: $\mathbf{q}(i) = 1/\gamma_i$.

The previous NEQs system is always invertible provided that $\mathbf{q}$ and $\mathbf{d}$ are not co-linear with each other. Given that the elements of the auxiliary vector $\mathbf{q}$ retain an almost constant value (their relative deviation does not exceed 10^{-4} even in large-scale continental networks), the inversion of Eq. 31.13 is practically guaranteed as long as the m control points do not have the same height level!

The correlation coefficient between the VRF transformation parameters is always negative and it is given by the general expression:

$$\rho_{\delta\hat{W}_o,\delta\hat{s}} = -\frac{\mathbf{q}^T\mathbf{P}\mathbf{d}}{(\mathbf{d}^T\mathbf{P}\mathbf{d})^{1/2}(\mathbf{q}^T\mathbf{P}\mathbf{q})^{1/2}} \simeq -\frac{mean[\mathbf{d}]}{rms[\mathbf{d}]} \tag{31.14}$$

A useful algebraic relationship for the optimal estimates obtained from the inversion of Eq. 31.13, as a function of their correlation coefficient, is:

$$\delta\hat{W}_o = \frac{\mathbf{q}^T\mathbf{P}(\mathbf{d}' - \mathbf{d})}{\mathbf{q}^T\mathbf{P}\mathbf{q}} + \rho_{\delta\hat{W}_o,\delta\hat{s}} \frac{(\mathbf{d}^T\mathbf{P}\mathbf{d})^{1/2}}{(\mathbf{q}^T\mathbf{P}\mathbf{q})^{1/2}} \delta\hat{s} \tag{31.15}$$

The separability of the VRF transformation parameters depends on the vertical network configuration. In the context of the joint estimation of δW_o and δs, an optimal vertical network geometry is not related to a homogeneous coverage over the Earth's surface, but to the *height variability* among its control stations. Specifically, the dispersion of the data vector $\mathbf{d}$ must be sufficiently large (with respect to the average height of the control stations) in order for the correlation coefficient in Eq. 31.14 to retain a reasonably low value.

Table 31.1 Transformation parameters between EVRF00 and EVRF07 based on the normal heights at 13 UELN stations over Europe

$\mathbf{d}$	$\mathbf{d}'$	$\delta\hat{W}_o$ (gpu)	$\delta\hat{s}$ (ppm)
EVRF00	EVRF07	0.002 ± 0.004	−25.5 ± 27.7

The initial weight matrix $\mathbf{P}$ was set equal to a unit matrix, and the a-posteriori variance factor of unit weight was estimated at $\sigma_o = 9$ mm

Let us give a few examples from the LS inversion of the transformation model in Eq. 31.12 for a number of VRFs in Europe. The first example employs the EVRF00 and EVRF07 normal heights at the 13 UELN fiducial stations that were used for the primary definition of the zero-height level in the official EVRF07 solution (Sacher et al. 2008). Although the zero-height levels of these two frames were a-priori aligned at the particular stations through a single constraint within the EVRF07 adjustment (see Sacher et al. 2008), our results in Table 31.1 show a small (mm-level) offset between their corresponding reference surfaces. This is caused by the inherent correlation between the estimated parameters $\delta\hat{W}_o$ and $\delta\hat{s}$ ($\rho = -0.7$ in this case), representing an unavoidable 'leakage' effect that occurs in most adjustment problems with coordinate transformation models. Nevertheless, the values of the transformation parameters between EVRF00 and EVRF07 seem to be statistically insignificant, within the limits of their statistical precision, over the particular 13-station UELN continental network.

The second example uses several VRFs that are realized over a network of 20 Swiss leveling benchmarks which are part of the EUVN-DA network (Marti 2010). The tested frames were compared on the basis of Eq. 31.12 using normal heights from: the EVRF00 and EVRF07 continental solutions, the combination of GPS heights with the European gravimetric geoid model EGG08, the official Swiss national height system LN02, and the LHN95 rigorous adjustment of the Swiss national height network. The estimated transformation parameters are given in Table 31.2, whereas the standard deviation of the height residuals (before and after the VRF transformation) are listed in Table 31.3.

Some notable highlights of the previous results are: the considerable scale difference between LHN95 and EVRF07 and the significant origin discrepancy between LN02 and EVRF07, the superiority of the GPS/EGG08 height frame (compared to the Swiss national VRFs) regarding its agreement with the official EVRF07 heights, and finally the sub-cm consistency between EVRF00 and EVRF07 at the

Table 31.2 Transformation parameters between different VRFs in Switzerland based on 20 EUVN-DA Swiss stations

d	d′	$\delta\hat{W}_o$ (gpu)	$\delta\hat{s}$ (ppm)
EVRF00	EVRF07	0.025 ± 0.001	2.9 ± 0.8
GPS/EGG08	EVRF07	0.044 ± 0.012	−76.6 ± 10.7
LN02	EVRF07	−0.251 ± 0.030	35.7 ± 26.8
LHN95	EVRF07	−0.060 ± 0.026	−220.7 ± 22.9

Table 31.3 Standard deviation (in cm) of the height residuals as obtained before and after the adjustment of Eq. 31.12 at 20 EUVN-DA Swiss stations

d	d′	σ (before)	σ (after)
EVRF00	EVRF07	0.3	0.2
GPS/EGG08	EVRF07	5.2	2.6
LN02	EVRF07	6.9	6.6
LHN95	EVRF07	14.0	5.6

particular 20 EUVN-DA Swiss stations, even before the implementation of the height transformation model of Eq. 31.12.

5 Epilogue

A preparatory discussion on the use of a conventional transformation model for evaluating and comparing VRFs has been presented in this paper. Our analysis has been restricted only to a static (time-independent) height transformation setting, yet its generalization for cases of dynamic vertical frames is also necessary in view of the following key tasks: (1) the assessment of systematic discrepancies in vertical velocity models obtained by different geodetic techniques and modeling assumptions, and (2) the optimal combination of individual time-dependent VRF realizations over a global or continental control network.

Acknowledgements Urs Marti from the Swiss Federal Office of Topography (SwissTopo) is acknowledged for providing the Swiss height data from the EUVN-DA project.

References

Altamimi Z, Collilieux X, Legrand J, Garayt B, Boucher C (2007) ITRF2005: a new release of the international terrestrial reference frame based on time series of station positions and Earth orientation parameters. J Geophys Res 112:B09401. doi:10.1029/2007JB004949

Ardalan A, Grafarend E, Kakkuri J (2002) National height datum, the Gauss-Listing geoid level value w_o and its time variation. J Geod 76:1–28

Burša M, Kouba J, Müller A, Raděj K, True SA, Vatrt V, Vojtišková M (2001) Determination of geopotential differences between LVDs and realization of a world height system. Stud Geophys Geod 45:127–132

Grappo GA (1980) Determination of the Earth's mean equatorial radius and center of mass from Doppler-derived and gravimetric geoid heights. Manuscr Geod 5:201–216

Heiskanen W, Moritz H (1967) Physical geodesy. WH Freeman, San Francisco

Ihde J (2007) Conventions for the definition and realization of a conventional VRS. Report of the IAG inter-commission project on vertical reference frames

Ihde J, Augath W (2001) The European vertical reference system (EVRS), its relation to a World Height System and to ITRS. IAG symposia, vol 125. Springer-Verlag, Berlin, pp 78–83

Jekeli C (2000) Heights, the geopotential and vertical datums. Department of Geodetic Science, The Ohio State University, OSU Report no. 459, Columbus, OH

Kotsakis C (2008) Transforming ellipsoidal heights and geoid undulations between different geodetic reference frames. J Geod 82:249–26

Kotsakis C, Katsambalos K (2010) Quality analysis of global geopotential models at 1542 GPS/levelling benchmarks over the Hellenic mainland. Surv Rev 42(318):327–344

Kotsakis C, Sideris M (1999) On the adjustment of combined GPS/geoid/leveling networks. J Geod 73:412–421

Leick A, van Gelder BHW (1975) On similarity transformations and geodetic network distortions based on Doppler satellite observations. Department of Geodetic Science, The Ohio State University, OSU Report no. 235, Columbus, OH

Mäkinen J, Ihde J (2009) The permanent tide in height systems. IAG symposia, vol 133. Springer-Velag, Berlin, pp 81–87

Marti U (2010) Private communication

Sacher M, Ihde J, Liebsch G, Mäkinen J (2008) EVRF07 as realization of the European vertical reference system. Presented at the annual EUREF symposium, Brussels, 18–21 June 2008

Schaab H, Groten E (1979) Comparison of geocentric origins of global systems from uniformly distributed data. Bull Geod 53:11–17

Soler T, van Gelder BHW (1987) On differential scale changes and the satellite Doppler system z-shift. Geophys J R Astr Soc 91:639–656

Comparison of Latest Global and Regional Gravimetric Geoid Models with GPS/Leveling Geoidal Undulations Over Japan

32

Y. Kuroishi

Abstract

The success of the dedicated gravity satellite missions CHAMP and GRACE has enabled global geoid determination at unprecedented accuracy and has stimulated study on geoid modeling at a very high-resolution on a global scale as well as for local improvement. The paper first describes the latest determination of a gravimetric geoid model for Japan, obtained by combining a GRACE-derived global geopotential model, altimetry-derived ocean gravity model and local surface gravity measurements on land and at sea. Then, a comparison of the model is made with EGM2008 and GPS/leveling geoidal undulations over four main islands of Japan. Third, we determine sea surface dynamic heights around Japan from a combination of a regional geoid model, tidal records and GPS/leveling data at the coast and on isolated islands and compare them with an oceanographic model.

Keywords

Gravimetric geoid • GPS/leveling geoidal undulations • GRACE • Comparison

1 Introduction

The geoid or quasi-geoid is one of the geodetic reference surfaces in geodesy and oceanography; orthometric heights are defined as physical lengths above the geoid; and sea surface dynamic heights are heights of sea level above the geoid. Determination of these heights is realized by precise leveling or by altimetric measurements and oceanographic observations, both of which suffer from biases and systematic errors at long wavelengths.

The convenience and accuracy of space geodetic techniques in positioning necessitates geodesists to determine the geoid with an accuracy that enables the conversion of geometric heights to physical heights with compatible precision. Enhanced requirements for global climate and environment studies also demand accurate geoid models that can improve our understanding of ocean dynamics in an absolute sense.

The success of the dedicated gravity satellite missions CHAMP (CHAllenging Mini-satellite Payload) and GRACE (Gravity Recovery And Climate Experiment) has enabled global geoid determination at unprecedented accuracy and has stimulated study on geoid modeling at a very high-resolution on a global scale as well as for local improvement. Based on GRACE-derived results and other gravity-related observations, Pavlis et al. (2008) determined a global geopotential model (GGM), EGM2008, at very high resolution while Kuroishi (2009) developed an improved, regional gravimetric geoid model for Japan, JGEOID2008.

This paper presents the results of the comparison of two models (EGM2008 and JGEOID2008) for Japan and its surroundings and their evaluation in terms of quality of gravity recovery at coastal areas and of a reference surface to sea surface dynamic heights (SSDHs) at long wavelengths.

Y. Kuroishi (✉)
Geospatial Information Authority of Japan, 1 Kitasato, 305-0811 Tsukuba, Ibaraki, Japan
e-mail: yuki@gsi.go.jp

Z. Altamimi and X. Collilieux (eds.), *Reference Frames for Applications in Geosciences*, International Association of Geodesy Symposia 138, DOI 10.1007/978-3-642-32998-2_32,

2 Local Data Sets of Geoid Undulations

2.1 Gravimetric Geoid Model for Japan, JGEOID2008

Kuroishi (2009) developed a national gravimetric geoid model on a $1' \times 1.5'$ grid for Japan, JGEOID2008, by combining three different types of data sets: (1) a GRACE-based GGM, (2) surface (land and ship-borne) gravity measurements, and (3) altimetry-derived marine gravity model, KMS2002 (Andersen et al. 2005). The generalized Stokes/Helmert approach was applied to residual Faye anomalies by using the one-dimensional fast Fourier transform in a remove-restore manner (Haagmans et al. 1993) with 100 % zero padding in the longitudinal direction. The geocentric gravitational constant and geoidal potential are taken from IERS Convention 2003 (McCarthy 2003), both of which define the zero elevation of the geoid. The geoid heights are given in the tide-free system as far as the permanent tide is concerned.

For the combination, the semidiscrete wavelet analysis/ reconstruction method with two-dimensional wavelets (Kuroishi and Keller 2005) was employed for selecting the spatial wavelength signals of the highest quality out of the respective data sets. The foundational model used was a GRACE-based GGM, GGM02C (Tapley et al. 2005) to degree and order 200, appended with EGM96 (Lemoine et al. 1997) from degrees 201 to 360. At long wavelengths GGM02C is used exclusively up to degree 90. At intermediate wavelengths KMS2002 is dominant. Its residual signals with respect to GGM02C/EGM96 are low-pass-filtered at a cutoff wavelength of about 1.25 ° and employed exclusively. On the other hand, signals at wavelengths shorter than that are solely taken out of surface measurements.

2.2 GPS/Leveling Data

The Geospatial Information Authority of Japan (GSI) has conducted GPS observations at bench marks throughout the country to measure geoidal undulations. There are 816 sites over the four main islands (Hokkaido, Honshu, Shikoku, and Kyushu), and more sites on smaller isolated islands.

Three-dimensional geocentric positions are determined from GPS observations in the ITRF94/GRS80 reference frame at epoch 1997.0 by putting strong constraints to the GEONET (nationwide permanent GPS array) coordinates at reference stations (Kuroishi et al. 2002). The positions are conventional tide-free values as far as the permanent tide is concerned.

Helmert orthometric heights are calculated from leveling measurements with reference to the latest national adjustment results, JGD2000 (vertical). No astronomical correction was applied in the adjustment of leveling data and accordingly the permanent tidal system of the heights is neither mean-tide, zero-tide, nor tide-free. Instead in this study we applied to the given orthometric heights conversion terms from the mean-tide system to the tide-free one, based on Ekman (1989). Under an assumption that time-dependent parts of tidal effects on leveling are negligibly small, the application of conversion can be considered as partial correction resulting in supposedly conventional tide-free heights.

The orthometric heights measured on the four main islands are uniquely referenced to the mean sea level (MSL) of Tokyo Bay, but those on isolated islands are defined independently on their respective local MSL. Since the mean SSDHs can differ by 1 m around the Japanese coasts due to the Kuroshio Current (Kuragano and Shibata 1997), GPS/leveling geoidal undulations on isolated islands may contain biases due to variation of the mean SSDHs with respect to the MSL of Tokyo Bay.

3 Model Comparison

3.1 Gravimetric Model Comparison Between EGM2008 and JGEOID2008

Two independent high-resolution models of both the gravity field and the geoid are available for Japan: a global model, EGM2008, and a regional model, JGEOID2008. EGM2008 is developed from combination of a GRACE-based GGM, altimetric marine gravity model, surface gravity measurements, and detailed global elevation model, all of which are different from those of JGEOID2008. The spatial resolution of EGM2008, corresponding to a $5' \times 5'$ grid, is slightly larger than that of JGEOID2008, but detailed enough to capture geoidal features.

The two models are compared to each other. Figure 32.1a, b show difference between JGEOID2008 and EGM2008 for free-air gravity anomalies and geoid heights, respectively. Over the Eurasian continent we find very large differences, attributable to the lack of local gravity data in JGEOID2008. We exclude the continental area from the discussion.

In the plot of gravity anomaly differences, a number of hillock patterns are observed in mountainous areas, indicating the limitation in the representation at EGM2008 resolution. Moreover, there exist some areas along the coasts with systematic features: Seto-inland Sea, Suruga Bay (around 35°N 139°E), Off Ibaraki (around 36–37°N 141°E), and east of Hokkaido (around 43–45°N 145°E+). In the two former areas, the ship-borne measurements used only in JGEOID2008 capture real variations in the gravity field, but altimetric models do not, leading to errors in EGM2008. On the other hand, the lack of surface gravity

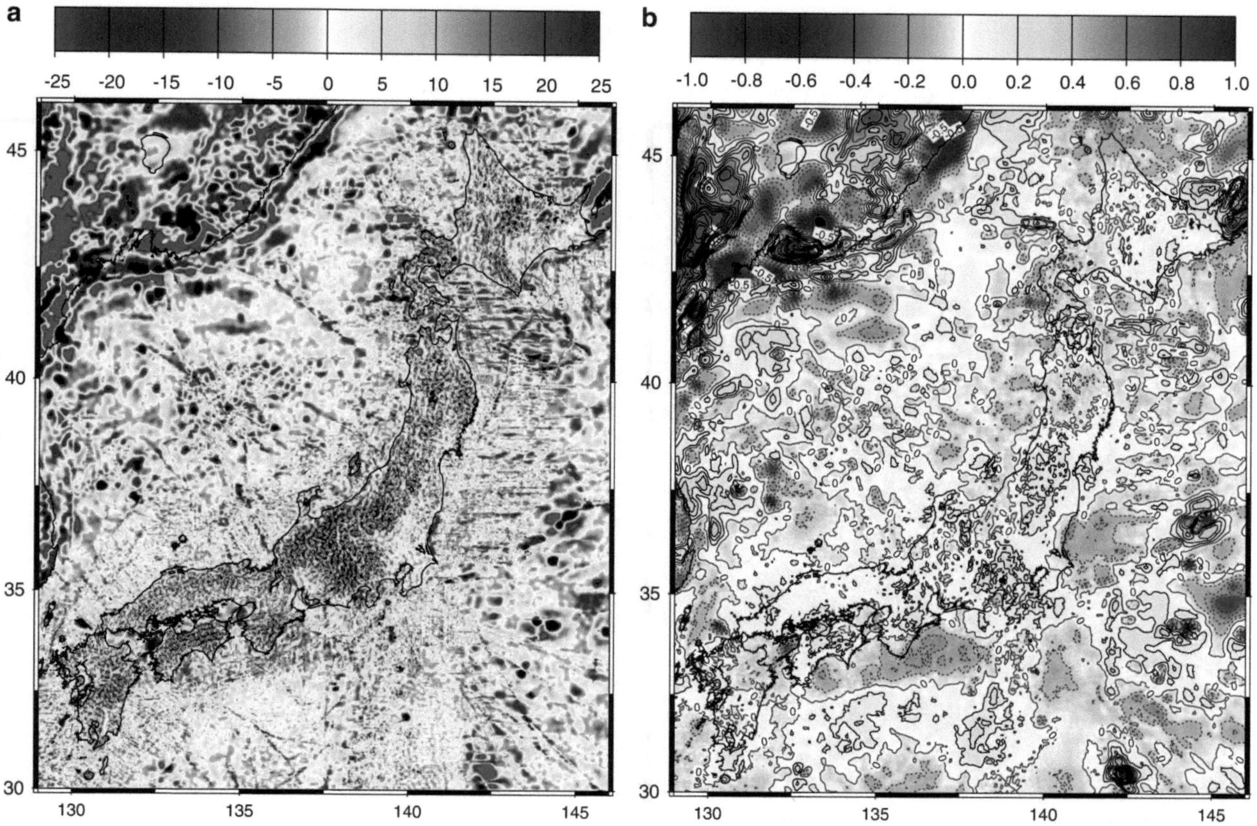

Fig. 32.1 Difference between JGEOID2008 and EGM2008: (**a**) free-air gravity anomalies in mGal and (**b**) geoid heights in meters with contour intervals of 0.1 m

data east of Hokkaido, sitting on the continental side of Chishima-Kamchatka (Kuril-Kamchatka) trench, where wide blocks of large gravity anomalies run parallel to the trench, does not allow JGEOID2008 to resolve features at smaller scales.

The plot of geoid differences depicts integrated effects of gravity differences. The disagreement is generally minor over Japan's landmass, but noticeable in some mountain ranges and coastal areas discussed above. It is seen that substantial errors of gravity anomaly in coastal seas propagate into the adjacent coastal lands, indicating the importance of accurate knowledge of gravity at shallow waters near coasts for precise geoid modeling on land.

3.2 Comparison with Respect to GPS/Leveling-Derived Geoid Undulations

The performance of the two gravimetric geoid models over Japan is evaluated by comparing them to 816 GPS/leveling geoidal undulations data. Post-fit residuals of geoid heights with respect to GPS/leveling data are shown in Fig. 32.2a, b for EGM2008 and JGEOID2008, respectively. Best-fit planar trends are removed from both figures.

For EGM2008 and JGEOID2008, the geoid discrepancies are generally smooth features in the range of several hundreds of kilometers. In some coastal areas, we notice the existence of dissimilar patterns between the two figures.

For EGM2008 (left panel) there are two areas with large positive discrepancies: Sata-misaki Peninsula (about 33.5°N 132°E) on the western edge of Shikoku Island and the coasts of Ibaraki (around 36–37°N 141°E). The former corresponds to the location where the altimetric model used could not correctly resolve the anomalous gravity feature in the western Seto-inland Sea. The latter matches to one of the regions where we observe large differences in the gravity field between the two models.

For JGEOID2008 (right panel) two areas are conspicuous with positive discrepancies: Shiretoko Peninsula of eastern Hokkaido (around 44°N 145°E) and the southern tip of Kii Peninsula (around 33.5°N 135.5°E). It is clear that the lack of gravity information contributes to errors in the former area. The latter should be associated to biases in the gravity field model at the coastal sea.

All of these prominent discrepancies found in both models demonstrate the importance of accurate gravity model along the coast for geoid determination on land. It

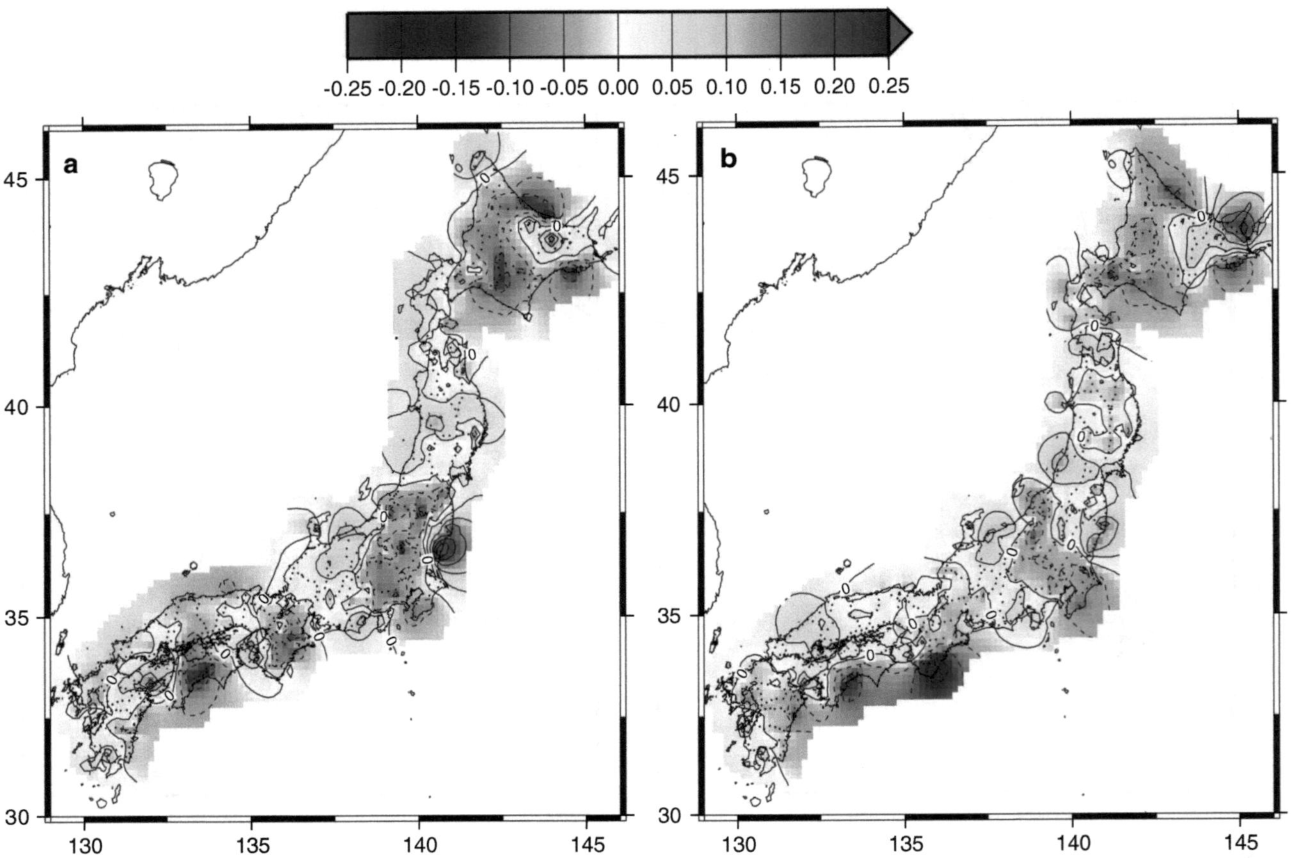

Fig. 32.2 Discrepancies of gravimetric geoid models against GPS/leveling geoidal undulations: (**a**) EGM2008 and (**b**) JGEOID2008. Best-fit planar trends are removed from both plots. Contour interval is 5 cm and dotted contours correspond to negative values

reveals that altimetric marine gravity modeling is still not satisfactory in some coastal areas.

Kuroishi (2009) provides statistics of geoid residuals. The standard deviation (SD) is 6.6 cm for EGM2008 about the best-fitted plane, whose maximum tilt is 0.11 ppm in the slope direction of N43°E, and the SD is 6.0 cm for JGEOID2008 about the best-fitted plane with the maximum tilt of 0.18 ppm in the slope direction of S83°E. Even if the magnitudes of the tilts are different by more than 60 %, the ranges of planar correction on the main islands are compatible to each other: 21 cm for EGM2008 and 24 cm for JGEOID2008. Though the difference in SD is small, these results show that JGEOID2008 is superior to EGM2008 over Japan.

3.3 Comparison with Oceanographic Estimates of Sea Surface Dynamic Height

The discrepancies of GPS-leveling geoidal undulations against JGEOID2008 are compared with orthometric heights of the local MSL at tide gauge stations and SSDHs estimated from a (physical) oceanographic approach. Such comparisons indicate the long-wavelength reliability of JGEOID2008 from a viewpoint of ocean dynamics.

The orthometric heights of the local MSL (the local separation between the MSL and the vertical datum) can be determined by precise leveling of the tidal stations in question. The Coastal Movement Data Center of Japan (secretariat at GSI) publishes such results at more than 140 registered tidal stations. A map of the heights in Fig. 32.3a shows the mean SSDHs around Japan for the period from 1990 to 1999 (reproduced from Kuroishi 2009). The orthometric heights of the MSL at the stations facing the Sea of Japan are about +20 cm.

Discrepancies of GPS-leveling geoidal undulations against JGEOID2008 are calculated for the entire Japanese dataset and the average is taken for each isolated island relative to that around the datum benchmark in Tokyo. The average can be considered as the difference SSDH from the MSL of Tokyo Bay with reference to JGEOID2008. Readers refer to Fig. 32.4a, b for the locations corresponding to the geographic names cited in the following text.

The averages at the isolated islands in the Sea of Japan are about +20 cm throughout, which agrees at the sub-decimeter

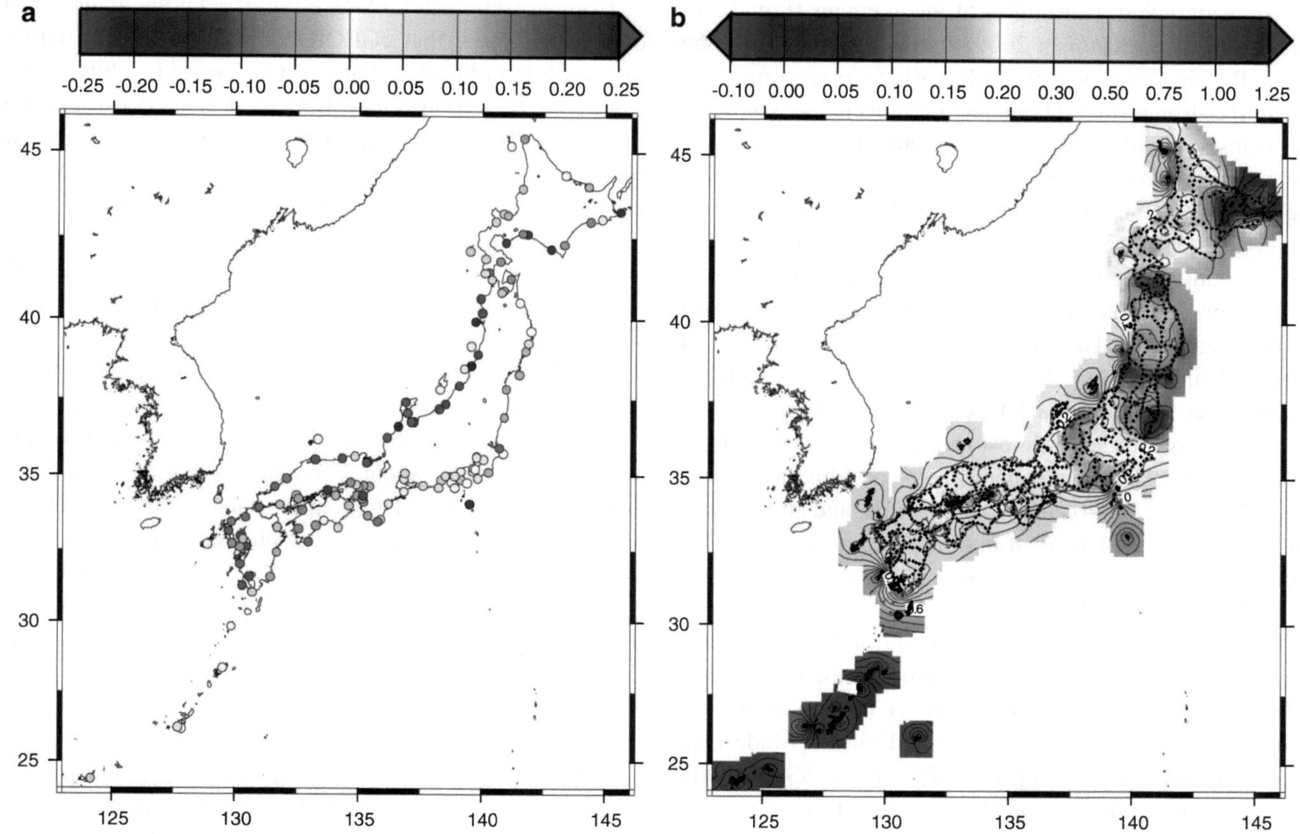

Fig. 32.3 (**a**) Orthometric heights of the local mean sea level at tidal stations for the period from 1990 to 1999 and (**b**) geoid differences between GPS/leveling geoid undulations and JGEOID2008 (both reproduced from Kuroishi 2009). Units of the scale above the images are meters in both figures, contour interval is 5 cm

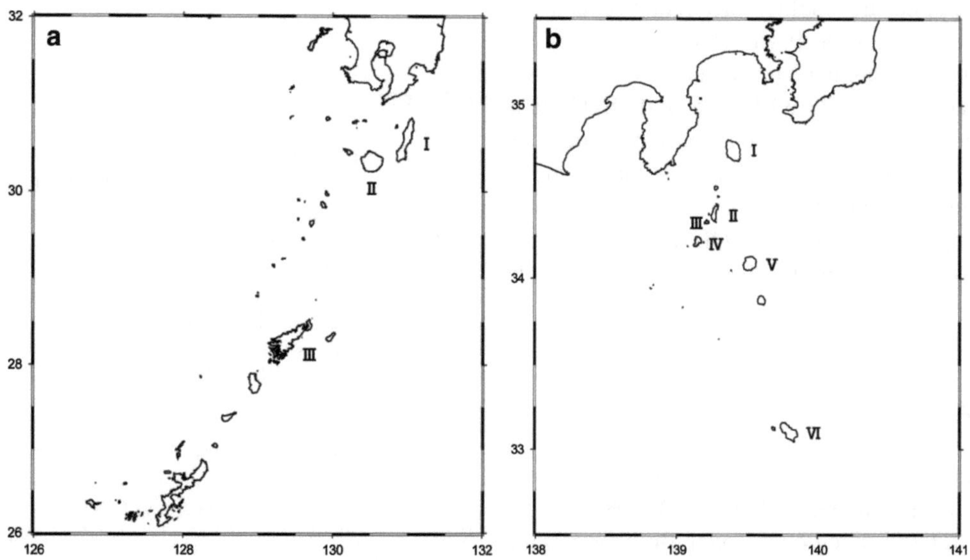

Fig. 32.4 The locations for the geographical names cited in the text. (**a**) Nansei Islands. *I*: Tanegashima, *II*: Yakushima, and *III*: Amami-Oshima Islands. (**b**) Izu Island chain. *I*: Oshima, *II*: Niijima, *III*: Shikinejima, *IV*: Kozushima, *V*: Miyakejima, and *VI*: Hachijojima Islands

level with the numeric values of the orthometric heights of the MSL along the Sea of Japan already discussed. These islands are scattered widely along the coast of the four main islands, from off Hokkaido to off Kyushu, and the agreement therefore suggests that JGEOID2008 is reliable with sub-decimeter precision on a nationwide scale.

The geographic distribution of the averages (Fig. 32.3b, reproduced from Kuroishi 2009) shows systematic features in the Pacific Ocean where the Kuroshio Current runs. It is one of the strongest west boundary currents in the world. For the Nansei (South-western) Islands and Izu Island chain, the averages show drastic changes around the mean axis of the Current. Along the Nansei Islands, the averages are larger than 1 m on, and to the south of, Amami-Oshima Island south of the mean axis, while they are smaller than +50 cm on Tanegashima and Yakushima Islands just north of the mean axis. The mean axis passes through the Izu Island chain (Oshima, Niijima, Shikinejima, Kozushima, Miyakejima, Hachijojima Islands are listed from north to south). The averages are almost zero on Oshima and Niijima Islands north of the axis, about 25 cm on Shikinejima and Kozushima Islands, 46 cm on Miyakejima, and about 63 cm on Hachijojima. These distribution patterns correspond to locations relative to the mean axis and reflect spatial changes in SSDHs across the Current.

Next, the averages are compared with SSDHs relative to that of Tokyo Bay from oceanographic estimates. We consider here a SSDH model processed by the Japan Oceanographic Data Center as the Regional Delayed Mode Data Base. Kuroishi (2009) has prepared the mean SSDH model for the period from 1993 to 1996 and shown its map.

The comparison reveals that the averages from JGEOID2008, tidal records, and GPS/leveling observations agree with the relative SSDHs from oceanographic estimates along the Nansei Islands and the Izu Island chain. Although the corrections applied are slightly different between the two, the agreement reaches the 10-cm level.

From the two types of comparisons in terms of SSDHs we can estimate the accuracy of JGEOID2008. JGEOID2008 is reliable at an accuracy of 10 cm on a nationwide scale even on the open seas. This accuracy is achievable thanks to the GRACE-based geopotential model (as constraints to the geoid model at long wavelengths). JGEOID2008 is expected to serve as a reference level for ocean dynamics study around Japan.

4 Conclusions

Two high-resolution models of the gravity field and geoid are compared over Japan to each other: EGM2008, a global model, and JGEOID2008, a regional model. Both models are developed distinctively from different sets of GRACE-based global geopotential model, altimetry-derived marine gravity model, surface gravity measurements, and detailed digital elevation model. The spatial resolution of EGM2008 is slightly coarser than JGEOID2008, producing a number of hillock patterns of gravity differences in mountainous areas at scales smaller than the resolution for the gravity field.

Comparisons with GPS/leveling geoidal undulation data over Japan show that JGEOID2008 has a slightly larger planar-tilt than EGM08, but that the former is a little superior in precision to the latter. Both models are consistent at a level better than 10 cm with GPS/leveling data in both raw (direct) and post-fit comparisons. There are a couple of coastal areas with systematic errors in both models, indicating respective deficiency in gravity field recovery in the adjacent coastal seas. This demonstrates that we still need to improve the recovery of the gravity field at coastal seas for accurate geoid modeling.

Discrepancies of GPS/leveling geoidal undulations against JGEOID2008 indicate differences of the local mean sea level (namely sea surface dynamic heights) between Tokyo Bay and isolated islands: about 20 cm higher in the Sea of Japan and about 1 m higher in the Nansei islands. The discrepancies south of Japan reflect the geographic distribution of SSDHs across the Kuroshio Current and these patterns match at a precision of about 10 cm with that of an oceanographic model. The regional geoid model can contribute to ocean dynamics studies as a reference surface model.

Acknowledgements The global high-resolution geopotential model, EGM2008, and its handling software are made available by the U.S. National Geospatial-Intelligence Agency. The Regional Delayed Mode Data Base used is provided by the Japan Oceanographic Data Center. The figures are created by GMT (Wessel and Smith 1998). The author's thanks are extended to the institutes and the researchers. The author is also grateful to two anonymous reviewers for their helpful comments and suggestions.

References

Andersen OB, Knudsen K, Trimmer R (2005) Improved high resolution altimetric gravity field mapping (KMS2002 Global Marine Gravity Field). In: Sanso F (ed) A Window on the future of geodesy: Proceedings IUGG 23rd general assembly, 2003, IAG symposium, vol 128. Springer, Sapporo, Japan, pp 326–331

Ekman M (1989) Impacts on geodynamic phenomena on systems for height and gravity. Bull Geod 63:281–296

Haagmans R, de Min E, van Gelderen M (1993) Fast evaluation of convolution integrals on the sphere using 1D FFT and a comparison with existing methods of stokes' integral. Manuscr Geod 18: 227–241

Kuragano T, Shibata A (1997) Sea surface dynamic height of the Pacific Ocean derived from TOPEX/POSEIDON altimeter data: calculation method and accuracy. J Oceanogr 53:585–599

Kuroishi Y (2009) Improved geoid model determination for Japan from GRACE and a regional gravity field model. Earth Planets Space 61:807–813

Kuroishi Y, Keller W (2005) Wavelet approach to improvement of gravity field-geoid modeling for Japan. J Geophys Res 110:B03402. doi:10.1029/2004JB003371

Kuroishi Y, Ando H, Fukuda Y (2002) A new hybrid geoid model for Japan, GSIGEO2000. J Geod 76:428–436

Lemoine FG, Smith DE, Kunz L, Smith R, Pavlis EC, Pavlis NK, Klosko SM, Chinn DS, Torrence MH, Williamson RG, Cox CM,

Rachlin KE, Wang YM, Kenyon SC, Salman R, Trimmer R, Rapp RH, Nerem RS (1997) The development of the NASA GSFC and NIMA joint geopotential model. In: Segawa J et al. (eds) Geoid and Marine Geodesy, Sep 30–Oct, 5 1996, IAG symposium, vol 117. Springer, Tokyo, Japan, pp 461–469

McCarthy DD (2003) IERS conventions (2003). IERS technical note 32

Pavlis NK, Holmes SA, Kenyon SC, Factor JK (2008) An Earth gravitational model to degree 2160: EGM2008. Presentation given at the 2008 General Assembly of the European Geosciences Union, Vienna, Austria, 13–18 April 2008

Tapley B, Ries J, Bettadpur S, Chambers D, Cheng M, Condi F, Gunter B, Kang A, Nagel P, Pastor R, Pekker T, Poole S, Wang F (2005) GGM02C – an improved Earth gravity field model from GRACE. J Geod 79. doi:10.1007/s00190-005-0480-z

Wessel P, Smith WHF (1998) New, improved version of generic mapping tools released. EOS Trans Am Geophys Union 79(47):579

Creation of Vertical Reference Surfaces at Sea Using Altimetry and GPS

33

L. Pineau-Guillou and L. Dorst

Abstract

The creation of vertical reference surfaces at sea, related to a reference ellipsoid, is a necessary step to enable the use of GPS (Global Positioning System) for referencing depth measurements at sea. Several projects exist for specific parts of the oceans, resulting in surfaces that partly overlap. As an example, we will present the French BATHYELLI project in detail, followed by a comparison of results for the North Sea area.

Keywords

Vertical references • Tide • Hydrography • Altimetry • GPS • Mean sea level • Lowest astronomical tide • Chart datum

1 Introduction

The BATHYELLI project (for BATHYmetry referred to the ELLIpsoid) is a set of models of reference surfaces at sea around the French coasts, and a computer program allowing for the transformation of data between one reference and another. The vertical reference surfaces included in this project are: Mean Sea Level (MSL) on the period of tide gauge measurements and on the period of spatial altimetry, Lowest Astronomical Tide level (LAT), Chart Datum (CD), ellipsoid, Land Datum (IGN69 in France) and geoid. In the BATHYELLI project, the surfaces MSL, LAT and CD have been computed with respect to the GRS80 ellipsoid.

The BATHYELLI project results will allow for Ellipsoidally Referenced Hydrographic Surveys (ERS) with GPS around the French coasts, avoiding tidal and meteorological corrections. It will also allow for the merging of land and sea, moving from a marine to a terrestrial reference. And, finally, it will allow for a comparison with similar surfaces realised by other countries.

L. Pineau-Guillou (✉)
French Research Institute for Exploration of the Sea, Technopôle de Brest Iroise, BP70, 29280 Plouzané, France

Former at French Hydrographic and Oceanographic Service, 13 rue du Chatellier, 29200 Brest, France
e-mail: Lucia.Pineau.Guillou@ifremer.fr

L. Dorst
Hydrographic Service of the Royal Netherlands Navy, Van der Burchlaan 31, The Hague 2597 PC, Netherlands

2 Technical Developments

2.1 Methodology

The methodology is based on Mean Sea Level computation; other surfaces (LAT, CD) are easily deduced by tidal modelling and the defined relations between LAT and CD.

The Mean Sea Level is obtained using several techniques (see Fig. 33.1):

- Spatial altimetry provides the Mean Sea Level far off the coasts. Close to the coasts, the Mean Sea Surface is not well calculated due to technical limitations. Because of excessive errors, these results cannot be used at less than 10 miles offshore.
- Tidal observations analysis provides Mean Sea Level on coastal sites. SHOM is responsible for the French Sea Level Observation Network, which includes 30 permanent tidal gauges along French metropolitan coasts, in addition to temporary tidal gauges.

Z. Altamimi and X. Collilieux (eds.), *Reference Frames for Applications in Geosciences*, International Association of Geodesy Symposia 138, DOI 10.1007/978-3-642-32998-2_33,

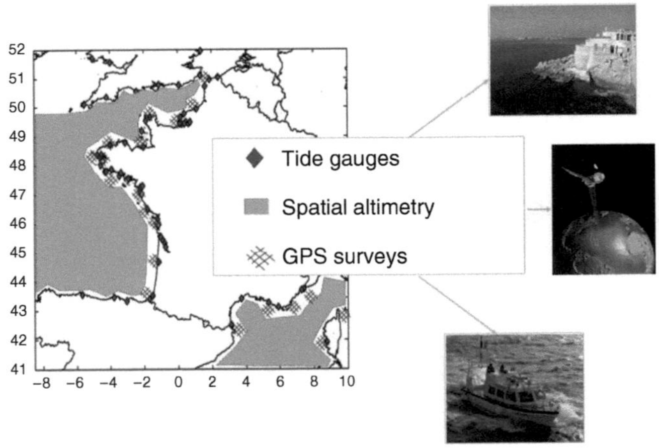

Fig. 33.1 Methodology to compute mean sea level: interpolation of tide gauges, spatial altimetry and GPS data

- To fill the gap between altimetry and tide gauge data, SHOM has planned surveys to measure the Mean Sea Level with reference to the ellipsoid using kinematic GPS.

The data sets resulting from altimetry, GPS surveys and tide gauges have been interpolated in a joint study carried out by SHOM and a company called Noveltis, for computing Mean Sea Level with a high associated precision, calculated explicitly. The data sets have been merged by a least squares method with covariance functions, see Jan (2009).

2.2 Spatial Altimetry

The currently available Mean Sea Surface is not fully satisfying for hydrographers, because the global tidal model used is not precise enough, and the water levels are corrected for the inverse barometer effect. Therefore, an alternative surface has been computed, called Hydrographic Mean Sea Surface.

The differences between Hydrographic Mean Sea Surface and usual Mean Sea Surface are:

- The tidal model used is the SHOM tidal model, instead of a global model, showing differences up to 25 cm,
- The Hydrographic Mean Sea Surface does not coincide with the geoid,
- Water levels are not corrected for the inverse barometer effect, for consistency with tide gauge processing; indeed, meteorological cyclic effects are a part of the tides, called radiational tide (contributing to harmonic constituents like Sa, Ssa or S_1).

The Hydrographic Mean Sea Surface, and associated precision, were computed in 2007 in a joint study carried out by SHOM, the Toulouse-based company CLS, and La Rochelle University, see Lefèvre et al. (2007). Altimetry data were processed between 1992 and 2005, from Topex/Poseidon, ERS1, ERS2 and GFO. The Hydrographic Mean Sea Surface is shown in Fig. 33.2.

2.3 GPS Surveys

A GPS survey campaign has been run by SHOM in about 20 sites along the French coasts and is now complete. The surveys were planned between 2006 and 2008. The survey locations are shown in Fig. 33.3.

For each site:

- A precise determination of Chart Datum at the tide gauge (fixed mark nearby the tide gauge) is made using a long GPS acquisition (at least 24 h, 48 h recommended); this operation is very important to ensure a precise MSL referred to the ellipsoid at the tide gauge;
- A GPS reference station is installed as close as possible to the tide gauge;
- Another GPS station is installed on board of a SHOM ship or launch (see Fig. 33.4), followed by a calibration procedure that allows for a precise computation of antenna height to take dynamic draught into account;
- The surveys are conducted using PPK differential GPS (centimetre accuracy); the survey generally extends about 15 nautical miles around the tide gauge and takes about 3 days; an example of a GPS survey is shown on Fig. 33.5;
- GPS data and attitude data are registered during the surveys.

The survey procedure is described in more detail in Pineau-Guillou and Dupont (2007).

The GPS data are post-processed, filtered, corrected for ship motions (pitch, roll and yaw), antenna height, surges, and the tides in order to compute the Mean Sea Surface related to the ellipsoid. An example of the processed GPS data is presented in Fig. 33.6.

The accuracy of each survey is evaluated:

- The "consistency" of the survey is the difference between the MSL at tide gauge and the MSL determined by GPS nearby the tide gauge. This difference must be at the

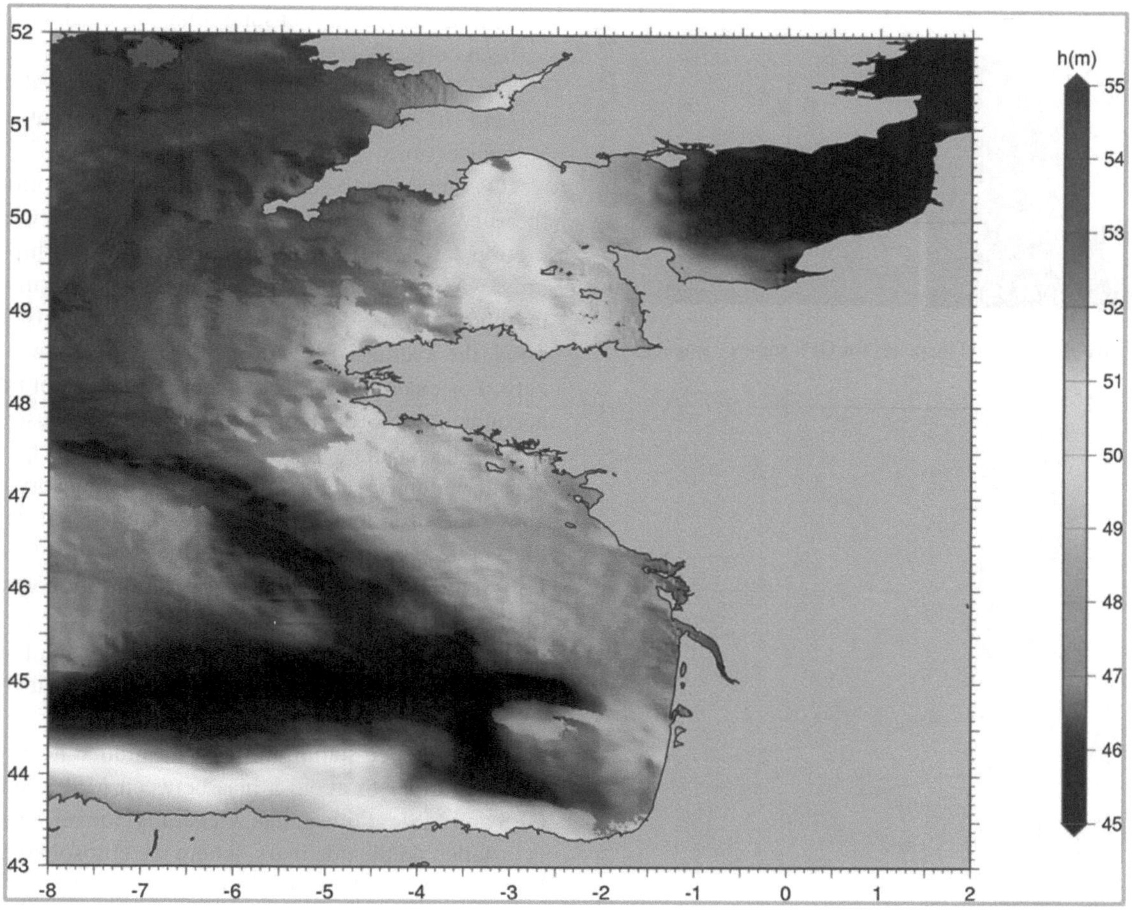

Fig. 33.2 Hydrographic mean sea surface with respect to the GRS80 ellipsoid

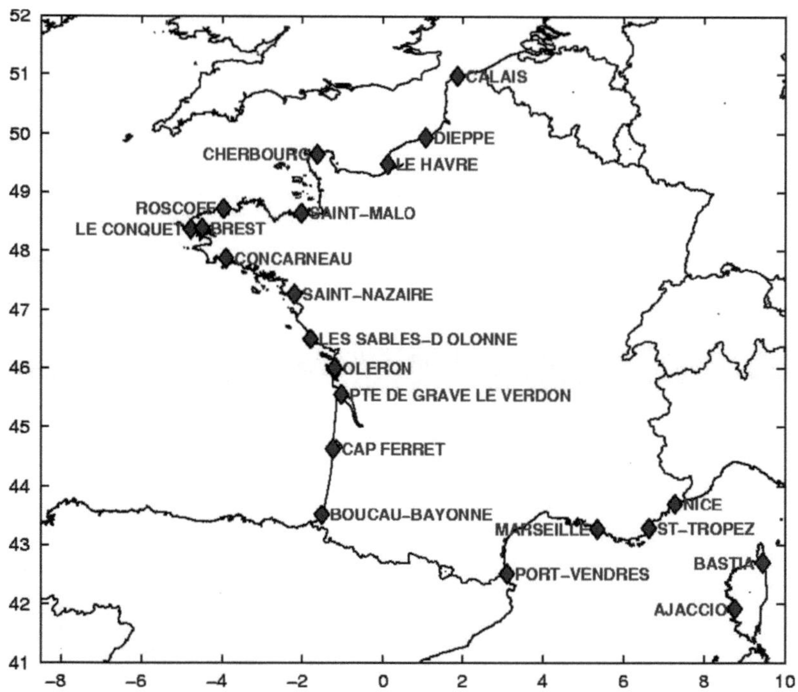

Fig. 33.3 2006–2008 GPS survey campaign locations

Fig. 33.4 SHOM ships and launches for GPS survey campaign

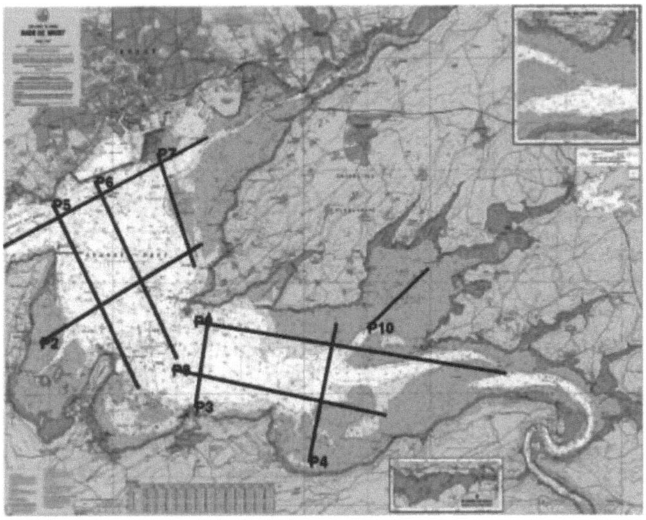

Fig. 33.5 GPS survey around Brest

centimeter level, it ensures that the antenna height has been correctly computed;

- The "precision" of the survey is the averaged difference (and its associated standard deviation) at the track intersections.

It is important to correct the data for ship motions during the survey. Attitude data are systematically recorded by a motion sensor. GPS data are corrected using ship motions, only when it is more appropriate than filtering (default method). Even if meteorological conditions are very good, the average roll and pitch of a vessel are not always zero. For example, they can reach 3 or 4 °. Not taking into account attitude data in such a case can cause a 15 cm bias in the results.

The antenna height must be calculated very precisely above the instantaneous sea level. This height depends on the vessel speed, as it modifies the draught of the vessel. Different tests show a draught variation of about 1 cm per knot, even at low speeds.

The survey must be conducted at a constant speed and the antenna height must be determined at this very speed.

The reference station must be positioned nearby the tide gauge in order to avoid a bias between MSL computed by GPS and MSL at tide gauge. Generally, the reference station is installed on the tide mark, where the ellipsoidal height has been precisely determined in ITRS.

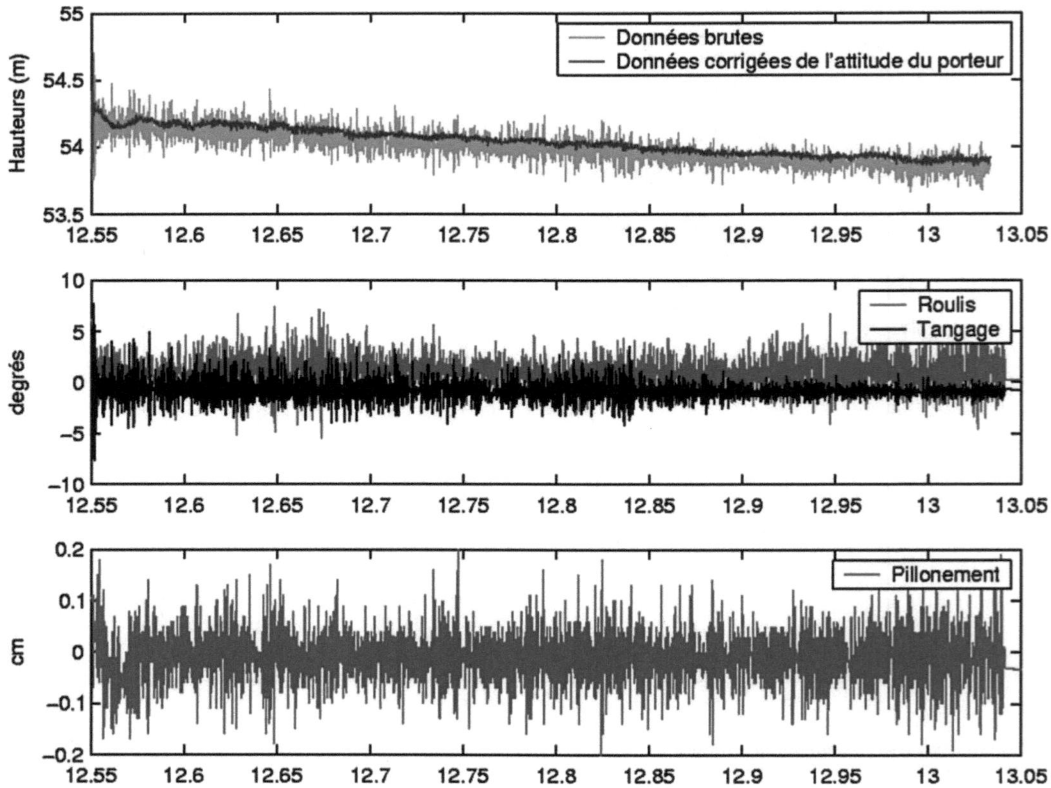

Fig. 33.6 Example of processed GPS data

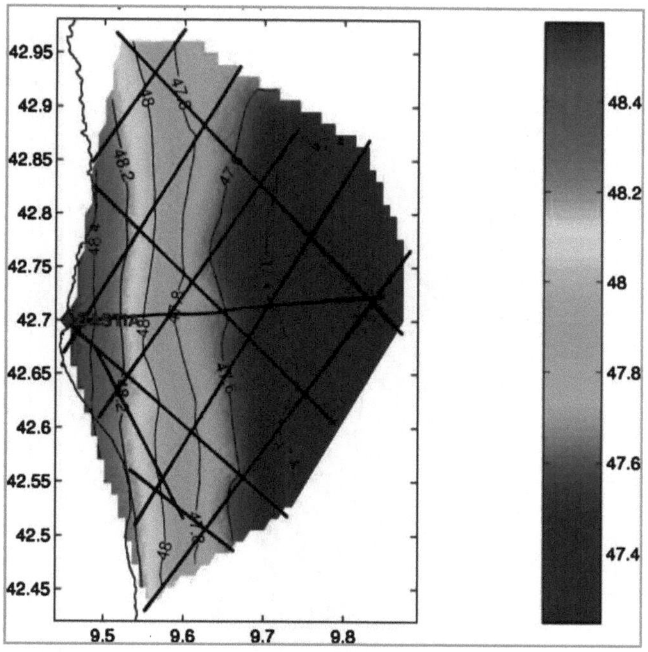

Fig. 33.7 Mean sea level related to ellipsoid around Bastia, France

3 Results

3.1 GPS Surveys

An example of Mean Sea Level surface from GPS around Bastia is shown in Fig. 33.7. MSL varies from 47.3 to 48.5 m. The survey "precision" is 3 cm, with a standard deviation of 2 cm, which is really satisfying. The survey "consistency" is better than 1 cm, which ensures that antenna height was correctly computed.

The MSL-variation is larger than 1 m on 15 nautical miles, and this variation is nonlinear. This confirms that GPS surveys will really help the interpolation between tide gauges and altimetry.

The values of precision, consistency, and survey period for all the surveys are shown in Table 33.1.

3.2 Vertical Reference Surfaces

The data have been interpolated on a finite element grid, see Jan (2009). The LAT surface, which is seamless, is shown in Fig. 33.8. CD, which is not seamless, and its associated error are shown in Figs. 33.9 and 33.10.

3.3 Comparison with Results in the North Sea Area

The Tidal Working Group of the IHO (International Hydrographic Organization) North Sea Hydrographic Committee

Table 33.1 Survey locations, precision (cm), consistency (cm) and period

Location	Precision mean/Std (cm)	Consistency (cm)	Period
Bastia	3.0/2.4	<1	May 16–18, 2007
Le Havre	8.4/6.8	3, 7	May 21–23, 2007
Saint-Malo	7.6/6.6	<1	July 17–18, 2007
Marseille	4.7/3.7	1	July 23–25, 2007
Brest	2.5/1.8	3	July 9–11, 2007
Sables d'Olonne	6.2/3.7	9	Sep 20–23, 2007
Saint-Nazaire	6.8/2.9	2	Sep 18–19, 2007
Oléron	6.8/4.7	1.5	June 6–8, 2007
Port-Vendres	8.0/4.6	3	Oct 25–27, 2007
Concarneau	6.3/3.5	11 (for 5 Nm)	Sep 17–19, 2007
Bayonne	3.0/1.8	Not computed	July 13–17, 2006
Cap Ferret	5.6/4.9	Not computed	Feb 23–24, 2008
Pointe de Grave 2007	5.0/0.4	7.4	Sep 27–28, 2007
Pointe de Grave 2008	11.5/9.6	Not computed	Feb 15–21, 2008
Dieppe	4.3/3.1	<1	May 25–26, 2007
Roscoff	4.0/4.0	1	Nov 09–11, 2007
Cherbourg	8.9/5.2	6, 3	Nov 03–08, 2007
Saint-Tropez	5.5/4.2	<1	Oct 08–11, 2007
Le Conquet	6.2/4.1	2	Nov 11–13, 2007; June 30, 2008; July 1, 2008
Mean	6.6/4.6	2.9	

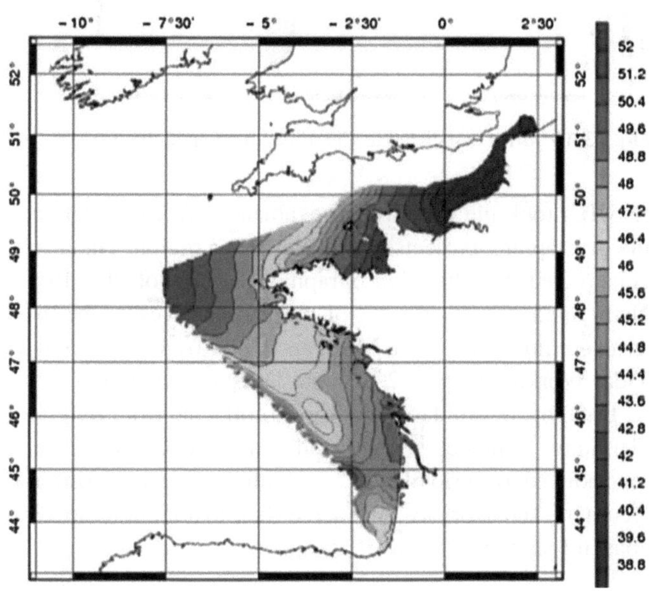

Fig. 33.8 LAT referred to ellipsoid (m)

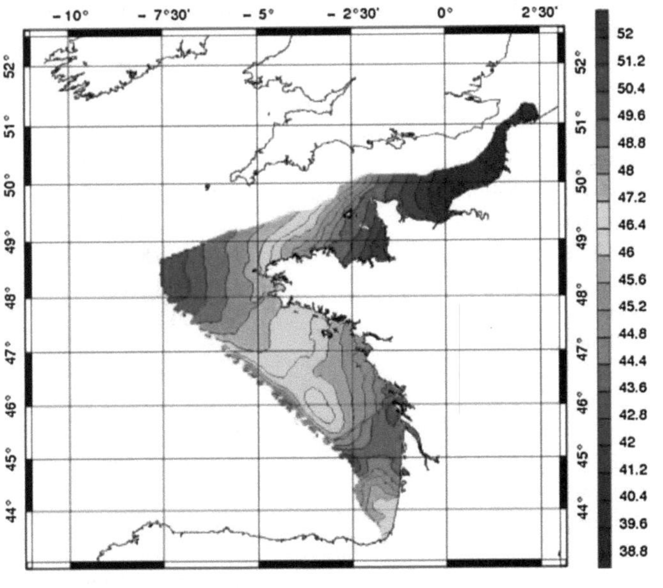

Fig. 33.9 Chart datum referred to ellipsoid (m)

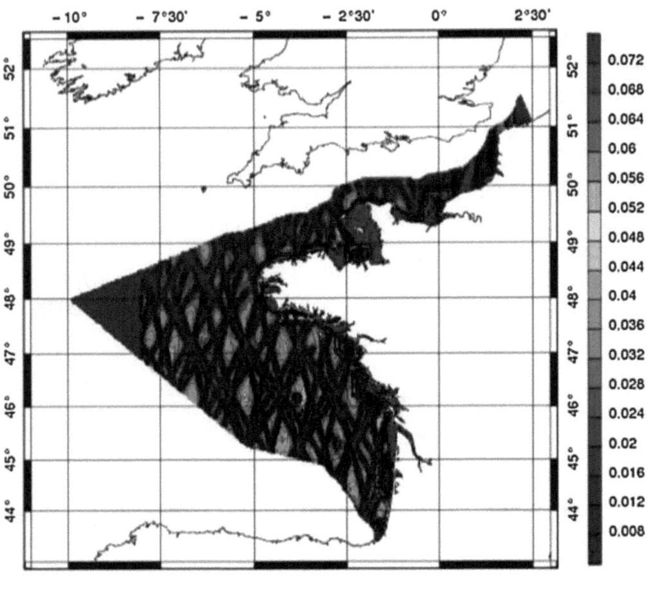

Fig. 33.10 Error associated to Chart datum computation (m)

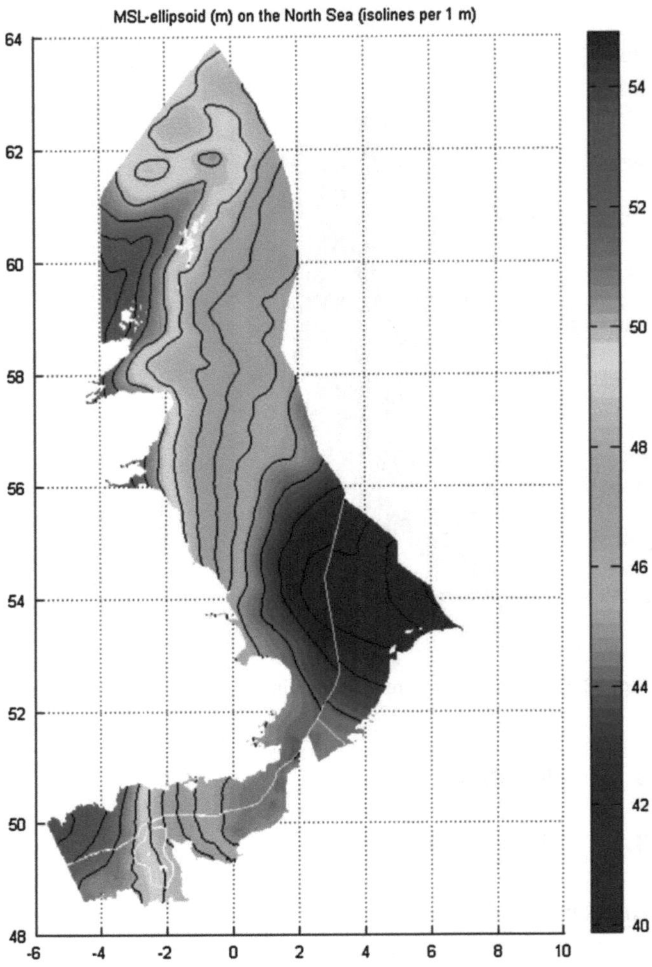

Fig. 33.11 MSL surface in relation to the ellipsoid, including one metre isolines in *black* and maritime boundaries in *white*

(NSHC) took the initiative to evaluate differences between the overlapping projects for the North Sea region. The result, constructed by the Hydrographic Service of the Royal Netherlands Navy, is a set of merged surfaces, to be used as a common LAT-level, MSL, and Chart Datum, see NSHC Tidal Working Group (2010) and Dorst et al. (2010).

Data sets referred to the ellipsoid collected from Belgium, Germany, France, Netherlands and UK have been interpolated to a grid of 0.02 ° for both Easting and Northing. Denmark and Norway only have data available at specific coastal locations.

MSL, LAT and Chart Datum referred to the ellipsoid are shown in Figs. 33.11, 33.12, and 33.13.

At maritime boundaries, differences for MSL and LAT are less than 0.6; for Chart Datum, differences are less than 0.8 m.

4 Conclusion

As a result of the BATHYELLI project, reference surfaces have become available since 2009, for studies only, not yet for "operational" purpose (Ellipsoidally Referenced Hydrographic Surveys).

The next step is the improvement and validation of these reference surfaces (addition of seven GPS surveys, validation by a comparison of a classical survey and a GPS survey in different areas). The next version should be an "operational" product.

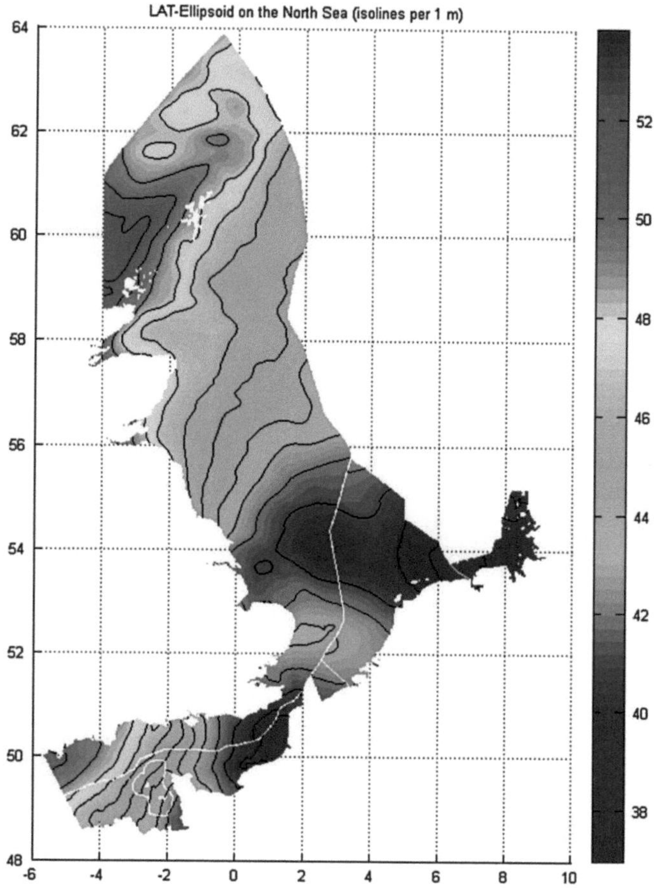

Fig. 33.12 LAT surface in relation to the ellipsoid, including 1 m isolines in *black* and maritime boundaries in *white*

A computer program that allows users to change easily from one vertical reference to another, is planned.

At the end of the project, SHOM plans to realize Ellipsoidally Referenced Hydrographic Surveys.

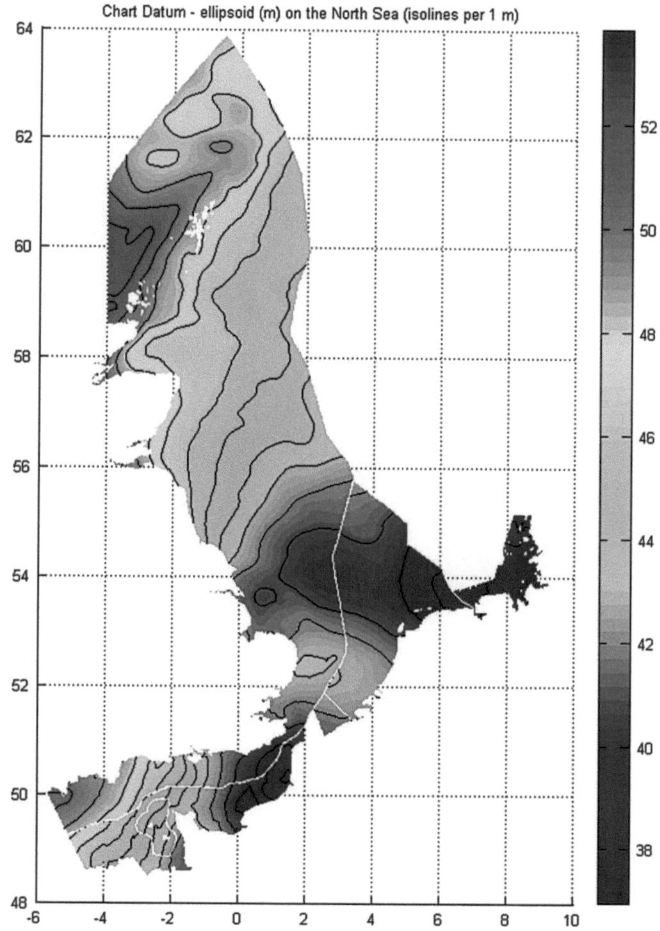

Fig. 33.13 Chart datum surface in relation to the ellipsoid, including 1 m isolines in *black* and maritime boundaries in *white*

References

Dorst L, Slobbe C, Klees R, Verlaan M, Ligteringen T (2010) Unifying vertical reference surfaces in the North Sea: an overview of developments. In: Proceedings of hydro 2010

Jan G (2009) Détermination des surfaces de reference pour l'hydrographie. Noveltis report for SHOM

Lefèvre F, Schaeffer P, Wöppelman G (2007) Etude et fourniture de données nécessaires au calcul d'une surface moyenne océanique issue de l'altimétrie spatiale – Rapport final, CLS report for SHOM

NSHC Tidal Working Group (2010) Merging and comparison of reference surfaces for the North Sea area, Annex D to the report of the 17th meeting of the NSHC-TWG to the 29th NSHC conference. Available at http://www.iho.int

Pineau-Guillou L, Dupont Y (2007) Guide technique pour la réalisation des levés BATHYELLI. SHOM report GU2007-020

Combined Adjustment of GRACE and Geodetic Observations of Vertical Crustal Motion in the Great Lakes Region

34

E. Rangelova and M.G. Sideris

Abstract

We combine GRACE-derived rates of vertical crustal motion, joint water gauge and satellite altimetry and GPS vertical velocities in the Great Lakes region. The combined vertical motion model is realized via a least-squares adjustment procedure, including variance-component estimation and robust outlier detection. This is necessary to ensure reliable estimates of the relative errors in the least-squares adjustment (via re-scaling of data variance-covariance matrices) and to ensure that the vertical motion model is not distorted by erroneous data. The combined vertical motion model shows a subsidence of 1–2 mm/year along the southern shores and an uplift of 3–4 mm/year along the northern shores generally consistent with the models of postglacial rebound in North America.

Keywords

Vertical crustal motion • Postglacial rebound • GRACE • Great lakes • Robust least-squares

1 Introduction

In North America, a new height reference system based on the most accurate continental geoid model and a model of the temporal geoid variations is currently being developed (Véronneau and Héroux 2007; NGS 2008). This new height datum will replace the traditionally defined and realized height datums in Canada (CGVD28) and the US (NAVD88) and will define the vertical component of an accurate geocentric spatial reference system for North America. In addition to the modernization of the continental datum, a local height datum with a centimetre-level of accuracy in the Great Lakes (GL) is currently under study.

This unprecedented one-centimetre accuracy of the new height datum requires that the time variable component of the datum be taken into account. Rangelova et al. (2009a) showed that the geoid change should be accounted for on a decadal time scale while the vertical crustal motion should be corrected for every second year. The secular temporal geoid changes and crustal motion in North America are due to the ongoing viscoelastic deformation of the Earth's crust and interior masses. This so-called postglacial rebound (PGR) change in the geoid height peaks at $1.3-1.5 \pm 0.1$ mm/year in Hudson Bay. An order of magnitude larger vertical uplift is observed. Being at the Laurentide ice sheet margin, the GL area has a geoid change of only few tenths of a mm/year while the crustal motion varies from few mm uplift in the northeast to 1–2 mm subsidence in the south, with the line of zero motion positioned through the lakes.

The PGR crustal motion can be described by either a model or an empirical velocity surface. Braun et al. (2008) provide a comprehensive assessment of the performance of a large number of PGR models in the GL area. The abundance of local geodetic measurements of crustal motion (GPS, water tide gauge and joint tide gauge/altimetry data) allows attempts to combine these heterogeneous data. The GRACE

E. Rangelova (✉) • M.G. Sideris
Department of Geomatics Engineering, Schulich School of Engineering, University of Calgary, 2500 University Drive NW, Calgary, AB T2N 1N4, Canada
e-mail: evrangel@ucalgary.ca

Z. Altamimi and X. Collilieux (eds.), *Reference Frames for Applications in Geosciences*, International Association of Geodesy Symposia 138, DOI 10.1007/978-3-642-32998-2_34,

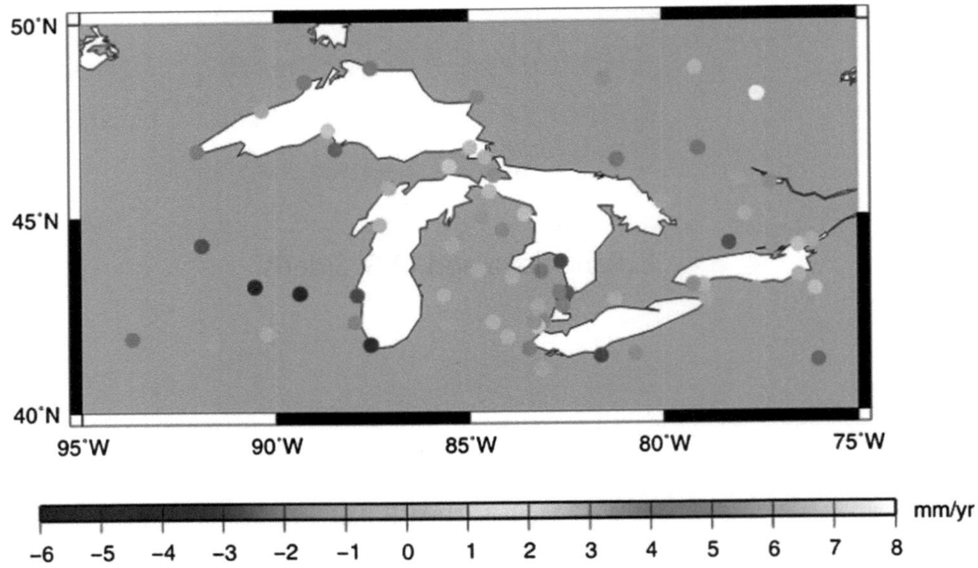

Fig. 34.1 Seventy-one GPS (IGb00) vertical velocities from the North American velocity data by Sella et al. (2007)

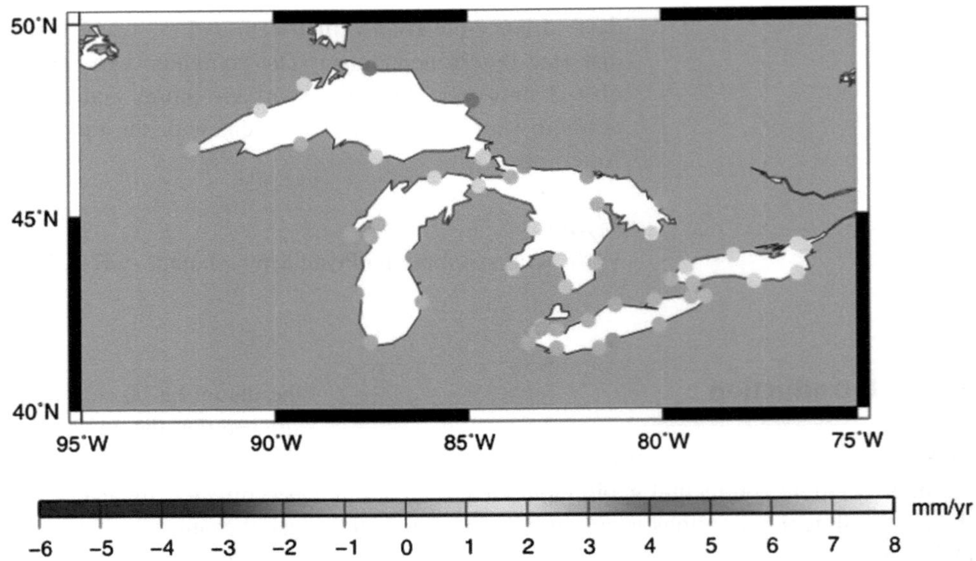

Fig. 34.2 Fifty-one joint tide gauge and TOPEX/Poseidon (TGA) absolute vertical motion estimates (Kuo et al. 2008)

mission provides a different kind of geodetic constraints, i. e., surfaces of vertical crustal motion derived from observed mass anomalies. Accurate empirical models of the crustal motion and reliable data errors are needed to define the line of zero motion as a constraint for the PGR models that can be used to maintain the height datum. Therefore, the objectives of this paper are to (1) derive a combined model of the vertical motion in GL using GPS and joint tide gauge/altimetry vertical velocities and GRACE-derived rates, (2) test methods for dealing with erroneous vertical motion observations, and (3) calibrate the variance-covariance matrices of the data sets.

2 Data

2.1 GPS and Joint Tide Gauge/Altimetry Data

Seventy-one vertical GPS velocities in the GL area (Fig. 34.1) are extracted from the published crustal velocities (with respect to IGb00) in North America by Sella et al. (2007). Generally, the GPS velocities show a large northeast/southwest slope and sharply decrease from 7 to 8 mm/year in the northeast to zero in the lakes area.

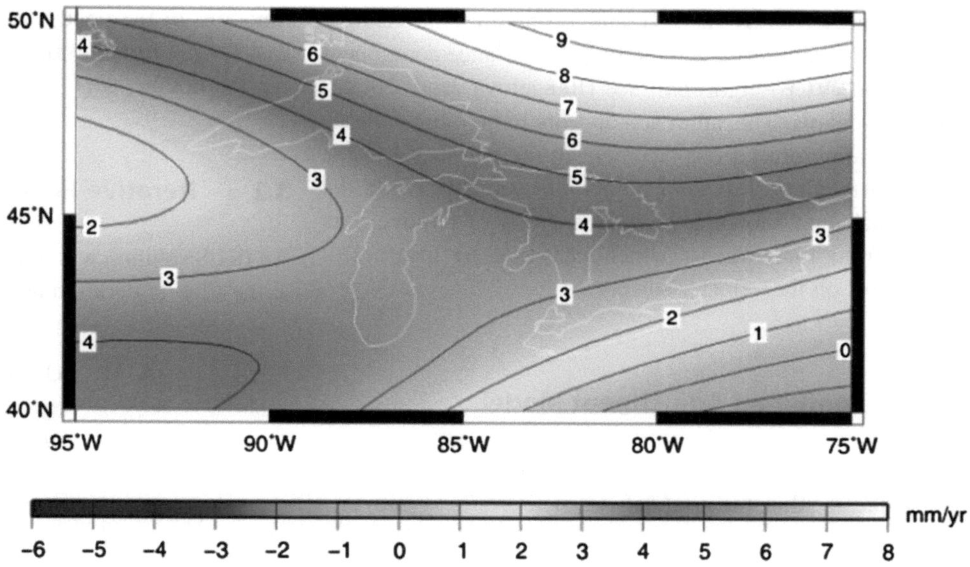

Fig. 34.3 CSR RL-04 GRACE-derived vertical motion rates for the Great Lakes

Figure 34.2 depicts the absolute vertical motion obtained from the optimally combined tide gauge and TOPEX/Poseidon altimetry (TGA) water heights by Kuo et al. (2008). The TGA data repeat the general pattern of the GPS velocities, but the zero velocities are observed farther south than the GPS ones. For both data sets, only diagonal variance-covariance matrices are available. Only eight GPS points are collocated with water tide gauges, but these are not used as constraints in the combined least-squares adjustment herein because of the different velocities estimated from the initial least-squares adjustments of the two data sets.

2.2 GRACE-Derived Uplift Rates

The GRACE data (Tapley et al. 2004) comprise 89 CSR-produced RL-04 monthly gravity solutions from January 2003 to June 2010 with the static gravity model GGM03S (epoch 2005.0) subtracted. The C_{20} coefficient is replaced with the SLR-derived monthly values. The GRACE solutions are corrected for the hydrology signal by means of the spherical harmonic coefficients derived from the GLDAS/Noah monthly water mass anomalies (Rodell et al. 2004) referenced to 2005.0 for consistency with GRACE. The standard de-striping (Swenson and Wahr 2006) is applied on the hydrology-corrected GRACE coefficients. Vertical rates of maximum spherical degree 50 are calculated by the approximate formula of Wahr et al. (2000) and are given in Fig. 34.3 on a 1-degree grid for the GL area. This transformation formula was derived using the fact that the PGR geoid change is mainly a result of the mass anomalies associated with the deformation of the lithosphere, which results in vertical crustal displacements. An isotropic Gaussian filter with a 300-km radius is applied to smooth any random and residual de-striping errors. Finally, the GRACE-derived rates are calculated at the 71 GPS station locations.

A common adjustment of the GRACE, GPS and TGA data is challenging due to a variety of factors. One factor is the (still) insufficient accuracy of the GRACE data partly due to the relatively short time span to deliver reliable rates of a secular signal. Another error source is the hydrology correction applied to the GRACE data. Imperfections in the GLDAS/Noah model likely corrupt the GRACE-derived rates. A PC/EOF analysis of the hydrology model data for the entire North American continent shows trend-like changes over the time period of the GRACE series with an apparent acceleration in 2009 and the first half of 2010. This trend signal leaks from the adjacent areas over the GL and through the hydrology correction contributes approximately 2 mm/year to the GRACE-derived rates in the GL area where the line of zero motion is placed by the GPS observations. Changes in the vertical motion pattern are observed if the time frame of the GRACE data is varied or the parameters of the de-correlation and Gaussian filters are changed. Other potential error contributions come from the different reference epochs of the geodetic data sets, differences from the degree one harmonic not present in the GRACE gravity solutions, the variable length of the time series (the tide gauge data time series are in most cases longer than the GPS time series), local tide gauge or monument instabilities (assumed to be largely filtered out in the original data adjustments), and unknown scale factors of the data variance-covariance matrices.

3 Methodology

An iterative procedure is designed which consists of identifying outliers and minimizing their effect on the combined velocity surface by means of the iterative re-weighting least-squares (IRLS) method and estimating the relative errors of the data sets using the BIQUE variance component estimation (VCE) algorithm; see (Rangelova et al. 2009b) for more details. This approach appears to be more suitable for estimating crustal displacement in the peripheral to the lakes areas where less data points are available while the traditional data snooping technique tends to remove data points from these areas, which can lead to insufficient constraints on the velocity surface.

3.1 Combined Adjustment Model

The GRACE, GPS and TGA data sets are combined via the least-squares adjustment model

$$\min \|\mathbf{l} - \mathbf{A}\mathbf{x}\|^2, \quad \mathbf{C}_l \tag{34.1}$$

where the observation vector $\mathbf{l} = \mathbf{A}\mathbf{x} + \mathbf{v}$ contains the observations $\mathbf{l} = \begin{bmatrix} \mathbf{l}_{GRACE}^T & \mathbf{l}_{GPS}^T & \mathbf{l}_{TGA}^T \end{bmatrix}^T$ with sub-vectors containing the vertical velocities in each of the data sets. The stochastic model is given by the block-diagonal variance-covariance matrix

$$\mathbf{C}_l = \begin{bmatrix} (\sigma^2\mathbf{Q})_{GRACE} & 0 & 0 \\ 0 & (\sigma^2\mathbf{Q})_{GPS} & 0 \\ 0 & 0 & (\sigma^2\mathbf{Q})_{TGA} \end{bmatrix} \tag{34.2}$$

with a scale factor σ^2 and a co-factor matrix $\mathbf{Q}$ (diagonal or fully-populated) for each data set. The calibrated scale factors are estimated by the VCE method.

The parametric model contains the coefficient matrix

$$\mathbf{A} = \begin{bmatrix} \mathbf{A}_0 & 0 & \mathbf{A}_{GRACE} \\ 0 & 0 & \mathbf{A}_{GPS} \\ 0 & 1 & \mathbf{A}_{TGA} \end{bmatrix} \tag{34.3}$$

where $\mathbf{A}_0$ defines the plane of the location-dependent offset of the GRACE-derived rates with respect to the GPS data while the TGA data set has a constant off-set. The coefficient matrices $\mathbf{A}_{GRACE}$, $\mathbf{A}_{GPS}$ and $\mathbf{A}_{TGA}$ are formed by means of the base function $\Phi(\rho)$

$$\mathbf{A}_k = [\Phi(\rho)], k = GRACE, GPS, TGA \tag{34.4}$$

where ρ is the distance between a grid node (from a grid that covers the entire area) where the base function is placed and the observation point. The velocity field is parameterized by the inverse multiquadrics $\Phi(\rho) = (\rho^2 + c^2)^{-1/2}$. In addition to providing an excellent approximation by adapting the shape parameter c^2 to the data set(s), the inverse multiquadrics have a particular physical meaning. They are base functions of the point mass method, which was used in the past for interpreting mass redistributions related to vertical crustal motion.

3.2 Iterative Re-Weighting Least-Squares

An IRLS solution with uncorrelated observations is obtained at the $(k + 1)$ iteration as (Hekimoğlu and Berber 2003)

$$\hat{\mathbf{x}}^{(k+1)} = \left(\mathbf{A}^T\overline{\mathbf{W}}^{(k)}\mathbf{A}\right)^{-1}\mathbf{A}^T\overline{\mathbf{W}}^{(k)}\mathbf{l}, \; \hat{\mathbf{v}}^{(k)} = \mathbf{l} - \mathbf{A}\hat{\mathbf{x}}^{(k)} \tag{34.5}$$

$$\overline{\mathbf{W}}^{(k)} = \mathbf{C}_l^{-1}\mathbf{W}^{(k)} \tag{34.6a}$$

$$\mathbf{W}^{(k)} = diag(w_1^{(k)}, \ldots, w_i^{(k)}, \ldots, w_n^{(k)}) \tag{34.6b}$$

$$\mathbf{C}_l = diag(\sigma_1^2, \ldots, \sigma_i^2, \ldots, \sigma_n^2) \tag{34.6c}$$

where $\mathbf{W}^{(k)}$is the weight matrix at the kth iteration and $\overline{\mathbf{W}}^{(k)}$is the so-called equivalent weight matrix (Yang 1994). The weight $w_i^{(k)}$for each observation i is computed as follows:

$$w_i^{(k)} = \Psi\left(\hat{\bar{v}}_i^{(k)}\right)/\hat{\bar{v}}_i^{(k)}, \tag{34.7}$$

where $\hat{\bar{v}}_i^{(k)}$is the standardized residual and, in our case, Ψ is the Fair influence function (Dollinger and Staudte 1991):

$$\Psi\left(\hat{\bar{v}}_i^{(k)}\right) = \hat{\bar{v}}_i^{(k)}/\left(1 + \left|\hat{\bar{v}}_i^{(k)}\right|/F\right) \tag{34.8}$$

with the parameter $F = 1.4$. The standardized residual is computed as follows:

$$\hat{\bar{v}}_i^{(k)} = \hat{v}_i^{(k)}/s \tag{34.9}$$

where s is the median absolute deviation of the residuals about the median, known as the MAD estimator (Rousseeuw and Croux 1993):

$$s = med\{|\hat{\mathbf{v}} - med\{\hat{\mathbf{v}}\}|\}. \tag{34.10}$$

4 Analysis of Results

We studied the optimal combination of the three data sets using a diagonal or a fully-populated variance-covariance (VC) matrix of the GRACE-derived vertical rates. The fully-populated VC matrix was computed by propagating the variances of the least-squares estimated trend in the GRACE spherical harmonic coefficients, thus we only accounted for the spatial scale of the GRACE errors. The standard errors of the monthly GRACE coefficients were computed from the distributed calibrated standard errors of the monthly solutions and GGM03S. Before applying the IRLS algorithm, the VC matrix was diagonalized via eigenvalue decompostion. This is necessary because the algorithm is available for uncorrelated data. The GPS and TGA VC matrices are diagonal as made available from the original studies.

The estimated biases of the TGA data and the GRACE-derived rates are given in Table 34.1 for both cases of the GRACE VC matrix and for the conventional least-squares adjustment (LSA) and IRLS. The estimates vary only slightly for all four combinations. A large bias of more than 3 mm/year is estimated for the GRACE-derived rates. As it was discussed, likely this large number is mostly due to the hydrology correction applied but it also includes the effect of the degree one harmonics not present in the GRACE-derived rates, as well as leakage of geophysical signals over the GL area. The TGA data set is less biased with respect to the GPS data set. The discrepancy of 1 mm/year is a combined effect of the different reference systems, possible systematic errors in the altimetry data, the different time spans and reference epochs of the data sets.

The estimated scale factors of the VC matrices in the LSA case are given in Table 34.2. Both GRACE and GPS VC matrices need to be up-weighted, which is not the case for TGA. The TGA matrix needs to be re-scaled such that the overly optimistic a-priori accuracy is compensated for in the combined least-squares adjustment.

The larger a-posteriori errors of the TGA data set are also seen in the statistics given in Table 34.3. In the case of the diagonal VC matrix of the GRACE rates, the mean a-posteriori TGA error increases more than twice. The increase is three times for LSA and four times for IRLS in the case of the fully-populated GRACE VC matrix with the maximum error reaching 1.1 and 1.9 mm/year, respectively.

The geographical distribution of the TGA and GPS a-priori (plotted in black) and a-posteriori (in red) errors from the IRLS solution is depicted in Figs. 34.4 and 34.5, respectively. An overall decrease in the a-posteriori GPS errors is evident, but also a reduction in the largest errors is observed. The a-posteriori errors of the TGA rates have an apparent geographical pattern with maximum values estimated for Lake Michigan. The vertical motion surface plotted on the two figures only slightly differs from the LSA-derived surface over the lakes. The line of zero motion is well constrained by the data, and less than 1 mm/year displacement is observed between the zero lines of the LSA- and IRLS-derived surfaces. The most noticeable differences are in the peripheral to the lakes areas that are less constrained by the data.

Table 34.1 Estimated biases, in mm/year

Method	GRACE	TGA
Diagonal VC matrix of the GRACE rates		
LSA	**3.5** ± 0.1	**1.0** ± 0.2
IRLS	**3.2** ± 0.1	**0.8** ± 0.1
Fully-populated VC matrix of the GRACE rates		
LSA	**3.3** ± 0.1	**1.0** ± 0.1
IRLS	**3.5** ± 0.1	**1.2** ± 0.1

Table 34.2 Estimated scale factors of the VC matrices in LSA

GRACE	GPS	TGA
Diagonal VC matrix of the GRACE rates		
0.2	0.6	6.7
Fully-populated VC matrix of the GRACE rates		
0.4	0.6	12.5

Table 34.3 Statistics of the a-posteriori errors obtained with the diagonal/fully-populated VC matrix of the GRACE rates, in mm/year

Data set	Min	Max	Mean
A priori *errors*			
GRACE	0.6	0.6	0.6
GPS	0.5	5.3	2.0
TGA	0.1	0.3	0.2
LSA			
GRACE	0.2/0.4	0.2/0.4	0.2/0.4
GPS	0.4/0.4	4.1/4.1	1.5/1.5
TGA	0.3/0.4	0.8/1.1	0.5/0.6
IRLS			
GRACE	0.2/0.6	0.3/0.6	0.2/0.6
GPS	0.2/0.5	5.2/4.6	1.4/1.5
TGA	0.2/0.3	1.1/1.9	0.4/0.8

Finally, using the calibrated data errors, we show by means of a simple example in Fig. 34.6 that the GPS and TGA data can be useful for constraining PGR models. A series of 8 models computed with the ICE-4G de-glaciation history and varied mantle parameters (van der Wal et al. 2004) are used to compute the standard deviation of the model values at the GPS and TGA points. The a-priori errors are plotted in black and the a-posteriori errors from the IRLS are in red. In the case of GPS, the increase in accuracy (smaller a-posteriori errors) leads to increased data resolution, and in approximately 30 % of the data points the a-posteriori error bars are within the PGR error bars. Thus, these points can be useful for constraining the PGR models. In the TGA case, the a-posteriori errors increase but are still within the model error bars and therefore provide better

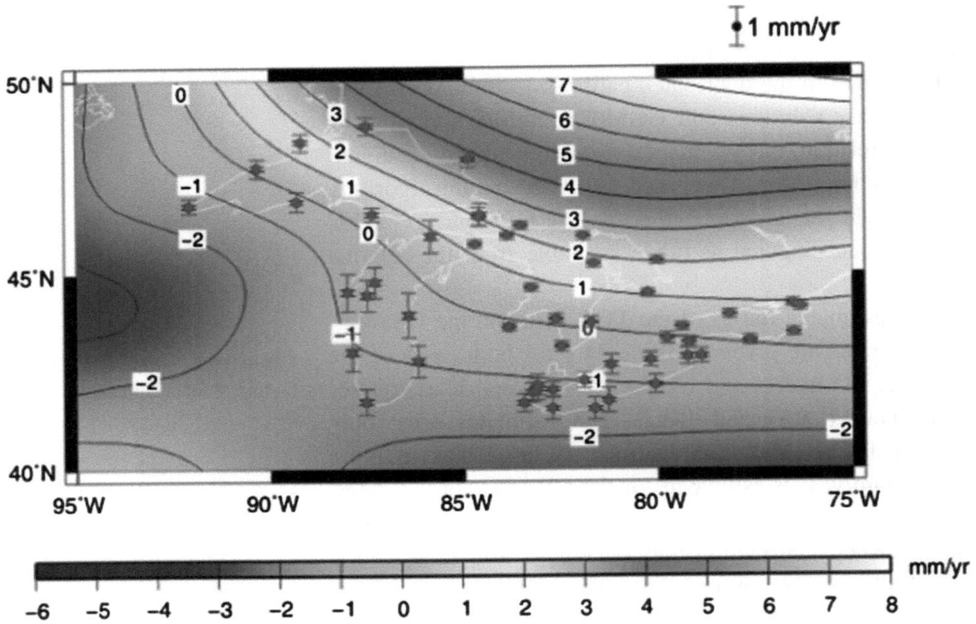

Fig. 34.4 Empirical model of the vertical motion surface with the a-priori (*in black*) and a-posteriori (*in red*) errors of the joint tide gauge and altimetry vertical velocities

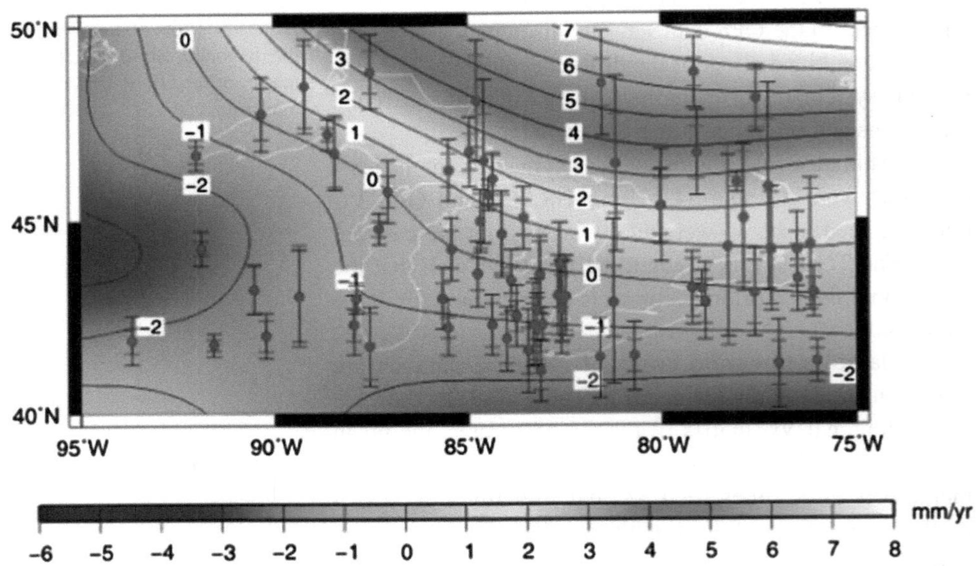

Fig. 34.5 Empirical model of the vertical motion surface with the a-priori (*in black*) and a-posteriori (*in red*) errors of the GPS

resolution compared to the GPS velocities. Of course, the bias of 1 mm/year in the TGA vertical velocity surface with respect to the GPS velocity surface should also be taken into account.

5 Conclusions

We have derived an empirical model of the vertical motion surface in the Great Lakes (Figs. 34.4 and 34.5) by means of an optimal combination of GPS vertical velocities, joint water tide gauge/altimetry absolute vertical motion data and vertical motion rates derived from almost 8 years of GRACE observations of the temporal variations of the global gravity field. The increasing time span of the GRACE mission leads to more accurate secular postglacial rebound rates that converge to the GPS velocities as observed for Fennoscandia by W. van der Wal (personal communication). If this is the case for North America is yet to be studied as the hydrology correction, which may in fact interrupt this convergence, appears to be of more importance for North America than for Fennoscandia.

We demonstrated the use of the iterative re-weighting least-squares method for combining vertical motion heterogeneous data. The advantage is that if outliers are present in the data (see Rangelova et al. 2009b), these data points are

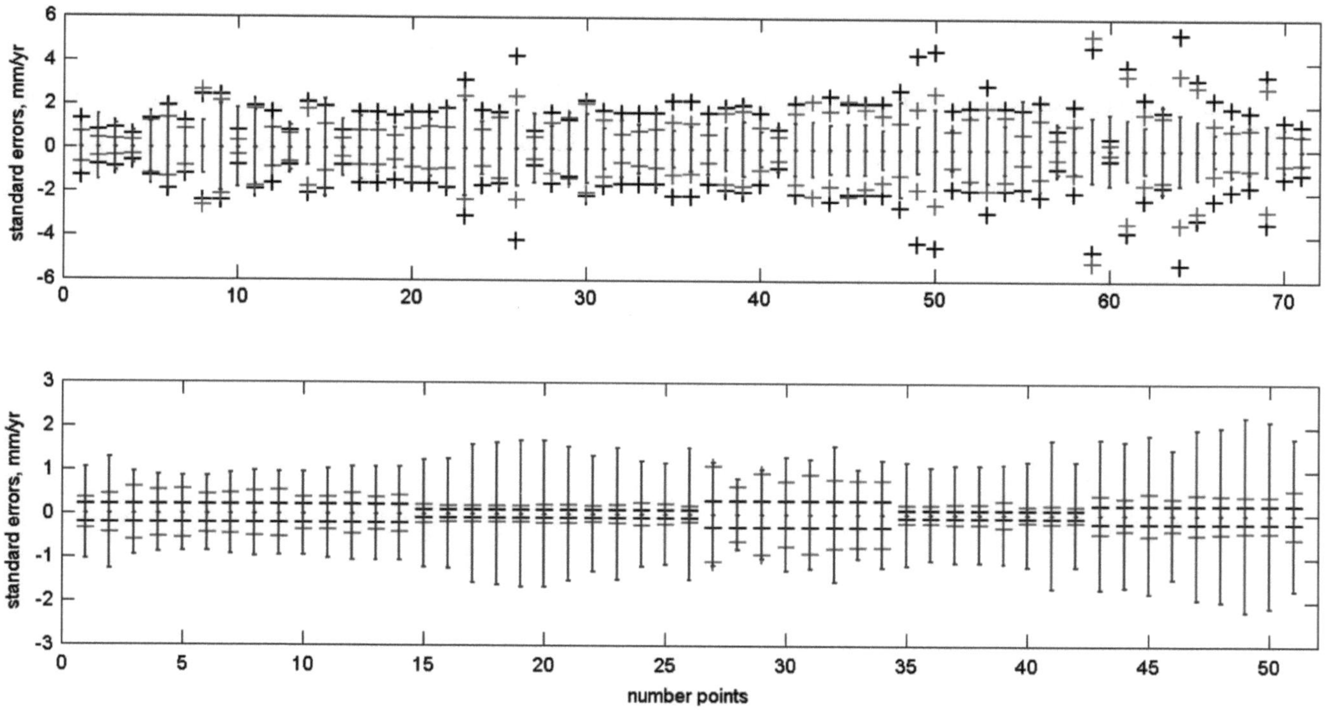

Fig. 34.6 A-priori (*black*) and a-posteriori (*red*) errors compared to PGR model variations (*blue*) at the GPS (*upper plot*) and TGA (*lower plot*) data points. Note the vertical axes are different for the two data sets

down-weighted and preserved in the optimal combination, thus allowing for more data constraints mostly in the peripheral areas, which are weakly constrained by the data. The line of zero motion, particularly over the lakes, is well constrained, and the presence of the data outliers (though detected) is not of importance.

By means of a comparison of the calibrated GPS and joint tide gauge/altimetry data errors and postglacial rebound model variability, we showed that both geodetic data sets can be useful for constraining the mantle profile in the model simulations. Our combined empirical models or the refined postglacial rebound models by means of the same empirical models are of ultimate importance for realizing and maintaining the modern, 1-cm accurate height datum in North America.

Acknowledgements The authors gratefully acknowledge Dr. W. van der Wal for providing the output of the postglacial rebound model simulations. The anonymous reviewers are acknowledged for their helpful comments. Financial support is provided by the GEOIDE NCE and NSERC to the second author.

References

Braun A, Kuo CY, Shum CK, Wu P, van der Wal W, Fotopoulos G (2008) Glacial isostatic adjustment at the Laurentide ice sheet margin: models and observations in the Great Lakes region. J Geodyn 46 (3–5):165–173

Dollinger MB, Staudte RG (1991) Influence functions of iteratively reweighted least squares estimators. J Am Stat Assoc 86 (415):709–716

Hekimoglu S, Berber M (2003) Effectiveness of robust methods in heterogeneous linear models. J Geod 76:706–713

Kuo CY, Shum CK, Braun A, Cheng KC, Yi Y (2008) Vertical motion determined by using satellite altimetry and tide gauges. Terr Atmos Ocean Sci 19(1–2):21–35

NGS (2008) The national geodetic survey ten-year plan: mission, vision, and strategy 2008–2018. Silver Spring, MD, U.S. Department of Commerce, National Oceanic and Atmospheric Administration, National Ocean Service. http://www.ngs.noaa.gov/INFO/NGS10yearplan.pdf

Rangelova E, Fotopoulos G, Sideris MG (2009a) A dynamic reference surface for heights in Canada. Geomatica 63(4):333–340

Rangelova E, Fotopoulos G, Sideris MG (2009b) On the use of iterative re-weighting least-squares and outlier detection for empirically modelling rates of vertical displacement. J Geodesy 83:523–535

Rodell M et al (2004) The global land data assimilation system. Bull Am Meteorol Soc 85(3):381–394

Rousseeuw PJ, Croux C (1993) Alternatives to the median absolute deviation. J Am Stat Assoc 88(424):1273–1283

Sella GF, Stein S, Dixon TH, Craymer M, James TS, Mazzotti S, Dokka RK (2007) Observation of glacial isostatic adjustment in "stable" North America with GPS. Geophys Res Lett 34:L02306. doi:10.1029/2006GL027081

Swenson S, Wahr J (2006) Post-processing removal of correlated errors in GRACE data. Geophys Res Lett 33. doi:10.1029/2005GL025285

Tapley BD, Bettadpur S, Watkins M, Reigber C (2004) The gravity recovery and climate experiment: mission overview and early results. Geophys Res Lett 31:L09607. doi:10.1029/2004GL019920

van der Wal W, Schotman HHA, Vermeersen LLA (2004) Geoid heights due to a crustal low viscosity zone in glacial isostatic

adjustment modeling: a sensitivity analysis. GOCE Geophys Res Lett 31:L05608. doi:10.1029/2003GL019139
Véronneau M, Héroux P (2007) Canadian height reference system modernization: rational, status and plans. Report Natural Resources of Canada
Wahr J, Wingham D, Bentley C (2000) A method of combining ICESat and GRACE satellite data to constrain Antarctic mass balance. J Geophys Res 105(B7):16,279–16,294
Yang Y (1994) Robust estimation for dependent observations. Manuscr Geod 19:10–17

Some Features of TOPEX/POSEIDON Data Application in Gravimetry

35

O. Yu. Vinogradova and E.A. Spiridonov

Abstract

The values of Δg corrections for vertical component of gravity produced by TOPEX/POSEIDON sea level variations were estimated. With this aim, the M. Molodensky boundary problem for deformation of elastic gravitating compressible sphere was solved. An equation connecting boundary conditions on the mantle surface and bottom was obtained. The values of load Love numbers for several models of upper layers of the Earth were calculated. We demonstrate that magnitude of Δg corrections appreciably depend from used models of the upper mantle and lithosphere structure.

Keywords

Load love numbers • Load oceanic effect

1 Equations and Boundary Conditions

In this work we are based on a system of three second-order differential equations described deformation of elastic gravitating compressible sphere. The system was obtained by Mikhail S. Molodensky in 1953 (Molodensky 1953):

$$-\left[\mu\left(T' + H - \frac{2}{r}T\right)\right]' = \rho(R + V'H) + \lambda f + +2\frac{\mu}{r}\left[2H + T' - \frac{n^2+n+1}{r}T\right], \qquad \text{(I)}$$

$$-(\lambda f + 2\mu H')' = \rho(R + V'H)' - \rho V' f + +4\frac{\mu}{r}\left(H' - \frac{H}{r}\right) - \frac{n(n+1)}{r^2}\mu\left(T' + H - 4\frac{T}{r}\right), \qquad \text{(II)}$$

$$R'' = -\frac{2}{r}R' + \frac{n(n+1)}{r^2}R + 4\pi G(\rho f + \rho' H), \qquad \text{(III)}$$

where μ, λ and ρ are Lame parameters and density as functions of r; $f = H' + \frac{2}{r}H - \frac{n(n+1)}{r^2}T$.

The equations enclose three functions of depth: function H for the radial displacement, T for the tangential displacement and R for the change in potential.

It is known six first-order equations can be obtained from (I, II, and III) by introduction the next variables:

$$L = r^2(R' - 4\pi G\rho H),$$
$$M = r^2\mu\left(T' + H - \frac{2}{r}T\right), \qquad (35.1)$$
$$N = (\lambda + 2\mu)H' + \frac{2}{r}\lambda H - \frac{n(n+1)}{r^2}\lambda T.$$

Then from Eq. 35.1 to (I, II, and III) we have:

$$H' = -\frac{2}{r}\frac{\lambda}{\lambda+2\mu}H + \frac{n(n+1)}{r^2}\frac{\lambda}{\lambda+2\mu}T + \frac{1}{\lambda+2\mu}N, \qquad (35.2)$$

O. Yu. Vinogradova (✉) • E.A. Spiridonov
Russian Academy of Science, Institute for Physics of the Earth, 10 Bolshaya Gruzinskaya street, Moscow, Russia
e-mail: sp287@mail.ru

Z. Altamimi and X. Collilieux (eds.), *Reference Frames for Applications in Geosciences*, International Association of Geodesy Symposia 138, DOI 10.1007/978-3-642-32998-2_35,

$$T' = -H + \frac{2}{r}T + \frac{1}{r^2\mu}M, \tag{35.3}$$

$$R' = 4\pi G\rho H + \frac{1}{r^2}L, \tag{35.4}$$

$$L' = -n(n+1)\cdot 4\pi G\rho T + n(n+1)R \tag{35.5}$$

$$\begin{aligned} M' = &-\left[2r\mu\frac{2\mu+3\lambda}{\lambda+2\mu} + \rho r^2 V'\right]H - \rho r^2 R + \\ &+\left[n(n+1)\frac{2\mu\lambda}{\lambda+2\mu} + 2\mu(n^2+n-1)\right]T \\ &-r^2\frac{\lambda}{\lambda+2\mu}N \end{aligned} \tag{35.6}$$

$$\begin{aligned} N' = &\left[\frac{4}{r^2}\mu + \frac{4}{r}\rho V' + \frac{8}{r^2}\frac{\lambda\mu}{\lambda+2\mu}\right]H - \\ &-\left[\rho V' + \frac{2}{r}\mu + \frac{4}{r}\frac{\lambda\mu}{\lambda+2\mu}\right]\frac{n(n+1)}{r^2}T - \\ &-\frac{\rho}{r^2}L + \frac{n(n+1)}{r^4}M - \frac{4}{r}\frac{\mu}{\lambda+2\mu}N \end{aligned} \tag{35.7}$$

If the order $n \geq 2$, boundary conditions on the surface of the sphere of the radius a will be (Pertsev 1976):

$$N = -\frac{2n+1}{3}\rho_m g_0 a, \tag{35.8}$$

$$M = 0, \tag{35.9}$$

$$L = (2n+1)a^2 g_0 - \frac{n+1}{a}R \tag{35.10}$$

where ρ_m is the mean density of the Earth, g_0 is the gravity on its surface.

Three conditions extra are determined on the core-mantle boundary:

$$L - r(n+rv)R + 4\pi G\rho_i r^2 H = 0, \tag{35.11}$$

$$M = 0, \tag{35.12}$$

$$N + \rho_i(R + V'H) = 0. \tag{35.13}$$

Here ρ_i is the core density on the core-mantle boundary, G is the gravity constant.

Function v is calculated from equation:

$$v' + v^2 + \frac{2(n+1)}{r}v + 4\pi G\frac{\rho'}{V'} = 0. \tag{35.14}$$

This equation follows from (III) where the change of variable $R' = R\left(v + \frac{n}{r}\right)$ was made. The integrating is carried out from the Earth's mass center ($v = 0$) up to the outer core surface.

In the process of the load coefficients calculation the following normalizations had been adopted: the mean Earth radius as unit of the length; the mean density of the Earth as unit of the density; the surface gravity g_0 as unit of acceleration.

Besides, a normalization had been realized due to the desired functions grow rapidly with order n (Pertsev 1976):

$$\begin{aligned} &H = (2n+1)\bar{H}/r^n, \quad T = \bar{T}/r^n, \quad R = \bar{R}/r^n, \\ &L = (n+1)\bar{L}/r^n, \quad M = n\bar{M}/r^n, \\ &N = n(2n+1)\bar{N}/r^n. \end{aligned} \tag{35.15}$$

Subject to adopted scales and normalizations (35.15) the boundary problem takes the form:

$$\begin{aligned} \bar{H}' = &\frac{1}{r}\left(n - 2\frac{\lambda}{\lambda+2\mu}\right)\bar{H} + \\ &+\frac{n(n+1)}{2n+1}\frac{1}{r^2}\frac{\lambda}{\lambda+2\mu}\bar{T} + \frac{n}{\lambda+2\mu}\bar{N}, \end{aligned} \tag{35.16}$$

$$\bar{T}' = -(2n+1)\bar{H} + \frac{1}{r}(n+2)\bar{T} + \frac{n}{r^2\mu}\bar{M}, \tag{35.17}$$

$$\bar{R}' = 4\pi G\rho(2n+1)\bar{H} + \frac{n}{r}\bar{R} + \frac{1}{r^2}(n+1)\bar{L}, \tag{35.18}$$

$$\bar{L}' = -n\cdot 4\pi G\rho\bar{T} + n\bar{R} + \frac{n}{r}\bar{L}, \tag{35.19}$$

$$\begin{aligned} \bar{M}' = &-\left(2r\mu\frac{2\mu+3\lambda}{\lambda+2\mu} + \rho r^2 V'\right)\frac{2n+1}{n}\bar{H} + \\ &+\left((n+1)\frac{2\mu\lambda}{\lambda+2\mu} + 2\frac{n^2+n-1}{n}\mu\right)\bar{T} - \\ &-\frac{\rho r^2}{n}\bar{R} + \frac{n}{r}\bar{M} - r^2\frac{\lambda}{\lambda+2\mu}(2n+1)\bar{N}, \end{aligned} \tag{35.20}$$

$$\begin{aligned} \bar{N}' = &\left(\frac{4}{r^2}\mu + \frac{4}{r}\rho V' + \frac{8}{r^2}\frac{\lambda\mu}{\lambda+2\mu}\right)\cdot\frac{1}{n}\bar{H} \\ &-\frac{1}{r^2}\left(\rho V' + \frac{2}{r}\mu + \frac{4}{r}\frac{\lambda\mu}{\lambda+2\mu}\right)\frac{n+1}{2n+1}\bar{T} - \\ &-\frac{\rho}{r^2}\frac{n+1}{n(2n+1)}\bar{L} + \frac{1}{r^4}\frac{n(n+1)}{2n+1}\bar{M} + \\ &+\frac{1}{r}\left(n - 4\frac{\mu}{\lambda+2\mu}\right)\bar{N}. \end{aligned} \tag{35.21}$$

Conditions on the mantle surface are

$$\bar{N}_0 = D, \quad (35.22)$$

$$\bar{L}_0 = A - B\bar{R}_0, \quad (35.23)$$

$$\bar{M}_0 = 0. \quad (35.24)$$

Conditions on the core-mantle boundary:

$$\bar{N}_k = -W\bar{R}_k - S\bar{H}_k, \quad (35.25)$$

$$\bar{L}_k = F\bar{R}_k - K\bar{H}_k, \quad (35.26)$$

$$\bar{M}_k = 0. \quad (35.27)$$

Here

$$D = -\frac{1}{3n}\left(\bar{g}_0^m\right)^2 \bar{r}_m^{n+1}, A = \frac{2n+1}{n+1}\bar{r}_m^{n+2}\bar{g}_0^m, B = \bar{r}_m$$
$$F = \frac{r_c(n + r_c v)}{n+1}, K = 4\pi\bar{G}\rho_i r_c^2 \frac{2n+1}{n+1}, \quad (35.28)$$

$$W = \frac{\rho_c}{n\,(2n+1)}, S = \frac{\rho_i V'}{n}.$$

In Eq. 35.28 $\bar{g}_0^m$, $\bar{r}_m$ and $\bar{G}$ are normalized for the PREM model values of the gravity, radius and gravity constant on mantle surface, i.e. under 3-km sea layer. For another continental models the first two values equal to 1 and $4\pi G \approx 3$.

In case of $n = 0$ instead of Eqs. 35.2, 35.3, 35.4, 35.5, 35.6 and 35.7 we have a system:

$$H' = -\frac{2}{r}\frac{\lambda}{\lambda + 2\mu}H + \frac{1}{\lambda + 2\mu}N, \quad (35.29)$$

$$N' = \left[\frac{4}{r^2}\mu + \frac{4}{r}\rho V' + \frac{8}{r^2}\frac{\lambda\mu}{\lambda + 2\mu}\right]H - \frac{4}{r}\frac{\mu}{\lambda + 2\mu}N. \quad (35.30)$$

The integration is carried out from mantle surface, whereon $N = -\frac{1}{3}\left(\bar{g}_0^m\right)^2\bar{r}_m$, up to incompressible inner core surface, where $H = 0$.

In case of $n = 1$ the problem (35.16), (35.17), (35.18), (35.19), (35.20) and (35.21) is solved. Here, conditions (35.22), (35.23) and (35.24) on boundaries are not independent. Therefore, condition (35.23) was replaced by independent condition: $\bar{R}_0 = \bar{g}_0^m\bar{r}_m$.

2 Method of Boundary Problem Solution

In every examined the Earth's models the dependences of parameters on depth were approximated by parabolas. The values of resulted functions were calculated with the step $t = 0.1$ km. Thereafter the values as well as the radius and step were normalized, and the variable coefficients ingressed into the solved equations were calculated. The integration of equations was carried out by the Euler method.

The boundary problem was solved in the following way. Column vectors of the six required functions on upper ($\bar{Y}_0$) and lower (Y_N) boundaries were connected by the equation:

$$\bar{Y}_N = C \cdot \bar{Y}_0, \quad (35.31)$$

where $\bar{Y}_N = (\bar{H}_N, \bar{T}_N, \bar{R}_N, \bar{L}_N, \bar{M}_N, \bar{N}_N)^T$, $\bar{Y}_0 = (\bar{H}_0, \bar{T}_0, \bar{R}_0, L0, \bar{M}_0, \bar{N}_0)^T$ while the matrix C corresponds with the product of matrices P the elements that express sequentially the variable coefficients of Eqs. 35.16, 35.17, 35.18, 35.19, 35.20 and 35.21 multiplied by the integration step t. The unit was added to diagonal elements of matrix P. Thus:

$$C_{N+1} = \prod_{i=0}^{N} P_i, \quad (35.32)$$

where i is the step number, N is the total amount of steps (in our case from surface to lower boundary of the mantle). As before, the letter n denotes an order of required functions. From Eqs. 35.16, 35.17, 35.18, 35.19, 35.20 and 35.21 we can easily write out the elements of matrix P:

$$p_{11} = \frac{1}{r}\left(n - 2\frac{\lambda}{\lambda + 2\mu}\right) \cdot t + 1;$$
$$p_{12} = \frac{n\,(n+1)}{2n+1}\frac{1}{r^2}\frac{\lambda}{\lambda + 2\mu} \cdot t;$$
$$p_{13} = 0; \quad p_{14} = 0; \quad p_{15} = 0;$$
$$p_{16} = \frac{n}{\lambda + 2\mu} \cdot t \text{ etc.}$$

After the matrix C calculation we substitute boundary conditions (35.22), (35.23), (35.24), (35.25), (35.26), and (35.27) into Eq. 35.32 and search out values of three required functions H_0, T_0, R_0 on the mantle surface for the case $n \geq 2$ from the system of equations:

$$A \cdot \tilde{\bar{Y}}_0 = \bar{B}, \quad (35.33)$$

where $\tilde{\bar{Y}}_0 = (\bar{H}_0, \bar{T}_0, \bar{R}_0)^T$, the elements of the matrix A and vector $\bar{B}$ are expressed in term of the matrix C elements and constants ingressed in boundary conditions (35.28) in the following manner:

$$A_{11} = F \cdot C_{31} - K \cdot C_{11} - C_{41};$$
$$A_{12} = F \cdot C_{32} - K \cdot C_{12} - C_{42};$$

$$A_{13} = F \cdot (C_{33} - C_{34} \cdot B) - K \cdot (C_{13} - C_{14} \cdot B) + C_{44} \cdot B - C_{43};$$
$$B_1 = A \cdot (C_{44} + K \cdot C_{14} - F \cdot C_{34}) + {} + D \cdot (C_{46} + K \cdot C_{16} - F \cdot C_{36});$$

$$A_{21} = C_{51};\ A_{22} = C_{52};$$
$$A_{23} = C_{53} - C_{54} \cdot B;$$
$$B_2 = -C_{54} \cdot A - C_{56} \cdot D;$$

$$A_{31} = -(W \cdot C_{31} + S \cdot C_{11} + C_{61});$$
$$A_{32} = -(W \cdot C_{32} + S \cdot C_{12} + C_{62});$$

$$A_{33} = -W \cdot (C_{33} - C_{34} \cdot B) - S \cdot (C_{13} - C_{14} \cdot B) - C_{63} + C_{64} \cdot B;$$

$$B_3 = A \cdot (C_{64} + S \cdot C_{14} + W \cdot C_{34}) + {} + D \cdot (C_{66} + S \cdot C_{16} + W \cdot C_{36}).$$

Thus, searching the load Love numbers $k' = R_0 - 1$; $h' = H_0$; $l' = T_0$ boils down to the matrices multiplication.

In the case of $n = 0$ the matrices P and C are of the second order and are being searched by the scheme given above. The matrix P elements will be respectively (see Eqs. 35.29 and 35.30):

$$p_{11} = -\frac{2}{r}\frac{\lambda}{\lambda + 2\mu} \cdot t + 1;\ p_{12} = \frac{1}{\lambda + 2\mu} \cdot t;$$

$$p_{21} = \left[\frac{4}{r^2}\mu + \frac{4}{r}\rho V' + \frac{8}{r^2}\frac{\lambda\mu}{\lambda + 2\mu}\right] \cdot t;$$
$$p_{22} = -\frac{4}{r}\frac{\mu}{\lambda + 2\mu} \cdot t + 1$$

The values of the required functions of zero order on the Earth surface and on inner-outer core boundary will respectively be equal:

$$H_0^0 = -\frac{C_{12}D}{C_{11}},\quad N_N = D\left(C_{22} - \frac{C_{21}C_{12}}{C_{11}}\right)$$

When $n = 1$ we are solving Eqs. 35.2, 35.3, 35.4, 35.5, 35.6 and 35.7. The matrices P and C can be calculated similarly based on coefficients of the equations. So long as the value of the function R on the Earth surface is known (equals to 1), Eq. 35.33 is solved relative unknowns H_0, T_0 and L_0. The A matrix elements of the linear equations system will be:

$$A_{11} = F \cdot C_{31} - K \cdot C_{11} - C_{41};$$
$$A_{12} = F \cdot C_{32} - K \cdot C_{12} - C_{42};$$

$$A_{13} = F \cdot C_{34} - K \cdot C_{14} - C_{44};$$

$$B_1 = D \cdot (C_{46} + K \cdot C_{16} - F \cdot C_{36}) + C_{43} - {} - F \cdot C_{33} + K \cdot C_{13};$$

$$A_{21} = C_{51};\ A_{22} = C_{52};\ A_{23} = C_{54};\ B_2 = -(C_{53} + C_{56} \cdot D);$$

$$A_{31} = -(W \cdot C_{31} + S \cdot C_{11} + C_{61});$$
$$A_{32} = -(W \cdot C_{32} + S \cdot C_{12} + C_{62});$$

$$A_{33} = -[W \cdot C_{34} + S \cdot C_{14} + C_{64}];$$

$$B_3 = D \cdot (C_{66} + S \cdot C_{16} + W \cdot C_{36}) + {} + W \cdot C_{33} + S \cdot C_{13} + C_{63}.$$

Table 35.1 Load Love numbers for PREM and Southern Africa NA model (Larson 2004)

	PREM		S_Africa	
n	−k·n·1,000	−h·1,000	−k·n·1,000	−h·1,000
2	612	990	612	993
3	589	1,048	588	1,053
4	535	1,050	534	1,055
5	524	1,082	522	1,087
6	542	1,138	540	1,144
7	575	1,206	571	1,212
8	613	1,276	607	1,283
10	691	1,414	683	1,420
18	965	1,858	942	1,865
32	1,250	2,311	1,192	2,313
100	1,470	2,900	1,365	2,926
150	1,489	3,119	1,442	3,210

3 The Gravity Effect Calculation

The TOPEX/POSEIDON sea height at any point on the Earth is given by the following expression:

$$H_n^m = \left(A_n^m \cos m\lambda + B_n^m \sin m\lambda\right) \cdot P_n^m(\theta)$$

In our calculations we have $n = 150$. A comparison with results obtained by Green's functions for $n = 10,000$gives in (Vinogradova and Spiridonov 2012).

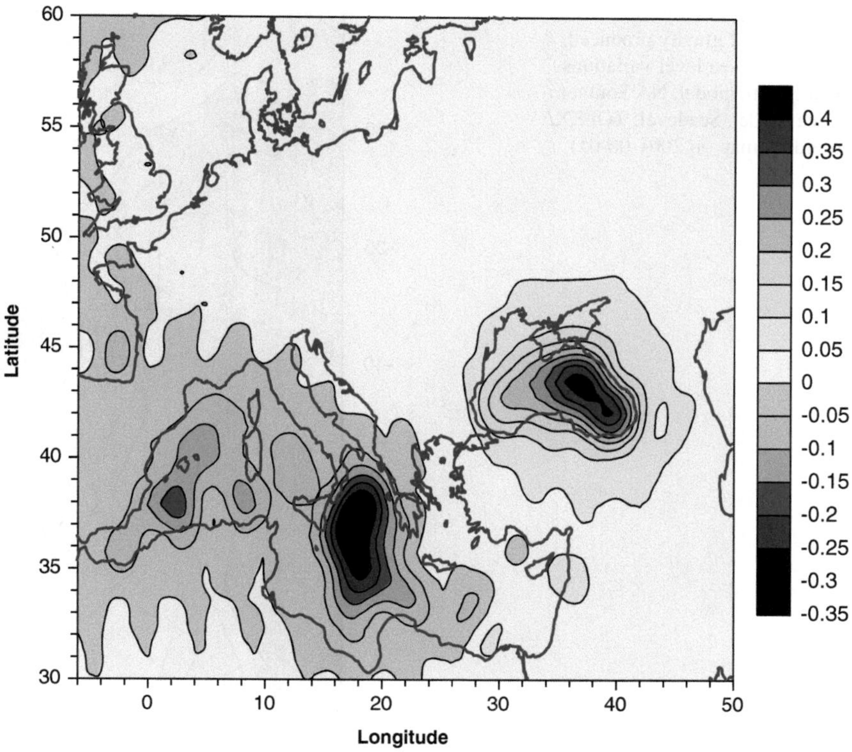

Fig. 35.1 Corrections to vertical component of gravity produced by non-tidal sea level variations (μGal). Earth model: PREM. Sea level: TOPEX/Poseidon (map on 2001-06-01)

The load gravity effect Δg is given by formula (Pertsev 1967):

$$\Delta g = -8\pi G\rho_0 \sum_n \frac{h'_n - \frac{n+1}{2}k'_n}{2n+1} \sum_m H_n^m,$$

where ρ_0 is the sea water density, h'_n and k'_n are load Love numbers calculated above (see Table 35.1).

4 Data

Sea level: TOPEX/Poseidon Sea Level Grids, version WOCE-PODAAC-v3.0-PF9.0 0.5 °. The values at each bin represent the difference between the average sea level measured in that space-time bin, and the 9-year average (1993–2001) of the measurements in that spatial bin (see http://woce.nodc.noaa.gov/).

Earth Models: (1) PREM (Anderson 1989); (2) The IASP91 Earth Reference Model (Wolfram et al. 2008, *Table A2*); (3) NA model for Southern Africa (*see* Larson 2004). The models have a different structure of the crust and upper mantle. Therefore, calculated by them load love numbers and gravity effect are different too.

5 Numerical Results

The results of calculations are presented in Table 35.1 and on Figs. 35.1, 35.2, and 35.3.

In Table 35.1 the load Love numbers values for two Earth models (PREM and NA Southern Africa Model (Larson 2004)) are given. The difference between the Love numbers values for two models growths with order n from part of percent to 10 %. The relative error of calculations is here within 0.01 %.

We calculated the load effect over globe and for different times.

For instance, on Figs. 35.1 and 35.2 we present the gravity anomalies Δg for EUROPE (Fig. 35.1) and Southern Africa (Fig. 35.2). The anomalies were calculated by TOPEX/POSEIDON sea heights (map date: 2001-06-01) and PREM data. In Europe the anomalies are not exceed 0.4 μGal and in Southern Africa amount to 1 μGal.

On Fig. 35.3 the difference between Δg values calculated for PREM and NA Earth model in Southern Africa is presented. This difference amounts to 0.03 μGal.

It is necessary to note that the significant distinctions between NA and PREM models extend to depth 400 km

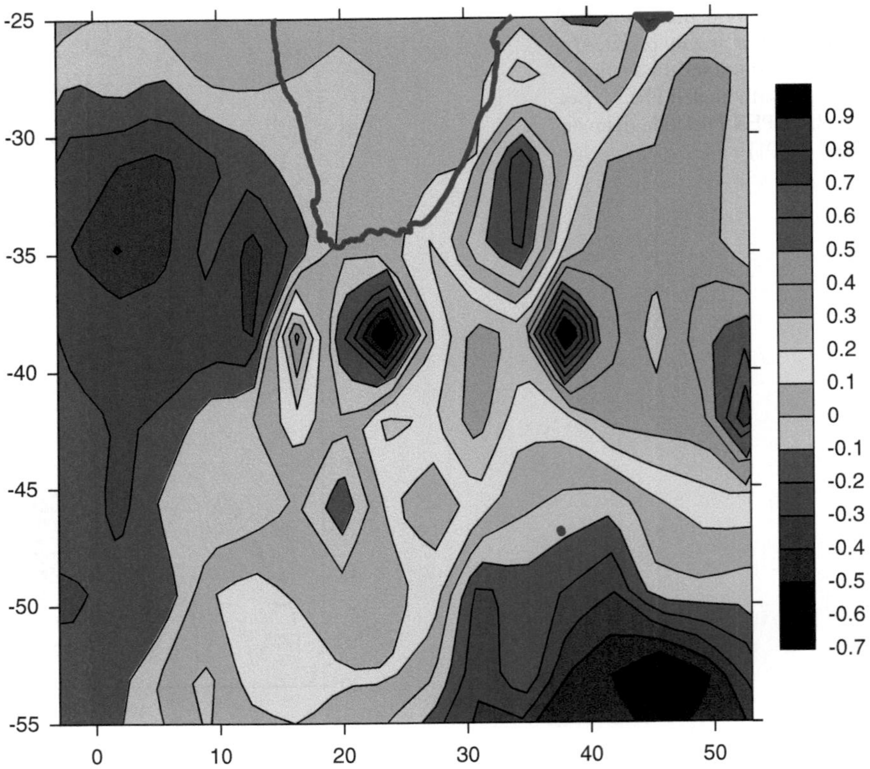

Fig. 35.2 Corrections to vertical component of gravity produced by non-tidal sea level variations (μGal) Earth model: NA Southern Africa Model. Sea level: TOPEX/Poseidon (map on 2001-06-01)

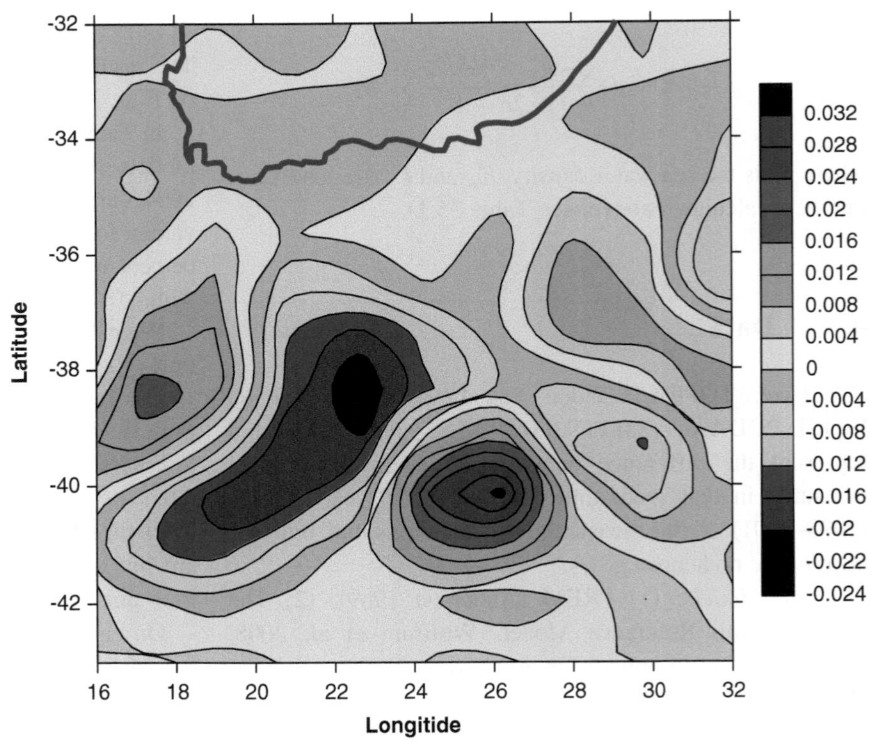

Fig. 35.3 Corrections to vertical component of gravity produced by non-tidal sea level variations (μGal) Earth model: difference between NA and PREM. Sea level: TOPEX/Poseidon (map on 2001-06-01)

If we use European model IASP91 for the Southern Africa observed sea level on 2001-06-01 then the differs by 0.1 μGal from PREM derived results. The reason is the differences of IASP and PREM models extend to depth 700 km. It leads to differences between the load Love numbers of high orders.

According to all our data, in different zones of the cean and different moments of time the discussed gravity correction, calculated by TOPEX/POSEIDON heights, does not exceed 1–2 μGal. An additional consideration of the crust local features and the upper mantle structure can make this correction more accurate by ±0.1 μGal.

This quantity is significant for the ocean gravity corrections.

6 Conclusion

Low-frequency sea level T/P perturbations give the magnitude of the gravimetric load effect of order 1 μGal. It is comparable with superconducting gravimeters precision.

In addition in the process of introduce of the ocean corrections into gravity observations data it is necessary to take into account the local features of the crust and upper mantle structure as the accounting of lateral irregularities allow to improve the value of the oceanic load effect in the range of 10 %.

References

Anderson DL (1989) Theory of the Earth. Blackwell Scientific Publications, Boston, http://resolver.caltech.edu/CaltechBOOK:1989.001

Larson AM (2004) S-wave velocity structure beneath the Kaapvaal Craton from surface-wave inversions compared with estimates from mantle xenoliths. Thesis submitted to the Faculty of the Virginia Polytechnic Institute and State University in partial fullment of the requirements for the degree of Master of Science in Geosciences, p 70

Molodensky MS (1953) Elastic tides, free nutation and some questions of Earth's structure. Trudi Geophizicheskogo Instituta Ac Sci USSR N19(146):3–52

Pertsev BP (1967) Estimation of the effect of marine tides on ground tides at points remote from the oceans, in book Zemnye prilivyi I Vnutrenneye Stroeniye Zemli, Moscow, IFZ, pp 10–22

Pertsev BP (1976) Influence of the near field sea tides on the Earth tidal observations. Phizika Zemli N1:13–22

Vinogradova O, Spiridonov E (2012) The comparative analysis of the oceanic corrections for gravity calculated for PREM and IASP91 models, Physika Zemli, N1 (in print)

Wolfram HG, Rainer K, Xiaohui Y (2008) Upper mantle and lithospheric heterogeneities in central and eastern Europe as observed by teleseismic receiver functions. Geophys J Int 174:351–376

Towards the Unification of the Vertical Datum Over the North American Continent

36

D.A. Smith, M. Véronneau, D.R. Roman, J. Huang, Y.M. Wang, and M.G. Sideris

Abstract

The United States adopted the North American Vertical Datum of 1988 (NAVD 88) for its official vertical datum in the 1990s. Canada has been using the Canadian Geodetic Vertical Datum (CGVD28) for its height applications since the 1930s. The use of the different datums causes inconsistent heights across the border between the two countries, and the topographic height data from the two countries are not compatible. Both datums rely on passive control and significant pre-modern survey data, yielding not only misalignment of the datums to the best known global geoid at approximately 1–2 m, but also local uplift and subsidence issues which may significantly exceed 1–2 m in extreme cases.

Today, the GNSS provides the geometric (ellipsoidal) height to an accuracy of 1–2 cm globally. The use of current inaccurate vertical datums no longer serves the purpose it once did. Because of this, users have begun to demand a physical height system that is closely related to the Earth's gravity field to a comparable accuracy. To address this need, government agencies of both countries are preparing the next generation of vertical datums. Even if the new datums are based on the same concepts and parameters, it is possible to have inconsistent heights along the borders due to the differences in the realization of the datums. To avoid inconsistency, it is in the interest of both countries to have a united, seamless, highly accurate vertical datum. The proposed replacements for CGVD28 and NAVD88 shall be based on GNSS positioning and a high accuracy gravimetric geoid that covers the territories of the United States, Canada, Mexico and the surrounding waters (to include all of Alaska, Hawaii, the Caribbean and Central America). To account for the effect of the sea level change, postglacial rebound, earthquakes and subsidence, this datum will also provide information on these changes. Detailed description of the definition, realization and maintenance of the datum is proposed. The challenges in realization and maintaining the datum are also discussed.

Keywords

Unification of the vertical datum • Gravity and geoid

1 Background

The governments of Canada and the USA have proposed improvements to their vertical datums, and a desire to unify the datum between the two countries, through the adoption of a new vertical datum based on GNSS positioning and a high accuracy geoid model. In Canada, a Height Modernization project was initiated in 2002 with the objective to adopt

D.A. Smith (✉) • D.R. Roman • Y.M. Wang
National Geodetic Survey, Silver Spring, MD, USA

M. Véronneau • J. Huang
Geodetic Survey Division, CCRS, Natural Resources Canada, Ottawa, Ontario, Canada

M.G. Sideris
University of Calgary, Calgary, Canada

Z. Altamimi and X. Collilieux (eds.), *Reference Frames for Applications in Geosciences*,
International Association of Geodesy Symposia 138, DOI 10.1007/978-3-642-32998-2_36,

a geoid model as the realization of the new vertical datum by 2013. The National Geodetic Survey (NGS) initiated more recently a comparable project under the Gravity for the Re-definition of the American Vertical Datum (GRAV-D) project, which includes extensive airborne gravity surveys across the USA. The implementation target date in the USA is currently 2022. Canada's Geodetic Survey Division (GSD) and NGS are engaged in ongoing discussions on how to define, realize and maintain this new vertical reference system to assure consistency and coherence across the two countries and to tie it properly to a global vertical reference system. Furthermore, it is desired to make this datum available to surrounding regions including nearby Greenland, Caribbean countries, Central America and the state of Hawaii. The USA and Canada have no plans to improve gravity coverage outside of the countries. The possibility exists that GRAV-D will extend 100 km or so to surrounding USA territories, but no formal plans exist. The National Geospatial-Intelligence Agency conducted an airborne survey providing a good quality dataset for Greenland. Furthermore, Greenland and Denmark are ready to work with North America for the development of a precise geoid model. However, they may not necessarily adopt the same geoid model. The expected accuracy in the unimproved regions will be what it is today, more or less what the EGM08 accuracy is in those regions.

The paper will give a brief history of the vertical datums in North America and provide insight into existing discrepancies and problems in these datums. It will define and discuss the fundamental components that form the basis of a modern vertical reference system and present the current options and challenges in defining and maintaining a North American vertical datum. It will investigate advantages and disadvantages for each option in terms of long-term consistency and needs in determining heights. The time dimension will be included to address the dynamic nature of the geoid and topography and future requirements in association with climate change and water resource management.

2 Status of the North America Vertical Datums

The NAVD88 (Zilkoski 1986; Zilkoski et al. 1992) is the vertical datum of the United States. It is realized through the publication of geopotential numbers and Helmert orthometric heights, based on decades of spirit leveling and surface gravity data. It was established in 1991 by the adjustment of geodetic leveling observations in Canada, the United States, and Mexico by holding the mean water level at Rimouski in Québec, Canada as the minimum constraint. The height information is delivered through passive benchmarks whose positions are subjected constantly to the changing environment, such as postglacial rebound, earthquakes and subsidence. Benchmarks have also been damaged or distorted by human activities such as construction.

The spirit leveling technique is the most accurate method for determining height differences over short distances. However, for nationwide networks, leveling errors may accumulate causing regional and nationwide network distortion. A comparison with the most recent GRACE-based satellite gravity model GGM03S (Tapley et al. 2007) reveals a 1.2 m Southeast-to-Northwest tilt of the NAVD88 (Fig. 36.1), which is in agreement with the 1.5 m height difference between the east and west coasts of Canada (Véronneau et al. 2006). Figure 36.1 shows a spherical harmonic expansion the geoid difference between GRACE and GPS/leveling-implied geoidal heights at 20,446 bench marks to degree and order 50 (Wang et al. 2012).

Assuming that the GRACE-derived geoid heights and GPS-derived ellipsoidal heights are free of long wavelength errors, the trend seen in Fig. 36.1 is purely due to long wavelength NAVD88 errors. However the actual source of these errors could be multi-fold, including field operations, data reduction and second order corrections. Because of that, the NAVD88 was not adopted by Canada. The current Canadian vertical datum is the Canadian Geodetic Vertical Datum (CGVD28), adopted in 1935. Like NAVD88, CGVD28 is based on a spirit leveling network, but does not incorporate actual surface gravity observations. The datum reference level was defined as mean sea-level determined from data collected at five tide gauges on the east and west coasts of Canada. Sea surface topography was not taken into consideration when using the mean sea levels at the five gauges, introducing a distortion of several decimeters into the CGVD28.

The currently published benchmark heights are a construct that results from annual survey observations dating back to 1904. Despite the great care administered to minimize potential error sources, the network is maintained in a piece-wise fashion by combining observations made over successive years and adjusted locally. This resulted in significant regional distortions in the current published heights. Further degradation of the accuracy of the published heights is attributable to vertical crustal motion over time. Due to the extent of the network and the related time required to carry out a full inspection, it is difficult to assess the exact physical state of the network at this time. We can only extrapolate based on statistics available in our databases or derived from the most recent inspections of small sections of the network. We can estimate that the degradation rate of the network across Canada is probably in the range of 15–20 % per 20 years. The continuous maintenance of the monumented vertical network on a 25-year cycle would require the leveling of approximately 5,600 km annually. This would still leave most of Canada without easy access to benchmarks. To meet current needs of GNSS users to access CGVD28, a geoid model has been used to provide a continuous and fully covered datum surface by fitting it to CGVD28 at

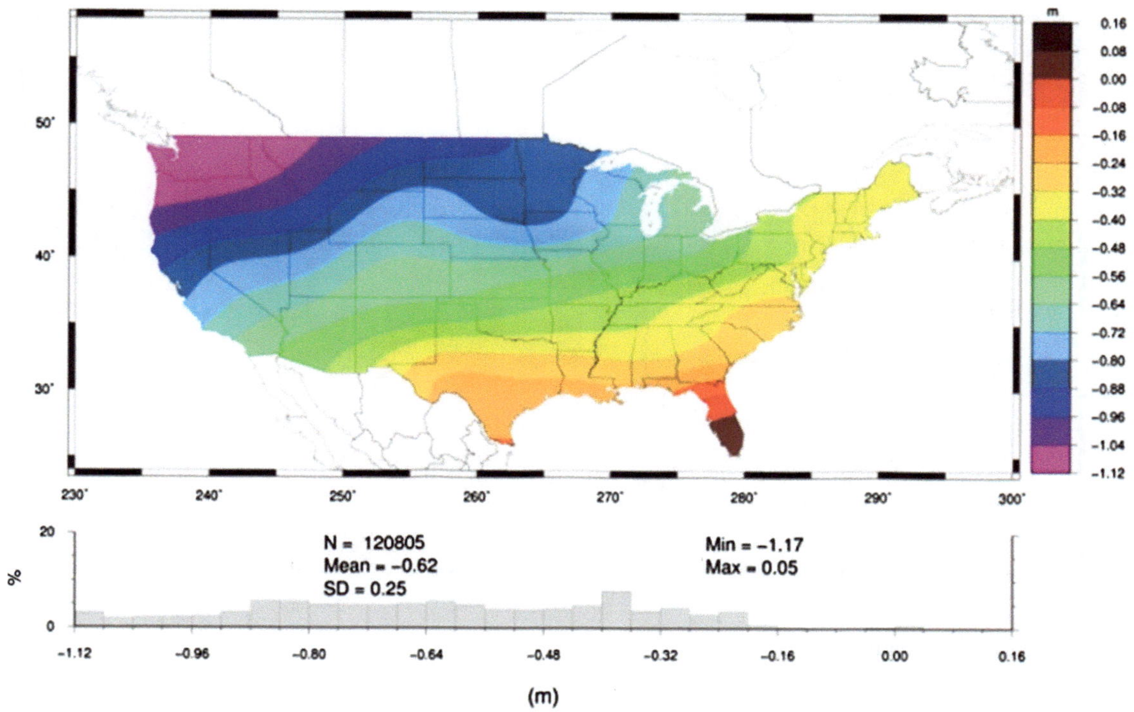

Fig. 36.1 Long wavelength errors (400 km resolution) of the USA vertical datum

GPS-occupied benchmarks. Therefore, the distortion in CGVD28 has been translated into the fitted geoid surface.

3 Height Modernization Programs of the USA and Canada

The vertical datums NAVD88 and CGVD28 were established by the classical levelling technique. This method is costly and the heights of the benchmarks are subjected to nature induced changes and human activities.

Both government agencies, NGS and GSD of Natural Resources Canada, are working independently on their next generation vertical datum (Smith 2007, 2008, 2010; Véronneau et al. 2006). In the United States, the decision to replace NAVD88 was codified as policy in the 2008 NGS 10 year plan (Smith 2008), which stated that the new geoid would be based upon a proposed aerial gravity survey of the United States (Smith 2007). Due to the time necessary to collect and process this new data source and the realities of budgeting, the current date for replacing NAVD88 in the USA is 2022.

Long before a replacement for NAVD88 was proposed, NGS recognized both the inherent power of GPS and ongoing deterioration of the benchmark network of NAVD88. Thus the "Height Modernization" program began in 1999. Under Height Modernization, NGS developed guidelines for the establishment of NAVD88 benchmarks using GPS technology. "Height Mod Surveys" were performed in many states over the last decade to shore up NAVD88. However, in 2008 NGS officially decided to replace NAVD88, but since this replacement is not implemented until 2022, the use of Height Mod surveys is expected to continue for the foreseeable future.

In Canada, a Height Modernization project was initiated in 2002 with the objective of adopting a geoid model as the realization of the new vertical datum by 2013. A major challenge is the development of a geoid model with an accuracy of 1–3 cm homogenously across Canada. The latest official geoid model CGG05 is estimated at an accuracy of better than 6 cm nationally and generally about 4 cm regionally (Huang et al. 2007). A new geoid model CGG10 is under development by taking advantage of the early release of GOCE models and new gravity and DEM data. It will be a prototype model for the new height datum. One key issue to be discussed and decided is to choose a potential value W_0 for the zero-height surface which is a definitive parameter toward unification of the vertical datums between USA and Canada.

4 Definition, Creation and Maintaining the NAVDXX

4.1 Definition of the NAVDXX

The new vertical datum (referred herein as "NAVDXX" though no formal name has yet been chosen) will use as its zero-height reference surface an equipotential surface with potential number W_0 (to be determined). NAVDXX will

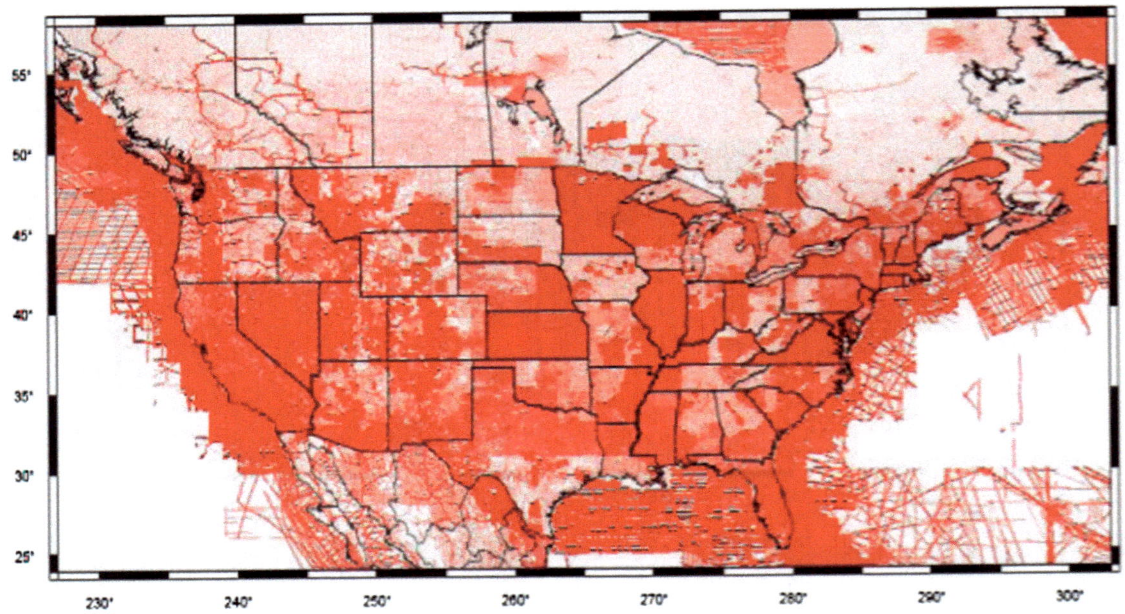

Fig. 36.2 Gravity data density over the North America

consist of a high accuracy gravimetric geoid model that covers North America, and all the surrounding regions. Additionally, the United States territories in the Pacific Ocean (American Samoa, Guam and the Commonwealth of the Northern Marianas) will have separate geoid models and associated vertical datums. In combination with the GNSS positioning, the orthometric height is determined to an accuracy compatible with the GNSS positioning at any given time and location. In addition to the conventional orthometric height, the accuracy of the height, height change rate and associated accuracy shall be provided to users. Furthermore, conversion from orthometric height to geopotential number will be provided, using the latest terrestrial gravity measurements. The geoid model is targeted to have an absolute accuracy of ± 1–2 cm in coastal and flat areas and ± 3–5 cm in mountainous regions.

4.2 Creation of NAVDXX

The key component of the NAVDXX is the geoid. There are still many challenges for computing the geoid with accuracy of a few cm. The challenges can be summarised in the following:

(a) Theory. The geoid theory is based on the fundamental geodetic boundary value problem. Lately, research in this field has been transferring from the free-boundary to the fix-boundary value problem, now that so much of the Earth's surface has been observed by GNSS and other remote sensing techniques. Sansò and Venuti (2008) showed that there is a unique solution for the fixed-boundary value problem, even if the inclination of the surface is as large as 89.6 °. In principle, the height anomaly can be accurately determined to any accuracy without knowing the density of the topographic masses by using Molodensky theory. However, the knowledge of the mass density of the topography is needed for geoid computations. Fortunately, the constant density assumption has its largest impact in the mountainous areas (e.g. Huang et al. 2001).

(b) Data. To reach a cm accurate geoid, the coverage of high quality data is the key. NGS and NRCan have historical gravity data dated back to 1960s. The quality of the data itself is believed to be good (around 50 to 100 μGal of measurement accuracy). However, the positions of the data, especially the vertical positions, are not accurate for non-benchmark data collected before the GPS era (yielding processed gravity anomaly error at the 1 mGal level or so). Because the gravity anomaly or gravity disturbance is used in geoid computations, the inaccurate height introduces artificial gravity signals. Another problem with the existing gravity database is data distribution. Figure 36.2 shows the gravity data density over the USA and Canada.

The gravity data is unevenly distributed. The data is densely measured in many states, but several mountainous and other states, Canada and Mexico have data density around one point per 81 to 225 km^2. The gravity data density does not support one arcminute data grid, which is currently the grid size needed to avoid <1 cm interpolation errors and omission errors in the most rugged geoid regions. Statistics show that around 70 % of $1'$ cells do not have any data. The insufficient data density will induce an omission error in the geoid (Wang 2012). This missing high frequency power of the gravity field can be compensated by that computed from high resolution digital elevation models (DEMs). Currently, the DEMs over North America come in 3 arcsecond resolution or finer. DEMs as fine as a few meter-resolution are also available for certain areas.

Since the density of the topographic masses is needed for geoid computations, it is a requirement to have at least some sampling of this type of data. Such an effort has been made in Canada (Huang et al. 2001) and is being investigated at NGS.

(c) Selection of the geoid computation method. There are different ways to compute the geoid, e.g., least squares collection and Stokes integration. In the Stokes integration method, there are kernel modifications, optimal combinations and many numerical methods. Li and Wang (2011) tested a few kernel modification methods, showing that the resulting geoid difference can be as large as a few decimeters. Therefore, the choice of geoid computation method is also a challenge for cm-geoid accuracy.

4.3 Validation

After the geoid is computed, it should be validated against adequate independent data sets. The validation is important since it should reveal the accuracy of the geoid. GPS/levelling data provide accurate geoid height differences, thus can be used to check the geoid in a relative sense. If we have lines of GPS/leveling in coastal (low) areas, and in medium and high variation topography areas, the relative accuracy of the geoid in low, medium and rugged topography areas can be reasonably estimated. In combination with the accuracy of the satellite gravity models, the formal error of the geoid can be estimated (see e.g. Huang et al. 2007). NGS is working on such a first test line in the Texas area (Smith et al. 2010). The survey is planned to start in the summer of 2011. Additional planned validations include comparisons with astro-geodetic deflections of the vertical, tide gauge data, and ocean dynamic topography models.

4.4 Maintenance

The Earth is a dynamic planet. The physical surface of the Earth changes every moment due to tectonics, postglacial rebound and subsidence attributed to the extraction of the fluids from the Earth, earthquakes and human activities. These changes can be monitored using GNSS positioning at Continuously Operating Reference Stations (CORS). In the same way, the geoid of the NAVDXX should change as mass changes cause geoid changes. As a rule of thumb, the geoid changes are one order of magnitude smaller than the Earth's physical surface changes.

It has been reported that the global sea level is rising at 2–3 mm annually. However, due to regional issues, local sea level may have much larger changes. The sea level change, unavoidably changes the W_0 value. To maintain the NAVDXX, the governments of the USA and Canada must jointly adopt a policy which addresses how sea level changes will be related to "zero height surface" changes, if at all. Such policy discussions are underway.

5 The Relationship Between the NAVDXX and the World Vertical Datum

A World Height System (WHS) has been proposed recently (Ihde et al. 2010). Since the WHS would be defined by a global geoid, the transition from the NAVDXX and WHS should be simply represented by a conversion surface, namely the difference map between the WHS and NAVDXX.

6 Conclusions

The vertical datums used by the USA and Canada are based on spirit-leveling networks. They are distorted, inconsistent, and have about 1-m error revealed by GRACE satellite only gravity model. To satisfy today's accurate height requirements, government agencies of both countries are working on their next generation of vertical datums. Canada plans to have its new datum in 2013 and the USA will replace the NAVD88 in 2022, which will likely coincide with an update to Canada's 2013 datum. It is in the interest of both countries to have a united, seamless, high accuracy vertical datum. The proposed new North America Vertical Datum (NAVDXX) shall be based on GNSS positioning and a high accuracy geoid that covers the territories of the United States, Canada, Mexico and the surrounding waters. NAVDXX is to be realized through a geoid model with well established parameters. The geoid model is expected to have an absolute accuracy of ±1–2 cm in coastal and flat areas and ±3–5 cm in mountainous regions.

References

Huang J, Vaníček P, Pagiatakis SD, Brink W (2001) Effect of topographical density on geoid in the Canadian Rocky Mountains. J Geodesy 74:805–815

Huang J, Fotopoulos G, Cheng MK, Véronneau M, Sideris MG (2007) On the estimation of the regional geoid error in Canada. IAG Sym 130(III):272–279

Ihde JM, Sideris G, Sánchez L (2010) Concepts for the realization of a World height system – theme 1 of the global geodetic observing system (GGOS). In: Proceedings of reference frames for applications in geosciences 2010, France

Li X, Wang YM (2011) Comparisons of geoid models over Alaska computed by different Stokes's kernel modifications. J Geod Sci 1(2):136–142

Sansò F, Venuti G (2008) On the explicit determination of stability constants for linearized geodetic boundary value problems. J Geodesy 82:909–916

Smith D (2007) The GRAV-D project: gravity for the redefinition of the American Vertical Datum. Available online at: http://www.ngs.noaa.gov/GRAV-D/pubs/GRAV-D_v2007_12_19.pdf

Smith D (2008) The NGS 10 year plan (2008–2018). Available online at: http://www.ngs.noaa.gov/INFO/NGS10yearplan.pdf

Smith D (2010) Improving the national spatial reference system. White paper prepared for the 2010 Federal Geospatial Summit on replacing NAD 83 and NAVD 88. Silver Spring, MD, May 2010. Available online at: http://www.ngs.noaa.gov/2010Summit/Improving_the_NSRS.pdf

Smith D, Eckl M, Fancher DRK, Kuper S, Smith C, Wang YM, Schmerge D, Winester D, Hirt C, Burki B, Guillame S (2010) A proposed geoid slope validation survey in the United States. In: Second international gravity field symposium, Fairbanks, Alaska

Tapley B, Ries J, Bettadpur S, Chambers D, Cheng M, Condi F, Poole S (2007) The GGM03 mean Earth gravity model from GRACE. Eos Trans AGU 88(52), Fall Meeting Suppl, Abstract G42A-03, 2007

Véronneau M, Duval R, Huang J (2006) A gravimetric geoid model as a vertical datum in Canada. Geomatica 60(2):165–172

Wang YM (2012) On the Omission Errors Due to Limited Grid Size in Geoid Computations, VII Hotine-Marussi Symposium on Mathematical Geodesy, IAG Symposia Vol. 137, pp 221–226

Wang YM, Saleh J, Li XP, Roman D (2012) The US Gravimetric Geoid of 2009 (USGG2009): model development and evaluation, J Geod, 86:165–180

Zilkoski DB (1986) The new adjustment of the North American datum. ACSM Bulletin, April, pp 35–36

Zilkoski DB, Richards JH, Young GM (1992) Results of the general adjustment of the North American vertical datum of 1988. Am Congr Surv Mapp Surv Land Inf Syst 52(3):133–149

Part VI

Usage and Applications of Reference Frames in Geosciences

Improving SIRGAS Ionospheric Model

37

C. Brunini, F. Azpilicueta, M. Gende, E. Camilion, and E. Gularte

Abstract

The IAG Sub-Commission 1.3b, SIRGAS (Sistema de Referencia Geocéntrico para las Américas), operates a service for computing regional ionospheric maps based on GNSS observations from its Continuously Operating Network (SIRGAS-CON). The ionospheric model used by SIRGAS (named La Plata Ionopsheric Model, LPIM), has continuously evolved from a "thin layer" simplification for computing the vTEC distribution to a formulation that approximates the electron density (ED) distributions of the E, F1, F2 and top-side ionospheric layers.

This contribution presents the newest improvements in the model formulation and validates the obtained results by comparing the computed vTEC to experimental values provided by the ocean altimetry Jason 1 mission. Comparisons showed a small underestimation of the Jason 1 vTEC by about 1.3 TECu on average and rather small differences ranging from −0.5 to −3.4 TECu (at 95 % probability level). The results are encouraging given that comparisons were made in the open ocean regions (far away from the SIRGAS-CON stations).

Keywords

Ionosphere • GNSS • SIRGAS

1 Introduction

Not long ago, the IAG Sub-Commission 1.3b called SIRGAS (Sistema de Referencia Geocéntrico para las Américas), i.e. the Regional Reference Frame for Central and South America, operates a service for computing regional ionospheric maps based on the GNSS observations provided by its Continuously Operating Network (SIRGAS-CON) (Brunini et al. 2011a). Since 2008, a continuous time series of maps describing the vertical Total Electron Content (vTEC) distribution for the SIRGAS region, with time resolution of 1 h, is available at the SIRGAS web page (www.sirgas.org).

As other vTEC maps computed within the geodetic community (e.g.: Hernández-Pajarez et al. 2009), SIRGAS maps were originally based on the so called thin layer ionospheric model (Brunini et al. 2011b). According to this model, the vertical structure of the ionosphere (from about 50 to 1,000 km above the Earth´s surface), is approximated with a spherical shell of infinitesimal thickness with equivalent vTEC (located somewhere between 350 and 450 km height). Within this approximation, the satellite-to-receiver slant Total Electron Content (sTEC) is converted into an equivalent vTEC on the shell, by means of a geometrical mapping function that only depends on the satellite elevation and the

C. Brunini (✉) • F. Azpilicueta • E. Camilion • E. Gularte
GESA, Facultad de Ciencias Astronómicas y Geofísicas, Universidad Nacional de La Plata, Paseo del Bosque s/n, La Plata 1900, Argentina
e-mail: claudiobrunini@yahoo.com

M. Gende
GESA, Facultad de Ciencias Astronómicas y Geofísicas, Universidad Nacional de La Plata, Paseo del Bosque s/n, La Plata 1900, Argentina

Consejo Nacional de Investigaciones Científicas y Técnicas, Buenos Aires, Argentina

Z. Altamimi and X. Collilieux (eds.), *Reference Frames for Applications in Geosciences*, International Association of Geodesy Symposia 138, DOI 10.1007/978-3-642-32998-2_37,

height of the shell. Spatial and temporal variations of the equivalent vTEC are represented on the shell by means of different kinds of 3-D (latitude, longitude and time) mathematical functions (e.g.: spherical harmonics expansion). The parameters of these functions are estimated from the GNSS observations, along with the inter frequency biases (IFB) that account for the frequency-dependent delays produced by the GNSS satellite and receiver hardware and firmware.

The ionospheric model used by SIRGAS, known as La Plata Ionospheric Model (LPIM) (Azpilicueta et al. 2005), has continuously evolved from the initial "thin layer with equivalent vTEC" simplification, to the present formulation in which the E, F1, F2 and top-side ionospheric layers are considered, and their vertical electron density (ED) distributions are approximated with empirical functions in similar way than the International Reference Ionosphere (IRI) (Bilitza and Reinisch 2008) or NeQuick (Nava et al. 2008) models.

At present, the SIRGAS ionospheric model is based on the geometry free observable computed from dual frequency carrier phase observations (Ciraolo et al. 2007) collected from either, ground-based or space-borne GNSS receivers on low Earth orbiting satellites (e.g.: SAC-C, GRACE, FORMOSAT-3 /COSMIC). After correcting carrier phase cycle slips and ambiguities, the geometry free observations are used (in connection with the geometrical description of the observed satellite-to-receiver line of sights; LOS) to estimate the parameters of the empirical functions that describe the 4-D (latitude, longitude, height and time) ED distribution of the different ionospheric layers. Satellite and receivers IFB are also estimated together with the function parameters.

The combination of GNSS observations collected from the Earth (with prevailing vertical LOS geometry) and from LEO (with prevailing horizontal LOS geometry) helps to solve the horizontal and vertical structure of the ED distribution (Hernández-Pajares et al. 2000).

The evolution of the SIRGAS ionospheric model has been reported in Brunini et al. (2011a) and (2011b). Those papers may be useful to clarify some points that are briefly presented in the following two sections. This contribution presents the newest improvements in the model formulation and validates the output of the model by comparing to experimental values provided by the ocean altimetry Jason 1 mission.

2 Model Formulation

A simple vertical ED profile is predicted by the Chapman theory (Chapman 1931), which assumes monochromatic radiation, photoionization of a single species neutral gas, and neglects transport processes:

$$N(h) = N_m \cdot exp[k \cdot (1 - z - exp(-z))]$$
$$z = \frac{h - h_m}{H}, \tag{37.1}$$

where $N(h)$ is the ED at a given height h, N_m and h_m are the ED and height of the peak of the Chapman function, and H is the scale height. Depending on the description of electron loss process, the factor k can be 0.5 or 1, defining respectively an α or β Chapman layer. According to Bilitza (2002), most modellers have found 0.5 to provide a closer match with observations and, according to that experience, 0.5 is used in the case of SIRGAS. In order to adapt the model to the different chemical and physical properties prevailing in the different ionospheric layers, SIRGAS uses four Chapman functions (37.1) with different parameters $N_{m,i}$, $h_{m,i}$, and H_i, $i = 1, \ldots, 4$, to represent the E and F1 layers, the bottom-side of the F2 layer and the top-side:

$$N(h) = \begin{cases} \sum_{i=1}^{3} N_i(h), if\ h \leq h_{m,3} \\ N_4(h),\ if\ h > h_{m,3} \end{cases}. \tag{37.2}$$

The ED, $N_{m,3}$, and height, $h_{m,3}$, of the F2 layer peak are computed from the corresponding critical frequency, f_3, and propagation factor, M_3. The critical frequency is converted into ED using the relation of proportionality between ED and squared critical frequency. The propagation factor is converted into height using the Dudeney (1974) formulae as modified by Bilitza et al. (1979) (which requires the use of the $N_{m,3}/N_{m,1}$ ratio).

As other ionospheric models (e.g.: IRI or NeQuick), SIRGAS uses the Jones and Gallet (1962) mathematical technique to represent the geographical (latitude and longitude) and daily (UT) variation of the critical frequency and propagation factor of the F2 layer peak; symbolically: $f_3(\varphi, \lambda, t; \mathbf{U}_f)$ and $M_3(\varphi, \lambda, t; \mathbf{U}_M)$, where $\mathbf{U}_f$ and $\mathbf{U}_M$ are vectors of 998 and 441 constant parameters respectively.

The critical frequencies and heights of the E and F1 layer peaks are computed according to the Comité Consultatif International des Radiocommunications (CCIR 1991) recommendations:

$$\left.\begin{aligned} N_{m,1} &= q_1 \cdot A_1 \cdot cos^{n_1}\chi \\ h_{m,1} &= (1 + r_1) \cdot 120\ km \end{aligned}\right\} E\ layer$$
$$\left.\begin{aligned} N_{m,2} &= q_2 \cdot A_2 \cdot cos^{n_2}\chi \\ h_{m,2} &= 165 + (0.6 + r_2) \cdot \chi \end{aligned}\right\} F1\ layer, \tag{37.3}$$

where χ is the solar zenith angle; A_1, A_2, n_1 and n_2 values depend on the geographic latitude and solar activity, and q_1, q_2, r_1 and r_2 are constant parameters.

The scale heights of the E, F1 and F2 layers are computed according to the CCIR (1991) recommendations:

$$H_3 = \frac{(1+s_1) \cdot 38.5 \cdot N_{m,3}}{exp\left[-15.2+2.0 \cdot \left(ln(N_{m,3}) + ln(M_3)\right)\right]}$$
$$H_2 = (1+s_2) \cdot 0.4 \cdot \left(h_{m,3} - h_{m,1}\right)$$
$$H_1 = (0.5 + s_3) \cdot \left(5.0 + h_{m,3} - h_{m,1}\right) \tag{37.4}$$

being s_1, s_2 and s_3 constant parameters.

The top-side profile of the SIRGAS model is represented with an α-Chapman function with height-varying scale height, $H_4(h)$ (Reinisch and Huang 2001):

$$N_4(h) = N_{m,3} \cdot \sqrt{\frac{H_4(h_{m,3})}{H_4(h)}} \times exp\frac{1}{2} \cdot [1 - z_4(h) - exp(-z_4(h))],$$
$$z_4(h) = \int_{h_6}^{h} \frac{d\zeta}{H_4(\zeta)}, \tag{37.5}$$
$$H_4(h) = H_4(h_T) + \frac{H_4(h) - H_4(h_T)}{tanh(p)} \times tanh\left(p \cdot \frac{h - h_T}{h_{m,3} - h_T}\right),$$

h_T is the transition height where the dominant ion species changes from O+ to H+, and p is a steepness parameter of the topside profile.

In summary, the SIRGAS ED profile is given by a function that depends on three sets of constant but unknown parameters: 1,000 in the vector $\mathbf{U}_N$ (including q_1, and q_2) and 443 in the vector $\mathbf{U}_h$ (including r_1 and r_2), needed to compute the ED and heights of the E, F1 and F2 layer peaks; and 5 parameters in the vector $\mathbf{U}_H = (s_1, s_2, s_3, h_T, p)^T$, required to compute the scale heights of the different layers. Symbolically, the ED profile at any point (latitude longitude and height) within the global ionosphere, and at any time (UT), is:

$$N(\varphi, \lambda, h, t; \mathbf{U}_N, \mathbf{U}_h, \mathbf{U}_H). \tag{37.6}$$

3 Data Assimilation

The observation equation for the problem reads as:

$$l + \varepsilon = sTEC + \beta_S + \beta_R, \tag{37.7}$$

where l is the geometry free observable from dual frequency carrier phase observations (collected from either, ground-based or space-borne GNSS receivers), already corrected by carrier phase cycle slips and ambiguities; ε is the observational error; *sTEC* is a function to be estimated; and β_S and β_R are the satellite and receiver IFB.

In terms of the expression (37.6), the *sTEC* function of Eq. 37.7 is written as:

$$sTEC = \int_{\Gamma} N(\varphi, \lambda, h, t; \mathbf{U}_N, \mathbf{U}_h, \mathbf{U}_H) \cdot d\Gamma, \tag{37.8}$$

where, is the LOS from the satellite to the ground-based or space-born receiver.

The function given in expression (37.6) is approxi mated with a linear expansion (with respect to the model parameters):

$$N(\varphi, \lambda, h, t; \mathbf{U}_N, \mathbf{U}_h, \mathbf{U}_H) = N(\varphi, \lambda, h, t)|_0 + \left.\frac{\partial N}{\partial \mathbf{U}_N}\right|_0 \cdot \Delta\mathbf{U}_N + \left.\frac{\partial N}{\partial \mathbf{U}_h}\right|_0 \cdot \Delta\mathbf{U}_h + \left.\frac{\partial N}{\partial \mathbf{U}_H}\right|_0 \cdot \Delta\mathbf{U}_H \tag{37.9}$$

where the notation $\cdot|_0$ is used to indicate evaluation of the function in the a-priory values of the $\mathbf{U}_N$, $\mathbf{U}_h$ and $\mathbf{U}_H$ parameters and the symbol $\Delta\cdot$ denotes correction to the corresponding a-priory value.

Finally, Eq. 37.7 is transformed into:

$$l + \varepsilon = sTEC|_0 + \Delta\mathbf{U}_N \cdot \int_{\Gamma} \left.\frac{\partial N}{\partial \mathbf{U}_N}\right|_0 \cdot d\Gamma + \Delta\mathbf{U}_h \cdot \int_{\Gamma} \left.\frac{\partial N}{\partial \mathbf{U}_h}\right|_0 \cdot d\Gamma + \Delta\mathbf{U}_H \cdot \int_{\Gamma} \left.\frac{\partial N}{\partial \mathbf{U}_H}\right|_0 \cdot d\Gamma + \beta_S + \beta_R \tag{37.10}$$

where $sTEC|_0 = \int_{\Gamma} N(\varphi, \lambda, h, t)|_0 \cdot d\Gamma$.

Around 5 × 105 observations per hour provided by ~200 SIRGAS-CON stations and 5 × 103 observations per hour derived from ~250 FORMOSAT-3/COSMIC radio occultations per day (over the SIRGAS region) are assimilated into the SIRGAS ionospheric model. The Least Square method is used to estimate daily sets of 1,000 ($\mathbf{U}_N$) + 443 ($\mathbf{U}_h$) + 5 ($\mathbf{U}_H$) model parameters, plus 31 IFB for the GPS satellites, plus ~200 IFB for the SIRGAS-CON receivers plus 6 IFB for the GPS receivers flying onboard the FORMOSAT-3/COSMIC satellites (Rocken et al. 2000).

4 Results

A complete year from November 1, 2006, to October 31, 2007, was processed in order to assess the performance of the SIRGAS model under low solar activity conditions.

Just to provide an example, Fig. 37.1 shows the computed electron density distribution for the fixed longitude of 60°W and fixed UT of 14 h, from −55° to +35° of geographic

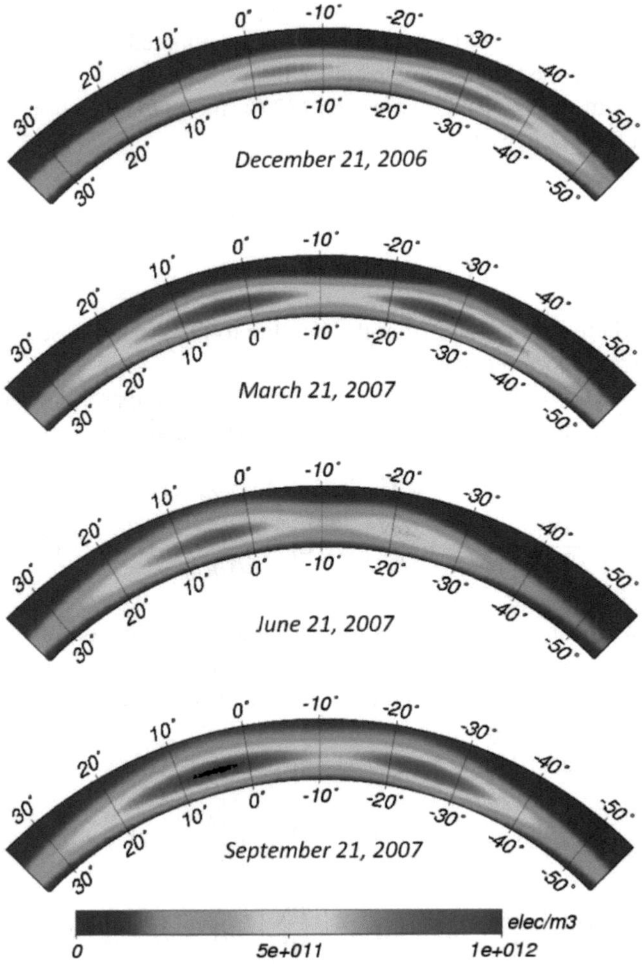

Fig. 37.1 Electron density distribution for fixed longitude and UT (60°W, 14 h), from −55° to +35° of geographic latitude and from 50 to 1,000 km height

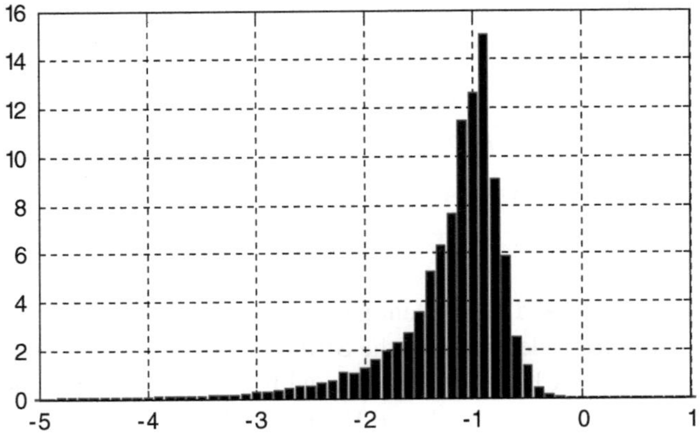

Fig. 37.2 Distribution of differences SIRGAS minus Jason 1 vTEC (x-value in TECu and y-value in percent)

latitude and from 50 to 1,000 km height, for the two solstices and equinoxes comprised in the analysed period. Seasonal changes in the morphology of the Appleton Anomaly are quite well represented by this figure.

In order to asses the accuracy of the obtained results, the vTEC distribution computed by integration of the SIRGAS ED distribution, is compared to experimental vTEC values derived form the dual frequency radar onboard the Jason 1 satellite altimetry mission (Menard and Haines 2001).

All Jason 1 observations for the whole year, within the ocean region surrounding the Latin America continent, from 0° to 120° W longitude and +40° to −60° modip latitude, are considered. In order to reduce the computational load, normal points every 30s are computed from the original 1 s sampling rate observations.

The SIRGAS vTEC is computed for the time and location of every smoothed Jason 1 vTEC, by integration of the ED up to the height of the Jason 1 satellite (~1,300 km). Then, the differences SIRGAS minus Jason 1 vTEC are evaluated.

The mean value of the differences for the whole year is −1.3 TECu, meaning that the SIRGAS model slightly underestimates the Jason 1 vTEC. According to Fig. 37.2, 95 % of the differences range from −0.5 to −3.4 TECu with a statistical distribution biased toward negative values.

Figure 37.3 shows a set of maps displaying the differences SIRGAS minus Jason 1 vTEC for the whole year, grouped into UT intervals of 2 h. The dashed lines represent the modip parallels of +30°, 0°, −30° and −60°. Within the mid latitude region ($|modip|>30°$) the differences stay very small (in absolute value), even far from the continent. Their values slightly increase in the low latitude region ($|modip|\leqslant 30°$) for the UT intervals when the Appleton Anomaly deploys over Latin America.

5 Conclusions

SIRGAS vTEC was compared to the corresponding values provided by the Jason 1 satellite altimetry mission, for a complete low solar activity year. According to this comparison, there is a bias of −1.3 TECu on average between the SIRGAS and Jason 1 vTEC (SIRGAS lower than Jason 1). This bias is compatible with other results reported in the literature regarding systematic differences between the vTEC estimated from GPS and from Jason 1–2 and TOPEX (Codrescu et al. 2001; Hernandez-Pajares 2003; Delay and Doherty 2004; Brunini et al. 2005). In fact, in a previous paper we pointed out the possibility that the TOPEX-GPS bias could be partially attributed to calibration errors on the TOPEX system biases.

Differences SIRGAS minus Jason 1 vTEC are very small (95 % differences range from −0.5 to −3.4 TECu). Within the mid latitude region the differences stay small even far

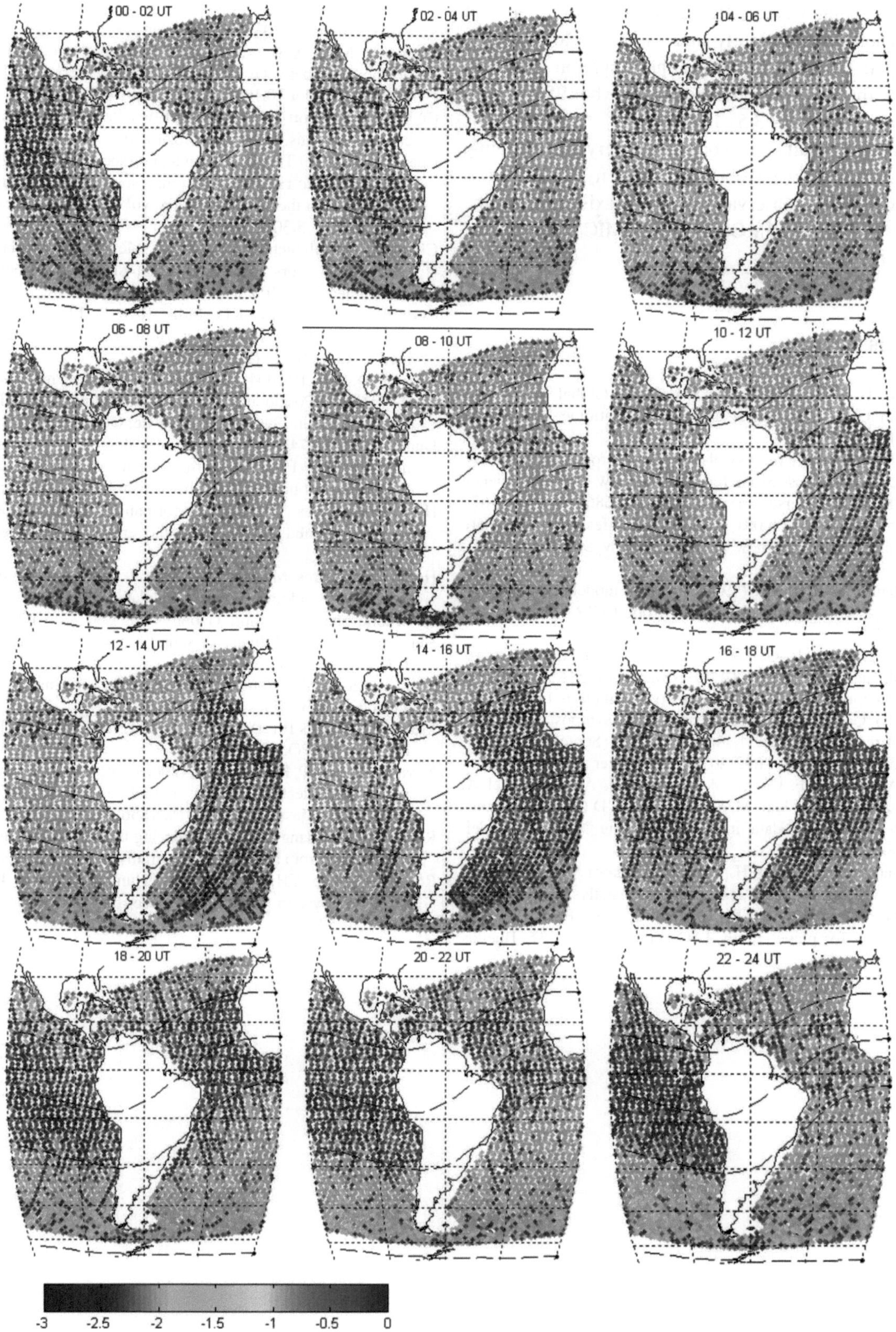

Fig. 37.3 Differences SIRGAS minus Jason 1 vTEC (TECu) for different UT intervals

from the continent, where SIRGAS-CON stations are not available. Their values slightly increase in the low latitude region for the UT intervals when the Appleton Anomaly evolves over the Latin American continent, but 95 % of the samples stay within (0, −3.5) TECu.

The obtained results are considered very encouraging, overall because they were obtained in open ocean regions where ground-based observations are inexistent and the model relies only on FORMOSAT-3/COSMIC observations.

References

Azpilicueta F, Brunini C, Radicella SM (2005) Global ionospheric maps from GPS observations using modip latitude. Adv Space Res 38(11):2324–2331

Bilitza D (2002) Ionospheric models for radio propagation studies. Chapter 28. In: W. Ross Stone (ed.) The Review of Radio Science 1999–2002. Wiley Interscience, ISBN 0-471-26866-6, pp 625–679

Bilitza D, Reinisch BW (2008) International reference ionosphere 2007: improvements and new parameters. Adv Space Res 42(4): 599–609

Bilitza D, Sheik NM, Eyfrig R (1979) A global model for the height of the F2-peak using M3000 values from CCIR. Telecommun J 46:549–553

Brunini C, Azpilicueta F, Gende M, Aragón-Ángel A, Hernández-Pajares M, Juan JM, Sanz J (2011a) Toward a SIRGAS service for mapping the ionosphere's electron density distribution. In: Pacino C et al. (eds) Geodesy for planet earth, IAG symposia, vol 135, Buenos Aires, Argentina, 31 August 31 – 4 September 2009, pp. 575–580, ISBN 978-3-642-20338-1, Springer

Brunini C, Azpilicueta F, Gende M, Camilion E, Aragón Ángel A, Hernandez-Pajares M, Juan M, Sanz J, Salazar D (2011b) Ground- and space-based GPS data ingestion into the NeQuick model. J Geod 18(12):931–939

Brunini C, Sánchez L, Drewes H, Costa S, Mackern V, Martínez W, Seemüller W, da Silva A (2011c) Improved analysis strategy and accessibility of the SIRGAS Reference Frame. In: Pacino C et al. (eds) Geodesy for planet Earth, IAG symposia, vol 135, Buenos Aires, Argentina, 31 August 31 – 4 September 2009, pp 3–8, ISBN 978-3-642-20338-1, Springer

Brunini C, Meza A, Bosch W (2005) Temporal and spatial variability of the bias between TOPEX and GPS derived TEC. J Geod 79:175–188. doi:10.1007/s00190-005-0448-z

CCIR (1991) Report 340–6. Comité Consultatif International des Radio communications, Genève, Switzerland

Chapman S (1931) The absorption and dissociative or ionizing effect of monochromatic radiation in an atmosphere on a rotating Earth. In: Proceedings of the physical society, vol 43, pp 483–501. doi:10.1088/0959-5309/43/5/302.

Ciraolo L, Azpilicueta F, Brunini C, Meza A, Radicela SM (2007) Calibration errors on experimental slant total electron content determined with GPS. J Geod 81(2):111–120

Codrescu MV, Beierle KL, Fuller-Rowell TJ (2001) More total electron content climatology from TOPEX/Poseidon measurements. Radio Sci 36(2):325–333

Delay S, Doherty P (2004) A decade of ionospheric measurements from the TOPEX/Poseidon mission. Contribution to the International Beacon Satellite Symposium, Trieste, October 18–22

Dudeney JR (1974) A simple empirical method for estimating the height of the F2-layer at the Argentine Islands Graham Land. Science report no. 88, London, British Antarctic Survey

Hernandez-Pajares M (2003) Performance of IGS ionosphere TEC maps. Presented at 22nd IGS Governing Board Meeting, Nice, 6 April 2003

Hernández-Pajares M, Juan JM, Sanz J (2000) Improving the Abel inversion by adding ground GPS data to LEO radio occultations in ionospheric sounding. Geophys Res Lett 27(16):2473–2476

Hernández-Pajarez M, Juan JM, Sanz J, Orus R, Garcia-Rigo A, Feltens J, Komjathy A, Schaer SC, Krankowski A (2009) The IGS VTEC maps: a reliable source of ionospheric information since 1998. J Geod 83:263–275

Menard Y, Haines B (2001) Jason-1 CALVAL plan, JPL Ref: TP2-J0-PL-974-CN (PO.DAAC)

Nava B, Coïsson P, Radicella SM (2008) A new version of the NeQuick ionosphere electron density model. J Atmos Sol Terr Phys. doi:10.1016/j.jastp. 2008.01.015, 1856-1862

Reinisch BW, Huang X (2001) Deducing topside profiles and total electron content from bottomside ionograms. Adv Space Res 27(1):23–30

Rocken C, Kuo YH, Schreiner W, Hunt D, Sokolovsky S (2000) COSMIC system description. Spec Iss Terr Atmos Ocean Sci 11(1):21–52

Use of Reference Frames for Interplanetary Navigation at JPL

38

Michael Heflin, Chris Jacobs, Ojars Sovers, Angelyn Moore, and Sue Owen

Abstract

Navigation of interplanetary spacecraft is typically based on range, Doppler, and differential interferometric measurements made by ground-based telescopes. Successful tracking requires knowledge of the telescope positions in the terrestrial reference frame. Spacecraft move against a background of extra-galactic radio sources and navigation depends upon precise knowledge of those background radio source positions in the celestial reference frame. Work is underway at JPL to reprocess historical VLBI and GPS data to improve realizations of the terrestrial and celestial frames. The purpose of this brief paper is to provide a snapshot of reference frame results.

Keywords

Reference frames • Navigation • VLBI • GPS • ITRF • ICRF

1 Introduction

Global VLBI data have been analyzed to estimate the coordinates of 1736 radio sources in the celestial reference frame ICRF2 (Ma et al. 2009). Global GPS data have been processed to estimate the positions and velocities of 909 stations in the terrestrial reference frame ITRF2005 (Altamimi et al. 2007). Analysis methods are described briefly and accuracy estimates are provided.

2 VLBI Celestial Frame Analyses

The VLBI source catalog was derived from 5.6 million delay measurements and an equal number of delay rates. Data were collected in the S/X band between 1978 and 2010. The wrms of delay residuals is 38 ps and the wrms of delay rate residuals is 96 femtoseconds/s.

Estimated parameters in the least squares fit included radio source positions, station locations, station clocks, wet tropospheric zenith delays, and tropospheric gradients. A no net rotation constraint was placed on the radio source positions relative to ICRF2 (Ma et al. 2009). Kinematic station positions were estimated daily for each site relative to ITRF2008 (Altamimi et al. 2011) apriori values and rates. Station clocks were estimated using a linear model reset about every 6 h. Wet zenith tropospheric delays were estimated roughly once per hour based on the VMF1 mapping function. Tropospheric gradients were estimated about once per day relative to apriori values from the Steigenberger equation (N-S = −0.5*sin(2*lat) mm) with 100 mm level a priori sigmas. The COMB2009 earth orientation solution (Ratcliff and Gross 2010) was used to fix UTPM values. Precession and nutation models were from MHB2000 (Matthews et al. 2002). Constant offsets for the PSIA and EPSA model parameters were estimated in order to avoid over constraint

M. Heflin (✉) • C. Jacobs • O. Sovers • A. Moore • S. Owen
Jet Propulsion Laboratory, California Institute of Technology, 4800 Oak Grove Drive, Pasadena, CA 91109, USA

Z. Altamimi and X. Collilieux (eds.), *Reference Frames for Applications in Geosciences*, International Association of Geodesy Symposia 138, DOI 10.1007/978-3-642-32998-2_38,

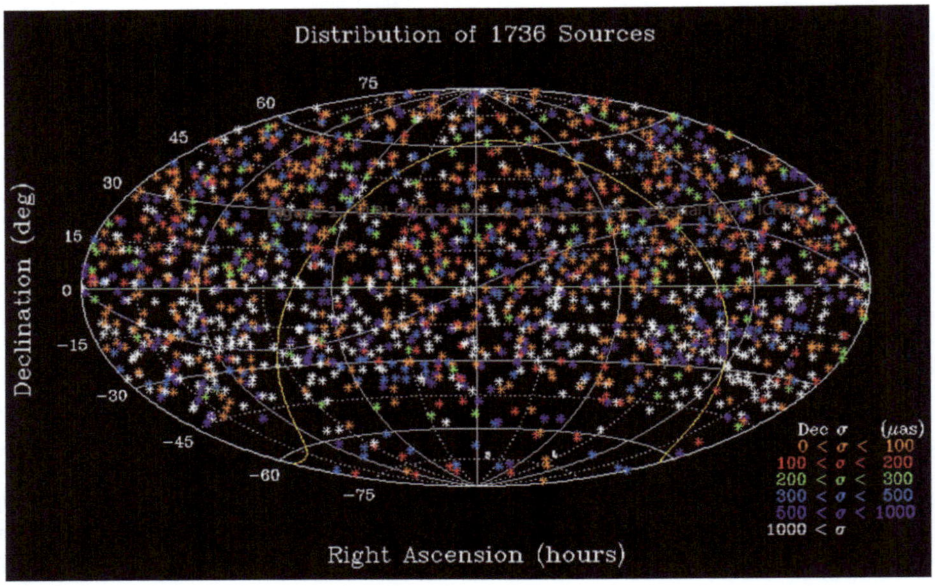

Fig. 38.1 VLBI radio source coordinates in the celestial frame ICRF2

Table 38.1 VLBI celestial frame alignment and residuals with respect to ICRF2

Parameter	Value (μas)	Sigma (μas)
Rx	4.8	5.3
Ry	2.4	5.5
Rz	1.0	4.4
Ra cos(dec) WRMS	57.3	
Dec WRMS	93.3	

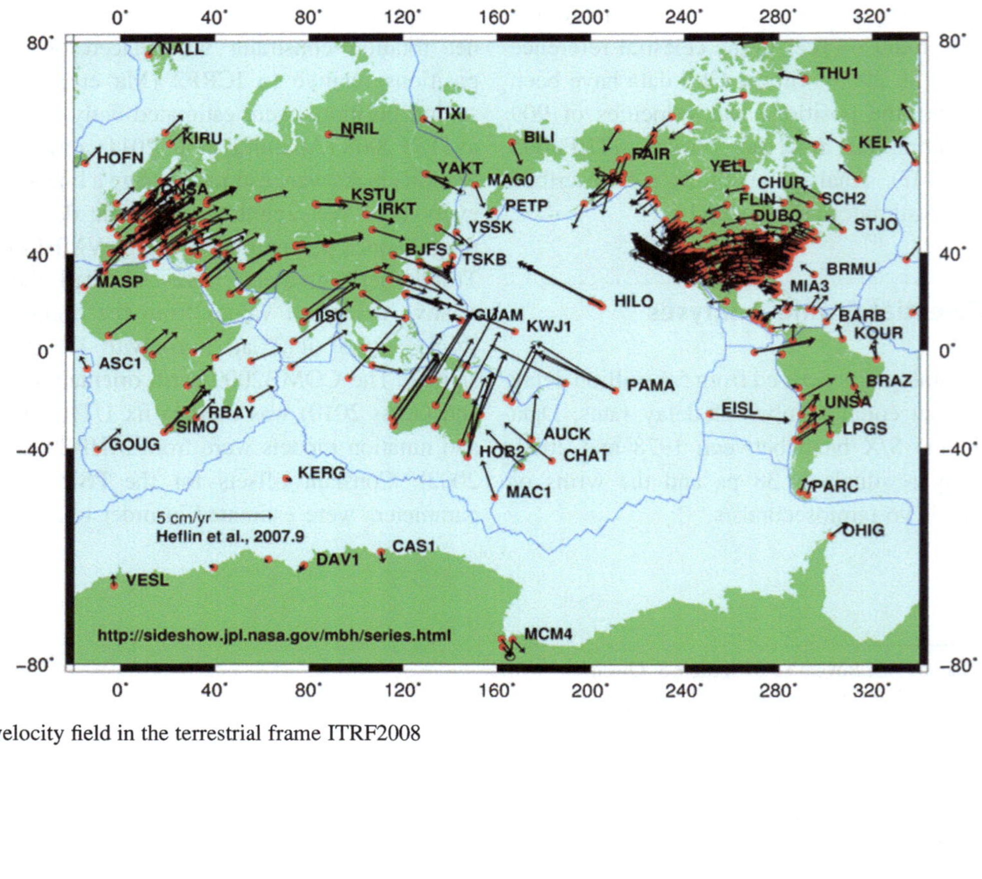

Fig. 38.2 GPS velocity field in the terrestrial frame ITRF2008

Table 38.2 GPS terrestrial frame alignment and residuals with respect to ITRF08

Parameter	Value	Sigma
Rx	−0.58 mm	0.13 mm
Ry	0.45 mm	0.14 mm
Rz	−1.42 mm	0.14 mm
Tx	−1.55 mm	0.12 mm
Ty	0.05 mm	0.12 mm
Tz	−3.40 mm	0.13 mm
Scale	2.51 mm	0.13 mm
N WRMS	1.6 mm	
E WRMS	2.1 mm	
V WRMS	4.5 mm	
Rx rate	−0.120 mm/year	0.016 mm/year
Ry rate	0.188 mm/year	0.017 mm/year
Rz rate	−0.340 mm/year	0.017 mm/year
Tx rate	0.236 mm/year	0.014 mm/year
Ty rate	0.045 mm/year	0.015 mm/year
Tz rate	0.048 mm/year	0.016 mm/year
Scale rate	−0.129 mm/year	0.016 mm/year
N rate WRMS	0.3 mm/year	
E rate WRMS	0.3 mm/year	
V rate WRMS	0.9 mm/year	

with respect to ICRF2.Celestial frame results are illustrated in Fig. 38.1 and summarized in Table 38.1.

3 GPS Terrestrial Frame Analyses

The GPS station catalog was derived from 2.1 million station-days of data from 909 sites. Orbits, clocks, and transformation parameters were used to compute precise point positions (Zumberge et al. 1997) in the ITRF2008 terrestrial frame (Altamimi et al. 2007) for each station-day based on 24 h of phase and range data. Estimated parameters included positions, tropospheric zenith delays, tropospheric gradients, clocks, and unresolved phase bias parameters. A segmented linear fit was computed for the time series of each site. Comparison with ITRF2008 was carried out by estimating 14 transformation parameters. Terrestrial frame results are illustrated in Fig. 38.2 and summarized in Table 38.2.

4 Conclusions

VLBI data have been analyzed to create a catalog of 1736 sources which agrees with ICRF2 defining coordinates at the 57–93 μas level. GPS data have been processed to create a catalog of 909 stations which agrees with ITRF2008 at the 1.6–4.5 mm level for positions and at the 0.3–0.9 mm/year level for velocities.

Acknowledgements The research described herein was performed at the Jet Propulsion Laboratory of the California Institute of Technology, under a contract with the National Aeronautics and Space Administration. Government sponsorship acknowledged.

References

Altamimi Z, Collileux X, Legrand J, Garayt B, Boucher C (2007) ITRF2005: a new release of the international terrestrial reference frame based on the time series of station positions and Earth orientation parameters. J Geophys Res 112:B09401. doi:10.1029/2007JB004949

Altamimi Z, Collileux X, Metivier L (2011) ITRF2008: an improved solution of the international terrestrial reference frame. J Geodesy 8(8):457–473. doi:10.1007/s00190-011-0444-4

Ma C et al. (2009) The 2nd realization of the ICRF by VLBI. In: Fey AL, Gordon D, Jacobs CS IERS Techinal note no. 35: IERS, Frankfurt

Matthews PM, Herring TA, Buffet BA (2002) Modeling of nutation and precession: new nutation series for nonrigid Earth and insights into the Earth's interior. J Geophys Res 107:B4. doi:10.1029/2001JB000390

Ratcliff JT, Gross RS (2010) Combinations of Earth orientation measurements: space 2009, COMB2009, and POLE 2009. JPL Publication, Pasadena, CA, pp 10–22

Zumberge J, Heflin M, Jefferson D, Watkins M, Webb F (1997) Precise point positioning for the efficient and robust analysis of GPS data from large networks. J Geophys Res 102(B3). doi:10.1029/96JB03860

Using Modified Allan Variance for Time Series Analysis

39

Z. Malkin

Abstract

Allan Variance (AVAR) was introduced more than 40 years ago as an estimator of the stability of frequency standards. Now it is also used for investigations of time series in astronomy and geodesy. However, there are several issues with this method that need special consideration. First, unlike frequency measurements, astronomical and geodetic time series usually consist of data points with unequal uncertainties. Thus one needs to apply data weighting during statistical analysis. Second, some sets of scalar time series naturally form multidimensional vector series. For example, Cartesian station coordinates form the 3D station position vector. The original AVAR definition does not allow one to process unevenly weighted and/or multidimensional data. To overcome these deficiencies, AVAR modifications were proposed in Malkin (2008. On the accuracy assessment of celestial reference frame realizations. J Geodesy 82: 325–329). In this paper, we give some examples of processing geodetic and astrometric time series using the classical and the modified AVAR approaches, and compare the results.

Keywords

Allan variance • Time series analysis

1 Introduction

The scatter of a geodetic time series provides a good measure of the series quality (and its more explicit statistical characteristics). One of the most effective approaches to analyze the noise component of a measured signal is the Allan Variance (AVAR) originally developed for the evaluation of the stability of time and frequency standards (Allan 1966). In particular, its advantage is the weak dependence of the noise parameter estimate on low-frequency components of the signal under study.

AVAR has already had a rather long history in geodesy and astrometry. It was used to study station coordinates time series (Malkin and Voinov 2001; Le Bail and Feissel-Vernier 2003; Le Bail 2006; Feissel-Vernier et al. 2007), radio source position stability (Feissel et al. 2000; Gontier et al. 2001; Feissel-Vernier 2003), the Earth orientation parameters (EOP) series (Gambis 2002), geocenter motion (Feissel-Vernier et al. 2006). AVAR can be also used to analyze the spectral characteristics of the signal (Feissel et al. 2000; Feissel-Vernier 2003; Feissel-Vernier et al. 2006, 2007) and Hurst parameter (Bregni and Primerano 2005). A detailed review and history of using AVAR in astrometry and geodesy can be found in Malkin (2011).

However, application of AVAR to analysis of the time series of geodetic and astrometric measurements yields sometimes unsatisfactory results because the analyzed series consists of the data points with unequal uncertainties. This necessitates proper weighting not provided by the original AVAR definition. Besides, we often deal with multidimensional values, e.g., terrestrial coordinates and/or velocities (3D or 6D), celestial coordinates and/or proper motions (2D or 4D). To provide adequate analysis of such kind of data,

Z. Malkin (✉)
Pulkovo Observatory, St. Petersburg, 196140, Russia
e-mail: malkin@gao.spb.ru

Z. Altamimi and X. Collilieux (eds.), *Reference Frames for Applications in Geosciences*,
International Association of Geodesy Symposia 138, DOI 10.1007/978-3-642-32998-2_39,

AVAR modifications WAVAR, MAVAR, and WMAVAR have been proposed by Malkin (2008a). They are explained in Sect. 2.

In Sect. 3, several practical examples are given to show how proposed AVAR modifications can be applied to investigation of real geodetic and astrometric data. These examples also allow us to better understand some practical features of various types of AVAR estimates.

We do not consider here the applications of AVAR to spectral or persistency analysis. Such applications require computation of modified AVAR on different averaging intervals, which should be straightforward.

2 AVAR and Its Modifications

In this section, we give definitions of AVAR modifications proposed in Malkin (2008a) to analyze real time series. The classical AVAR is defined as

$$\sigma^2 = \frac{1}{2(n-1)} \sum_{i=1}^{n-1} (y_i - y_{i+1})^2 . \quad (39.1)$$

where y_i are measurements, $i = 1,\ldots,n$.

This definition is not satisfactory for the analysis of real measurements, which generally have different uncertainties and thus should be unevenly weighted during analysis to obtain realistic and statistically meaningful estimates for the investigated parameters. For such kind of data, the weighted AVAR, WAVAR, is introduced in Malkin (2008a). It can be defined as

$$\sigma_1^2 = \frac{1}{2p} \sum_{i=1}^{n-1} p_i (y_i - y_{i+1})^2, \qquad p = \sum_{i=1}^{n-1} p_i, \qquad p_i = (s_i^2 + s_{i+1}^2)^{-1} . \quad (39.2)$$

where s_i are the measurement uncertainties.

Another modification of AVAR proposed by Malkin (2008a) is intended for processing of multidimensional data. The latter can be considered as a k-dimensional vector $y_i = (y_i^1, y_i^2, \ldots, y_i^k)$, $i = 1, \ldots, n$, with standard errors $s_i = \left(s_i^1, s_{i,}^2 \ldots, s_i^k\right)$. The corresponding weighted multidimensional AVAR, WMAVAR, is given by

$$\sigma_2^2 = \frac{1}{2p} \sum_{i=1}^{n-1} p_i d_i^2, \qquad d_i = |y_i - y_{i+1}|, \qquad p = \sum_{i=1}^{n-1} p_i . \quad (39.3)$$

Strictly speaking, the weights p_i should be computed in accordance with the error propagation law.

$$p_i = \left(\sum_{j=1}^{k} \left\{ \left[\left(y_i^j - y_{i+1}^j\right) / d_i \right]^2 \left[\left(s_i^j\right)^2 + \left(s_{i+1}^j\right)^2 \right] \right\} \right)^{-1} . \quad (39.4)$$

However, the expression Eq. 39.4 has a singular point in the case of $di = 0$, i.e. of equal adjacent measurements. Hence, this case require special treatment, e.g. assigning unit weight for close (near-)zero value of *di*. To avoid the singularity, the following simplified formula for p_i is recommended in Malkin (2008a) for practical use:

$$p_i = \left(\sum_{j=1}^{k} \left[\left(s_i^j\right)^2 + \left(s_{i+1}^j\right)^2 \right] \right)^{-1} . \quad (39.5)$$

No significant difference between results computed with weights given by Eqs. 39.4 and 39.5 was found during real data processing.

In the case of $p_i = 1$, we have unweighted multidimensional AVAR (MAVAR). It is easy to see that WMAVAR is a universal definition which includes all others as special cases. It can be also noted that for homogeneous time series with close s_i, MAVAR (WMAVAR) values should be equal to AVAR (MAVAR) ones computed for one of k vector dimensions and multiplied by $\sqrt{k}$.

In noise component analysis, one often uses Allan Deviation ADEV computed as the square root of AVAR. Correspondingly, ADEV modifications, WADEV, MADEV, and WMADEV can be used in most practical applications, including this study.

3 Practical Examples

In this section, we give several examples of using modified ADEV. Four examples will be considered: comparison of station displacement time series, investigation of source position stability, comparison of celestial pole offset (CPO) time series, and quality assessment of celestial reference frame (CRF) realization (radio source position catalogues).

3.1 Station Position Series

ADEV is an effective tool for investigation of noise characteristics of station position time series (Malkin and Voinov 2001; Le Bail and Feissel-Vernier 2003; Le Bail 2006; Feissel-Vernier et al. 2007). However, the classical ADEV applied to a series with heterogeneous

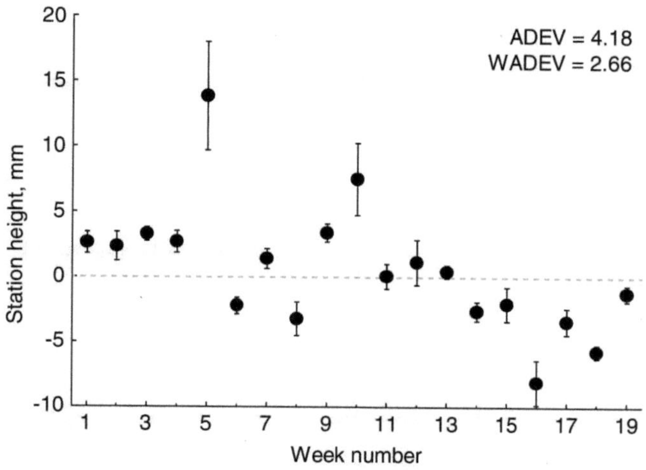

Fig. 39.1 Difference between ADEV and WADEV in case of station height time series with implied outliers

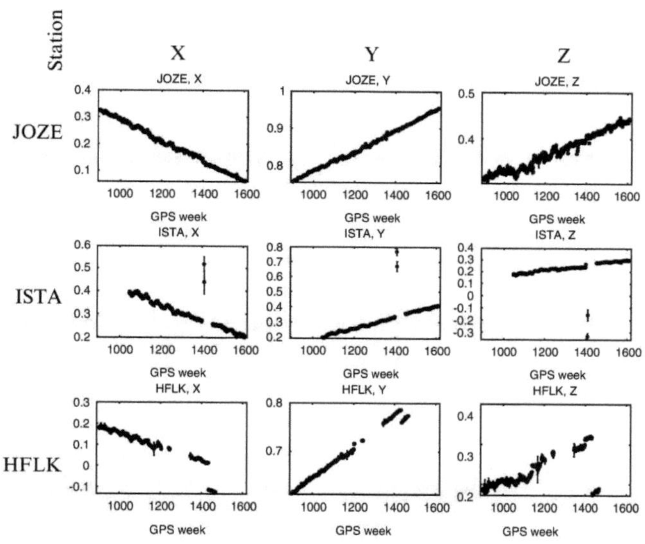

Fig. 39.2 EPN station position time series. X, Y, and Z are the station Cartesian geocentric coordinates. Unit: m

uncertainty can give a biased estimate of the actual signal characteristics.

As the first example, let us consider a series of weekly station height estimates depicted in Fig. 39.1. One can see that one to three points may be considered as excludable because of large bias. Their inclusion in detailed analysis (e.g. scatter statistics) depends on the method used to detect outliers. Using WADEV mitigates the impact of implied outliers without application of special statistics.

Now, let us consider three station position time series provided by the European Permanent GPS Network (EPN) Central Bureau.[1] These series are shown in Fig. 39.2, and

Table 39.1 Scatter characteristics for EPN station position time series shown in Fig. 39.2. Unit: mm

	ADEV			WADEV			
Station	X	Y	Z	X	Y	Z	3D WADEV
JOZE	2.9	1.1	2.7	1.6	1.0	2.0	2.8
ISTA	10.2	17.1	24.0	2.3	1.5	1.9	3.3
HFLK	5.2	1.9	5.6	20.2	2.9	21.2	28.9

scatter statistics is given in Table 39.1. Station JOZE shows stable behavior without jumps and outliers; for this station all the estimates give close results. Station ISTA has two outliers of several decimeters with large uncertainties; for this station using unweighted estimate gives unsatisfactory result, and using WAVAR allows us to practically eliminate outliers. The last case of station HFLK shows in contrast to the previous example unsatisfactory result obtained with WAVAR.

The reason is that the HFLK position uncertainties were relatively large in the period when the station showed position stability and became much smaller to the end of the series where the position jump occurred. As a result, the position estimates around the jump epoch were entered to the statistics with large weight.

3.2 Source Position Series

Investigation of the noise characteristics of source position time series are usually aimed at ranking of sources by time series statistics, and compiling list of sources that are not stable enough to be solved in VLBI global solution, and require special handling. In particular, computation of some quantitative source stability indices are desirable to make a selection of the International Celestial Reference Frame (ICRF) defining sources as objective as possible (Ma et al. 1998; Feissel-Vernier 2003; Gordon et al. 2008; Malkin 2008b).

In Malkin (2009) we investigated source position time series submitted by VLBI analysis centers in the framework of preparation of the Second ICRF realization ICRF2 (Ma et al. 2009). These series were computed using different software and/or analysis strategies, which makes their comparison especially interesting and instructive. Each analyzed series consists of source positions obtained from analysis of 24-h observing sessions (one source position estimate for each session).

Out of several scatter indices generally used for analysis of source position time series we used the following two: weighted root-mean-square (WRMS) residuals of the session source position with respect to weighted mean position (the weights are inversely proportional to the square of the reported position uncertainty), and WADEV. Our study

[1] http://epncb.oma.be/

(Malkin 2009) confirmed that both statistics have their own advantages and disadvantages. WRMS estimate is affected by trend-like and low-frequency signal components.

In contrast, WADEV does not depend on the slow position variations. However, it may give inadequate estimate of the scatter index in the case when the time series contains jumps but is stable between jumps. In order to get a more general measure of source position stability, a composite index of WRMS and WADEV can be used.

It can be noted that some examples given above may seem to be artificial. It is common practice to identify and reject outliers before further statistical analysis. However, as already mentioned above, it can be not a trivial in all the time series considered here, and others we meet in our work. So, our intention is to show that weighted ADEV (WADEV) provides more robust estimate in case of outliers, provided the outliers have large uncertainties, which is quite common, see e.g. Figs. 39.1 and 39.3 (next section). In such a case, the result obtained with WADEV is more independent of the quality of outlier detection procedure used.

3.3 CPO Series Comparison

Investigation of the noise characteristics of CPO time series is important for different tasks, such as quality assessment of EOP series and weighting EOP series during combination (Gambis 2002). Here we consider ten series of the CPO coordinates *dX* and *dY* computed at International VLBI Service for Geodesy and Astrometry (IVS) analysis centers,[2] including the IVS combined series. They are depicted in Fig. 39.3 in a deliberately small scale to show most of the outliers.

A visual inspection of Fig. 39.3 reveals, for example, that OPA series seems to be the noisiest, and IAA series seems to be practically noiseless. However, a deeper look at the data shows that the first impression is wrong. First of all, the series contain different number of sessions. One also notices that most of the outliers have enormously large uncertainties. Both factors should be taken into account. In particular, it requires to compute weighted scatter index and compare subsets of original series consisted of the sessions common for all compared CPO series.

In Tables 39.2 and 39.3, the results of computation of several statistics are presented. The former table shows the impact of data weighting only. The latter table provides scatter indices computed using the same data set for all the series, i.e. free of selection effect. The statistics used are WRMS, MADEV (unweighted 2D ADEV estimate) and WMADEV (weighted 2D ADEV estimate). WRMS values given in the table are computed as average of those for dX and dY. In fact, WRMS1 is just the WRMS value of the reported CPO estimates. Every statistics is computed in two variants: for

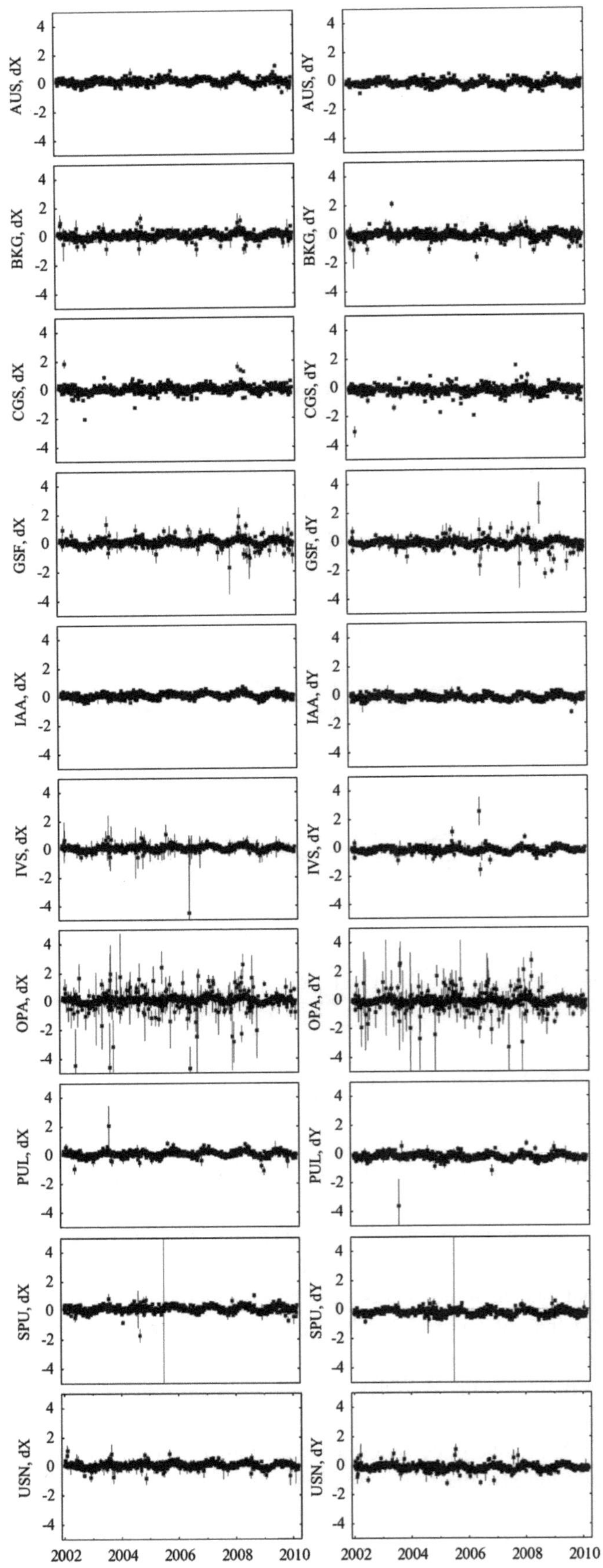

Fig. 39.3 VLBI CPO time series. Units: mas

Table 39.2 Statistics of the VLBI CPO time series depicted in Fig. 39.3. The statistics indexed with 1 are computed from raw data, the statistics indexed with 2 are computed after removing CPO model and linear trend. Units: μas

Series	No. of sessions	WRMS1	WRMS2	MADEV1	MADEV2	WMADEV1	WMADEV2
AUS	862	251	113	206	206	177	177
BKG	1212	191	101	245	245	163	163
CGS	1083	224	129	301	300	200	200
GSF	1276	197	86	279	279	137	137
IAA	1101	211	97	173	173	146	146
IVS	1158	202	80	236	236	125	126
OPA	1437	189	87	680	680	149	148
PUL	1167	213	87	215	215	137	137
SPU	854	253	115	227	227	180	180
USN	1208	195	86	200	200	132	132

Table 39.3 The same as Table 39.2, but computed for common sessions

Series	No of sessions	WRMS1	WRMS2	MADEV1	MADEV2	WMADEV1	WMADEV2
AUS	740	251	113	195	195	176	176
BKG	740	197	101	180	180	159	158
CGS	740	226	128	222	221	203	203
GSF	740	199	82	145	145	131	130
IAA	740	214	98	165	165	151	151
IVS	740	209	79	139	139	125	125
OPA	740	189	78	141	141	128	127
PUL	740	217	90	152	152	139	139
SPU	740	254	114	223	223	179	179
USN	740	192	85	146	146	132	132

Table 39.4 Statistics of the two CPO time series computed with two source catalogues ICRF1 and ICRF2. Unit: μas

CRF	WADEV(dX)	WADEV(dY)	WMADEV
ICRF1	102	107	149
ICRF2	92	90	129

original CPO estimates as reported by authors (indexed with 1), and for values corrected for CPO model ZM2 (Malkin 2007) and linear trend (indexed with 2). A detailed discussion of the results of these computations is beyond the scope of this paper. Let us just mention several conclusions.

- The comparison of "1" and "2" variants confirms that WRMS depends heavily on the model of systematic errors, whereas WADEV and WMADEV do not.
- Using weighted ADEV estimates allows us to severely mitigate the influence of outliers. It is especially visible for the OPA series which includes many CPO estimates having large bias and uncertainty (coming from sessions with poor geometry), evidently for completeness. However it affects less the CGS statistics because the latter includes CPO estimates with large bias but small uncertainty (see Fig. 39.3).
- Weighted ADEV estimate provides robust statistics for ranging CPO series which does not practically depend on low-frequency components of the time series. In particular, it can be mentioned that combined IVS series shows the least noise level.

One more result not included in these tables is that 2D WMADEV estimates are very close to the average of WADEV estimates computed for dX and dY multiplied by $\sqrt{2}$. Thus the multidimensional ADEV provides more compact expression for noise characteristics.

3.4 Comparison of CRF Realization

Noise level comparison of CPO time series computed with different radio source catalogues is one of very few absolute methods (maybe the only method proposed so far) for quality assessment of the CRF realizations. In Malkin (2008b), Sokolova and Malkin (2007), this method was used for the first time to compare several catalogues. Here we use it for comparison of the first and the second ICRF realizations. For this purpose, we computed two CPO series using ICRF1 (ICRF-Ext.2, Fey et al. 2004) and ICRF2 (Ma et al. 2009) source positions. Then we computed their scatter level using modified ADEV estimates WADEV and 2D WMADEV.

From the results presented in Table 39.4, one can see that the scatter of the CPO series computed with ICRF2 is substantially smaller than for series computed with ICRF1. We consider this result as clear evidence of better accuracy of the ICRF2 source positions as compared with the ICRF1.

4 Summary

The modified Allan variation (AVAR) and associated Allan deviation (ADEV) estimators proposed in Malkin (2008a) for processing unevenly weighted and/or multidimensional data is an effective and convenient tool for geodetic and astronomical time series scatter analysis. An important advantage of ADEV is its low sensitivity to low-frequency signal variations as compared to WRMS, which heavily depend on the model used to separate the stochastic and systematic signal components. However, both the original and the modified ADEV may inadequate estimate the noise level when jumps are present the time series.

The AVAR (ADEV) modifications, WAVAR (WADEV) and WMAVAR (WMADEV) developed to process unevenly weighted one- and multidimensional measurements, allow us to get noise characteristics that are less sensitive to outliers than the original ADEV estimate, provided the outliers have exaggerated uncertainties, which is usually the case. It is clear that the WMADEV (WMAVAR) definition is the most general as it includes the original AVAR definition and its modifications, WADEV and MADEV, as special cases.

For thoroughness' sake, it should be mentioned that the original AVAR is computed on evenly spaced time series. Unfortunately, it is not the case for many geodetic and astronomical applications. Many time series have regular time span however, e. g. daily or weekly station coordinates, daily troposphere parameters, etc. Other series can be made (near)-regular by averaging measured data over equal intervals. For example, such a method was used by Feissel-Vernier (2003) for radio source position time series. On the other hand, if AVAR and its modifications are used as a series scatter index, uneven spacing should not influence result of analsys, to our mind. And otherwise, averaging the original series may lead to loss of information.

Acknowledgements This work has made use of data and products provided by the International VLBI Service for Geodesy and Astrometry (IVS, Schlueter and Behrend 2007) and European Permanent GPS Network (EPN, Bruyninx and Roosbeek 2006). The author is grateful to two anonymous reviewers for careful reading of the manuscript and helpful comments and suggestions.

References

Allan DW (1966) Statistics of atomic frequency standards. Proc IEEE 54:221–230

Bregni S, Primerano L (2005) Using the modified allan variance for accurate estimation of the hurst parameter of long-range dependent traffic. In: arXiv: cs/0510006, Milano

Bruyninx C, Roosbeek F (2006) The EUREF permanent network: recent achievements. EUREF Publication No. 16, Mitteilungen des Bundesamtes für Kartographie und Geodäsie, Band, vol 40, pp 105–112

Feissel M, Gontier A-M, Eubanks TM (2000) Spatial variability of compact extragalactic radiosources. Astron Astr 359:1201–1204

Feissel-Vernier M (2003) Selecting stable extragalactic compact radio sources from the permanent astrogeodetic VLBI program. Astron Astr 403:105–110. doi:10.1051/0004-6361:20030348

Feissel-Vernier M, Le Bail K, Berio P et al (2006) Geocentre motion measured with DORIS and SLR, and predicted by geophysical models. J Geodesy 80:637–648. doi:10.1007/s00190-006-0079-z

Feissel-Vernier M, de Viron O, Le Bail K (2007) Stability of VLBI, SLR, DORIS, and GPS positioning. Earth Planets Space 59:475–497

Fey AL, Ma C, Arias EF et al (2004) The second extension of the international celestial reference frame: ICRF-Ext.2. Astron J 127:3587–3608. doi:10.1086/420998

Gambis D (2002) Allan variance in earth rotation time series analysis. Adv Space Res 30:207–212. doi:10.1016/S0273-1177(02)00286-7

Gontier A-M, Le Bail K, Feissel M, Eubanks T-M (2001) Stability of the extragalactic VLBI reference frame. Astron Astr 375:661–669. doi:10.1051/0004-6361:20010707

Gordon D, Ma C, Gipson J, Petrov L, MacMillan D (2008) On selection of "defining" sources for ICRF2. In: Finkelstein A, Behrend D (eds) Proceedings of fifth IVS general meeting, pp 261–264

Le Bail K (2006) Estimating the noise in space-geodetic positioning: the case of DORIS. J Geodesy 80:541–565. doi:10.1007/s00190-006-0088-y

Le Bail K, Feissel-Vernier M (2003) Time series statistics of the DORIS and GPS colocated observations. Geophysical Reseaech Abstracts, EGS-AGU-EUG Joint Assembly, vol 5, Nice, 6–11 April 2003, p 04078

Ma C, Arias EF, Eubanks TM et al (1998) The international celestial reference frame as realized by very long baseline interferometry. Astron J 116:516–546

Ma C, Arias EF, Bianko G et al. (2009) The second realization of the international celestial reference frame by very long baseline interferometry. In: Fey A, Gordon D, Jacobs CS (eds) Presented on behalf of the IERS/IVS working group, IERS technical note no. 35, Frankfurt am Main, Verlag des Bundesamts für Kartographie und Geodäsie

Malkin ZM (2007) Empiric models of the Earth's free core nutation. Solar Syst Res 41:492–497. doi:10.1134/S0038094607060044

Malkin Z (2008a) On the accuracy assessment of celestial reference frame realizations. J Geodesy 82:325–329. doi:10.1007/s00190-007-0181-x

Malkin Z (2008b) On construction of ICRF-2. In: Finkelstein A, Behrend D (eds) Proceedings of fifth IVS general meeting, pp 256–260

Malkin Z (2009) Some results of analysis of source position time series. IVS Memorandum 2009-001v01. http://ivscc.gsfc.nasa.gov/publications/memos/

Malkin ZM (2011) Study of astronomical and geodetic series using the allan variance. Kinemat Phys Celest Bodies 27:42–49. doi:10.3103/S0884591311010053

Malkin ZM, Voinov AV (2001) Preliminary results of processing EUREF network observations using a non-fiducial strategy. Phys Chem Earth (A) 26:579–583. doi:10.1016/S1464-1895(01)00104-1

Schlueter W, Behrend D (2007) The international VLBI service for geodesy and astrometry (IVS): current capabilities and future prospects. J Geodesy 81:379–387. doi:10.1007/s00190-006-0131-z

Sokolova J, Malkin Z (2007) On comparison and combination of catalogues of radio source positions. Astron Astr 474:665–670. doi:10.1051/0004-6361:20077450

The Role of the TRS in Precision Agriculture: DGPS with EGNOS and RTK Positioning Using Data from NTRIP Streams

40

I. Osório and M. Cunha

Abstract

For Precise Agriculture purposes, several steps of a maize crop-system were recorded by the use of a GPS receiver with EGNOS and RTK capabilities. The field is about 35 km far from two GNSS CORS, one from RENEP, operated by IGS, and the other from SERVIR, operated by IGEoE. Both networks disseminate real-time GNSS data streams over the Internet using the NTRIP protocol. The GNSS data streams from RENEP reference stations (including validated station coordinates) provide the user with a real-time access to the ETRS89 and, those same streams from IGEoE, a military institution, are in ITRS, allowing large scale scientific applications. The validation of the EGNOS and the RTK solutions, obtained in the two TRS systems, was achieved by the results from post-processed measurements. RTK solutions, when compared to the post-processed values in the same TRS, show sub-decimeter accuracy what is enough for many of the Precision Agriculture studies. However, the two RTK solutions have a translation with a magnitude of the order of 0.5 m that can be explained by the independence of the ETRS89 on the continental drift. Indeed, at the zone where the field is located, while the ETRFyy Cartesian coordinates have velocities less than 1 mm/year, the ITRFyy Cartesian coordinates have velocities greater than 1 cm/year, what give rise to a point position variation with a magnitude of 2.5 cm/year.

In order to correlate the tractor velocity, during a pre-emergence herbicide application, to the terrain slope, the field orthometric heights were obtained by the use of GRS80 ondulations, on a $1.5' \times 1.5'$grid, in the local Portuguese geoid model GeodPT08. The global precision of this model is estimated in 4 cm, which is within the error for the real time solutions obtained.

Keywords

Precision agriculture • TRS • RTK positioning • GNSS • SBAS • GBAS

1 Introduction

Precision Agriculture is a management system where crop production practices and inputs such as seed, fertilizers and pesticides are variably applied within a field, which is considered as a solution to keep a balance between sustainable development and environmental benefits.

The risk of agronomic and environmental damage associated to the maize-application of persistent herbicides is very high. Uniform applications of maize herbicides, very much dependent on the technologies and the operator, are an important way to minimize the environmental and agronomic risks. It was recognized that the maintenance of correct distance between contiguous runs and tractor velocity although repetitive, is one of the most difficult tasks for

I. Osório (✉) • M. Cunha
Faculdade de Ciências, Universidade do Porto, Rua do Campo Alegre, 687, Porto 4169-007, Portugal
e-mail: iposorio@fc.up.pt; mcunha@mail.icav.up.pt

Z. Altamimi and X. Collilieux (eds.), *Reference Frames for Applications in Geosciences*, International Association of Geodesy Symposia 138, DOI 10.1007/978-3-642-32998-2_40,

an operator. This is particularly so when wide equipment is used (McDougall et al. 2001; Easterly et al. 2010).

The use of GNSS receivers represents a rapidly expanding technology in precision agriculture to perform uniform applications of herbicides, as well as other crop inputs, despite the presence of heterogeneous physiographic field, farm equipment and environmental conditions (Pérez-Ruiz et al. 2010). However, in stand alone mode, even with a very precise GNSS receiver, the accuracy of the position solution is in a range of 10 m for planimetry and 30 m for altimetry. This accuracy is not enough for some steps of precision agriculture where it is needed the management of each crop production input, like fertilizers, herbicides, seeds, on a site-specific basis to reduce waste, increase profits and maintain the quality of the environment. The real-time GNSS positioning accuracy can be improved with SBAS and GBAS.

The EGNOS is anopen SBAS available since 1 October 2009. It allows users, in Europe, to find their positions to within 1.5 m against the mentioned10–30 m for the stand alone GNSS positioning.

In which concerns GBAS, at present, in Portugal, there are two official networks, RENEP and SERVIR, operated, respectively, by IGP and by IGEoE. By the use of NTRIP streams, both transmit GNSS real-time data via Mobile Internet allowing precisions of a few centimeters in RTK positioning. There are some differences between the ways as these Augmentation Systems work, but the main and crucial one is the TRS: positions from RENEP are in ETRS89, a system realized by stations over Europe, while, positions from SERVIR are in ITRS a system whose frames result from stations all over the world. As we are in Europe, these last ones real-time GNSS solutions are affected by the tectonic movement.

In this work, the geometric solutions accuracy obtained in real time with two TRS systems as well as the post processing data were analyzed. The field results were discussed in the context of the Precision Agriculture applications.

2 GNSS Data Acquisition for Precision Agriculture Purposes

In order to monitor several steps of a maize production, in a field of approximately 1.3 ha, located in the northern part of Portugal, it was decided to use a GPS receiver since GPS is the only GNSS with Full Operational Capability (FOC). So, a Trimble R7 GPS receiver, with a Zephyr Geodetic antenna and a Trimble TSC2 data collector were used. The R7 is a double frequency receiver that, among several features, has WAAS/EGNOS capability and an adaptive dual-frequency RTK engine. The Trimble TSC2 is a handheld field computer, with the Microsoft Windows Mobile operating system running the Trimble Survey Controller field software. All the measurements were obtained in continuous kinematic mode, with an elevation mask of 10 and at a cadency of 1 s, for the measurements with the antenna mounted on a tractor, and 5 or 10 s in the other cases.

The positioning solution given by a GNSS receiver in standalone mode is not accurate enough for precision agriculture purposes. Indeed, even after the removal of the Selective Availability voluntary degradation, introduced in the GPS signals until May2 2002,the GPS Control Segment policy dictates a predictable positioning accuracy of 10 m horizontally, at the 95 % confidence level, and 30 m (95 %) vertically. With such errors, accurate real time solutions require the appeal to a GNSS augmentation system. Several augmentation systems were used accordingly to the phase of the maize production.

The weed control was made with residual herbicides applied to the maize crop by tractor mounted sprayer-boom. This operation took place at May 15 2010. As the field is not a flat piece of land there is interest in the knowledge of the different heights in the field, for instance, to evaluate how this affects the tractor velocity and, consequently, the rate of herbicide application. For these measurements EGNOS was used for an easier manipulation of the material installed in a conventional tractor. During the tractor trajectory the data acquisition was stopped twice, for some minutes, for sprayer tank refill.

One and half months later the maize field limits, including the two strips for irrigation and other mechanical operations, were registered in RTK mode by the use of a Bluetooth connection between the data collector and a mobile phone, provided with mobile internet to allow the reception of data from NTRIP streams.

As said before, the RTK data corrections were got from two official GBAS broadcasting NTRIP streams: RENEP and SERVIR. The reception of data from these NTRIP broadcasters is tax free requiring only the user registration. The first one broadcast RTK corrections in single base station mode while the RTK positioning with the second one is based on the concept of VRS, a service that uses data from several CORS, around the user, to compute the broadcast corrections. This is a very convenient method in situations where the mobile receiver is far from all the Reference Stations. But, this is only a second order difference between the GNSS networks. What really is fundamental to take into account, is the Reference System.

2.1 The TRS Used by the Portuguese NTRIP Broadcasters

A TRS is a spatial reference system co-rotating with Earth in its diurnal motion in space. Its materialization is known as TRF, through the realization of its origin, scale, axes orientation and their time evolution. Seven parameters are needed to fix a TRF at a given epoch, to which are added another seven, their time derivatives, to define the TRF time evolution. These 14 parameters are called the datum definition and space geodesy techniques (LLR, SLR, VLBI, GPS, DORIS) are not sensitive to all the parameters needed for the completely establishment of the TRF. For this reason constraints must be, also, introduced in addition to the space geodesy observations.

SERVIR is a GNSS CORS network operated by IGeoE (http://www.igeoe.pt), a military institution that adopted the ITRS to allow large scale scientific applications.

ITRS is a geocentric system whose definition (IERS Technical Note No. 36, 4.1.4) fulfills, among others, the condition that the time evolution of the orientation is ensured by using a no-net-rotation condition with regards to horizontal tectonic motions over the whole Earth. This is a constraint that, until ITRF2005, was applied by aligning the orientation time evolution of the frame to that of the geophysical model NNR-NUVEL 1 or 1A. For instance, ITRF2000 provides an accurate estimation of relative motions for six major tectonic plates (Altamini et al. 2003). However, in ITRF2005, and for the first time, the input data are time series (weekly from satellite techniques and 24-h session-wise from Very Long Baseline Interferometry) of station positions and daily Earth Orientation Parameters (EOPs). A velocity field of 152 sites with an error less than 1.5 mm/year is used to estimate absolute rotation poles of 15 tectonic plates that are consistent with the ITRF2005 frame (Altamini et al. 2007).

So, an ITRS common user sees the Cartesian coordinates changing with a velocity of a few centimeters by year, what is not very convenient for some applications. However, it must be said that these velocities (v_x, v_y, v_z) are widely available for all the stations of the IERS network trough several official sites like those of IERS, EUREF or IGS. For instance, in the northern part of Portugal, a granitic zone, the vector velocity for ITRF2005 coordinates of the EPN station GAIA is (−0.0088, 0.0192, 0.0123)m/year being its magnitude of 2.5 cm/year.

RENEP is a GNSS CORS network operated by IGP (http://www.igeo.pt), the Portuguese Institute for Geodesy and Cartography. Some of its stations belong to EPN, so must follow EUREF guidelines. Accordingly to these guidelines the GNSS data streams must provide the user with a real-time access to the ETRS89.

The ETRS89 is a TRS whose definition specifies two main conditions: the ETRS89 should coincide with International Terrestrial Reference System (ITRS) at epoch 1989.0 and the ETRS89 is co-moving with the stable part of the Eurasian tectonic plate. This last condition has the objective of to avoid the influence of the tectonic plates motion on the European points coordinates. The realizations of this reference system are achieved with a set of European Reference Stations under the supervision of EUREF. The last frame for this TRS is ETRF2005. However, EUREF advises the adoption of ETRF2000 as the conventional frame of the ETRS89 system, since the coordinates of a point at this system, for a central epoch, are related, through the 14 parameters, with the corresponding coordinates in ITRFyy and there are Z-translation drift of 1.8 mm/year between ITRF2005 and ITRF2000 (Altamimi 2008).

From the point of view of a European user, not familiarized with TRS, this reference system is more comprehensive since the variation of the coordinates is less than 1 mm/year. So, for the most problems involving positioning techniques, the error of the solution is greater than the variation of the coordinates along several years.

2.2 Data for Post-Processed Solutions

In order to explore the use of GNSS real time measurements for precision agriculture studies it was decided to validate the measurements with the post-processed positioning solutions by the use of three Continuously Operating Reference Stations (CORS): GAIA, a EUREF station from RENEP, PVAR from SERVIR and FCUP from GEONET (http://geonet.fc.up.pt). This last permanent GNSS Reference Stations network is the positioning component of an extended National network for geophysical purposes (Osório 2010), supported by the National Scientific Foundation (FCT) under Scientific Equipment Renewal Programme. FCUP is a GEONET CORS 35 km away from the maize field and 6 km away from GAIA. Since its coordinates are under study, we know the Cartesian stations coordinates (2 s) in ETRS89/ETRF2000 and also in ITRS/ITRF2005 for the epoch 2009.0, and a preliminary solution for the velocities.

3 Exploitation of GPS Measurements

In a first step, all the real time kinematic positions were visualized with the TTC software, from Trimble. The raw data differential correction was obtained with the three reference stations considered independently and not in triangulation mode to avoid adjusted solutions. Also, as said before, for FCUP and for GAIA both coordinates, ETRS89 and ITRS, where known while for PVAR only ITRS/ITRF2005 coordinates are available.

The tractor trajectory, during the pre-emergence herbicide application was recovered. The maize field limits, including the two strips for crop practices, were surveyed one and half months later.

The broadcast NTRIP streams from SERVIR network gave rise to RTK vectors with origin in a base station located west of the field while those from RENEP network had its origin from a base station at south. This means real time coordinates of the same point in two Reference Systems. Several questions immediately arise: How different are these solutions? Can a farmer, not used to deal with variation of the coordinates, take both NTRIP broadcasters accordingly to the availability of the signal and the power of the received corrections? Is the price of mobile internet for the reception worthwhile or a SBAS like EGNOS is enough?

To explore conveniently the data, all the real time solutions and the post processed ones were exported, in text files, in two different modes: 3D coordinates in both TRS and 2D coordinates in the Portuguese system PT-TM06/ETRS89 (PorTuguese Transverse Mercator map projection system). The field orthometric heights, also exported, were obtained by the use of GRS80 ondulations, on a $1.5' \times 1.5'$ grid, in the local Portuguese geoid model GeodPT08. The global precision of this model is estimated in 4 cm, which is within the error for the real time solutions obtained. All the programs for data analysis were developed with MATLAB software.

3.1 Mean Sea Level Heights and Tractor Velocities

For each point of the field where a GPS measurement was recorded the respective orthometric height was estimated and a 3D plot of such values is presented in Fig. 40.1. Each raw that the tractor runs along has approximately 200 m and the height variation is of the order of 10 m.

As said before, there is interest to evaluate how this affects the tractor velocity, since it is related with the spray volume rate.

The real time positions, corrected by EGNOS, have errors of the order of 30 cm when compared to the post-processed ones. Since the velocities were calculated by numerical differentiation of the Cartesian coordinates, the real time values were used. The results of a second order method, $\frac{f(x)=f(x+h)-f(x-h)}{2h}$, for the derivatives, when compared to those obtained by a first order one, $\frac{f(x)=f(x+h)-f(x)}{h}$, show that the second method is enough. Moreover, by taking into account that the tractor velocity is not uniform and that we are looking for velocity variation with slops, the conclusions are much more realistic if, in the computations, only two points are considered for each epoch. Finally, before these computations, the points were split in two groups: the zones where the tractor was moving freely and the maneuver zones (Fig. 40.2).

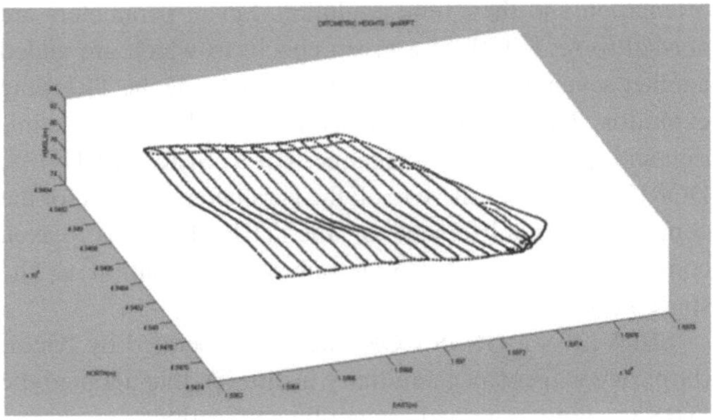

Fig. 40.1 Mean sea level heights

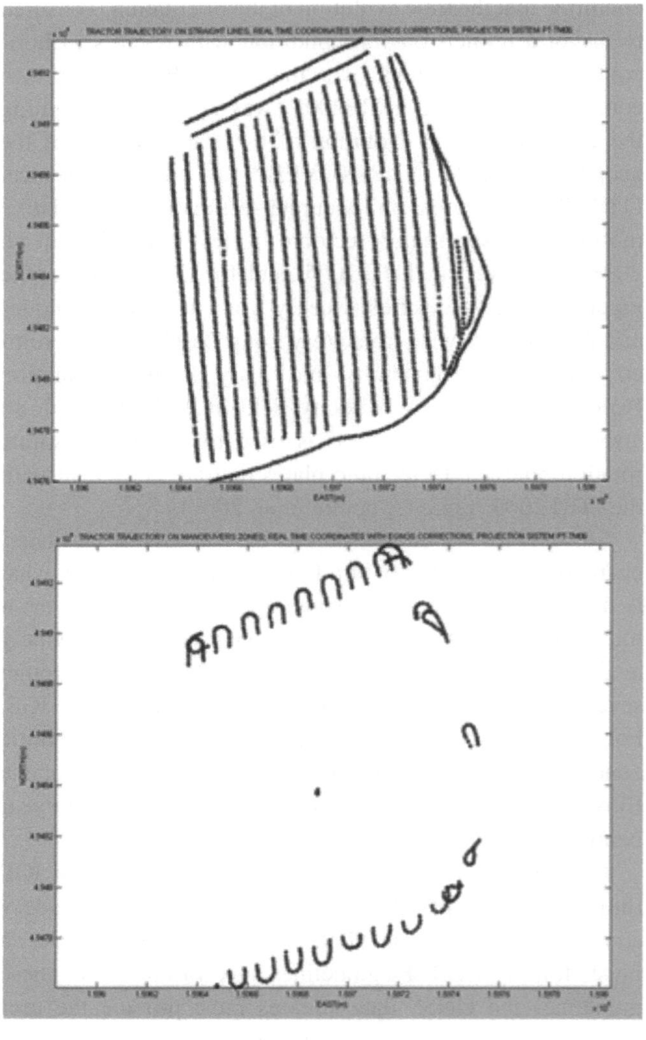

Fig. 40.2 Straight lines and maneuver zones

The influence of the slope in the tractor velocity was study along all the lines and, also, during all the maneuvers. Although the driver be an experimented farmer, the influence of the slope in the velocity of the tractor is easily detected in Figs. 40.3 and 40.4.

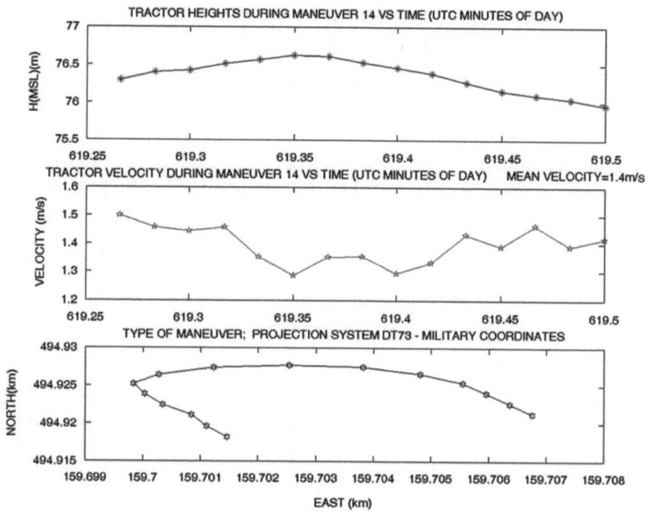

Fig. 40.3 Heights and velocities during a maneuver

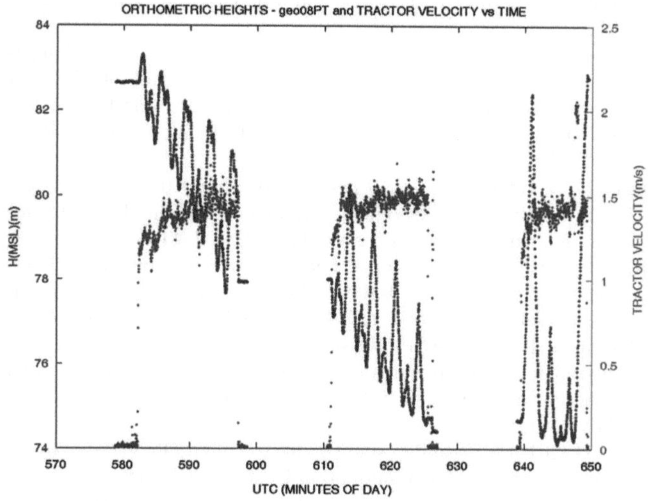

Fig. 40.4 Heights and corresponding velocities along the field. Time intervals without data correspond to the herbicide refill

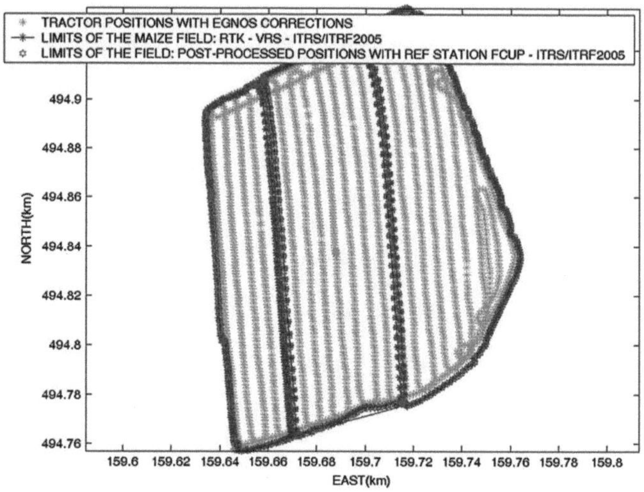

Fig. 40.5 Limits of the maize growing

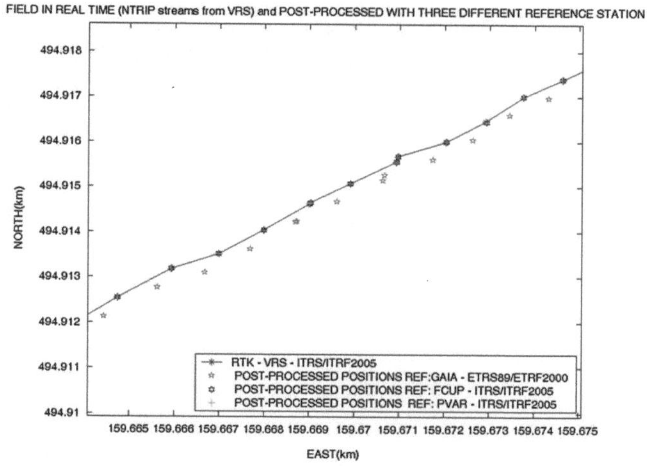

Fig. 40.6 RTK from SERVIR network and post processed solutions

The herbicide application was successful and, as it can be observed in Fig. 40.5, the limits of maize growing are around the tractor trajectory.

3.2 The Accuracy of RTK Positioning and the Influence of the TRS

During the RTK positioning the TSC2 was programmed for data acquisition of only fixed solutions with a precision of 2 cm. The calculated residuals between these real time positions and the post-processed ones show that the accuracy must have an order of 5–8 cm, when the reference station coordinates are in the same Reference System as that of the broadcast corrections. But the residuals became ten times bigger (0.5 m) when the real time solution from one of the NTRIP streams broadcaster is compared to the post-processed coordinates having as reference the station of the other broadcaster (Figs. 40.6 and 40.7).

The post-processed coordinates having as Reference Station FCUP with coordinates ITRS/ITRF2005 are very close to those solutions from PVAR, what is, by one hand, a nice result for the work that is being developed to coordinate GEONET CORS and, by the other hand, a good test in order to take into account that, for Precision Agriculture purposes, RTK positioning is enough for all the problems under study.

According to the Sect. 2.1, the 0.5 m difference between the ETRS89 and the ITRS solutions can be explained by the ITRS coordinates variation of 2.5 cm/year during 21 years (1989–2010). As already said, nowadays, these two TRS are very well defined and accurately realized by estimates of the coordinates and velocities of a very well chosen set of stations (Altamini et al. 2007), observed by different methods including GPS (Collilieux et al. 2007). From the scientific point of view the access to the transformation parameters between the different realizations is public as

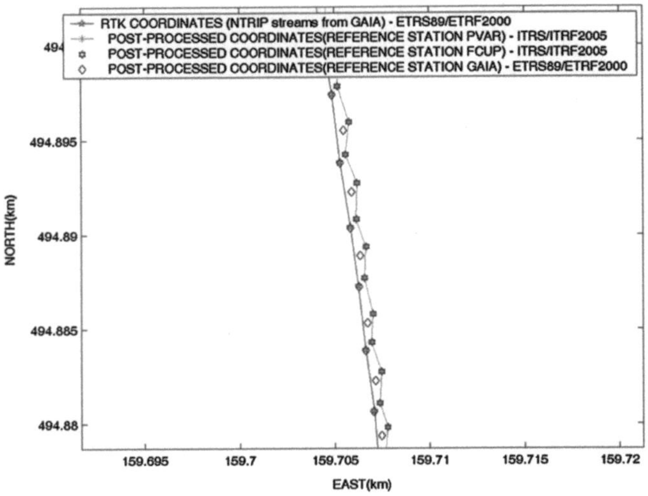

Fig. 40.7 RTK positions (at 10 s) with RENEP network and post processed ones (at 5 s)

well as the coordinates and velocities of the stations and everything is clear. But from the point of view of the applications it is not so easy.

4 Conclusions

Aided real time GNSS positioning may be an important tool in different phases of Precision Agriculture. Real time positioning data can be recorded during crop-operations like sowing, herbicide applications and so on, in order to derive outputs for future improvement of the procedures.

However, attention must be given to the TRS of the broadcasted data by a GNSS permanent network.

In the experiments presented here the effect was of the order of 0.5 m, what is contrasting with the centimeter level accuracy in the positioning with each one of the NTRIP broadcasters. This problem must be considered, mainly by taking into account that the users in this type of application, usually, will not be sensible to this type of situation.

Acknowledgements Geo-Space Sciences Research Centre, Faculty of Sciences, University of Porto, financially supported by the Foundation for Science and Technology (FCT). Thanks are also due to Mr. P. Veloso for helping with fieldwork at "Q^{ta} de Crujes".

References

Altamini Z, Sillard P, Boucher C (2003) The impact of a no-net-rotation condition on ITRF2000. Geophys Res Lett 30. doi:10.1029/2002GL016

Altamini Z, Collilieux X, Legrand J, Garayt B, Boucher C (2007) ITRF2005: a new release of the international terrestrial frame on the time series of station positions and Earth orientation parameters. J Geophys Res 112. doi:10.1029/2007JB004949

Altamimi Z (2008) EUREF symposium, Brussels, Belgium, 17–20 June 2008

Collilieux X, Altamimi Z, Coulot D, Ray J, Sillard P (2007) Comparison of very long baseline interferometry, GPS, and satellite laser ranging height residuals from ITRF2005 using spectral and correlation methods. J Geophys Res 112:B12403. doi:10.1029/2007JB004933

Easterly D, Adamchuk V, Kocher M, Hoy R (2010) Using a vision sensor system for performance testing of satellite-based tractor auto-guidance. Comput Electron Agric 72:107–118

IERS Conventions (2010) Gérard Petit and Brian Luzum (eds.). (IERS Technical Note ; 36) Frankfurt am Main: Verlag des Bundesamts für Kartographie und Geodäsie, 2010. p. 179 ISBN 3-89888-989-6

McDougall K, Gibbings P, Wolski I (2001) Comparison of a DGPS system and conventional guidance for spraying applications. In: 1st conference on geospatial information in agriculture, Sydney, 17–19 July 2001, p 13

Osório I (2010) A feasibility study on the use of GLONASS signals for meteorology in RTK positioning. EUP J Nav 8:10–15

Pérez-Ruiz M, Carballido J, Aguera J, Gil J (2010) Assessing GNSS correction signals for assisted guidance systems in agricultural vehicles. Prec Agric. doi:10.1007/s11119-010-9211-4

Author Index

Z. Altamimi and X. Collilieux (eds.), *Reference Frames for Applications in Geosciences*,
International Association of Geodesy Symposia 138, DOI 10.1007/978-3-642-32998-2,